Protocols of Actinomycetes

Many of the natural products have been isolated and characterized from actinobacteria, for example, polyketides, phenazines, peptides, indolocarbazoles, and sterols. To explore new antibiotics from actinobacteria, several bioinformatics and synthetic biology tools were developed. This book covers the basics of recent protocols for drug discovery from actinobacteria.

Features:

1. Discusses the benefits of the production of antibiotics and enzymes from actinomycetes in a large scale.
2. Covers the synthetic biology approach.
3. Describes the strain improvement of actinobacteria.
4. Provides information on basic isolation of actinobacteria and modern techniques.
5. Covers the applications and metabolic engineering strategies of actinomycetes.

This book will be helpful for pharmaceutical industries and researchers to develop new antibiotics from actinobacteria and can be used in support of future research in drug discovery.

Protocols of Actinomycetes
Microbiology to Gene Editing

Edited by
L. Karthik

CRC Press
Taylor & Francis Group
Boca Raton London New York

CRC Press is an imprint of the
Taylor & Francis Group, an **informa** business

First edition published 2025
by CRC Press
2385 NW Executive Center Drive, Suite 320, Boca Raton FL 33431

and by CRC Press
4 Park Square, Milton Park, Abingdon, Oxon, OX14 4RN

CRC Press is an imprint of Taylor & Francis Group, LLC

ISBN: 9781032504230 (hbk)
ISBN: 9781032504254 (pbk)
ISBN: 9781003398400 (ebk)

DOI: 10.1201/9781003398400

Typeset in Times
by codeMantra

Dedication

I dedicate this book to Almighty God, my father S. Loganathan, my mother L. Sasikala, my wife K. Ramapriya, and my daughter K.R. Dhanya Sri.

Contents

Foreword

It is a great pleasure to write a foreword for this new book from CRC Press – Taylor & Francis titled *Protocols of Actinomycetes: Microbiology to Gene Editing*, with Dr. Karthik Loganathan serving as an editor. This book includes 51 chapters that cover various aspects of research on actinomycetes from the basics of isolation of actinomycetes from various habitats to taxonomic characterization, screening for bioactivities, molecular techniques, OMICS tools, and CRISPR-Cas technology. Being passionately working with the research on the diversity and applications of actinomycetes from underexplored habitats for over two decades, I am also happy to contribute two chapters in this book.

Actinomycetes are Gram-positive bacteria; they are a member of the phylum Actinomycetota with a unique filamentous morphology and high guanine-cytosine (GC) content. They are one of the most studied microorganisms traditionally investigated for their secondary metabolites, especially antibiotics. Their capacity as prolific producers of bioactive compounds is second to none, with recent studies revealing their potential in additional applications, including agriculture and environment. However, they are generally slow-growing bacteria and difficult to isolate from environmental samples. Routine microbiological procedures are generally not suitable for actinomycetes. Books or protocol manuals specifically for actinomycetes are also not widely available. This collection of protocols from the contributing authors who have been working with actinomycetes will surely be valuable for those working with actinomycetes.

I strongly believe that this new book *Protocols of Actinomycetes: Microbiology to Gene Editing* will be a success in providing the readers, especially young researchers, with useful resources for both basic and advanced protocols necessary to continue the exploration of actinomycetes for the benefits of human society.

Wasu Pathom-aree

Wasu Pathom-aree, PhD.
Associate Professor
Head of Microbiology Division, Department of Biology, Faculty of Science
Chiang Mai University
Chiang Mai 50200, Thailand
Editor in Chief: Chiang Mai Journal of Science; Natural and Life Sciences Communications
Associate Editor: Current Microbiology
Section Editor: All Life

Preface

Antimicrobial resistance cases are increasing day by day globally. To solve this issue, finding new antibiotics is important. With the recent revolution in whole-genome sequence analysis, researchers found that actinobacteria are potential producers of new antibiotics.

There are only a few books available on the protocols of actinobacteria, and they also cover mostly basic experiments and do not cover molecular and gene-editing technologies. In this book, I majorly covered molecular biology and gene-editing techniques. Hence, this book will be useful for all researchers, academicians, and industry people.

Around 150 authors from all around the world contributed to this book, including India, the USA, Egypt, China, Turkey, and Thailand. I strongly believe that this book will be useful for current researchers and company people to find new antibiotics and other industrially important metabolites from actinobacteria.

I personally thank all authors for their active participation and timely submission. I extend my sincere thanks to Dr. Renu, Editor, CRC Press, for all her support. I also thank personally Dr. A. Vishnu Kirthi, Assistant Professor, Lovely Professional University, India, for his continuous support. I sincerely thank Dr. D. Dhanasekaran, Associate Professor, Bharathidasan University, TN, India, for his continuous motivation and encouragement.

About the Editor

Dr. L. Karthik has extensive expertise in actinobacteria, synthetic biology, and drug discovery strategies and is presently working as Project Leader, Synthetic Biology at Salem Microbes Pvt. Ltd, Salem, India. He is **Top 2% of all scientists in the world (the Stanford University report published by Elsevier on 10th October 2022 and 2023)**.

He is an elected fellow member of the Linnean Society of London. He did his first postdoctoral training at the Shanghai Jiao Tong University, China, under the guidance of eminent scientist Prof. Zhiyong Li. Further, he has completed his second postdoctoral fellowship in ECUST, Shanghai, China, under the guidance of eminent scientist Prof. Lixin Zhang. He received his Ph.D. degree in Microbiology from the VIT University in 2013. He had published around 75 research papers (h-index: 26), 5 books, and 6 patents and 6 products commercialized.

He is a communication committee member of the International Chemical Biology Society, USA, and President of the Society of Chemical and Synthetic Biology, India. In 2016, he received the Prestigious "Young Scientist Award" from the Indian Association of Applied Microbiologists for his outstanding research work and received the Best Research Scholar award 2012 from VIT University. He received 200 Young Scientist-2012 fellowship (DST) to attend international conferences. He received Best VIT ALUMNI AWARD-2022 from VIT University, TN, India. He is the Associate Editor of *Frontiers in Microbiology* and the Review Editor of *Frontiers in Marine Science.* He is also the Associate Editor of *Current Chinese Engineering Science* journal (Bentham Science). Recently, he proved the COVID-19 activity of Kabasura Kudineer (Siddha Herbal formulation). During the COVID-19 wave, this drug saved a lot of lives. For this contribution, he received the COVID Warrior Award from SEED Trust, TN, India.

Contributors

Omnia Mohamed Abdallah
Microbiology Department, Faculty of Science
Ain Shams University
Cairo, Egypt

Noor Afsha
Department of Biochemistry
Oriental Institute of Science and Technology
Bardhaman, India

Ramzan Ahmed
Department of Applied Biology
University of Science and Technology
Ri-Bhoi, India
and
Department of Physics, Faculty of Science
Kasetsart University
Bangkok, Thailand

A. R. Akhiya
CSIR-Institute of Genomics and Integrative Biology
New Delhi, India

Md Quasid Akhter
CSIR-Institute of Genomics and Integrative Biology
New Delhi, India

Dina Hatem Amin
Microbiology Department, Faculty of Science
Ain Shams University
Cairo, Egypt

K. Prem Anand
Marine Biotechnology and Bioproducts Lab, Department of Biomedical Sciences, School of Bioscience and Technology
Vellore Institute of Technology
Vellore, India

Khanna Asha
Avika Society
Biological Research Foundation
Jabalpur, India

Apsara S. Babu
Marine Biotechnology Laboratory
Vellore Institute of Technology
Vellore, India

Swarnendu Bag
CSIR-Institute of Genomics and Integrative Biology
New Delhi, India

Lavanya Balaji
Department of Biochemistry
Auxilium College (Autonomous)
Vellore, India

Uttra Chauhan
Department of Microbiology
Graphic Era (Deemed to be University)
Uttarakhand, India

D. M. Chetan
Dept. of Biotechnology Engineering
N.M.A.M. Institute of Technology (Deemed to be University)
Nitte, India

Shiela Chetri
Department of Microbiology
Thassim Beevi Abdul Kader College for Women
Kilakarai, India

John Godwin Christopher
VIT University
Vellore, India

Chitrakshi Consul
Department of Microbiology, School of Bioengineering and Biosciences
Lovely Professional University
Phagwara, India

Vipin Mohan Dan
Mycobacterium Research Laboratory
Rajiv Gandhi Centre for Biotechnology
and
Department of Microbiology
KSCSTE – Jawaharlal Nehru Tropical Botanical Garden and Research Institute, Palode
Trivandrum, India

Aishwarya Das
Department of Biotechnology
Dayananda Sagar College of Engineering
Bengaluru, India

Ipsita Debnath
School of Bioscience and Technology
Vellore Institute of Technology
Vellore, India

C. Subathra Devi
School of Bio Sciences and Technology
Vellore Institute of Technology
Vellore, India

Awalagaway Dhulappa
Department of Microbiology
Maharani Cluster University
Bangalore, India

Ankita J. Dobariya
UGC-CAS Department of Biosciences
Saurashtra University
Rajkot, India

Nalini Easwaran
School of Bio Sciences and Technology
Vellore Institute of Technology
Vellore, India

Lekshmi K. Edison
Department of Comparative, Diagnostic, and Population Medicine
College of Veterinary Medicine
University of Florida
Gainesville, Florida

Alaa F. Elsayed
Department of Microbiology, Faculty of Science
Ain Shams University
Cairo, Egypt

Sanjeev K Ganesh
School of Bio Sciences and Technology
Vellore Institute of Technology
Vellore, India

Shilpi Garg
Department of Biological Sciences
BITS-PILANI
Pilani, India

Sankaranarayanan Gomathinayagam
School of Bio Sciences and Technology
Vellore Institute of Technology
Vellore, India

Kodiveri Muthukaliannan Gothandam
School of Bio Sciences and Technology
Vellore Institute of Technology
Vellore, India

M. Mohamed Hasan
Department of Biotechnology
Vel Tech High Tech Dr. Rangarajan Dr. Sakunthala Engineering College
Chennai, India

Shabiha Nudrat Hazarika
Microbial Biotechnology Laboratory, Life Sciences Division
Institute of Advanced Study in Science and Technology
Guwahati, India

Birson Ingti
Department of Microbiology
The Assam Royal Global University
Guwahati, India

R. C. Jaysree
Department of Biotechnology, School of Life Sciences
Nehru Arts and Science College
Coimbatore, India

Loganathan Karthik
Research and Development
Salem Microbes Private limited
Salem, India

Tarun Keswani
Food Allergy Center and Center for Immunology and Inflammatory Diseases, Massachusetts General Hospital
Harvard Medical School
Boston, Massachusetts

Khushal Khambhati
Department of Biosciences, School of Science
Indrashil University
Rajpur, Gujarat, India

A. Vishnu Kirthi
Department of Microbiology, School of Bioengineering and Biosciences
Lovely Professional University
Phagwara, India

Benyapa Kitwetch
Interdisciplinary Program in Biotechnology, Graduate School
Chiang Mai University
Chiang Mai, Thailand

Uma Kizhuveetil
Department of Chemical Engineering
Indian Institute of Technology Madras
Chennai, India

Merja H. Kontro
Faculty of Biological and Environmental Sciences, Ecosystems and Environment Research Programme
University of Helsinki
Lahti, Finland

Ramesh K. Kothari
UGC-CAS Department of Biosciences
Saurashtra University
Rajkot, India

Gaurav Kumar
Department of Microbiology, School of Bioengineering and Biosciences
Lovely Professional University
Phagwara, India

K. Ajith Kumar
Department of Botany, School of Life Sciences
St. Joseph's University
Bengaluru, India

S. R. Sathish Kumar
Head-NRL
Neogen Food and Animal Security India Pvt. Ltd.
Cochin, India

V. Naveen Kumar
ImmuGenix Biosciences Pvt Ltd
Chennai, India

G. R. Shree Kumari
Department of Botany, School of Life Sciences
St. Joseph's University
Bengaluru, India

Soumen Kundu
CSIR-Institute of Genomics and Integrative Biology
New Delhi, India

Vartika Lohani
Department of Pharmacy
Banasthali Vidyapith
Jaipur, India

M. Maneesha
School of Bio Sciences and Technology
Vellore Institute of Technology
Vellore, India

Gira P. Mankad
HSc. College
Rajkot, India

Mathurin Meethangdee
Interdisciplinary Program in Biotechnology, Graduate School
Chiang Mai University
Chiang Mai, Thailand

Gangadhar Megeri
Dept. of Biotechnology
Basaveshwar Engineering College
Bagalkote, India

Andrea Michelle
Department of Botany
St Joseph's University
Bengaluru, India

V. Mohanasrinivasan
School of Bio Sciences and Technology
Vellore Institute of Technology
Vellore, India

Yugal Kishore Mohanta
Department of Applied Biology
University of Science and Technology
Ri-Bhoi, India

U. Monisha
Department of Botany
St Joseph's University
Bengaluru, India

Balaji Muralikrishnan
Department of Biotechnology
PSG College of Arts & Science
Coimbatore, India
and
Mycobacterium Research Laboratory
Rajiv Gandhi Centre for Biotechnology
Trivandrum, India

Ramamurthy Nagarajan
Department of Chemical Engineering
Indian Institute of Technology Madras
Chennai, India

S. Jemimah Naine
School of Bio Sciences and Technology
Vellore Institute of Technology
Vellore, India

Kothari Neeti
School of Bio Sciences and Technology
Vellore Institute of Technology
Vellore, India

Ayyadurai Niraikulm
Department of Biochemistry and Biotechnology
Council of Scientific and Industrial Research (CSIR) – Central Leather Research Institute (CLRI)
Chennai, India

M. Joyce Nirmala
Department of Chemical Engineering
Indian Institute of Technology Madras
Chennai, India

Nisha
Department of Microbiology, School of Bioengineering and Biosciences
Lovely Professional University
Phagwara, Punjab, India

Mustafa Oskay
Faculty of Sciences and Letters, Department of Biology
Section of Basic and Industrial Microbiology, Manisa Celal Bayar University
Manisa, Turkey

Gaurav Pant
Department of Microbiology
Graphic Era (Deemed to be University)
Uttarakhand, India

Wasu Pathom-aree
Department of Biology, Faculty of Science
Chiang Mai University
Chiang Mai, Thailand

Veilumuthu Pattaulavar
VIT University
Vellore, India

M. Pavithra
Department of Biotechnology
D G Vaishnav College
Chennai, India

T. S. Sai Pawan
Department of Biotechnology
Dayananda Sagar College of Engineering
Bengaluru, India

Saravanan Periasamy
Centre of Excellence in Biofilms, Department of Biotechnology
Rajalakshmi Engineering College (Autonomous)
Chennai, India

N. S. Pradeep
KSCSTE-Malabar Botanical Garden and Institute for Plant Sciences
Kozhikode, India

Bhatele Pranita
Department of Botany and Microbiology
Govt. M. H. College of Home Science and Science for Women
Jabalpur, India

Akshatha Prasanna
Department of Biotechnology
Dayananda Sagar College of Engineering
Bengaluru, India

Charles Lekhya Priya
Department of Biosciences
Mohan Babu University, Sree Vidyanikethan Engineering College
Tirupati, India

Hina N. Radadiya
UGC-CAS Department of Biosciences
Saurashtra University
Rajkot, India

R. Rajamathanghi
Department of Biotechnology
Vel Tech High Tech Dr. Rangarajan Dr. Sakunthala Engineering College
Chennai, India

Muniyan Rajiniraja
Department of Integrative Biology, School of Bio-Sciences and Technology
Vellore Institute of Technology (VIT)
Vellore, India

Hasti Ramavat
Department of Biosciences
Saurashtra University
Rajkot, India

A. Sai Ramesh
Department of Biotechnology
Prathyusha Engineering College
Thiruvallur, India

Pharada Rangseekaew
Department of Biology, Faculty of Science
Chiang Mai University
Chiang Mai, Thailand

K. V. Bhaskara Rao
Marine Biotechnology Laboratory
Vellore Institute of Technology
Vellore, India

Manik Prabhu Narsing Rao
Department of Genetics, Faculty of Science
Kasetsart University
Bangkok, Thailand
and
Omics Center for Agriculture, Bioresources, Food and Health
Kasetsart University (OmiKU)
Bangkok, Thailand

Lokesh Ravi
Department of Biomedical Sciences, School of Health Sciences
The Apollo University
Chittoor, Andhra Pradesh, India

Anand Ravichandran
Centre of Excellence in Biofilms, Department of Biotechnology
Rajalakshmi Engineering College (Autonomous)
Chennai, India

Bhagwan Narayan Rekadwad
National Centre for Cell Science, NCCS Complex
Savitribai Phule Pune University Campus
and
Institute of Bioinformatics and Biotechnology (IBB)
Savitribai Phule Pune University
and
Yenepoya Research Centre
Yenepoya (Deemed to be University), University Road
Mangalore, India

Kangkon Saikia
Bioinformatics Infrastructure Facility
Institute of Advanced Study in Science and Technology
Guwahati, India

Srivarshini Sankar
School of Bio Sciences and Technology
Vellore Institute of Technology
Vellore, India

P. Chandra Sekar
School of Bioscience and Technology
Vellore Institute of Technology
Vellore, India

Anirban Sengupta
Department of Biomedical & Clinical Science, Division of Molecular Medicine & Virology
Linkoping University
Linkoping, Sweden

B. Kiran Sharma
Department of Biotechnology
D G Vaishnav College
Chennai, India

Shasmeen Shora
Department of Chemical Engineering
National Institute of Technology
Srinagar, India

Satya P. Singh
UGC-CAS Department of Biosciences
Saurashtra University
Rajkot, India

Vijai Singh
Department of Biosciences, School of Science
Indrashil University
Rajpur, India

Ramamoorthy Siva
VIT University
Vellore, India

Veena Sreedharan
Department of Biomedical Sciences, School of Bio Sciences and Technology
VIT University
Vellore, India

Kasbe Prajapati Sudhakar
Lonza Biologics Inc.
Houston, USA

Ranjitha Dhevi V. Sundar
School of Biosciences and Technology
Vellore Institute of Technology
Vellore, India

K. Suthindhiran
Marine Biotechnology and Bioproducts Lab, Department of Biomedical Sciences, School of Bioscience and Technology
Vellore Institute of Technology
Vellore, India

K. Swetha
Department of Biotechnology
DKM College for Women
Vellore, India

Debajit Thakur
Microbial Biotechnology Laboratory, Life Sciences Division
Institute of Advanced Study in Science and Technology
Guwahati, India

Arinthip Thamchaipenet
Department of Genetics, Faculty of Science
Kasetsart University
and
Omics Center for Agriculture, Bioresources, Food and Health
Kasetsart University (OmiKU)
Bangkok, Thailand

Durairaj Thirumurugan
Department of Biotechnology
SRM Institute of Science and Technology
Kancheepuram, India

Sirikanjana Thongmee
Department of Physics, Faculty of Science
Kasetsart University
Bangkok, Thailand

Balendu Upmanyu
CSIR-Institute of Genomics and Integrative Biology
New Delhi, India

Abhitha Venkataramakrishnan
Centre of Excellence in Biofilms, Department of Biotechnology
Rajalakshmi Engineering College (Autonomous)
Chennai, India

Ramasamy Vijayakumar
Faculty of Microbiology
Centre for Distance and Online Education, Bharathidasan University
Tiruchirappalli, India

Jayachandra S. Yaradoddi
Dept. of Biotechnology
Basaveshwar Engineering College
Bagalkote, India

D. M. Yashaswini
Department of Botany
St Joseph's University
Bengaluru, India

1 Isolation of Actinomycetota by Different Methods

Ramzan Ahmed, Birson Ingti, Noor Afsha, Sirikanjana Thongmee, and Yugal Kishore Mohanta

1.1 INTRODUCTION

Secondary metabolites derived from microorganisms continue to be a variety of bioactive sources for the exploration and creation of pharmacological drugs, as well as biochemical probes for studying human disease processes (Igarashi et al., 2002). Over the past few decades, many Actinomycetota have been isolated and screened from soil, and they are responsible for 70%–80% of significant secondary metabolites that are being sold commercially (Igarashi et al., 2002). Actinomycetota are filamentous, branched, Gram-positive bacteria with a morphology resembling that of a fungal organism. Due to their contribution to the discovery of much more than half of all commonly used antibiotics to date, they constitute a significant source of new antibiotics. Antibiotics are essential in the generation of novel techniques for primary screening and tissue culture, particularly in microbiology, biochemistry, genetics, and molecular biology, as well as, to a limited extent, pharmaceuticals and organic chemistry (Nanjwade et al., 2010). Actinomycetota, which were once thought of as microbiological fascinations with little economic significance, are now the focus of extensive research into potential sources of novel, physiologically active chemicals. They are mostly found in soil but are also widely distributed in natural settings, giving rise to the fresh possibility of isolating various actinomycetota in the quest for novel metabolites.

Recently, it was discovered that marine habitats are one of the key places to look for new Actinomycetota with the ability to produce a large variety of chemically distinct substances with a broad spectrum of biological functions (Bredholt et al., 2008). The endophytic Actinomycetota that live on plants are crucial for defending their hosts against invasions from plant pathogens (Crawford et al., 1993). A number of endophytic Actinomycetota produce phytohormones, iron-chelating compounds, indole-3-acetic acid, and siderophores that stimulate plant development (Indananda et al., 2011).

The isolation of numerous Actinomycetota improves the finding of undiscovered microbial metabolites from these species. In order to develop Actinomycetota resources, acquiring new variants is a difficult condition; hence, innovative separation techniques are often essential. Actinomycetota genera have been isolated using a variety of techniques in the past, and new techniques have been used subsequently. Different techniques were devised by Hayakawa and Nonomura for the isolation of suitable, rare Actinomycetota species from their natural environments. These procedures use a number of preprocessing approaches in conjunction with enrichment procedures to properly add selective antibiotics to agar media (Hayakawa et al., 2008). Further innovative techniques had been used to isolate rare and unusual Actinomycetota. In this chapter, we looked at more advanced techniques for the isolation of different genera of Actinomycetota from soil, water, plants, and other environmental samples.

1.2 MATERIALS AND METHODS

1.2.1 Isolation of Actinomycetota

Actinomycetota are Gram-positive bacteria that can thrive on both simple and complicated substrates. Since these bacterial groups are prevalent in various environments, it is important to acquire

DOI: 10.1201/9781003398400-1

proper samples for a successful isolation. Actinomycetota can be isolated using general bacterial isolation techniques such as serial dilution to decrease overcrowding, pour plates, streaking, and centrifugation procedures.

1.2.1.1 Isolation of Actinomycetota from Soil

Since Actinomycetota are soil organisms, it is necessary to collect soil samples from the study region in order to isolate Actinomycetota.

1.2.1.2 Collection of Soil Samples

Four to five grams of soil samples should be taken from a depth of 4–5 cm, put in a sterile container, and occasionally combine them with around 1 g of $CaCO_3$. The soil should be allowed to dry out at room temperature for about 3 weeks in order to destroy the majority of the Gram-negative bacteria. To prevent the growth of other bacterial flora, heat treatment can also be used by heating the samples separately in a hot air oven at 120°C for 1 hour. The serial dilution procedure can then be used to minimize the quantity of microbial cells (Figure 1.1).

1.2.1.3 Pretreatment of Soil Samples

Pretreating the soil samples taken from various geographic locations is necessary because by either encouraging the growth of Actinomycetota or eradicating the majority of undesirable Gram-negative bacteria, pretreating soil can accelerate the isolation of Actinomycetota. Different pretreatment methods have been devised for various Actinomycetota genera. Streptomycetes are widespread in natural environments and typically make up a significant portion of the entire Actinomycetota population (Hayakawa, 2008). Physical pretreatment methods can be used to easily isolate streptomycetes; however, chemical procedures or methods that combine chemical and physical methods should be utilized to isolate other types of streptomycetes. The pretreatment of soils involves the following techniques.

1.2.1.4 Physical Treatments

Actinomycetota from various genera are regularly subjected to physical treatment, including centrifugation, pollen baiting, dry heat, wet incubation using radiation, glycerol, and air drying. As Gram-negative bacteria are less resistant to desiccation than Actinomycetota spores, a soil sample heated at 120°C for an hour promotes the development of streptomycetes and other uncommon species like *Actinomadura*, *Microbispora*, *Spirilliplanes*, and others on humic acid vitamin agar (HVA). Centrifugation can be used to preferentially isolate motile Actinomycetota by removing streptomycetes and other non-motile Actinomycetota while allowing motile Actinomycetota to be maintained in the supernatant and dispersed on a suitable culture medium that incorporates

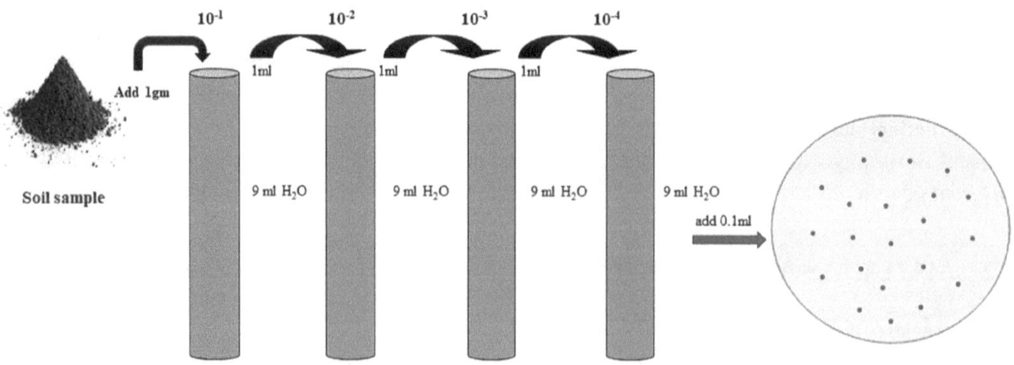

FIGURE 1.1 Serial dilution of soil sample.

trimethoprim and nalidixic acid (Hayakawa et al., 2008). In addition to these, UV-irradiation is excellent for the isolation of *Pseudonocardia* and *Nocardiopsis* spp., while ultra-high-frequency irradiation is favorable for the isolation of *Streptosporangium* and *Rhodococcus* spp. Extremely high-frequency irradiation was also effective for *Streptosporangium* and *Nocardia* spp.

1.2.1.5 Chemical Treatments

Actinomycetota can be isolated, especially using chemical treatments such as chitin and calcium carbonate treatments, sodium dodecyl sulfate (SDS), calcium chloride, phenol, germicides, yeast extract, chloramine-T, and chemotactic agents. Actinomycetota growth is supported by the carbon and nitrogen sources chitin and calcium carbonate. Treatment of the soil sample with 1.5% phenol will permit the growth of *Micromonospora* spp. and *Streptomyces violaceusniger* (Hayakawa et al., 2004, 1991a). Chloramine-T-treated soil samples were observed to encourage the growth of *Microbispora*, *Microtetraspora*, *Herbidospora*, *Nonomuraea*, and *Streptosporangium* on HV agar containing nalidixic acid (Hayakawa et al., 1997).

1.2.1.6 Physical and Chemical Treatments

It was revealed that combining physical and chemical treatments was the best way to successfully isolate different genera of Actinomycetota. *Actinomadura viridis* is best isolated on kanamycin, josamycin, lysozyme, and nalidixic acid-containing HV agar after being dry heated at 110°C for 1 hour and supplied with 1% phenol (Hayakawa et al., 1995). *Microbispora* on nalidixic acid-containing HV agar are promoted by dry heating a soil sample to 120°C for 1 hour with 1.5% phenol (Hayakawa et al., 1991a). *Dactylosporangium* or *Streptosporangium* genera on leucomycin and nalidixic acid contain HV agar that benefits from dry heat at 120°C together with 0.01% benzethonium chloride (Hayakawa et al., 1991b).

1.2.2 Isolation of Actinomycetota

Actinomycetota can be isolated using different methods.

1.2.2.1 Direct Inoculation Technique

In petri plates, 2–3 mg of soil samples are uniformly dispersed over the agar media and allowed to come into contact with the medium. At a temperature of 28°C, the incubation period lasts for 5–25 days. On malt extract and yeast extract agar, colonies exhibiting typical Actinomycetota characteristics are sub-cultured. M1 slant and 50% glycerol broth can be used to keep pure cultures alive at –4°C.

1.2.2.2 Spread Plate Method

One gram of soil sample should be serially diluted (see Figure 1.1) and shaken briskly with a vortex; 0.1 mL of each dilution is then placed on the isolation medium (see Table 1.1), and a sterile plastic spreader was used to distribute the inoculum uniformly. In order to allow the liquid to be incorporated, the inoculated plates were allowed to remain at room temperature for 5–10 minutes and then incubated at 28°C for 5–25 days.

1.2.2.3 Membrane Filter Technique

Actinomycetota can be isolated with greater accuracy using this method. This technique involves autoclaving an appropriate agar medium for 15 minutes at 121°C. When the medium has cooled down to 45°C, it is placed in petri plates and given time to settle. Aseptically, cellulose ester membrane filters with pore diameters ranging from 0.22 to 0.45 mm can be placed in the middle of the plates. After that, soil is sprayed onto the membrane filters. After approximately 4 days of incubation at 28°C, the membrane filters are removed. Actinomycetota have highly branching mycelial networks, which give them the ability to get inside membrane filter holes. Actinomycetota mycelium enters the filter holes during incubation to reach the underneath agar media, whereas non-Actinomycetota can only grow on the surface of the filter paper. The plates were re-incubated for 2–3 weeks.

1.2.2.4 Integrated Method

One of the most popular techniques for isolating Actinomycetota with accuracy is this one. Actinomycetota appear to be growing as appropriate when certain antibiotics and other selective methods are used in combination with physical and chemical methods. Following the soil's initial treatment with $CaCO_3$ and pouring it on a nalidixic acid-containing selective medium, the number of isolated pure cultures of Actinomycetota has increased significantly (Alferova et al., 1989). The growth of Actinoplanes is specifically encouraged by the pollen-baiting and drying technique that uses HVA with nalidixic acid (Hayakawa et al., 1991d). The growth of Actinoplanes, Actinosynnema, Actinokineospora, Cryptosporangium, Catenuloplanes, Geodermatophilus, Dactylosporangium, Sporichthya, and Kineosporia is supported by centrifuging the sample at 1,500 *g* for 20 minutes and rehydrating it at 30°C for 90 minutes.

1.2.2.5 Enrichment Treatment

For the purpose of selective isolation, the number of different genera of Actinomycetota has been enhanced by the inclusion of dietary or auxiliary ingredients. Enrichment is one of the best ways to increase the diversity and number of culturable microorganisms. Hayakawa et al. (1991c) devised an enhanced chemotactic approach that makes use of the potent chemotactic reaction of Actinomycetota zoospores to collidine and results in higher revival of *Actinoplanes* and *Dactylosporangium* spp. from different soils. The thermophilic Actinomycetota's phage susceptibility offered a selective method for lowering their populations on culture plates; this enhanced the populations of thermophilic *Streptomyces* spp., Thermomonospora, and *Saccharopolyspora rectivirgula* (Kurtboke et al., 1993). Kyun Man made a coal-vitamin medium in which the quantity of Actinomycetota had grown well while the development of other soil bacteria had been restricted. The preparation of a soil mixture with peptone (6%) and lauryl sulfate (0.05%) at 50°C for 10 minutes considerably enhanced the number of Actinomycetota in the soil before incubation with the new medium. Antonie Van Leeuwenhoek used differential centrifugation to isolate motile Actinomycetota from the soil and plant litter in 2000. Hayaka et al. established the rehydration and centrifugation (RC) method for the selective separation of various zoosporic actinomycete taxa (Kineosporia, Actinokineospora, and Catenuloplanes) from soil and plant litter. Streptomycetes and other non-motile Actinomycetota were significantly reduced during the centrifugation stage, allowing for the selective growth of uncommon, motile Actinomycetota on the culture plates after inoculation.

1.2.3 Isolation of Actinomycetota from Marine Environment

Actinomycetota are abundant in the marine environment, particularly in marine deposits, but they also coexist alongside a variety of marine species, including corals, fish, macroalgae, sponges, and tunicates (Abdelmohsen et al., 2014). Several methods were used to selectively isolate marine Actinomycetota, including (a) the preparation of seawater-based medium to isolate the first marine obligatory genus *Salinispora* (Maldonado et al., 2005a), (b) setting up somewhat acidic environments (Ramesh and Mathivanan, 2009; Goodfellow and Williams, 1983), (c) the abundance of a variety of marine Actinomycetota increases when sea sediments are heated at 50°C for 60 minutes, then diluted with a sterile saline solution of 0.5%, and treated with cycloheximide and nystatin, (d) the use of the heat treatment and serial dilution method, and (e) the application of various types of radiation (Bredholt et al., 2008).

Sediment from the deep ocean is retrieved using a sampler, and the samples need to be stored at 4°C in sterile glass bottles. After 1 week of air drying at room temperature, the sediment sample should be processed utilizing various pretreatment and enrichment techniques. You can serially dilute 1 g of sludge in 50% sterile saltwater. Then, 0.1 mL of each dilution should be plated onto various isolation and selective media, such as starch casein agar. The plates underwent a 4-week incubation period at room temperature.

1.2.4 Isolation of Plant Endophytic Actinomycetota

Actinomycetota are also abundantly distributed in plants. In recent years, endophytic Actinomycetota have been investigated and can either be isolated from the plant's internal tissues or from the tissues' disinfected surfaces (Siva et al., 2011). The surface-sterilized tissues need to be dried correctly under aseptic conditions and then dried continuously at 100°C for 15 minutes. After being cut into very small pieces using aseptic techniques, the surface-treated tissues are finally homogenized using a homogenizer. A total of 0.1 mL of the sample solution is then spread out on the agar plate at three different dilutions. The inoculation plates were kept in an incubator at 28°C for 2–4 weeks. In addition to surface sterilization, moist incubation and desiccation are utilized as enrichment techniques to isolate endophytic Actinomycetota.

1.2.5 Isolation of Thermophilic Actinomycetota

For the isolation of Actinomycetota, the samples from hot springs or other tropical settings were air-dried at room temperature for 7–10 days before being subjected to 120°C for 1 hour.

1.2.6 Isolation of Halophilic and Alkalophilic Actinomycetota

Isolating the alkalophilic Actinomycetota requires using the following medium (Table 1.1), with the pH raised to 10–11 using autoclaved Na_2CO_3 or NaOH prior to being dispersed on a petri plate. All media should contain 25–40 mg of nalidixic acid (for a 1 L medium) to suppress Gram-negative bacteria. Halophilic Actinomycetota usually grow very slowly. As a result, isolation media should be thicker, and isolation plate cultivation times should be extended to 20–35 days in order to maintain humidity. The same isolation medium is used to inoculate a slant with a single actinomycete colony.

1.2.7 Isolation of Acidophilic Actinomycetota

Only a few studies on acidophilic streptomycetes and acidophilic Actinomycetota have been conducted worldwide. The rapid development of bacteria and fungi in test materials when isolating plates with low pH makes it challenging to isolate these Actinomycetota. The following media can be used to isolate acidophilic Actinomycetota: YIM 17, YIM 6, YIM 47, and YIM 21. It is recommended to use 25 g of agar per 1 L and to set the pH between 4.0 and 4.5 using sterilized hydrochloric acid. Filter-sterilized mixtures are provided for all media (50 mg autoclaved $K_2Cr_2O_7$ or 50 mg nystatin + 50 mg cycloheximide + 20 mg nalidixic acid).

1.2.8 Isolation of Actinomycetota in Animal Feces

Fresh fecal samples were taken. As soon as possible, the samples were placed on sterilized glass plates to dry at 28°C for 10 days. Each dry sample was prepared for an hour at 80°C, and then, 2 g of each sample were dissolved in 18 mL of sterile water containing 0.1% $Na_4P_2O_5$ and then shaken for an hour at 220 rpm. Before coating, the suspension is subjected to a 40-second ultrasonic treatment at 150 W. Three dilutions, 10^{-5}, 10^{-6}, and 10^{-7}, were utilized to isolate Actinomycetota after diluting the sample from 10^{-1} to 10^{-7}.

1.2.9 Endophytic Actinomycetota Isolation from Roots

The plants were completely cleansed to eliminate any soil from the root mass and set aside to air dry for 48 hours. To remove any soil and organic materials from the roots, a sonication stage was included in the washing process. The roots were cut out and then sterilized in three steps, such as 99% ethanol for 60 seconds, 3.125% NaOCl for 6 minutes, and 99% ethanol for 30 seconds, and then autoclaved with reverse osmosis (RO) water for the last rinse. The root surfaces were cleaned before being sliced into 1-cm pieces and placed in the culture medium (Table 1.1), where they were subsequently incubated for up to 4 weeks at 27°C.

1.2.10 Endophytic Actinomycetota Isolation from Seeds

Additionally, a technique for separating endophytes from seeds was devised. To do this, it was necessary to slice up particular, surface-cleaned seeds into slices of 1–2 mm thickness and aseptically place them on the culture medium (Table 1.1). These slices were left at 27°C for up to 4 weeks before the endophytes began to develop, but they did not germinate.

1.2.11 Media Used for the Isolation of Actinomycetota

In order to isolate Actinomycetota, selective media have always been suggested because the majority of the sample contains other microbial genera. As a result, Actinomycetota propagules must not be negatively affected by the design of isolation media that inhibits the growth of competing bacteria (Goodfellow and O'Donnel, 1989; Cross, 1982). As nitrogen sources, proteins

TABLE 1.1
List of Few Media Used for Isolation of Actinomycetota

SL No.	Media	Composition
1.	MA	40 g Marine broth (Labscan Ltd., Thailand), 15 g agar, and 1 L distilled water
2.	Starch casein agar (SCA)	10 g soluble starch, 2 g K_2HPO_4, 2 g KNO_3, 2 g NaCl, 0.05 g $MgSO_4{\cdot}7H_2O$, 0.3 g casein, 0.02 g $CaCO_3$, 0.01 g $FeSO_4{\cdot}7H_2O$, 15 g agar, and 1 L distilled water
3.	GNO_2A	3 g tryptone, 10 g glucose, 5 g peptone, 5 g NaCl, 15 g agar, and 1 L distilled water
4.	Czapek Dox Agar	Sucrose 20 g, K_2HPO_4 1 g, $NaNO_3$ 2 g, $MgSO_4$ $7H_2O$ 0.5 g, $FeSO_4$ $7H_2O$ 0.01 g, KCl 0.5 g, and vitamin mixtures
5.	Luria Bertani (M1) agar	Luria Bertani (M1) medium (peptone 2.0 g, starch 10 g, yeast extract 4.0 g, agar 18.0 g, distilled water 1 L, and pH; 7.0±0.1)
6.	HV agar	1 g humic acid, 0.5 g Na_2HPO_4, 1.7 g KCl, 0.5 g $MgSO_4$, 0.01 g $FeSO_4$, 0.02 g $CaCO_3$, 1 mL VB stock solution (added after autoclaving), 10 g agar, and 1 L dH_2O
7.	Starch nitrate agar	Soluble starch 20.0 g, KNO_3 2.0 g, K_2HPO_4 1.0 g, $MgSO_4$ 0.5 g, $FeSO_4$ 0.1 g, $CaCO_3$ 3.0 g, NaCl 100 g, $ZnSO_4$ 0.1 g, $MnCl_2$ 0.1 g, distilled water 1 L, and pH 7.0±0.1
8.	Yeast extract agar	0.25 g of yeast extract (Oxoid), 0.5 g K_2HPO_4, and 18 g agar (Oxoid) per liter of distilled water
9.	Yeast extract-casein hydrolysate agar	0.3 g of yeast extract (Oxoid), 0.3 g of D-glucose, 2 g of K_2HPO_4, and 18 g of agar (Oxoid) per liter of RO water
10.	Glycerol asparagine agar	Glycerol 10 g, K_2HPO_4 1 g, sodium asparaginate 1 g, agar 15 g, and distilled water 1 L (add 3 g/ L $CaCO_3$ to adjust pH to 7.0)
11.	Benedict agar	Glycerol 20 g, *L. arginine* 2.5 g, $FeSO_4{\cdot}7H_2O$ 0.1 g, $MgSO_4{\cdot}7H_2O$ 0.1 g, $CaCO_3$ 0.1 g, pimaricin 50 mg, agar 20 g, NaCl 1 g, and distilled water 1 L

and amino acids are extremely important in the selective separation of Actinomycetota. For the specific isolation of Actinomycetota, a variety of selective media, such as Kusters's agar, HVA, starch nitrate agar, starch casein agar, glycerol glycine agar, chitin agar, and inorganic salt starch agar, are commonly used (Table 1.1).

In order to isolate certain genera of Actinomycetota, many researchers have utilized antibiotics (such as nalidixic acid, trimethoprim, leucomycin, and kanamycin) and antifungal drugs (such as nystatin, cycloheximide, and pimaricin) in media. Many experts use a combination of antibacterial and antifungal drugs to promote Actinomycetota development (Ravi Ranjan et al., 2015).

ACKNOWLEDGMENTS

Dr. Ramzan Ahmed acknowledges the Ministry of Higher Education, Science, Research and Innovation and the Thailand Science Research and Innovation Project under the Kasetsart University Reinventing University Program 2022 for Post-Doctoral Fellowship (Ref. No. 6501.0207/8434, dated August 26, 2022).

CONFLICT OF INTEREST

The authors have no conflicts of interest to declare that are associated with the subject matter of this chapter.

DATA AVAILABILITY

All the information generated or assessed during this assessment is contained in this chapter.

REFERENCES

Abdelmohsen UR, Bayer K, Hentschel U. 2014. Diversity, abundance and natural products of marine sponge-associated Actinomycetes. *Nat. Prod. Rep.*, 31, 381–399. DOI: 10.1039/C3NP70111E.

Alferova I, Terekhova L, Prauzer K. 1989. Selective medium with nalidixic acid for isolating antibioticproducing Actinomyces. *Antibiot. Khimioter.*, 34, 344–348.

Bredholt H, Fjaervik E, Jhonsen G, Zotechev SB. 2008. Actinomycetes from sediments in the Trondhein Fjrod, Norway: diversity and biological activity. *Mar. Drugs*, 6, 12–24.

Crawford D, Lynch J, Whipps J, Ousley M. 1993. Isolation and characterization of actinomycete antagonists of a fungal root pathogen. *Appl. Environ. Microbiol.*, 59, 3889–3905. DOI: 10.1128/aem.59.11.3899-3905.1993.

Cross T. 1982. Actinomycetes: a continuing source of new metabolites. *Dev. Ind. Microbiol.*, 23, 1–18.

Goodfellow M, O'Donnel AG. 1989. Search and discovery of industrially significant actinomycetes. In *Microbial Products: New Approaches, Society for General Microbiology Symposium,* No. 44, eds. Baumberg, S., Hunter, I. S., and Rhodes, P. M. pp. 343–383. Cambridge: Cambridge University Press.

Goodfellow M, Williams ST. 1983. Ecology of actinomycetes. *Annu. Rev. Microbiol.*, 37, 189–216.

Hayakawa M. 2008. Studies on the isolation and distribution of rare actinomycetes in soil. *Actinomycetologica*, 22, 12–19. DOI: 10.3209/saj.SAJ220103.

Hayakawa M, Iino H, Takeuchi S, Yamazaki T. 1997. Application of a method incorporating treatment with Chloramine-T for the selective isolation of Streptosporangiaceae from soil. *J. Ferment. Bioeng.*, 84, 599–60. DOI: 10.1016/S0922-338X(97)81919-8.

Hayakawa M, Momose Y, Kajiura T, Yamazaki T, Tamura T, Hatano K, Nonomura H. 1995. A selective isolation method for Actinomadura viridis in soil. *J. Ferment. Bioeng.*, 79, 287–289. DOI: 10.1016/0922-338X(95)90619-B.

Hayakawa M, Sadakata T, Kajiura, T, Nonomura H. 1991a. New methods for the highly selective isolation of Micromonospora and Microbispora from soil. *J. Ferment. Bioeng.*, 72, 320–326. DOI: 10.1016/0922-338X(91)90080-Z.

Hayakawa M, Tamura T, Iino H, Nonomura H. 1991b. New methods for the highly selective isolation of Streptosporangium and Dactylosporangium from soil. *J. Ferment. Bioeng.*, 72, 327–333. DOI: 10.1016/0922-338X(91)90081-Q.

Hayakawa M, Tamura T, Iino H, Nonomura H. 1991d. Pollen-baiting and drying method for highly selective isolation of *Actinoplanes* spp. *from soil. J. Ferment. Bioeng.*, 72, 433–438. DOI: 10.1016/0922-338X(91)90050-Q.

Hayakawa M, Tamura T, Nonomura H. 1991c. Selective isolation of Actinoplanes and Dactylosporangium from soil by using collidine as the chemoattractant. *J. Ferment. Bioeng.*, 72, 426–432.

Hayakawa M, Yoshida Y, Iimura Y. 2004. Selection of bioactive soil actinomycetes belonging to the Streptomyces violaceusniger phenotypic cluster. *J. Gen. Appl. Microbiol.*, 96, 973–981. DOI: 10.1111/j.1365-2672.2004.02230.x.

Igarashi Y, Iida T, Sasaki T, Saito N, Yoshida R, Furumai T. 2002. Isolation of actinomycetes from live plants and evaluation of antiphytopathogenic activity of their metabolites. *Actinomycetologica*, 16, 9–13. DOI: 10.3209/saj.16_9.

Indananda C, Matsumoto A, Inahashi Y, Takahashi Y, Duangmal K, Thamchaipenet A. 2011. Actinophyto-cola gen. nov., a new genus of the family Pseudono-cardiaceae and description of a new species, *Actinophytocoloa oryzae* sp. nov., isolated from root of Thai jasmine rice plant. *Int. J. Syst. Evol. Microbiol.*, 4, 737–741.

Kurtboke D, Wilson C, Sivasithamparam K. 1993. Occurrence of Actinomadura phage in organic mulches used for avocado plantations in Western Australia. *Can. J. Microbiol.*, 39, 389–394. DOI: 10.1139/m93-057.

Maldonado LA, Fenical W, Jensen PR, Kauffman CA, Mincer TJ, Ward AC, Bull AT, Goodfellow M. 2005a. Salinispora arenicola gen. nov., sp. nov. and Salinispora tropica sp. nov., obligate marine actinomycetes belonging to the family Micromonosporaceae. *Int. J. Syst. Evol. Microbiol.*, 55, 1759–1766. DOI: 10.1099/ijs.0.63625-0.

Nanjwade B, Chandrashekhara S, Goudanavar P, Shamarez A, Manvi F. 2010. Production of antibiotics from Soil-isolated actinomycetes and evaluation of their antimicrobial activities. *Trop. J. Pharma. Res.*, 9, 373–377. DOI: 10.4314/tjpr.v9i4.58933.

Ramesh S, Mathivanan N. 2009. Screening of marine actinomycetes isolated from the Bay of Bengal, India for antimicrobial activity and industrial enzymes. *World J. Microbiol. Biotechnol.*, 25, 2103–2111. DOI: 10.1007/s11274-009-0113-4.

Ravi Ranjan K, Vasantba J, Bhoomi M, Bonisha T, Bhumika C. 2015. Antibacterial potentials of Actinomycetes isolated from Gujarat. *Int. J. Pharm. Sci. Rev. Res.*, 30(1), 78–83.

Siva TK, Umesh K, Arjun S. 2011. Isolation and screening of endophytic actinomycetes from different parts of Emblica officinalis. *Ann. Biol. Res.*, 2(4), 423–434.

2 Recent Traditional Methods for Isolating Diverse Actinobacteria

Omnia Mohamed Abdallah and Dina Hatem Amin

2.1 CHAPTER 1: DIVERSITY OF ACTINOBACTERIA

Actinobacteria belong to the filamentous Gram-positive bacterial group that has extra guanine–cytosine (G + C) content [1]. High number of actinobacteria produce mycelium.

Actinobacteria communities are found frequently in soil, freshwater, and marine environments and habitats with extreme conditions.

There are pathogens from actinobacteria such as *Mycobacterium tuberculosis, Nocardia* spp., and *Actinomyces* spp. The actinobacterial pathogens are classified as human opportunistic pathogens, and they are the causative agents of tuberculosis, actinomycosis, and *nocardiosis*, respectively.

The analysis of genome sequence data revealed the genomic heterogeneity, which elucidates the ability of actinobacteria to produce metabolites with various applications [2].

Extreme environmental conditions encourage the unique actinobacterial species to evolve and to develop unique techniques to adapt to the limited nutrient conditions that significantly affect the structure of microbial communities [3], such as desert and marine habitats.

A large number of actinobacteria isolates from extreme habitats have unique capabilities: *Streptomyces bullii* [4] can grow in a medium with pH of up to 11; *Streptomyces aridus* [5] has tolerance to grow in the temperature range between 10 and 40°C; and *Streptomyces asenjonii* [6] and *Streptomyces leeuwenhoekii* [7] can grow in desert conditions with the capability of producing antibacterial metabolites. Genus *Geodermatophilus* is another desert actinobacterium that has nine new species—members of this genus have reported radiation-resistant capabilities.

2.2 SELECTIVE ISOLATION METHODS OF ACTINOBACTERIA

The materials used for the isolation of Actinobacteria are as follows:

70% ethanol, 3.25% NaOCl (NA, Difico), (TSA, Difico), Cycloheximide (50 µg/mL), nalidixic acid (10 µg/mL), nystatin (50 µg/mL), rifampicin (5 µg/mL), neomycin sulfate (10 µg/mL), calcium chloride, phenol, SDS, yeast extract, germicides, chemotactic agents, and chloramine-T, Ringer's solution (60 mg of sodium chloride, 31 mg of sodium lactate, 3 mg of potassium chloride, and 2 mg of calcium chloride per 50 mL of solution (pH 6.5)) (Figure 2.1).

Steps of actinobacteria isolation reporting different sample pretreatment techniques.

2.2.1 Isolation of Endophytic Actinobacteria

Separation of surface debris from plant tissue through surface sterilization is required for selective isolation of endophytic actinobacteria.

DOI: 10.1201/9781003398400-2

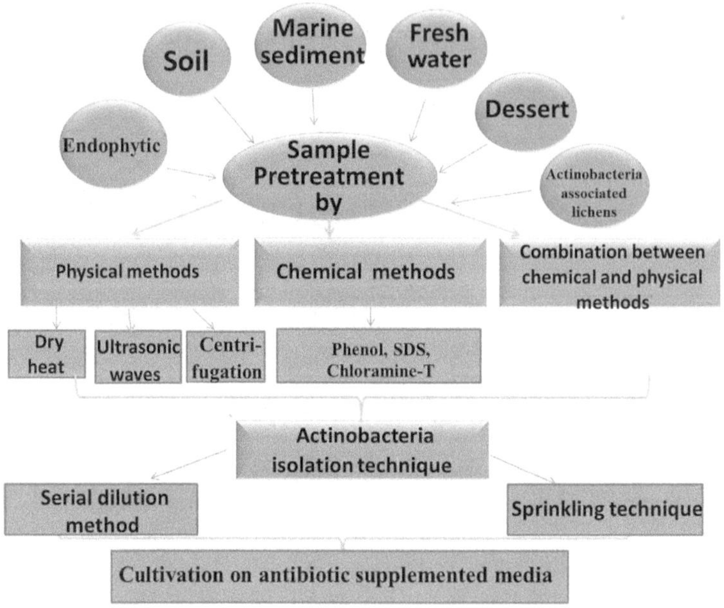

FIGURE 2.1 Diagram showing different methods of pretreatment of samples and isolation of actinobacteria.

2.2.2 Method 1

2.2.2.1 Surface Sterilization

- Soak plant tissue in 70% of ethanol for 1 minute.
- Soak in 3.25% NaOCl for 6 minutes, followed by a 1-minute soaking in 70% of ethanol and a 1-minute rinse with sterile water five times [8].

To confirm the surface sterilization, the sterilized tissues were overlaid onto nutrient agar and tryptic soy agar and incubated at 30°C for 1 week.

2.2.2.2 Isolation Technique

- Under aseptic conditions, 5 g surface-sterilized samples from three sampling branches were cut up and ground in a sterile mortar and then were diluted to 10^{-1}, 10^{-2}, and 10^{-3} with sterile water.

2.2.2.3 Incubation

An aliquot of 200-μL dilutions were spread over the surface of HV media supplemented with antibiotics and incubated at 30°C for 5 days.

2.2.3 Method 2

Another technique is used for selective isolation of endophytic actinobacteria; first you need to remove surface debris from plant samples using running tap water; then follow the procedure this procedure requires rinsing the samples three times with sterile water in between steps.

- The plant samples were segmented into stem, root, and leaf parts aseptically using sterilized scissors.
- Samples were subjected to sonification for 30 minutes at 45 kHz to remove the soil particles and organic matter.

2.2.3.1 The Sterilization of Surface

2.2.3.1.1 Five-Step Protocol

- Rinse plant segments in Tween 20 with concentration of 0.1% for 1 minute and then immerse the pieces in 5% sodium hypochlorite (NaOCl) for 4 minutes in case of leaf parts and 6 minutes in case of stem and root parts.
- Wash by 2.5% $Na_2S_2O_3$ for 10 minutes and soak the samples in ethanol with concentration of 70% for 4 minutes in case of leaves and for 6 minutes in case of stems and roots.
- Wash the segments by $NaHCO_3$ 10% for 10 minutes to inhibit the growth of fungi.

To test the success of the sterilization steps, inoculate 0.2 m/L of the sterile distilled water from the final wash process of each tissue onto ISP2 [9] and incubate at 28°C for 7 days.

- Dry the samples on sterilized filter paper and put them in a sterile Petri dish with subjecting to the laminar flow for one day.
- Grind the sterilized samples in a sterilized blender for 20 seconds in case of leaves and 30 seconds in case of stems and roots [10,11].

Pre-treatment was performed to eliminate bacteria in the plant tissues (roots, stems, and leaves) by one of following methods:

- **Freezing treatment:** the method involves freezing plant tissues at –80°C for 2 weeks; the other treatment involves drying plant tissues at 100°C for 20 minutes.
- Weigh aseptically 1 g of each treated tissue samples (roots, stems, and leaves) and grindi with a sterile mortar in the presence of 9 mL double-sterilized distilled water.

2.2.3.2 Isolation Media and Incubation

The treated samples solution was left at room temperature for 10 minutes for precipitation, the supernatant was serially diluted by concentration of (10^{-2} and 10^{-3}), and an inoculum of 0.2 mL was inoculated onto different selective isolation media supplemented with antibiotics like nutrient agar (Difico), tryptic soya agar (Difico), and tea-extract agar (tryptone 10 g, tea extract: 5.0 mL, NaCl 5 g, yeast extract 5 g, agar 18 g, 1 L distilled water, pH 7.3) were used for the selective isolation. Tea extract was prepared by leaf sterilization by 0.22 µm filter; tea-extract medium was prepared by adding sterilized tea extract to the sterilized basic minimum medium at 50°C and HV media supplemented with antibiotics (the concentrations are added in each medium by 50–100 mg cycloheximide, 100 mg nystatin, 25 mg nalidixic acid, and 5 mg penicillin per 1000 mL) [12].

- Isolate the separated colonies that formed in single form and have firm association to the medium and powdery appearance on IS4 media.

2.2.4 Isolation of Marine Actinobacteria

Pretreatment methods are used to prepare the samples for actinobacteria growth prior to the isolation by dilution and plate technique.

- Collect soil samples from primeval forest, saline, alkaline soils, and desert at depth of 5–20 cm and sediment in Deep Ocean with sampler.
- The samples are preserved at 4°C in sterilized glass container.

2.2.4.1 Pretreatment Can be Done Using Several Techniques

- Air drying of samples like soil, sediment, lichen, and fecal samples at room temperature.
- Heating of air-dried soil samples at 120°C for an hour.

- Using moist heat by heating samples at 60°C for 1 h in a water bath.
- Heating air-dried soil samples at 100°C for 15 minutes.
- Heating of soil suspensions or water samples at 45°C or 50°C for 10 minutes.
- Heating air-dried soil samples at 120°C for an hour.
- Heating air-dried soil samples at 28°C for a week.
- Heating soil suspension at 110°C for an hour.

Pretreatment with $CaCO_3$ promotes the growth and development of aerial mycelia in several actinobacteria and is done using the following procedure:

- The sediment samples were mixed with $CaCO_3$ by 1:1 ratio and incubated at 28°C for 10 days.
- After the treatment with calcium carbonate, the samples were sprinkled over one of the selective media for actinobacteria.

2.2.5 Isolation Technique

2.2.5.1 Method 1: Serial Dilution Method

2.2.5.1.1 Serial Dilution Method Procedure

- Adding aseptically 1 mL of the seawater or sediment samples into 9 mL of sterile distilled water (dilution 1:10).

2.2.5.2 Isolation Technique

- Further dilutions up to 1:1,000 were then prepared.
- Inoculate 0.1 mL from each dilution of the water and sediment sample over the plates of selective media: ryptone-yeast extract broth (ISP-1), yeast extract-malt extract agar (ISP-2), oatmeal agar (ISP-3), inorganic salts-starch agar (ISP-4), glycerol-asparagine agar (ISP-5), peptone-yeast extract iron agar (ISP-6), tyrosine agar (ISP-7), nutrient agar, starch casein agar (SCA), and actinobacteria isolation agar (AIA).
- Adjust the pH of all media to 9 by adding separately autoclaved Na_2CO_3 and salt concentrations in the range 0%–15%.
- Incubation was carried out at 28°C for 2 weeks reaching to 8 weeks.

2.2.5.3 Method 2: Enrichment Method

- Water and sediment samples were enriched with different concentrations of NaCl in the range of 0%–15% and incubated at 28°C for 7 days and then the enriched samples were then subjected to dilution and plating for the isolation of the actinobacteria [13].
- Pure actinobacterial colonies are maintained on media slant at 4°C and 20% (w/v) glycerol suspensions at –20°C.

2.2.6 Isolation of Actinobacteria from Fresh Water

2.2.6.1 Pretreatment

- The collected samples were subjected to air drying.
- The dried samples are treated by wet heat: heating in a water bath (55°C for 20 minutes) or treatment with phenol solution at conditions of (concentration 1.5%, for 30 minutes at 30°C).

Isolation technique

- One milliliter of treated sediments was added to 9 mL of 25% Ringer's solution in a test tube, and the solution was mixed for 40 minutes.

- The pretreated samples were serially diluted with Ringer's solution to 10^{-4}.
- Inoculate 100 µL of each dilution on selective media used for actinobacteria, including humic acid–vitamin agar (HV) [14], starch casein agar (SCA) [15], nocardia agar (NA) [16], M1 agar [17] supplements with concentrations of antibiotics.
- Isolate the single well separated colonies that have firmly attached to the medium and have a powdery appearance on IS4 media.

2.2.7 Isolation of Soil Actinobacteria

Pretreatment techniques of soil samples can support the selective isolation of actinobacteria by either promoting growth of actinobacteria or eliminating most undesired bacteria, especially Gram-negative bacteria. Using physical, chemical, and/or mixing of both methods are suggested options for selective isolation of actinobacteria species.

2.2.7.1 Physical Treatment

Heat drying of soil sample at 120°C for 1 hour is suitable for selective isolation of actinobacteria species because spores of actinobacteria can stand heat without desiccation more than Gram-negative bacteria and hence promote growth of *Streptomyces* and other rare genera on humic acid vitamin-supplemented (HV) media [18,19].

Heating of soil samples at 110°C for 10 minutes for inhibition of growth of bacteria and fungi [20].

Centrifugation is selective for isolation of motile actinobacteria and eliminates *Streptomycetes* and other non-motile actinobacteria; the motile species can be recovered in the supernatant and can be inoculated on suitable media containing nalidixic acid and trimethoprim [21].

Ultrasonic waves: Sample suspensions can be treated with ultrasonic waves at 180 W for 40 that can be used for pretreatment of sample suspensions. It helps in releasing the actinobacterial isolates associated with soil granules into the suspension [22].

2.2.7.2 Membrane Baiting Technique

Actinobacteria are characterized by the presence of mycelium; this feature is used for selective isolation of actinobacteria species from mixed microbial populations through:

- Overlaying of 0.22–0.45 µm pore size cellulose filter membrane on suitable prepared media.
- Spreading of the soil on the settled membrane and incubating for 3–4 days; the branched mycelium of actinobacteria has selective ability to penetrate the pores of membrane filter.

During incubation, the mycelium of actinobacteria penetrates the filter pores reaching to the agar medium, and the growth of other bacteria remains on the filter surface.

Finally, the filter membrane is eliminated and the culture is returned to incubation for 2–3 weeks to allow the well-developed isolated actinobacteria colonies to replicate [22].

2.2.8 Chemical Treatments

Calcium chloride, yeast extract, SDS, germicide, chemotactic agents, chloramine-T, and phenol can be used for selective isolation of actinobacteria (Ravi Rajan et al., 2016).

- **Phenol treatment** is applied by the addition of phenol solution by concentration of 1.4% (w/v) to soil suspensions but gave lower colonies count from untreated suspensions [21].
- **ChloramineT**-treated soils promoted the selective growth of *Streptosporangiaceae* species on HV agar supplemented with nalidixic acid [21].

- **SDS 0.05% and yeast extract 5% treated soil samples** induced the growth of *Streptomyces* and other genera on HV agar supplemented with nalidixic acid [22].
- Combination of physical and chemical treatment was also found to be a more suitable method to effectively isolate various genera of actinobacteria.
- **Dry heating at 110°C for 1 hour and treatment with 1% phenol** induce recovering of *Actinomadura* spp. on kanamycin, josamycin, lysozyme, and nalidixic acid-containing HV agar [23].
- **Addition of 0.01% benzethonium chloride to dry-heated soil samples at 120°C** induces isolation of *Streptosporangium* or *Dectylosporangium* genera on nalidixic acid- and leucomycin-containing HV agar [24].

2.2.8.1 Isolation Technique

- Use the vortex to suspend the soil and leave for 30 minutes; then, soil samples were serially diluted (10^{-3} 10^{-6}).
- One hundred microliters of each "dilution" was plated on nutrient agar (NA), actinomycetes isolation agar (AIA), yeast malt glucose agar (M6), starch casein agar, and Czapek Dox agar.
- Isolate the single well-separated colonies that they have a sticked firm attachment appearance to the medium and powdery appearance on IS4 media.

2.2.9 Isolation of Actinobacteria Associated Lichens

- Grind 2 g of each dried sample in a sterilized glass homogenizer.
- Dissolve the grinded samples in 18 mL of sterile water and add 0.1% $Na_4P_2O_5$.
- Shake the suspension at 220 rpm/minute for 60 minutes.
- Subject the suspension to ultrasound waves at 150 W for 40 seconds.

2.2.10 Isolation Technique

- Dilute the suspension by concentration of 10^{-1} to 10^{-5}, and 0.1 mL of three dilutions, 10^{-3}, 10^{-4}, and 10^{-5}, was inoculated on the plates and incubated for 10–25 days at 28°C on selective isolation media such as casein soybean peptone medium.
- Pick up the single actinobacteria colony and inoculate to a slant with the same isolation medium.

2.2.11 Isolation of Actinobacteria in Animal Feces

- Collect fresh waste samples and transfer immediately to sterile glass dishes and allow them to dry for 10 days at 28°C.
- Pretreat 2 g of each dried sample using heat at 80°C for 1 hour.
- Suspend the pretreated sample in 18 mL of sterile water containing 0.1% $Na_4P_2O_5$ and then shake at 220 rpm/minute for 60 minutes.
- Treat the suspension with ultrasound waves for 40 seconds at 150 W before coating.

Isolation technique:

Dilute the suspension from 10^{-1} to 10^{-7} and use the three dilutions, 10^{-5}, 10^{-6}, and 10^{-7}, for isolating actinobacteria.

Isolation media:

The selective media that can be suitable for isolation of actinobacteria from animal feces are modified starch casein, HV, modified glycerol asparagine media supplemented with potassium bichromate 50 mg, and antibacterial such as penicillin 5 mg or mix of antifungal and antibacterial like nystatin 50 mg, nalidixic acid 20 mg, and 5 mg penicillin per 1 L should be added into the isolation medium to inhibit the growth of Gram-negative bacteria and fungi [24].

- Isolate the separated colonies solely that have firm association to the medium and powdery appearance on IS4 media.

2.2.12 Isolation of Desert Actinobacteria

2.2.12.1 Isolation Technique

The sprinkling technique is proper for the isolation of different classes of actinobacteria from desert [25].

Sprinkling technique is done by inoculateing the pretreated soil granules directly over the solid selective media.

2.2.12.2 Isolation Media

The suitable media for selective isolation are enriched starch casein agar and actinobacterial isolation media supplemented with antibiotics to inhibit the growth of fungi, G+ve and G–ve bacteria:

- Amphotericin B (20 µg/mL), calcium propionate (30 mg/mL), nystatin (25 µg/mL), potassium dichromate (45 mg/L), and cycloheximide (25 µg/mL) have been used for fungi and yeast.
- Penicillin G and vancomycin can be used against Gram-positive bacteria [26]. Polymyxin B (25 mg/L), colistin (2 µg/mL), nalidixic acid (25 mg/L), and ceftazidime (8 µg/mL) can inhibit the growth of Gram-negative bacteria.
- A number of antibiotics are effective against both Gram-positive and Gram-negative bacteria, including cefixime, trimethoprim, chloramphenicol, rifampicin (5 µg/mL), erythromycin, gentamicin, neomycin, cefamandole (4 µg/mL), fosfomycin, and novobiocin (25 µg/mL) [26].

2.2.13 Incubation Conditions

- Incubation is applied at different temperatures to recover different actinobacteria taxa.
- The optimum temperature for numerous number of actinobacteria species is at 25°C–30°C, so it can be the incubation temperature for growing desert actinobacteria.
- Some species in desert microbial community can grow in a temperature range of (19°C–21°C),
- Incubate at a temperature of 45°C for selective isolation of thermophilic species.
- Isolate the single purified colonies that have firm association to the medium and powdery appearance on IS4 media (Figure 2.2).

Cultivated Streptomyces on starch casein agar media showing colonies morphology.

2.3 CONCLUSION

Various methods of pretreatment are used to eliminate fast-growing microorganisms and encourage the growth of actinobacteria. This chapter shows a number of pretreatment and isolation methods that can be used to isolate actinobacterial species from extreme conditions like desert, endophytic actinobacteria, and actinobacteria associated lichens with suitable selective media for the selective isolation of actinobacteria.

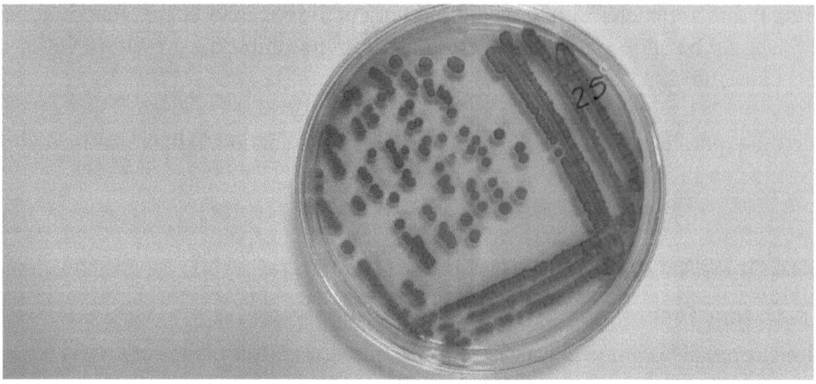

FIGURE 2.2 Representative of pure culture of actinobacteria isolated from Egyptian Sinai desert.

2.4 MEDIA COMPONENTS

2.4.1 Selective Isolation Media

Casein soybean peptone medium: casein 1.5 g, soybean peptone 0.5 g, K_2HPO_4 H_2O 1 g, $MgSO_4$ $7H_2O$ 0.5 g, $CaCO_3$ 0.3 g, NaCl 5 g, vitamin mixtures, agar 15 g, pH 7.5.

Glycerol-asparagine medium: glycerol 10 g, asparagine 1 g, K_2HPO_4 H_2O 1 g, $MgSO_4$ $7H_2O$ 0.5 g, $CaCO_3$ 0.3 g, vitamin mixture of HV medium 3.7 mg, and agar 15 g, pH 7.2.

Starch casein agar: starch 10 g, casein 0.3 g, KNO_3 2 g, $MgSO_4$ $7H_2O$ 0.05 g, NaCl 2 g, K_2HPO_4 2 g, $CaCO_3$ 0.02 g, $FeSO_4$ 10 mg, agar 15 g, pH 7.2–7.4.

Oatmeal medium: oatmeal 20 g (cook or steam 20 g oatmeal in 1,000 mL distilled water for 20 minutes, filter through cheese cloth, and add distilled water to restore volume of filtrate to 1,000 mL).

Czapek medium: sucrose 20 g, $NaNO_3$ 2 g, K_2HPO_4 1 g, $MgSO_4$ $7H_2O$ 0.5 g, KCl 0.5 g, $FeSO_4$ $7H_2O$ 0.01 g, vitamin mixtures [26] 3.7 mg, agar 25 g, pH 7.2.

HV medium: humic acid 1.0 g, Na_2HPO_4 0.5 g, KCl 1.7 g, $MgSO_4$ $7H_2O$ 0.05 g, $FeSO_4$ $7H_2O$ 0.01 g, $CaCl_2$ 1 g, B-vitamins (0.5 mg each of thiamine–HCl, riboflavin, niacin, pyridoxine, Ca-pantothenate, inositol, p-aminobenzoic acid, and 0.25 mg of biotin), agar 18 g, pH 7.2.

Actinobacteria isolation agar (2.0 g sodium caseinate, 0.1 g asparagine, 4.0 g sodium propionate, 0.5 g K_2HPO_4, 0.1 g $MgSO_4{\cdot}7H_2O$, 0.001 g $FeSO_4{\cdot}7H_2O$, and 20.0 g agar in 1,000 mL distilled water; pH 7.0) [26].

REFERENCES

1. Barka, E. A., Vatsa, P., Sanchez, L., Gaveau-Vaillant, N., Jacquard, C., MeierKolthoff, J. P., et al. 2016. Taxonomy, physiology, and natural products of actinobacteria. *Microbiology and Molecular Biology Reviews*. 80, 1–43.
2. Jose, P. A., Maharshi, A., Jha, B. 2021. Actinobacteria in natural products research: progress and prospects. *Microbiological Research*. 246, 126708.
3. Hozzein, W. N., Goodfellow, M. 2007. *Streptomyces synnematoformans* sp. nov., a novel actinomycete isolated from a sand dune soil in Egypt. *International Journal of Systematic and Evolutionary Microbiology*. 57, 2009–2013.
4. Santhanam, R., Rong, X., Huang, Y., Andrews, B. A., Asenjo, J. A., Goodfellow, M. 2013. *Streptomyces bullii* sp. nov., isolated from a hyper-arid AtacamaDesert soil. *Antonie van Leeuwenhoek*. 103, 367–373.
5. Idris, H., Goodfellow, M., Sanderson, R., Asenjo, J. A., Bull, A. T. 2017. Actinobacterial rare biospheres and dark matter revealed in habitats of the Chilean Atacama Desert. *Scientific Reports*. 7, 110, 705–717.
6. Abdelkader, M., Philippon, T., Asenjo, J. A., Bull, A. T., Goodfellow, M., Ebel, R., et al. 2018. Asenjonamides A-C, antibacterial metabolites isolated from Streptomyces asenjonii strain KNN 42.f from an extreme-hyper arid Atacama Desert soil. *The Journal of Antibiotics*. 71, 425–431.

7. Gran-Scheuch, A., Trajkovic, M., Parra, L., Fraaije, M. W. 2018. Mining the Genome of Streptomyces leeuwenhoekii: two new type I baeyer-villiger Monooxygenases from Atacama Desert. *Frontiers in Microbiology.* 9, 1609.
8. Qin, S., Li, J., Chen, H. H., Zhao, G. Z., Zhu, W. Y., Jiang, C. L., et al. 2009. Isolation, diversity, and antimicrobial activity of rare actinobacteria from medicinal plants of tropical rain forests in Xishuangbanna, China. *Applied and Environmental Microbiology.* 75, 6176–6186.
9. Shirling, E. T., Gottlieb, D. 1966. Methods for characterization of *Streptomyces* species. *International Journal of Systematic and Evolutionary Microbiology.* 16, 313–340.
10. Qin, S., Wang, H. B., Chen, H. H., Zhang, Y. Q., Jiang, C. L., Xu, L. H., et al. 2008. *Glycomyces endophyticus* sp. nov., an endophytic actinomycete isolated from the root of Carex baccans Nees. *International Journal of Systematic and Evolutionary Microbiology.* 58, 2525–2528.
11. Li, L., Mohamad, O. A. A., Ma, J., Friel, A. D., Su, Y., Wang, Y., et al. 2018. Synergistic plant-microbe interactions between endophytic bacterial communities and the medicinal plant *Glycyrrhiza uralensis* F. *Antonie Van Leeuwenhoek.* 111, 1735–1748.
12. Sheikh, M., Rathore, D. S., Gohel, S. D., Singh, S. 2019. Cultivation and characteristics of the Marine Actinobacteria from the Sea water of Alang, Bhavnagar. *Indian Journal of Marine Sciences.* 48, 12, 1896–1901.
13. Kuester, E., Williams, S. T. 1964. Selection of media for isolation of *streptomycetes. Nature.* 202, 928–929.
14. Sanglier, J. J., Whitehead, D., Saddler, G. S., Elayne V. Ferguson, Michael Goodfellow. 1992. Pyrolysis mass spectrometry as a method for the classification, identification and selection of actinomycetes. *Gene.* 115, 1–2, 235–242.
15. Mincer, T. J., Jensen, P. R., Kauffman, C. A., William Fenical. 2002. Widespread and persistent populations of a major new marine actinomycete taxon in ocean sediments. *Applied and Environmental Microbiology.* 68, 10, 5005–5011.
16. Hayakawa, M., Sadakata, T., Kajiura, T., Nonomura, H. 1991. New methods for the highly selective isolation of *Micromonospora* and *Microbispora* from soil. *Journal of Fermentation and Bioengineering.* 72, 320–326.
17. Tamura, T., Hayakawa, M., Hatano, K. 1997. A new genus of the order *Actinomycetales, Spirilliplanes* gen. nov., with description of Spirilliplane *Syamanashiensis* sp. nov. *International Journal of Systematic and Evolutionary Microbiology.* 47, 97–102.
18. Agate, A. D., Bhat, J. V. 1963. A method for the preferential isolation of actinomycetes from soils. *Antonie Van Leeuwenhoek.* 29, 297–304.
19. Hayakawa, M. 2008. Studies on the isolation and distribution of rare actinomycetes in soil. *Actinomycetologica.* 22, 12–19.
20. Jiang, Y., Cao, Y. R., Zhao, L. X., Wang, Q., Jin, R. X., He, W. X., Xue, Q. H. 2010. Treatment of ultrasonic to soil sample for increase of the kind of rare actinomycetes. Acta Microbiologica Sinica, 50(8):1094–7.
21. Hirsch, C., Christensen, D. 1983. Novel method for selective isolation of Actinomycetes. *Applied and Environmental Microbiology.* 46, 925–929.
22. Lawrence, C. 1956. A method of isolating actinomycetes from scabby potato tissue and soil with minimal contamination. *Canadian Journal of Botany.* 34, 44–47.
23. Jiang, Y., Li, Q., Chen, X., Jiang, C. 2016. Isolation and cultivation methods of Actinobacteria. In D. Dhanasekaran & Y. Jiang (Eds.), *Actinobacteria – Basics and Biotechnological Applications,* IntechOpen, 40–52.
24. Xie, F., Pathom-aree, W. 2021. Actinobacteria from desert: Diversity and biotechnological applications. *Frontiers in Microbiology.* 12, 765531.
25. Bonnet, M., Lagier, J. C., Raoult, D., Khelaifia, S. 2019. Bacterial culture through selective and non-selective conditions: the evolution of culture media in clinical microbiology. *New Microbes and New Infections.* 34, 100622.
26. Masand, M., Sivakala, K. K., Menghani, E., Thinesh, T., Anandham, R., Sharma, G., Sivakumar, N., Jebakumar, S. R. D., Jose, P. A. 2018. Biosynthetic potential of bioactive streptomycetes isolated from arid region of the Thar Desert, Rajasthan (India). *Frontiers in Microbiology.* 18, 9, 687.

3 Isolation of Actinobacteria from Caves and Limestone Habitats

Pharada Rangseekaew and Wasu Pathom-aree

3.1 INTRODUCTION

Extreme environments are unique and special ecosystems which can be characterized by the presence of physio-chemical stressors such as high salt concentration, limited availability, low organic matter, high pressure, and high/low temperature. Extremophilic microorganisms which are able to grow under these conditions require specific adaptations that are absent in other species or same species from normal environments (Begon et al., 2014; Gómez, 2011; Sapkota, 2020, Tobler, 2008). Extreme environments are good sources for discovery of novel microbial taxa and useful microorganisms for biotechnological applications. Actinobacteria are prolific producers of bioactive compounds and show potential in various applications, including agriculture and environment. They have been discovered in various extreme environments such as deserts (Xie and Pathom-aree, 2021), marine habitats (Kamjam et al., 2017), caves (Niyomvong et al., 2012; Cheeptham et al., 2013; Wiseschart et al., 2018; Rangseekaew and Pathom-aree, 2019; Farda et al., 2022), and limestone quarry (Nimaichand et al., 2015).

Limestone habitats are the area rich in calcium carbonate. In this research, we focus on limestone caves, karst landscapes, and quarries. They represent an extreme environment in terms of light, humidity, nutrients, and temperature (Gómez, 2011). Karst is a unique landscape which forms from the dissolution of soluble rocks, especially carbonate rocks (limestone and dolomite) by surface or groundwater (Alpha et al., 1997; Bakalowicz, 2005; Waele, 2017). It is estimated that karst landscape covers around 7–10% of the earth's surface (Alpha et al., 1997; Bakalowicz, 2005). Cave is one of the ecosystems in karst landscape which possess unique conditions, including low organic matter content, low light intensity, low stable temperature, and high humidity (Tomczyk-Żak and Zielenkiewicz, 2016). The special features of these limestone habitats would promote special properties for survival of microorganisms that live in such extreme environments. Few studies have been reported on actinobacterial diversity in limestone habitats, partly due to the difficulty in isolating actinobacteria from such habitats (Niyomvong et al., 2012; Rangseekaew and Pathom-aree, 2019; Nimaichand et al., 2015; Fang et al., 2017). This chapter is designed to introduce and share the protocols for isolation of actinobacteria from caves and other limestone habitats in particular dripping water and speleothems to interested microbiologists.

3.2 MATERIALS

3.2.1 Selective Media

The following selective media were used for isolating actinobacteria from different types of samples taken from caves and other limestone habitats. Starch casein agar (Küster and Williams, 1964), water proline agar (Matsumoto et al., 2006), humic acid vitamin agar (Hayakawa and Nonomura, 1987), raffinose-histidine agar (Vickers et al., 1984), and R2A agar (Reasoner and Geld Reich, 1985).

 DOI: 10.1201/9781003398400-3

3.2.2 Antibiotics Supplemented in Selective Media

All selective media for actinobacterial isolation were supplemented with antibiotics to inhibit or suppress unwanted microorganisms. Generally, nalidixic acid and ketoconazole were used to prevent the growth of Gram-negative bacteria and fungi at a concentration of 25 and 100 μg/mL, respectively.

3.3 METHODS

3.3.1 Isolation of Actinobacteria from Cave Soils

1. Top soil samples (5 cm) were collected using a sterile spoon and put in a sterile zip-lock bag or bottle. The samples were air-dried for 3–7 days in a sterile Petri dish at room temperature.
2. Soil (1 g) was suspended in 0.85% (w/v) sodium chloride solution (9 mL) and mixed by vortexing.
3. The soil suspension was treated with microwave radiation at a power setting of 800 W for 30 seconds at a frequency of 2,460 MHz (Niyomvong et al., 2012).
4. The treated soil suspension was left at room temperature for 5 minutes to precipitate the soil particles.
5. The top layer was transferred to a new tube and subjected to a serial 10-fold dilution to make up 10^{-5} dilution.
6. An aliquot of 100 μL of suspension was inoculated on selective media amended with 25 μg mL^{-1} nalidixic acid and 100 μg mL^{-1} ketoconazole in triplicates using spread plate technique.
7. All inoculated plates were cultured in an incubator for up to 4 weeks at 28°C.
8. Presumptive colonies of actinobacteria were counted every week and the number of colonies recorded as colony-forming unit per gram of soil dry weight (CFU/g).
9. Presumptive colonies were purified by restreaking on glucose peptone meat extract (GPM) agar (Takahashi et al., 2003) and Seino agar (Intra et al., 2011).
10. Pure colonies were subcultured on International *Streptomyces* Project 2 (ISP2) agar or other suitable culture media and used as a working stock.
11. Each pure culture of actinobacteria was stored as a mixture of spores and mycelium fragments at –20°C/–80°C in 20% (v/v) sterile glycerol solution containing cryotubes for long-term preservation.

3.3.2 Isolation of Actinobacteria from Dripping Water Samples

1. Dripping water sites were selected inside the cave. The sampling site should be devoid of human activities.
2. Dripping water samples were collected using sterile tubes such as 15 or 50 mL Falcon tubes.
3. Samples should be kept at 4°C during transportation to the laboratory and preferably subjected to selective isolation. The samples can also be stored under refrigerated condition until use.
4. Dripping water samples were pretreated with microwave radiation with a power setting of 800 W for 30 seconds at 2,460 MHz of frequency (Niyomvong et al., 2012).
5. Pretreated dripping water sample (100 μL) was inoculated on selective media supplemented with 25 μg mL^{-1} nalidixic acid and 100 μg/mL ketoconazole in triplicates using spread plate technique.
6. All inoculated plates were cultured in an incubator for up to 4 weeks at 28°C.
7. Presumptive colonies of actinobacteria were recorded every week and the number of colonies reported as colony-forming unit per mL of dripping water (CFU/mL).

8. Actinobacterial colonies with different morphology were selected and purified by a series of streak plates on GPM agar (Takahashi et al., 2003) and Seino agar (Intra et al., 2011). Pure culture of each actinobacteria was grown on the suitable culture media such as International *Streptomyces* Project 2 (ISP2) agar (Shirling and Gottlieb, 1966).
9. A mixture of spores and mycelium fragments of each pure culture of actinobacteria were stored in 20% (v/v) sterile glycerol solution at –20°C/–80°C for long-term preservation.

3.3.3 Isolation of Actinobacteria from Speleothem Samples (Stalactites and Stalagmites)

1. Speleothem samples of at least 1 g (stalactites and stalagmites) were collected from undisturbed location from human and anthropogenic activities using sterile forceps, scalpel, geologic hammer, and chisel. All collecting equipments were cleaned with 70% (v/v) ethanol before use (Dhami et al., 2018; Mudgil et al., 2022).
2. Collected samples were placed in a sterile plastic zip-lock bag or sterile bottle using aseptic technique.
3. Samples were kept on ice at 4°C until they arrived at the laboratory and stored at –20°C until use.
4. Speleothem samples were aseptically ground to powder in a pre-sterilized mortar using sterile pestle.
5. Speleothem powder (1 g) was mixed with 9 mL of normal saline solution (0.85% (w/v) NaCl) to make 10^{-1} dilution and further diluted 10-fold using normal saline to 10^{-5}.

TABLE 3.1
Media for Selective Isolation and Cultivation of Actinobacteria from Caves and Limestone Habitats

Selective Media	Ingredients	References
Humic acid vitamin agar	Humic acid 1.0 g/L; KCl 1.71 g/L; Na_2HPO_4 0.5 g/L; $MgSO_4 \cdot 7H_2O$ 0.05 g/L; $CaCO_3$ 0.02 g/L; $FeSO_4 \cdot 7H_2O$ 0.01 g/L; B-vitamins (100×) 10 mL; Agar 15.0 g/L	Hayakawa and Nonomura (1987)
Raffinose-histidine agar	Raffinose 10.0 g/L; L-Histidine 1.0 g/L; $MgSO_4 \cdot 7H_2O$ 0.5 g/L; $FeSO_4 \cdot 7H_2O$ 0.01 g/L; K_2HPO_4 1.0 g/L; Agar 15.0 g/L	Vickers et al. (1984)
R2A agar	Casamino acids 0.5 g/L; dextrose 0.5 g/L; proteose peptone No. 3 0.5 g/L; soluble starch 0.5 g/L; yeast extract 0.5 g/L; sodium pyruvate 0.3 g/L; dipotassium phosphate 0.3 g/L; Magnesium sulfate 0.05 g/L; Agar 15.0 g/L	Reasoner and Geld Reich (1985)
Starch casein agar	Starch 10.0 g/L; KNO_3 2.0 g/L; NaCl 2.0 g/L; Difco vitamin-free casein 0.3 g/L; $MgSO_4 \cdot 7H_2O$ 0.05 g/L; $CaCO_3$ 0.02 g/L; $FeSO_4 \cdot 7H_2O$ 0.01 g/L; Agar 18.0 g/L	Küster and Williams (1964)
Water proline agar	Proline 10.0 g/L; Agar 15.0 g/L	Matsumoto et al. (2006)
General Cultivation Media	**Ingredients**	**References**
International *Streptomyces* Project 2 (ISP2) agar	Malt extract 10.0 g/l; dextrose 4.0 g/L; yeast extract 4.0 g/L; agar 20.0 g/L	Shirling and Gottlieb (1966)
Glucose peptone meat extract (GPM) agar	D-glucose 10.0 g/L; meat extract 5.0 g/L; peptone 5.0 g/L; NaCl 3.0 g/L; agar 15.0 g/L	Takahashi et al. (2003)
Seino agar	Starch 10.0 g/L; NZ amine 3.0 g/L; $CaCO_3$ 3.0 g/L; meat extract 1.0 g/L; yeast extract 1.0 g/L; agar 15.0 g/L	Intra et al. (2011)

6. Each diluted sample (100 µL) was spread on selective media amended with 25 µg/mL nalidixic acid and 100 µg/mL ketoconazole in triplicates and incubated at 28°C for up to 4 weeks.
7. The number of presumptive colony of actinobacteria was recorded weekly and reported as colony-forming unit per gram of speleothem sample (CFU/g).
8. The representative actinobacterial colonies were selected based on their colony morphology and re-streaked on GPM agar (Takahashi et al., 2003) and Seino agar (Intra et al., 2011) to obtain pure culture. Pure culture of each actinobacteria was grown on the suitable culture media such as International *Streptomyces* Project 2 (ISP2) agar (Shirling and Gottlieb, 1966).
9. A mixture of spores and mycelium fragments of each pure culture of actinobacteria were kept at –20°C/–80°C for long-term preservation in 20% (v/v) sterile glycerol solution (Table 3.1).

3.4 CONCLUSION

The selective isolation procedures for actinobacteria from limestone habitats were proposed in this chapter. The proposed methods have been used with success from samples ranging from cave soils, dripping water, and speleothems which will be useful for investigation of cultivable actinobacteria from limestone habitats.

REFERENCES

Alpha, T. R., Galloway, J. P., and Tinsley, I. C. 1997. Karst Topography Karst topography; computer animations and paper model. Geological survey. USGS ESIC open-file report Section [distributor]. https://doi.org/10.3133/ofr97536A (accessed December 15, 2022).

Bakalowicz, M. 2005. Karst groundwater: A challenge for new resources. *Hydrogeology Journal* 13: 148–60. https://doi.org/10.1007/s10040-004-0402-9.

Begon, M., Howarth, R. W., and Colin, R. T. 2014. *Essentials of Ecology* (4th ed.). New York: John Wiley & Sons.

Cheeptham, N., Sadoway, T., Rule, D., Watson, K., Moote, P., Soliman, L. C., Azad, N., Donkor, K. K., and Horne, D. 2013. Cure from the cave: Volcanic cave actinomycetes and their potential in drug discovery. *International Journal of Speleology* 42(1): 35–47. https://doi.org/10.5038/1827-806X.42.1.5.

Dhami, N. K., Mukherjee, A., and Watkin, E. L. 2018. Microbial diversity and mineralogical-mechanical properties of calcitic cave speleothems in natural and in vitro biomineralization conditions. *Frontier in Microbiology* 9: 40. https://doi.org/10.3389/fmicb.2018.00040.

Fang, B. Z., Salam, N., Han, M. X., Jiao, J. Y., Cheng, J., Wei, D. Q., Xiao, M., and Li, W. J. 2017. Insights on the effects of heat pretreatment, pH, and calcium salts on isolation of rare actinobacteria from karstic caves. *Frontier in Microbiology* 8: 1535. https://doi.org/10.3389/fmicb.2017.01535.

Farda, B., Djebaili, R., Vaccarelli, I., Del Gallo, M., and Pellegrini, M. 2022. Actinomycetes from caves: An overview of their diversity, biotechnological properties, and insights for their use in soil environments. *Microorganisms* 10: 453. https://doi.org/10.3390/microorganisms10020453.

Gómez, F. 2011. Extreme environment. In *Encyclopedia of Astrobiology*, eds. M. Gargaud, R. Amils, J. C. Quintanilla, H. J. Cleaves II, W. M. Irvine, D. L. Pinti, and M. Viso, pp. 570–572. Berlin, Heidelberg: Springer Berlin Heidelberg. https://doi.org/10.1007/978-3-642-11274-4_566.

Hayakawa, M., and Nonomura, H. 1987. Humic acid-vitamin agar, a new medium for the selective isolation of soil actinomycetes. *Journal of Fermentation Technology* 65(5): 501–509. https://doi.org/10.1016/0385-6380(87)90108-7.

Intra, B., Mungsuntisuk, I., Nihira, T., Igarashi, Y., and Panbangred, W. 2011. Identification of actinomycetes from plant rhizospheric soils with inhibitory activity against *Colletotrichum* spp., the causative agent of anthracnose disease. *BMC Researh Notes* 4(98): 1–9. https://doi.org/10.1186/1756-0500-4-98.

Kamjam, M., Sivalingam, P., Deng, Z., and Hong, K. 2017. Deep sea actinomycetes and their secondary metabolites. *Frontier in Microbiology* 8: 760. https://doi.org/10.3389/fmicb.2017.00760.

Küster, E., and Williams, S. T. 1964. Selection of media for isolation of streptomycetes. *Nature* 202: 928–929. https://doi.org/https://doi.org/10.1038/202928a0.

Matsumoto, A., Takahashi, Y., Iwai, Y., and Omura, S. 2006. Isolation of Gram-positive bacteria with high G+C from inside soil aggregates. *Actinomycetologica* 20(2): 30–34. https://doi.org/10.3209/saj.20.30.

Mudgil, D., Paul, D., Baskar, S., Baskar, R., and Shouche, Y. S. 2022. Cultivable microbial diversity in speleothems using MALDI-TOF spectrometry and DNA sequencing from Krem Soitan, Krem Lawbah, Krem Mawpun, Khasi Hills, Meghalaya, India. *Archives of Microbiology* 204(8): 1–15.

Nimaichand, S., Devi, A. M., Tamreihao, K., Ningthoujam, D. S., and Li, W. J. 2015. Actinobacterial diversity in limestone deposit sites in Hundung, Manipur (India) and their antimicrobial activities. *Frontier in Microbiology* 6: 413. https://doi.org/10.3389/fmicb.2015.00413.

Niyomvong, N., Pathom-aree, W., Thamchaipenet, A., and Duangmal, K. 2012. Actinomycetes from tropical limestone caves. *Chiang Mai Journal of Science* 39(3): 373–388.

Rangseekaew, P., and Pathom-aree, W. 2019. Cave actinobacteria as producers of bioactive metabolites. *Frontier in Microbiology* 10: 387. https://doi.org/10.3389/fmicb.2019.00387.

Reasoner, D. J., and Geld Reich, E. E. 1985. A new medium for the enumeration and subculture of bacteria from potable water. *Applied and Environmental Microbiology* 49(1): 1–7.

Sapkota, A. 2020. "Microbiology of extreme environments (types and examples)". Microbe Notes. December 15, 2022. https://microbenotes.com/microbiology-of-extreme-environments/#references.

Shirling, E. B., and Gottlieb, D. 1966. Methods for characterization of *Streptomyces* species. *International Journal of Bacteriology* 16(3): 313–340. https://doi.org/10.1099/00207713-16-3-313.

Takahashi, Y., Katoh, S., Shikura, N., Tomoda, H., and Omura, S. 2003. Superoxide dismutase produced by soil bacteria increases bacterial colony growth from soil samples. *The Journal of General and Applied Microbiology* 49(4): 263–266. https://doi.org/10.2323/jgam.49.263.

Tobler, M. 2008. Divergence in trophic ecology characterizes colonization of extreme habitats. *Biological Journal of the Linnean Society* 95(3): 517–528. https://doi.org/10.1111/j.1095-8312.2008.01063.x.

Tomczyk-Żak, K., and Zielenkiewicz, U. 2016. Microbial diversity in caves. *Geomicrobiology Journal* 33(1): 20–38. https://doi.org/10.1080/01490451.2014.1003341.

Vickers, J. C., Williams, S. T., and Ross, G. W. 1984. A taxonomic approach to selective isolation of Streptomycetes form soil. *Biological, Biochemical, and Biomedical Aspects of Actinomycetes* 1984: 553–561. https://doi.org/10.1016/B978-0-12-528620-6.50045-8.

Waele, J. De. 2017. Karst processes and landforms. In *International Encyclopedia of Geography*, eds. D. Richardson, N. Castree, M. F. Goodchild, A. Kobayashi, W. Liu, and R. A. Marston. New York: John Wiley & Sons. https://doi.org/10.1002/9781118786352.wbieg0968.

Wiseschart, A., Mhuanthong, W., Thongkam, P., Tangphatsornruang, S., Chantasingh, D., and Pootanakit, K. 2018. Bacterial diversity and phylogenetic analysis of type II polyketide synthase gene from Manao-Pee cave, Thailand. *Geomicrobiology Journal* 35(6): 518–27. https://doi.org/10.1080/01490451.2017.1411993.

Xie, F., and Pathom-aree, W. 2021. Actinobacteria from desert: Diversity and biotechnological applications. *Frontier in Microbiology* 12: 765531. https://doi.org/10.3389/fmicb.2021.765531.

4 Isolation of Endophytic Actinobacteria from Different Sources

Wasu Pathom-aree, Benyapa Kitwetch, Mathurin Meethangdee, and Pharada Rangseekaew

4.1 INTRODUCTION

Actinomycetota are a phylum of high G+C% Gram-positive bacteria which are widely distributed in nature as they can adapt to live in diverse ecological habitats. Soil is generally a major source of actinobacteria and a zone of active interactions between the actinobacteria and the root system of plants (Singh and Dubey, 2018). As a result, plant roots can be considered as an attractive source of endophytic actinobacteria. Endophytic actinobacteria are actinobacteria that live in the internal tissues of healthy plants without causing any negative effects or diseases to plants (Singh and Dubey, 2018; Dinesh et al., 2017). Actinobacteria can live inside plants during some growth stages or the entire life cycle in a mutualistic relationship; plants provide nutrients and shelter for endophytes and in turn obtain secondary metabolites from actinobacteria to enhance fitness and resilience against abiotic (drought and salinity) and biotic stress (insects, pests, and pathogens) (Singh and Dubey, 2018; Dinesh et al., 2017; Ek-Ramos et al., 2019). Endophytes can live in many locations within plants, namely leaves, roots, stems, and a few in flowers, fruits, and seeds (Ek-Ramos et al., 2019). Various plant species provide home for endophytic actinobacteria such as medicinal plants (Qin et al., 2009; Musa et al., 2020), crop plants, woody plants, and mangroves (Singh and Dubey, 2018). Almost 300,000 plant species around the world are hosts of actinobacteria that extremely contribute to biodiversity (Strobel et al., 2004).

Endophytic actinobacteria produce various secondary metabolites with diverse applications in agriculture, industrial, and pharmaceutical sectors (Ganapathy and Sivakumar, 2018; Singh et al., 2022). Plants benefit from endophytic actinobacteria via the production of phytohormones to improve plant growth, and they also protect plants from phytopathogens by producing antibiotics and siderophores (Ayswaria et al., 2020). Endophytic actinobacteria are also able to produce hydrolytic enzymes (chitinase, cellulase, hemicellulose, and lignin-degrading enzymes) that can be used in biorefinery and waste management (Ayswaria et al., 2020). In addition, various antimicrobial, anticancer, and anti-inflammatory substances have been reported from endophytic actinobacteria. Due to their remarkable potential, endophytic actinobacteria are regarded as valuable sources for search and discovery programs for various biotechnological applications. An effective approach to isolate endophytic actinobacteria from their host plants is essentially needed. In this chapter, we aim to present protocols useful for the isolation of endophytic actinobacteria from bryophytes, lettuce, maize, and tomato.

4.2 MATERIALS

4.2.1 Selective Media

Media specifically designed for selective isolation of actinobacteria are generally being used to recover endophytic actinobacteria from several plant species. Details of media composition and related references are summarized in Table 4.1.

DOI: 10.1201/9781003398400-4

TABLE 4.1
Media for Selective Isolation and Cultivation of Endophytic Actinobacteria

Selective Media	Ingredients	References
Actinomycete isolation agar (AIA)	Sodium caseinate 2.0 g/L; L-asparagine 0.1 g/L; sodium propionate 4.0 g/L; K_2HPO_4 0.5 g/L; $MgSO_4$ 0.1 g/L; FeSO4 0.001 g/L; agar 15 g/L	Gandhimathi et al. (2009)
Humic acid vitamin agar	Humic acid*1.0 g/L; Na_2HPO_4 0.5 g/L; KCl 1.71 g/L; $MgSO_4{\cdot}7H_2O$ 0.05 g/L; $FeSO_4{\cdot}7H_2O$ 0.01 g/L; $CaCO_3$ 0.02 g/L; B-vitamins (100×) 10 mL; Agar 15.0 g/L	Hayakawa and Nonomura (1987)
Rhizosphere mimicking agar		Brescia et al. (2020)
Raffinose-histidine agar	Raffinose 10.0 g/L; L-Histidine 1.0 g/L; $MgSO_4{\cdot}7H_2O$ 0.5 g/L; $FeSO_4{\cdot}7H_2O$ 0.01 g/L; K_2HPO_4 1.0 g/L; Agar 15.0 g/L	Vickers et al. (1984)
R2A agar	Yeast extract 0.5 g/L; proteose peptone No. 3 0.5 g/L; casamino acid 0.5 g/L; dextrose 0.5 g/L; soluble starch 0.5 g/L; sodium pyruvate 0.3 g/L; dipotassium phosphate 0.3 g/L; Magnesium sulfate 0.05 g/L; agar 15.0 g/L	Reasoner and Geld Reich (1985)
S-medium	Dextrose 10.0 g/L, casein hydrolysate 4.0 g/L; K_2HPO_4 0.5 g/L; $MgSO_4{\cdot}7H_2O$ 0.2 g/L; $CaCl_2{\cdot}2H_2O$ 0.1 g/L; ferric citrate 10 mg/L; CoSO4·$7H_2O$ 0.01 mg/L; CuSO4·$5H_2O$ 0.1 mg/L; H_3BO_3 1.5 mg/L; $MnSO_4{\cdot}H_2O$ 0.8 mg/L; $(NH_4)_6Mo_7O_{24}{\cdot}4H_2O$ 0.2 mg/L; $ZnSO_4{\cdot}7H_2O$ 0.6 mg/L	Cao et al. (2004) and Baker (1990)
Starch casein agar	Starch 10.0 g/L; Difco vitamin-free casein 0.3 g/L; KNO_3 2.0 g/L; NaCl 2.0 g/L; $MgSO_4{\cdot}7H_2O$ 0.05 g/L; $CaCO_3$ 0.02 g/L; $FeSO_4{\cdot}7H_2O$ 0.01 g/L; agar 18.0 g/L	Küster and Williams (1964)
Water agar	Agar 15.0 g/L	Bonaldi et al. (2015)
Water proline agar	Proline 10.0 g/L; Agar 15.0 g/L	Matsumoto et al. (2006)
General Culture Media	**Ingredients**	**References**
International *Streptomyces* Project 2 (ISP2) agar	Dextrose 4.0 g/L; yeast extract 4.0 g/L; malt extract 10.0 g/l; agar 20.0 g/L	Shirling and Gottlieb (1966)
Glucose peptone meat extract (GPM) agar	D-glucose 10.0 g/L; peptone 5.0 g/L; meat extract 5.0 g/L; NaCl 3.0 g/L; agar 15.0 g/L	Takahashi et al. (2003)
Nutrient agar	Peptone 5.0 g/L; NaCl 5.0 g/L; yeast extract 2.0 g/L; beef extract 1.0 g/L; agar 15 g/L	Passari et al. (2016)
Seino agar	Starch 10.0 g/L; NZ amine 3.0 g/L; yeast extract 1.0 g/L; meat extract 1.0 g/L; $CaCO_3$ 3.0 g/L; Agar 15.0 g/L	Intra et al. (2011)

4.2.2 Antibiotics Supplemented in Selective Media

All selective isolation media for endophytic actinobacteria were generally supplemented with antibiotics to prevent the growth of unwanted microorganisms. In general, nalidixic acid (25 μg/mL) was added to suppress the growth of fast-growing Gram-negative bacteria and ketoconazole (100 μg/ mL^{-1}) or cycloheximide (10 mg/mL) or nystatin (25 mg/mL) or benomyl (25 μg/mL) for fungi.

4.3 METHODS

4.3.1 Isolation of Endophytic Actinobacteria from Bryophytes

4.3.1.1 Sample Collection

Bryophyte samples (mosses and liverwort) were collected from their natural habitats. Only healthy plants without disease lesions were selected. Bryophyte samples were stored in zip-lock bags using an aseptic technique to prevent the plants from losing moisture and stored under refrigerated conditions at 4°C for longer preservation and processed within 48 hours.

Method 1 (Boonsnongcheep et al., 2016; Insuk et al., 2020)

1. The plant samples were disinfected using 70% (v/v) ethyl alcohol for 1 minute and 3% (v/v) sodium hypochlorite solution for 10 minutes. Samples were rinsed three times with sterile distilled water.
2. Sterile samples (1 g fresh weight) were crushed in a sterile mortar with 1 mL of sterile normal saline solution (0.85% (w/v) NaCl) and used to prepare serial dilutions up to 10^{-3}.
3. An aliquot of 100 µL from each dilution was spread on starch casein agar (Küster and Williams, 1964), humic acid vitamin B agar (Hayakawa and Nonomura, 1987), Reasoner's 2A (R2A) agar (Reasoner and Geld Reich, 1985), and water proline agar (Matsumoto et al., 2006) in triplicates. All media were supplemented with 25 mg/mL nystatin and 10 mg/mL cycloheximide.
4. All cultures were incubated for 30 days at 28°C. A colony of actinobacteria was subcultured on glucose peptone meat extract (GPM) agar (Takahashi et al., 2003) and Seino agar (Intra et al., 2011) to obtain pure culture.
5. The pure culture of each actinobacteria was grown on suitable culture media such as International *Streptomyces* Project 2 (ISP2) agar (Shirling and Gottlieb, 1966).

Method 2

1. The plant samples were washed with 30 mL of sterile distilled water two times.
2. An aliquot of 200 µL from the first washed water was inoculated in triplicates on water proline agar (Matsumoto et al., 2006) and 343 V8 juice agar amended with 25 µg/mL each of nalidixic acid and benomyl using pour plate technique.
3. All isolating plates were incubated for 7 days at 27°C.
4. For surface sterilization, plants were soaked in 70% (v/v) ethyl alcohol for 1 minute, followed by 1% (v/v) sodium hypochlorite for 1 minute. Samples were rinsed in sterile distilled water twice.
5. The disinfected plant samples were transferred to a sterile filter paper in a sterile glass dish and kept in a silica gel jar for 3 days.
6. Dried plant sample (1 g) was ground in 1 mL of sterile distilled water using a sterile mortar and used to prepare a series of tenfold dilutions up to 10^{-4}.
7. An aliquot of 200 µL from each dilution was mixed with water proline agar and 343 V8 juice agar amended with 25 µg/mL nalidixic acid and 25 µg/mL benomyl in triplicates using the pour plate technique.
8. Isolation plates were cultured in an incubator for 30 days at 27°C.
9. Presumptive colonies were subcultured by restreaking on GPM agar (Takahashi et al., 2003) and Seino agar (Intra et al., 2011) to obtain a pure culture.
10. Pure culture of each actinobacteria was grown on the suitable culture media such as International *Streptomyces* Project 2 (ISP2) agar (Shirling and Gottlieb, 1966).

Method 3 (Insuk et al., 2022)

1. The bryophyte sample was transferred into the sterile 50 mL centrifugation tube with sterile forceps. The soil surrounding the bryophyte was removed by shaking.
2. The soil was kept in an oven at 60°C for sample drying for 2 hours, and 1 g was used to prepare suspension in sterile Winogradsky's salt solution (K_2HPO_4 3.8 g/L; KH_2PO_4 1.2 g/L; $MgSO_4{\cdot}7H_2O$ 5.1 g/L; NaCl 2.5 g/L; $Fe_2(SO_4)_3$ 0.05 g/L; $MnSO_4{\cdot}5H_2O$ 0.05 g/L) and continued with a 10-fold serial dilution to make up the 10^{-4} dilution.
3. An aliquot of 200 μL from each dilution was mixed with water proline agar (Matsumoto et al., 2006) or 1/10 V8 Juice agar amended with 25 μg/mL each of nalidixic acid and benomyl in triplicates using the pour plate technique.
4. Plates were incubated at 27°C for 7 days.
5. The surface of plant samples was decontaminated in 70% (v/v) ethanol for 1 minute and 1% (v/v) sodium hypochlorite for 1 minute. Samples were further washed twice with sterile distilled water.
6. The plant samples were dried for 2 hours in an oven at 60°C.
7. Dried samples (1 g) were crushed in 1 mL of sterile Winogradsky's salt solution in sterile mortars and then the homogenized solution was serially diluted to make up 10^{-4} dilution.
8. An aliquot of 200 μL from each dilution was mixed with water proline agar (Matsumoto et al., 2006) or 1/10 V8 juice agar amended with nalidixic acid and benomyl each at 25 μg/mL.
9. Plates were cultured in an incubator for 30 days at 27°C.
10. Presumptive actinobacteria were restreaked on GPM agar (Takahashi et al., 2003) and Seino agar (Intra et al., 2011) to obtain a pure culture. The pure culture of each actinobacteria was grown on suitable culture media such as International *Streptomyces* Project 2 (ISP2) agar (Shirling and Gottlieb, 1966).

Method 4 (Insuk et al., 2022)

1. The plant samples were washed twice with sterile distilled water (30 mL).
2. An aliquot of 100 μL from the first washed water was inoculated on humic acid vitamin agar (Hayakawa and Nonomura, 1987), raffinose-histidine agar (Shirling and Gottlieb, 1966), starch casein agar (Küster and Williams, 1964), water proline agar (1% L-proline) (Matsumoto et al., 2006), and rhizosphere mimicking agar (Brescia et al., 2020) in triplicates. All media were supplemented with nalidixic acid and nystatin at 25 μg/mL.
3. All cultured plates were incubated for 7 days at 27°C.
4. The surface of plant samples was decontaminated by treating with 70% (v/v) ethanol for 1 minute and with 1% (v/v) sodium hypochlorite solution for an additional 1 minute. Samples were washed with sterile distilled water twice.
5. One gram of sterilized samples was crushed in 1 mL of sterile Winogradsky's salt solution in sterile mortars and used to prepare a serial dilution to make up 10^{-4} dilution.
6. Each diluted solution (100 μL) was plated on humic acid vitamin agar (Hayakawa and Nonomura, 1987), raffinose-histidine agar (Vickers et al., 1984), starch casein agar (Küster and Williams, 1964), water proline agar (Matsumoto et al., 2006), and rhizosphere mimicking agar (Brescia et al., 2020) in triplicates. All media were supplemented with nalidixic acid and nystatin at 25 μg/mL amended.
7. Plates were incubated for 30 days at 28°C.
8. Presumptive colonies were restreaked on GPM agar (Takahashi et al., 2003) and Seino agar (Intra et al., 2011). The pure culture of each actinobacteria was grown on suitable culture media such as International *Streptomyces* Project 2 (ISP2) agar (Shirling and Gottlieb, 1966).

4.3.2 Isolation of Endophytic Actinobacteria from Lettuce

1. Lettuce samples were harvested from the field and kept at 4°C in a sterile plastic bag during transportation to the laboratory and processed within 48 hours.
2. The roots were cleaned with sterile distilled water to remove soil and other growth substrate.
3. Lettuce roots were decontaminated by soaking in 1% (v/v) sodium hypochlorite (NaHClO) for 5 seconds followed by a dip in 70% (v/v) ethyl alcohol for 5 seconds and washed twice using sterile distilled water (Franz et al., 2007).
4. Surface sterile lettuce roots were placed on nutrient agar to check sterilization efficiency.
5. Sterile lettuce roots were ground in a sterilized mortar using a sterile pestle.
6. One gram of ground lettuce root was added into 9 mL of normal saline solution (0.85% (w/v) NaCl) to make 10^{-1} dilution and further diluted to 10^{-6} using normal saline.
7. An aliquot of 100 µL from each dilution was spread in triplicates on humic acid vitamin agar (Hayakawa and Nonomura, 1987), R2A agar (Reasoner and Geld Reich, 1985), starch casein agar (Küster and Williams, 1964), water proline agar (Matsumoto et al., 2006), and water agar (Bonaldi et al., 2015) amended with nalidixic acid (25 µg/mL) and ketoconazole (25 µg/mL) and incubated at 28°C for up to 4 weeks.
8. Morphologically distinct actinobacterial colonies were selected and purified by a series of streak plates on GPM agar (Takahashi et al., 2003) and Seino agar (Intra et al., 2011). Pure culture of each actinobacteria was grown on suitable culture media such as International *Streptomyces* Project 2 (ISP2) agar (Shirling and Gottlieb, 1966) for routine use.
9. Spores and mycelium fragments from each pure actinobacteria were mixed in 20% (v/v) sterile glycerol solution and stored at –20°C/–80°C for long-term preservation.

4.3.3 Isolation of Endophytic Actinobacteria from Maize

1. Maize root samples were harvested from the field and kept in a sterile plastic bag. Samples should be processed within 24 hours after harvest or kept at 4°C in a refrigerator.
2. The roots were cleaned in sterile distilled water with an ultrasonic step to remove soil and other growth substrates.
3. The roots were surface sterilized by sequential soaking in a series of chemical solutions: 5% (v/v) sodium hypochlorite (NaHClO) for 5 minutes, 2.5% (w/v) sodium thiosulfate ($Na_2S_2O_3$) for 10 minutes, and 75% (v/v) ethanol for 5 minutes. The samples were further rinsed in sterile distilled water twice and finally soaked in a solution of 10% (v/v) sodium hypochlorite for an additional 10 minutes (Shen et al., 2014).
4. The surface sterile roots were air-dried and placed on a nutrient agar plate to check sterilization efficiency.
5. Surface sterile roots were crushed in a sterile mortar using a sterile pestle.
6. One gram of ground root was used to prepare 10^{-1} dilution by mixing with 9 mL of normal saline solution (0.85% (w/v) NaCl) and further diluted to 10^{-6} using normal saline.
7. An aliquot of 100 µL from each dilution was spread in triplicates on humic acid vitamin agar (Hayakawa and Nonomura, 1987), R2A agar (Reasoner and Geld Reich, 1985), starch casein agar (Küster and Williams, 1964), water proline agar (Matsumoto et al., 2006), and water agar (Bonaldi et al., 2015) amended with nalidixic acid (25 µg/mL) and ketoconazole (25 µg/mL) and incubated at 28°C for up to 4 weeks.
8. Morphologically distinct actinobacterial colonies were selected and purified by a series of streak plates on GPM agar (Takahashi et al., 2003) and Seino agar (Intra et al., 2011). Pure culture of each actinobacteria was grown on suitable culture media

such as International *Streptomyces* Project 2 (ISP2) agar (Shirling and Gottlieb, 1966) for routine use.
9. A mixture of spores and mycelium fragments from each pure culture in 20% (v/v) sterile glycerol solution was stored at –20°C/–80°C for long-term preservation.

4.3.4 Isolation of Endophytic Actinobacteria from Tomato

4.3.4.1 Isolation of Endophytic Actinobacteria from Tomato Roots

1. Tomato plants were harvested and loosely soil particles were removed from the roots, and then roots were aseptically cut by sterile scissors or scalpel.
2. Tomato roots were thoroughly washed with running tap water to eliminate any remaining soil particles.
3. Tomato roots were sequentially soaked in 70% (v/v) ethanol for 5 minutes, 2.5% (v/v) sodium hypochlorite solution for 10 minutes, and subsequently rinsed thrice with sterile distilled water to eliminate surface sterilizing agents (de Oliveira et al., 2010).
4. For validation of the surface sterilization method, decontaminated roots were randomly selected and placed on nutrient agar without antibiotics. In addition, 0.1 mL aliquot of water from the final washing step was spread on nutrient agar, and inoculated plates were cultured in an incubator at 28°C±2°C for 24 hours. No microbial growth on the agar plates indicates effective sterilization.
5. Isolation of endophytic actinobacteria can be divided into **two methods**.

   Method 1: Surface disinfected root samples were cut into small fragments and directly placed on selective media amended with nalidixic acid (25 µg/mL) and ketoconazole (25 µg/mL) such as starch casein agar (SCA) (Küster and Williams, 1964), S-medium (Cao et al., 2004), and ISP2 agar (Shirling and Gottlieb, 1966) in triplicates.

   Method 2: One gram of surface disinfected root was homogenized in 9 mL of normal saline (0.85% (w/v) sodium chloride) solution using sterile mortar and pestle.

   1. 1 mL of homogenized sample was 10-fold diluted up to 10–6 using normal saline solution.
   2. Diluted sample (100 µL) was inoculated on selective media such as starch casein agar (SCA) (Küster and Williams, 1964), S-medium (Cao et al., 2004), and ISP2 agar (Shirling and Gottlieb, 1966) using the spread plate technique. All selective media were supplemented with 25 µg/mL nalidixic acid and 100 µg/mL ketoconazole.
6. All plates were incubated at 28°C±2°C for 4 weeks.
7. Presumptive colonies of actinobacteria were counted every week and reported as colony-forming units (CFU/g) per gram of root samples.
8. Presumptive actinobacteria were repeatedly streaked on GPM agar (Takahashi et al., 2003) and Seino agar (Intra et al., 2011) to obtain pure culture.
9. Pure culture of each actinobacteria was grown on suitable culture media such as International *Streptomyces* Project 2 (ISP2) agar (Shirling and Gottlieb, 1966) for routine use.
10. The purified actinobacterial isolates were stored in 20% (v/v) sterile glycerol solution at –20°C for long-term preservation.

4.3.4.2 Isolation of Endophytic Actinobacteria from Tomato Stems

1. After tomato plant harvesting, stems were separated from roots and leaves by sterile scissors or scalpel.
2. Stems were washed with sterile distilled water and surface sterilized with 2% (v/v) sodium hypochlorite for 2 minutes with constant agitation. The samples were washed with sterile distilled water four times prior to selective isolation (Bernabeu et al., 2015).
3. Surface-sterilized stems were randomly selected and placed on nutrient agar without antibiotics to check sterilization efficiency. In addition, 0.1 mL aliquot of water from the final

washing step was spread on nutrient agar which was incubated at 28°C±2°C for 24 hours. No microbial growth on the agar plates indicates effective sterilization.

4. Isolation of endophytic actinobacteria can be divided into **two methods**.

   Method 1: The outer tissue layer was removed from the surface-sterilized stem using a sterile scalpel, next the inner tissue was carefully dissected into small fragments (around 0.5 cm in size) (Passari et al., 2016) and directly placed on selective media amended with nalidixic acid (25 µg/mL) and ketoconazole (25 µg/mL) such as starch casein agar (SCA) (Küster and Williams, 1964), S-medium (Cao et al., 2004), and ISP2 agar (Shirling and Gottlieb, 1966) in triplicates.

   Method 2: One gram of surface-sterilized stems was homogenized in 9 mL of normal saline solution using a sterile mortar and pestle.

   1. Homogenized sample (0.1 mL) was serially diluted up to 10^{-6} using a normal saline solution.
   2. Diluted sample (100 µL) was inoculated on selective media such as starch casein agar (SCA) (Küster and Williams, 1964), S-medium (Cao et al., 2004), and ISP2 agar (Shirling and Gottlieb, 1966) using the spread plate technique. All selective media were supplemented with 25 µg/mL nalidixic acid and 100 µg/mL ketoconazole.

5. All isolating plates were incubated at 28°C±2°C for 4 weeks.
6. Every week, presumptive actinobacterial colonies were repeatedly streaked on plates with GPM agar (Takahashi et al., 2003) and Seino agar (Intra et al., 2011) to obtain pure culture.
7. Pure actinobacterial culture was preserved in 20% (v/v) sterile glycerol solution and stored at –20°C for long-term preservation.

4.3.4.3 Isolation of Endophytic Actinobacteria from Tomato Leaves

1. Plants were harvested and washed with tap water for 3–5 minutes to remove unwanted debris.
2. Leaf samples were cut into small pieces of 1 cm^3 and rinsed with 0.1% (v/v) Tween 20 for a few seconds.
3. Leaf samples were surface decontaminated by soaking in a solution of 70% (v/v) ethanol for 3 minutes, followed by 0.4% (v/v) sodium hypochlorite (NaOCl) for 1 minute, 70% (v/v) ethanol for 2 minutes, and finally washed with sterile distilled water three times and dried in a biosafety cabinet (Passari et al., 2016).
4. Surface sterilization procedure was validated by placing decontaminated leaves on nutrient agar and incubated at 28°C±2°C, and growth was observed after 24 hours. In addition, 0.1 mL aliquot of water from the final washing step was spread on nutrient agar and plates were incubated at 28°C±2°C for 24 hours. No microbial growth on the agar plates indicates effective sterilization.
5. The surface-sterilized leaves were spread on two selective media, namely starch casein nitrate agar (SCNA) and actinomycete isolation agar (AIA) amended with 25 µg/mL nalidixic acid and 100 µg/mL ketoconazole as antibacterial and antifungal agents, respectively.
6. The inoculated plates were incubated at 28°C±2°C for 4 weeks and observed the appearance of actinobacterial growth.
7. The pure isolates were obtained by repeated streaking on GPM agar (Takahashi et al., 2003) and Seino agar (Intra et al., 2011).
8. Pure isolates of actinobacteria were maintained in 20% (v/v) sterile glycerol solution and stored at –20°C for long-term preservation.

4.4 CONCLUSION

The selective isolation procedures designed for cultivable actinobacteria from several plant species were proposed in this chapter. The proposed methods have been successfully used with samples ranging from bryophytes, lettuce, tomato, and maize which will be useful for the investigation of cultivable endophytic actinobacteria from plants.

REFERENCES

Ayswaria, R., Vineeth, V., and Rijesh, K. 2020. Diverse endophytic *Streptomyces* species with dynamic metabolites and their meritorious applications: A critical review. *Critical Reviews in Microbiology* 46(6): 750–758. https://doi.org/10.1080/1040841X.2020.1828816.

Baker, D. (1990). Methods for the isolation, culture and characterization of the Frankiaceae: soil actinomycetes and symbionts of actinorhizal plants. In: Labeda, D.P. (ed.) *Isolation of biotechnological organisms from nature*. McGraw-Hill Publishing Company, New York, pp. 213–236.

Bernabeu, P.R., Pistorio, M., Torres-Tejerizo, G., Estrada-De los Santos, P., Galar, M.L., Boiardi, J.L., and Luna, M.F. 2015. Colonization and plant growth-promotion of tomato by *Burkholderia tropica. Scientia Horticulturae* 191: 113–120.

Bonaldi, M., Chen, X., Kunova, A., Pizzatti, C., Saracchi, M., and Cortesi, P. 2015. Colonization of lettuce rhizosphere and roots by tagged *Streptomyces. Frontier in Microbiology* 6: 25. https://doi.org/10.3389/fmicb.2015.00025.

Boonsnongcheep, P., Nakashima, T., Takahashi, Y., and Prathanturarug, S. 2016. Diversity of endophytic actinomycetes isolated from roots and root nodules of *Pueraria candollei* Grah. ex Benth. and the analyses of their secondary metabolites. *Chiang Mai Journal of Science* 44: 1–14.

Brescia, F., Marchetti-Deschmann, M., Musetti, R., Perazzolli, M., Pertot, I., and Puopolo, G. 2020. The rhizosphere signature on the cell motility, biofilm formation and secondary metabolite production of a plant- associated *Lysobacter* strain. *Microbiological Research* 234: 126424. https://doi.org/10.1016/j.micres.2020.126424.

Cao, L., Qiu, Z., You, J., Tan, H., and Zhou, S. 2004. Isolation and characterization of endophytic *Streptomyces* strains from surface-sterilized tomato (*Lycopersicon esculentum*) roots. *Letters in Applied Microbiology* 39(5): 425–430.

de Oliveira, M.F., da Silva, M.G., and Van Der Sand, S.T. 2010. Anti-phytopathogen potential of endophytic actinobacteria isolated from tomato plants (*Lycopersicon esculentum*) in southern Brazil, and characterization of *Streptomyces* sp. R18 (6), a potential biocontrol agent. *Research in Microbiology* 161(7): 565–572.

Dinesh, R., Veeraraghavan, S., Sheeja, T.E., Muthuswamy, A., and Hamza, S. 2017. Endophytic actinobacteria: diversity, secondary metabolism and mechanisms to unsilence biosynthetic gene clusters. *Critical Reviews in Microbiology* 43(5): 546–566. https://doi.org/10.1080/1040841X.2016.1270895.

Ek-Ramos, M.J., Gomez-Flores, R., Orozco-Flores, A.A., Rodríguez-Padilla, C., González-Ochoa, G., and Tamez-Guerra, P. 2019. Bioactive products from plant-endophytic Gram-positive bacteria. *Frontiers in Microbiology* 10: 463. https://doi.org/10.3389/fmicb.2019.00463.

Franz, E., Visser, A.A., Diepeningen, A.D. Van., Klerks, M.M., Termorshuizen, A.J., and Bruggen, A.H.C. 2007. Quantification of contamination of lettuce by GFP-expressing *Escherichia coli* O157:H7 and *Salmonella enterica* serovar Typhimurium. *Food Microbiology* 24(1): 106–112. https://doi.org/10.1016/j.fm.2006.03.002.

Ganapathy, A. and Sivakumar, N. 2018. Metabolic potential and biotechnological importance of plant associated endophytic actinobacteria. In *New and Future Developments in Microbial Biotechnology and Bioengineering*, eds. B. P. Singh, V. K. Gupta, A. K. Passari, pp. 207–224. Elsevier. https://doi.org/10.1016/B978-0-444-63994-3.00014-X.

Gandhimathi, R., Seghal Kiran, G., Hema, T.A., Selvin, J., Rajeetha Raviji, T., and Shanmughapriya, S. 2009. Production and characterization of lipopeptide biosurfactant by a sponge-associated marine actinomycetes *Nocardiopsis alba* MSA10. *Bioprocess and Biosystems Engineering* 32: 825–835.

Hayakawa, M. and Nonomura, H. 1987. Humic acid-vitamin agar, a new medium for the selective isolation of soil actinomycetes. *Journal of Fermentation Technology* 65(5): 501–509. https://doi.org/10.1016/0385-6380(87)90108-7.

Insuk, C., Kuncharoen, N., Cheeptham, N., Tanasupawat, S., and Pathom-aree, W. 2020. Bryophytes harbor cultivable actinobacteria with plant growth promoting potential. *Frontiers in Microbiology* 11: 563047. https://doi.org/10.3389/fmicb.2020.563047.

Insuk, C., Pongpamorn, P., Forsythe, A., Matsumoto, A., Omura, S., Pathom-aree, W., Cheeptham, N., and Xu, J. 2022. Taxonomic and metabolite diversities of moss-associated actinobacteria from Thailand. *Metabolites* 12: 22. https://doi.org/10.3390/metabo12010022.

Intra, B., Mungsuntisuk, I., Nihira, T., Igarashi, Y., and Panbangred, W. 2011. Identification of actinomycetes from plant rhizospheric soils with inhibitory activity against *Colletotrichum* spp., the causative agent of anthracnose disease. *BMC Research Notes* 4(1): 1–9. https://doi.org/10.1186/1756-0500-4-98/TABLES/4.

Küster, E. and Williams, S.T. 1964. Selection of media for isolation of streptomycetes. *Nature* 202: 928–929. https://doi.org/10.1038/202928a0.

Matsumoto, A., Takahashi, Y., Iwai, Y., and Omura, S. 2006. Isolation of Gram-positive bacteria with high G+C from inside soil aggregates. *Actinomycetologica* 20(2): 30–34. https://doi.org/10.3209/saj.20.30.

Musa, Z., Ma, J., Egamberdieva, D., Abdelshafy, M.O.A., Abaydulla, G., Liu, Y., Li, W.J., and Li, L. 2020. Diversity and antimicrobial potential of cultivable endophytic actinobacteria associated with the medicinal plant *Thymus Roseus*. *Frontiers in Microbiology* 11: 191. https://doi.org/10.3389/fmicb.2020.00191.

Passari, A.K., Chandra, P., Mishra, V.K., Leo, V.V., Gupta, V.K., Kumar, B., and Singh, B.P. 2016. Detection of biosynthetic gene and phytohormone production by endophytic actinobacteria associated with *Solanum lycopersicum* and their plant-growth-promoting effect. *Research in Microbiology* 167(8): 692–705.

Qin, S., Li, J., Chen, H.H., Zhao, G.Z., Zhu, W.Y., Jiang, C.L., Xu, L.H., and Li, W.J. 2009. Isolation, diversity, and antimicrobial activity of rare actinobacteria from medicinal plants of tropical rain forests in Xishuangbanna, China. *Applied and Environmental Microbiology* 75(19): 6176–6186. https://doi.org/10.1128/AEM.01034-09.

Reasoner, D.J. and Geld Reich, E.E. 1985. A new medium for the enumeration and subculture of bacteria from potable water. *Applied and Environmental Microbiology* 49(1): 1–7.

Shen, Y., Zhang, Y., Liu, C. Wang, X., Zhao, J., Jia, F., Yang, L., Yang, D., and Xiang, W. 2014. *Micromonospora zeae* sp. nov., a novel endophytic actinomycete isolated from corn root (*Zea* mays L.). *The Journal of Antibiotics* 67: 739–743. https://doi.org/10.1038/ja.2014.54.

Shirling, E.B., and Gottlieb, D. 1966. Methods for characterization of *Streptomyces* species. *International Journal of Systematic Bacteriology* 16(3): 313–340. https://doi.org/10.1099/00207713-16-3-313.

Singh, D., Thapa, S., Mahawar, H., Kumar, D., Geat, N., and Singh, S.K. 2022. Prospecting potential of endophytes for modulation of biosynthesis of therapeutic bioactive secondary metabolites and plant growth promotion of medicinal and aromatic plants. *Antonie van Leeuwenhoek* 115(6): 699–730. https://doi.org/10.1007/s10482-022-01736-6.

Singh, R. and Dubey, A.K. 2018. Diversity and applications of endophytic actinobacteria of plants in special and other ecological niches. *Frontiers in Microbiology* 9: 1767. https://doi.org/10.3389/fmicb.2018.01767.

Strobel, G., Daisy, B., Castillo, U., and Harper, J. 2004. Natural products from endophytic microorganisms. *Journal of Natural Products* 67(2): 257–268. https://doi.org/10.1021/np030397v.

Takahashi, Y., Katoh, S., Shikura, N., Tomoda, H., and Omura S. 2003. Superoxide dismutase produced by soil bacteria increases bacterial colony growth from soil samples. *Journal of General and Applied Microbiology* 49(4): 263–266. https://doi.org/10.2323/JGAM.49.263.

Vickers, J.C., Williams, S.T., and Ross, G.W. 1984. A taxonomic approach to selective isolation of Streptomycetes form soil. Biological, biochemical, and biomedical aspects of actinomycetes 1984: 553–561. https://doi.org/10.1016/B978-0-12-528620-6.50045-8.

5 Sample Treatment and Protocols for the Cultivation of Marine Actinomycetes

Hasti Ramavat, Ankita Dobariya, and Satya P. Singh

5.1 INTRODUCTION

Extremophiles grow and survive in extreme environments, which are usually considered too hostile to support life. They are adapted at various levels to withstand harsh conditions. Recent studies have shown that marine environments are the richest source of high G + C-containing bacteria known as actinomycetes [1,4,21,22]. They have evolved various unique features to survive and live under various extremities. Many unique properties reflected by these organisms and their metabolites are important in various applications and industrial processes [2,14,15,22,23]. Actinomycetes from saline habitats are largely unexplored for their novel biocatalysts and bioactive compounds [2,3]. Halophilic actinomycetes are found in marine salterns, marine basins, Soda Lake, and many other such habitats across the globe. Among the actinomycetes cultivated from saline habitats using unique culturing techniques, the majority represent novel genera with unique adaptation strategies and stability of biomolecules. Different isolation strategies have been employed to cultivate and capture the diversity of marine actinomycetes [4,5,6]. Some of these actinomycetes are rare and difficult to grow using conventional cultivation methods due to the requirement of specific nutritional conditions and isolation and cultivation methods. The isolation of the actinomycetes can be enhanced by (a) appropriate selective media containing macromolecules to promote the growth of actinomycetes and suppress the contaminant organisms [8,9], (b) the addition of different antibacterial and antifungal antibiotics in the medium [10], (c) selective isolation of sporulating actinomycetes by using chemicals that attract the spores of the actinomycetes species [11], (d) selective isolation of certain specific genera by chloramine treatment to suppress the contaminants while promoting the growth of the rare actinomycete in humic acid and vitamin-rich medium [12], (e) application of other radiation, such as super high frequency (SHF), extreme high frequency (EHF), and UV to favor the selective isolation of various genera of actinomycetes [13], and (f) using seawater-based media for the isolation of various genera of actinomycetes. Thus, applying different selective media can promote cultivating a diverse group of actinomycetes [7,20]. Besides, pretreatment of the source sample in combination with the enrichment techniques appears useful for the selective isolation of diverse actinobacterial genera.

5.2 MATERIALS

Based on the methods of cultivation, the requirements of the material may vary.

5.3 METHODS

The cultivation of marine actinomycetes [16] (Figure 5.1).

 DOI: 10.1201/9781003398400-5

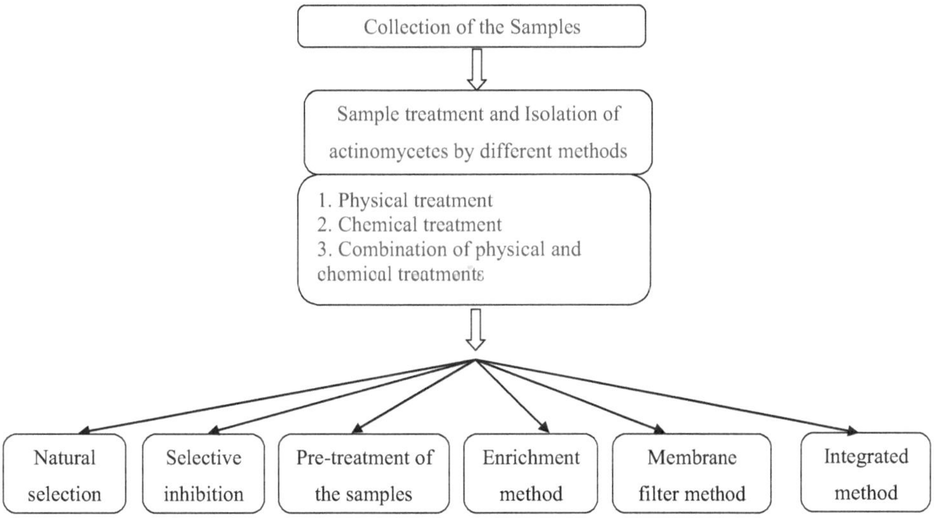

FIGURE 5.1 Steps for the cultivation of marine actinomycetes.

5.3.1 Sample Collection and Determination of the Physicochemical Properties

Soil samples are collected from the pristine, polluted, and marine sediment at a depth of 5–20 cm in the sampling bags. Water samples from different parts of the saline water bodies are collected in pre-sterilized glass bottles. The samples are stored at 4°C till further treatment and isolation of actinomycetes.

pH, salinity, temperature, and other physical and chemical parameters of the collected samples are determined.

5.4 ISOLATION OF ACTINOMYCETES BY DIFFERENT METHODS

5.4.1 Nutritional Selection Method

The medium for the isolation with specific carbon, nitrogen, and complex substances is designed to inhibit other fast-growing organisms while supporting the growth of actinomycetes. Table 5.1 represents the specific carbon and nitrogen sources that favor the growth of actinomycetes over other fast-growing bacteria.

5.4.2 Selective Inhibition by the Antibacterial/Antifungal Agents

Various antibacterial and antifungal antibiotics are used to suppress the growth of bacteria and fungi and thus allowing the growth of actinomycetes.

Various antibiotics used for the selective inhibition of bacteria and fungi are listed in Table 5.2.

5.4.3 Pretreatment of Samples

5.4.3.1 Physical Treatment

Physical treatments include air drying, dry heat, moist incubation using radiation, cellulose filtration, centrifugation, and pollen baiting [16] (Table 5.3).

TABLE 5.1
Different Carbon and Nitrogen Sources Used in the Isolation Medium

Sr. No.	Nutritional Requirement/Isolation Media	References
1.	Nitrogen source: L-arginine	[17,29]
2.	Starch/glycerol, casein, and nitrate	[30]
3.	Nitrogen source: glycine in place of L-arginine (for the isolation of non-Streptomyces actinomycetes)	[30]
4.	Chitin as the sole source of carbon and nitrogen	[31]
5.	Actinomycetes isolation agar Yeast extract-malt extract agar Oatmeal agar Inorganic salt-starch agar Glycerol-asparagine agar Starch casein agar Starch agar ISP medium (All media with 5%–10% w/v NaCl and nalidixic acid 15 μg/mL, pH 8–9)	[15,18,32]
6.	Glucose as a carbon source Different amino acids as the sole source of carbon and nitrogen	[24]
7.	Amino acids as the sole source of carbon and nitrogen	[25]

TABLE 5.2
Antibacterial/Antifungal Agents Used in the Isolation Medium

Sr. No.	Antibacterial/Antifungal Agent	Inhibits	References
1.	Nalidixic acid and cycloheximide	Gram-negative bacteria and fungi	[33,34]
2.	Nalidixic acid	Gram-negative bacteria and fungi	[18,27]
3.	Nystatin	Fungi	[35]
4.	Nalidixic acid, cycloheximide, and nystatin	Bacteria and fungi	[36]

TABLE 5.3
Various Sample Pretreatment Methods Used for Isolating Actinomycetes

Sr. No.	Pretreatment Type	Pretreatment	Organisms Isolated	References
1.	Physical treatments	Dry heating at 120°C for 1 hour	*Streptosporangium* spp.	[37]
		Air drying at 100°C for 15 minutes	*Actinomadura* and *Nocardiopsis*	[37]
		Air drying at 110°C for 10 minutes	*Nocardia* spp.	[38]
		Heat treatment at 50°C–60°C for 30–60 minutes	*Nocardiopsis* spp.	[19,39]
		Differential centrifugation: isolation of motile actinomycetes is done	*Nocardiopsis, Micromonospora,* and *Actinomadura*	[40]
		Radiation: ultrasonic waves, high-frequency radiation, super high-frequency radiation, electromagnetic radiation, etc.	*Micromonospora, Micropolyspora, Nocardia,* and *Lentzea* spp.	[43]

(Continued)

TABLE 5.3 (*Continued*)
Various Sample Pretreatment Methods Used for Isolating Actinomycetes

Sr. No.	Pretreatment Type	Pretreatment	Organisms Isolated	References
2.	Chemical treatment	Pre-treating the sample with 1.5% phenol followed with and then plating on HVA	*Polymorphospora, Nonomuraea,* and *Nocardia*	[41]
		Use of antimicrobial agents: novobiocin, gentamicin along with nalidixic acid, and nystatin	*Micromonospora*	[26]
		Calcium carbonate treatment alters pH in favoring the growth of actinomycetes, while calcium ion favors the growth of aerial mycelia	*Actinokineospora* spp., *Saccharopolyspora, Dietzia, Blastococcus, Dactylosporangium, Promicromonospora, Oerskovia,* and *Actinocorallia*	[42]
		Chemoattractants: xylose, chloride, bromide, vanillin, and γ-collidine	*Actinoplanes, Dactylosporangium,* and *Catenuloplanes*	[11]
		Chloramine treatment (chlorination)	*Herbidospora, Microbispora, Microtetraspora,* and *Streptosporangium*	[12]
		Heat treatment (60°C for 1 hour) + calcium carbonate treatment (sediment sample: $CaCO_3$ 1:1) followed by the enrichment (media supplemented with 0%–15% NaCl)	Rare Actinomycetes species (from seawater of Alang)	[4]
		Heat treatment (60°C for 30 minutes) + calcium carbonate treatment followed by the enrichment (media supplemented with 0%–15% NaCl)	Marine actinomycetes from seawater and sediment of Kachhigadhi coast	[15]

5.4.4 Enrichment Techniques

The nutritional or non-nutritional components are added into the medium to enhance the selective isolation of actinomycetes.

5.4.5 Membrane Filter Method

Nutrient agar medium is overlayed with cellulose ester membrane filter of different pore sizes. The mycelial structure of actinomycetes penetrates into the pores of the membrane to facilitate growth (Hersch and Christensen [28]).

5.4.6 Integrated Method

A combination of the physicochemical methods with antibiotics and other selective isolation methods is used for the isolation of actinomycetes.

5.4.6.1 A General Protocol for the Cultivation of Actinomycetes from Saline Habitats [18]

Seawater sample is treated at 50°C–60°C for 30–60 minutes.

⇩

Serial dilution of the sample is carried out up to 10^{-10}, and 0.1 mL of the appropriate dilution is spread over the media mentioned in [18].

⇩

The inoculated plates are incubated at 27°C–30°C and periodically observed for the isolated actinobacterial colonies.

Overall, this chapter imparts knowledge about the different approaches/protocols used for the sample pretreatment and cultivation of marine actinomycetes. The diverse cultivation methods signify the ability of marine actinomycetes to tolerate various environmental stresses and grow irrespective of other invasive species. The sample treatment, therefore, plays an important role in enhancing the cultivation of a wider range of marine actinomycetes, particularly the rare and slow-growing species.

ACKNOWLEDGMENTS

AD and HR received the SHODH Fellowships from the Government of Gujarat. SPS acknowledges the award of the DBT Multi-Institutional Project, MoES Net Working Project, UGC-BSR Faculty Fellowship, and International Travel Fellowships from DST-SERB, CSIR, DBT, and UGC. Research facilities under the UGC-CAS and DST-FIST are duly acknowledged, besides the support and facilities created by Saurashtra University. The authors also wish to mention the contributions of Dr. Sangeeta Gohel (Co-PI), Dr. Dalip Singh Rathore, and Dr. Mahejbin Sheikh as research fellows in the MoES Project on marine actinomycetes.

REFERENCES

1. Poli, A., Finore, I., Romano, I., Gioiello, A., Lama, L. and Nicolaus, B., 2017. Microbial diversity in extreme marine habitats and their biomolecules. *Microorganisms*, *5*(2), p. 25.
2. Singh, S.P., Raval, V.H., Purohit, M.K., Thumar, J.T., Gohel, S.D., Pandey, S., Akbari, V.G. and Rawal, C.M., 2012. Haloalkaliphilic bacteria and actinobacteria from the saline habitats: new opportunities for biocatalysis and bioremediation. In Satyanarayan, T., editor. *Microorganisms in environmental management* (pp. 415–429). Springer, Dordrecht.
3. Purohit, M.K., Raval, V.H. and Singh, S.P., 2014. Haloalkaliphilic bacteria: molecular diversity and biotechnological applications. In Parmar, N., Singh, A., editors. *Geomicrobiology and biogeochemistry* (pp. 61–79). Springer, Berlin, Heidelberg.
4. Sheikh, M., Rathore, D.S., Gohel, S. and Singh, S.P., 2019. Cultivation and characteristics of the Marine Actinobacteria from the Sea water of Alang, Bhavnagar. *Indian Journal of Geo Marine Sciences*, *48*(12), pp. 1896–1901.
5. Singh, S.P., Raval, V.H., Purohit, M.K., Thumar, J.T., Gohel, S.D., Pandey, S., Akbari, V.G. and Rawal, C.M., 2012. Haloalkaliphilic bacteria and actinobacteria from the saline habitats: new opportunities for biocatalysis and bioremediation. In Satyanarayan, T., editor. *Microorganisms in environmental management* (pp. 415–429). Springer, Dordrecht.
6. Magarvey, N.A., Keller, J.M., Bernan, V., Dworkin, M. and Sherman, D.H., 2004. Isolation and characterization of novel marine-derived actinomycete taxa rich in bioactive metabolites. *Applied and Environmental Microbiology*, *70*(12), pp. 7520–7529.
7. Subramani, R. and Aalbersberg, W., 2012. Marine actinomycetes: an ongoing source of novel bioactive metabolites. *Microbiological Research*, *167*(10), pp. 571–580.
8. Zhang, J. and Zhang, L., 2011. Improvement of an isolation medium for actinomycetes. *Modern Applied Science*, *5*, pp. 124–127.
9. Zhang, L., 2005. Integrated approaches for discovering novel drugs from microbial natural products. In: Zhang L, Demain A, editors. *Natural products: drug discovery and therapeutics medicines* (pp. 33–56). Human Press, Totowa, NJ.
10. Hong, K., Gao, A.H., Xie, Q.Y., Gao, H., Zhuang, L., Lin, H.P., et al. 2009. Actinomycetes for marine drug discovery isolated from mangrove soils and plants in China. *Marine Drugs*, *7*, pp. 24–44.
11. Hayakawa, M. 2008. Studies on the isolation and distribution of rare actinomycetes in soil. *Actinomycetologica*, *22*, pp. 12–19.
12. Hong, K., Gao, A.H., Xie, Q.Y., Gao, H., Zhuang, L., Lin, H.P., et al. 2009. Actinomycetes for marine drug discovery isolated from mangrove soils and plants in China. *Marine Drugs*, *7*, pp. 24–44.

13. Bredholdt, H., Galatenko, O.A., Engelhardt, K., Fjaervik, E., Terekhova, L.P., and Zotchev, S.B. 2007. Rare actinomycete bacteria from the shallow water sediments of the Trondheim fjord, Norway: isolation, diversity and biological activity. *Environmental Microbiology*, 9, pp. 2756–2764.
14. Tendler, M.D. and Burkholder, P.R., 1961. Studies on the thermophilic Actinomycetes: I. Methods of cultivation. *Applied Microbiology*, *9*(5), pp.394–399.
15. Rathore, D.S., Sheikh, M., Gohel, S. and Singh, S.P., 2019. Isolation strategies, abundance and characteristics of the marine actinomycetes of Kachhighadi, Gujarat, India. *Journal of the Marine Biological Association of India*, *61*(1), pp. 71–78.
16. Kumar, R.R. and Jadeja, V.J., 2016. Isolation of actinomycetes: a complete approach. *International Journal of Current Microbiology and Applied Sciences*, *5*(5), pp. 606–618.
17. Porter, J., Wilhelm, J. and Tresner, H. 1960. Method for the preferential isolation of actinomycetes from soil. *Applied Microbiology*, *8*, pp. 174–178.
18. Sharma, A.K., Kikani, B.A. and Singh, S.P., 2021. Diversity and phylogeny of actinomycetes of Arabian Sea along the Gujarat Coast. *Geomicrobiology Journal*, *38*(4), pp. 347–364.
19. Sharma, A.K. and Singh, S.P., 2016. Effect of amino acids on the repression of alkaline protease synthesis in haloalkaliphilic *Nocardiopsis dassonvillei*. *Biotechnology Reports*, *12*, pp. 40–51.
20 Ezeobiora, C.E., Igbokwe, N.H., Amin, D.H., Enwuru, N.V., Okpalanwa, C.F. and Mendie, U.E., 2022. Uncovering the biodiversity and biosynthetic potentials of rare actinomycetes. *Future Journal of Pharmaceutical Sciences*, *8*(1), pp. 1–19.
21. Chen, X., Zheng, Z., Li, F., Ma, X., Chen, F., Chen, M. and Tuo, L., 2023. Description and genomic characterization of *Nocardioides bruguierae* sp. nov., isolated from Bruguiera gymnorhiza. *Systematic and Applied Microbiology*, *46*(2), p. 126391.
22. Kuzmanovic, N., Bunk, B., Sproeer, C., Fruehling, A., Neumann-Schaal, M., Overmann, J. and Smalla, K., 2023. Genomics of the "tumorigenes" clade of the family Rhizobiaceae and description of Rhizobium rhododendri sp. nov. *bioRxiv*.
23. Ibrahimi, M., Loqman, S., Jemo, M., Hafidi, M., Lemee, L. and Ouhdouch, Y., 2023. A facultative predatory potential Actinobacteria spp. and prospects in the plant disease biocontrol's advances. *Frontiers in Microbiology*, *13*, p. 5402.
24. Thumar, J.T. and Singh, S.P., 2011. Repression of alkaline protease in salt-tolerant alkaliphilic *Streptomyces clavuligerus* strain Mit-1 under the influence of amino acids in minimal medium. *Biotechnology and Bioprocess Engineering*, *16*(6), pp. 1180–1186.
25. Sharma, A.K. and Singh, S.P., 2016. Effect of amino acids on the repression of alkaline protease synthesis in haloalkaliphilic *Nocardiopsis dassonvillei*. *Biotechnology Reports*, *12*, pp. 40–51.
26. Sarmiento-Vizcaíno, A., Martín, J., Ortiz-López, F.J., Reyes, F., García, L.A. and Blanco, G., 2022. Natural products, including a new caboxamycin, from Streptomyces and other Actinobacteria isolated in Spain from storm clouds transported by Northern winds of Arctic origin. *Frontiers in Chemistry*, *10*, pp. 1–18.
27. Raval, V.H., Joshi, R.H., Bhatt, H.B. and Singh, S.P., 2022. Phenotypic characteristics, phylogenetic analysis and characterization of alkaline proteases of marine bacteria Geomicrobium halophilum, Oceanobacillus oncorhynchi, and Oceanobacillus khimchii. *Biologia*, *77*(4), p. 2405.
28. Hirsch, C.F. and Christensen, D.L., 1983. Novel method for selective isolation of actinomycetes. *Applied and Environmental Microbiology*, *46*(4), pp. 925–929.
29. El-Nakeeb, M.A. and Lechevalier, H.A., 1963. Selective isolation of aerobic actinomycetes. *Applied Microbiology*, *11*(2), pp. 75–77.
30. Küster, E. and Williams, S.T., 1964. Selection of media for isolation of streptomycetes. *Nature*, *202*(4935), pp. 928–929.
31. *Lingappa*, Y. and Lockwood, J.L., 1962. Chitin media for selective isolation and culture of actinomycetes.
32. Gohel, S.D., Sharma, A.K., Dangar, K.G., Thakrar, F.J. and Singh, S.P., 2015. Antimicrobial and biocatalytic potential of haloalkaliphilic actinobacteria. In Maheshwari, D., Saraf, M., editors. *Halophiles: Biodiversity and Sustainable Exploitation* (pp. 29–55), Springer, Cham.
33. Ravi, R.K., Vasantba, J., Bhoomi, M., Bonisha, T. and Bhumika, C., 2015. Antibacterial potentials of actinomycetes isolated from Gujarat. *International Journal of Pharmaceutical Sciences Review and Research*, *30*(1), pp. 78–83.
34. Sarmiento-Vizcaíno, A., Martín, J., Ortiz-López, F.J., Reyes, F., García, L.A. and Blanco, G., 2022. Natural products, including a new caboxamycin, from Streptomyces and other Actinobacteria isolated in Spain from storm clouds transported by Northern winds of Arctic origin. *Frontiers in Chemistry*, *10*, p. 948795.

35. Daquioag, J.E.L. and Penuliar, G.M., 2021. Isolation of actinomycetes with cellulolytic and antimicrobial activities from soils collected from an urban green space in the Philippines. *International Journal of Microbiology*, *2021*(1), p. 6699430.
36. Shamikh, Y.I., El Shamy, A.A., Gaber, Y., Abdelmohsen, U.R., Madkour, H.A., Horn, H., Hassan, H.M., Elmaidomy, A.H., Alkhalifah, D.H.M. and Hozzein, W.N., 2020. Actinomycetes from the Red Sea sponge Coscinoderma mathewsi: Isolation, diversity, and potential for bioactive compounds discovery. *Microorganisms*, *8*(5), p. 783.
37. Qin, S., Li, J., Chen, H.H., Zhao, G.Z., Zhu, W.Y., Jiang, C.L., Xu, L.H. and Li, W.J., 2009. Isolation, diversity, and antimicrobial activity of rare actinobacteria from medicinal plants of tropical rain forests in Xishuangbanna, China. *Applied and Environmental Microbiology*, *75*(19), pp. 6176–6186.
38. Yamamura, H., Hayakawa, M., Nakagawa, Y., Tamura, T., Kohno, T., Komatsu, F. and Iimura, Y., 2005. Nocardia takedensis sp. nov., isolated from moat sediment and scumming activated sludge. *International Journal of Systematic and Evolutionary Microbiology*, *55*(1), pp. 433–436.
39. Sharma, A. K., Kikani, B. A. and Singh, S. P. (2020) Diversity and Phylogeny of Actinomycetes of Arabian Sea Along the Gujarat Coast. *Geomicrobiology Journal*, *38*(4), pp. 347–364. doi: 10.1080/01490451.2020.1860165.
40. Hayakawa, M., Otoguro, M., Takeuchi, T., Yamazaki, T. and Iimura, Y., 2000. Application of a method incorporating differential centrifugation for selective isolation of motile actinomycetes in soil and plant litter. *Antonie van Leeuwenhoek*, *78*, pp. 171–185.
41. Istianto, Y., Koesoemowidodo, R.S.A., Watanabe, Y., Pranamuda, H. and Marwoto, B., 2012. Application of phenol pretreatment for the isolation of rare Actinomycetes from Indonesian soil. *Microbiology Indonesia*, *6*(1), p. 7.
42. Tiwari, K. and Gupta, R.K., 2012. Rare actinomycetes: a potential storehouse for novel antibiotics. *Critical Reviews in Biotechnology*, *32*(2), pp. 108–132.
43. Jiang, Y., Cao, Y.R., Zhao, L.X., Wang, Q., Jin, R.X., He, W.X. and Xue, Q.H., 2010. Treatment of ultrasonic to soil sample for increase of the kind of rare actinomycetes. *Acta Microbiol Sinica*, 50(8), pp.1094–1097.

6 A Protocol for Screening, Cultivation, and Molecular Identification of Rare and Slow-Growing Actinomycetes with Biosynthetic Gene Clusters

Bhagwan Narayan Rekadwad and Vikas Balajirao Maske

6.1 INTRODUCTION

Microbial bioactive compounds are an important part of the drug bank mainly produced by actinomycetes. They are playing a keen role in curing diseases and being lifesavers since the 20th century. The fact that known metabolites are being found again and again is making people less interested (Hifnawy et al., 2020). There is an urgent need for new drugs which can fight microbes that are resistant to antibiotics (multidrug-resistant superbugs) and that cause life-threatening diseases (WHO, 2021). The low success of chemical medicine or its side effects has brought natural product research, especially from microbial sources, back to life (Hifnawy et al., 2020). In this regard, actinomycetes continue to be an important group of bacteria with a diverse range of roles and applications. Currently, actinomycetes are the preferred bacteria in some areas of interest because of their importance. These include antibiotic production, biotechnology, ecological roles, and human health.

Actinomycetes species such as *Micromonospora* sp. LHW63014T (Figure 6.1a), *Streptomyces coelicolor*, *Actinomadura* sp., *Streptomyces albus*, *Saccharothrix algeriensis*, and *Nonomuraea* sp. (Mohammadipanah et al., 2017) are important producers of antibiotics, including erythromycin, chloramphenicol, tetracycline, nucleosides, polyenes, and macrolides (Lo Grasso et al., 2016; Mast and Stegmann, 2019; De Simeis and Serra, 2021). However, the overuse and misuse of antibiotics have led to the emergence of antibiotic-resistant bacteria, which is a growing public health concern (Ventola, 2015; Rahman et al., 2022; Boyanova et al., 2023). Researchers are continuing to study actinomycetes to identify new antibiotics and alternative approaches to combating bacterial infections (De La Hoz-Romo et al., 2022; Ibnouf et al., 2022; Helmy et al., 2023). In order to cultivate rare actinomycetes, it may be necessary to use specialized media, as these microorganisms have unique nutritional and environmental requirements that may not be met by conventional growth media. Actinomycetes are a diverse group of bacteria known for their ability to produce a variety of secondary metabolites, such as antibiotics, antifungals, and anticancer agents. Numerous uncommon actinomycetes have the potential to produce novel bioactive compounds with considerable therapeutic value, but they may not grow well on standard media. List of cultivation media specific for *Micromonospora* sp. is not limited to, e.g., ISP-4 medium, Gauze's medium No. 1, humic

DOI: 10.1201/9781003398400-6

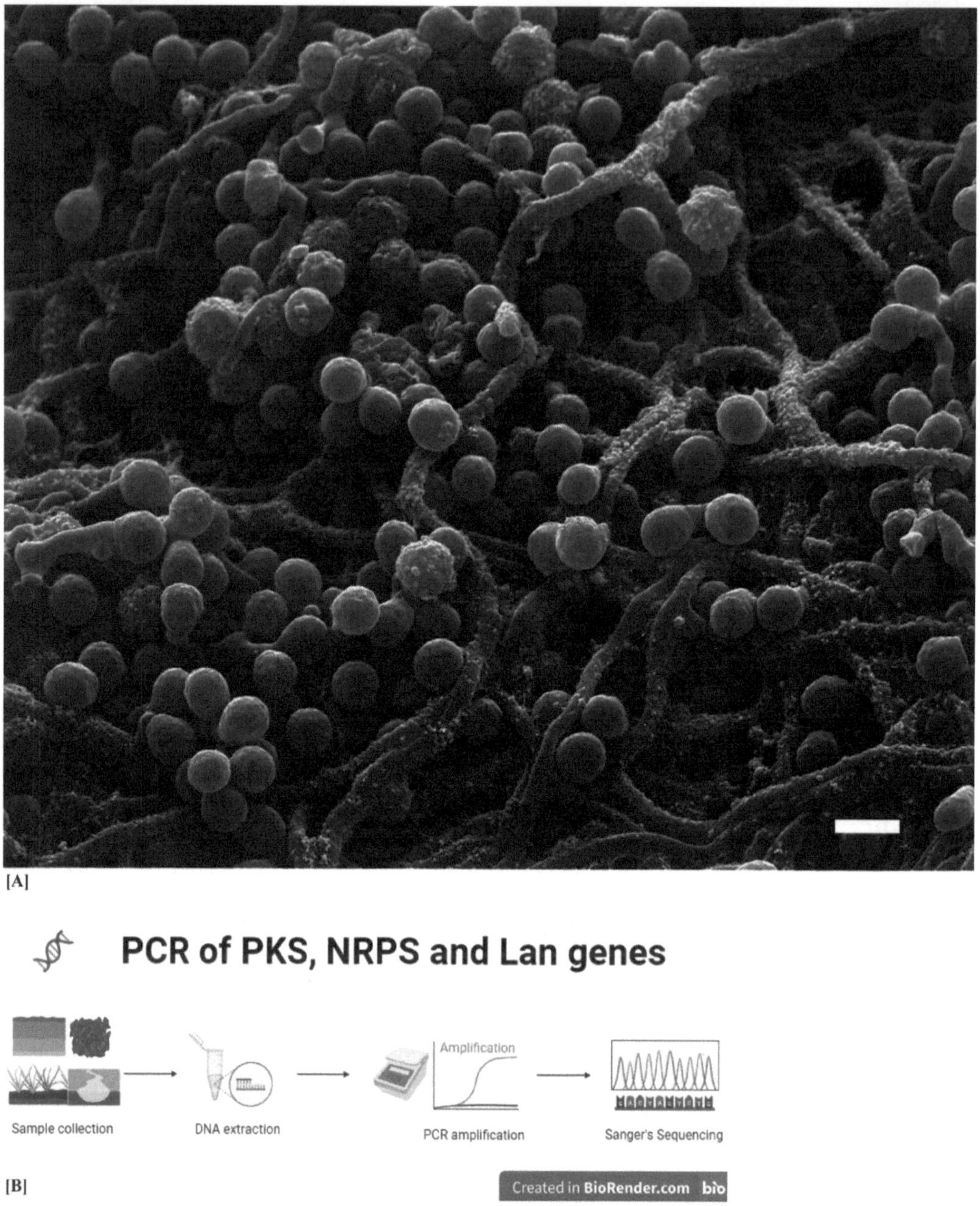

FIGURE 6.1 (a) A representative scanning electron microscopy (SEM) image of rare actinomycete *Micromonospora* sp. LHW63014T at 1 μm resolution. LHW63014T was incubated on TSP-3 agar media at 28°C for 28 consecutive days. (This SEM image is adopted from Li et al., 2019, under CC BY 4.0 License.) (b)] Overview and graphical representation of methodology enumeration and PCR amplification of specific genes of rare actinomycetes.

acid-vitamin agar (HVA), R2A medium, Bennett's agar, starch casein agar (SCA), and glycerol asparagine agar (GAA), which are selectively used for the cultivation of rare and new actinomycetes strains (Seong et al., 2001). There are many rare actinomycetes species with the ability to produce a wide range of bioactive compounds, including antibiotics and anticancer agents. For example, *Streptomyces calvus*, *Micromonospora marina*, and *Actinoplanes missouriensis* produce the antibiotics calvacin, marinocine, and actinomycin, respectively. These are just a few examples of the many rare actinomycetes species that exist in nature. There is an intense need to find novel antibiotics from rare actinomycetes to combat antibiotic resistance. This chapter describes how to grow, identify, and sequence the whole genome and use bioinformatics to study rare and slow-growing actinomycetes by using a typical actinomycetes species as an example.

6.2 MATERIALS

6.2.1 Source of Samples

Soil samples are from environment, compost, or pristine places.

6.2.2 Growth Media for Screening and Cultivation

Some important features of the following media for the cultivation of actinomycetes, e.g., *Micromonospora*, are given as follows:

6.2.2.1 ISP-4 Medium (M359, HiMedia Lab.)

- Low-nutrient medium that contains glucose, yeast extract, and soluble starch.
- Ideal for the cultivation of slow-growing or sporadic actinomycetes.
- Provides a source of simple carbohydrates and nitrogen compounds for growth.

6.2.2.2 Gauze's No. 1 Medium (DSMZ)

- Contains 20 g soluble starch, 1 g KNO_3, 0.5 g NaCl, 0.5 g K_2HPO_4, 0.5 g $MgSO_4$, 0.01 g $FeSO_4$, 20 g agar, and 1 L H_2O.
- Ideal for the cultivation of actinomycetes that produce spores that are difficult to dislodge.

6.2.2.3 Humic Acid-Vitamin Agar (HVA) (Hayakawa and Nonomura, 1987)

- Contains humic acid, vitamins, and minerals.
- Ideal for the cultivation of rare actinomycetes that are difficult to grow on other media.
- Humic acid is thought to mimic the natural environment of actinomycetes and provide a source of organic matter.

6.2.2.4 R2A Medium (M1743, HiMedia Lab)

- Low-nutrient medium that contains a variety of organic compounds.
- Ideal for the isolation of slow-growing bacteria, including actinomycetes.
- Provides a variety of nutrients, including peptides, amino acids, and organic acids.

6.2.2.5 Bennett's Agar (ATCC Medium 174)

- Contains 1 g yeast extract, 1 g beef extract, 2 g N-Z amine type A, 10 g glucose, 20 g agar, and 1 L H_2O.
- Ideal for the cultivation of alkaliphilic actinomycetes.
- Calcium carbonate helps neutralize acidic compounds that may inhibit the growth of some actinomycetes.

6.2.2.6 Casein Agar/Starch Casein Agar (SCA) (Revised as Starch M-Protein Agar) (M801, HiMedia Lab.)

- Contains casein as a source of nitrogen.
- Ideal for the cultivation of *Streptomyces* species.
- Casein is a complex protein that provides a variety of amino acids for growth.

6.2.2.7 Glycerol Asparagine Agar (GAA)/Glycerol Asparagine Agar Base (ISP Medium No. 5) (M360, HiMedia Lab)

- Contains glycerol and asparagine as sources of carbon and nitrogen.
- Ideal for the cultivation of *Streptomyces* species.
- Provides a source of simple carbohydrates and nitrogen compounds for growth.

Note: Autoclave media at 121°C for 15 minutes.

6.2.3 Requirements for Spread Plate Method, Streak Plate Method, Isolation, and Preservation

The spread plate and streak plate methods are two commonly used techniques for isolating pure cultures of microorganisms, including actinomycetes.

- Cooling incubator
- Petri plate (glass or disposable)
- Wire loop, sterile cotton, alcohol (70%), and test tubes
- Sterile spreader or glass rod
- Sterile diluent (such as sterile saline)
- Soil samples
- Incubator set to the optimal temperature for the growth of the actinomycetes
- Laminar flow hood
- Phosphate-buffered saline with 20% glycerol
- Sterile micropipette with sterilized micropipette tips (1 mL and 200 μL capacity) (refer to Sanders, 2012, for Sections 6.2.1–6.2.3)

6.2.3.1 DNA Extraction

DNA from pure culture can be extracted using a QIAGEN DNA extraction kit (QIAamp DNA Micro Kit (50 preps)) according to the manufacturer's instructions.

6.2.3.2 General Requirement for Polymerase Chain Reaction

- Template DNA from *Micromonospora* species
- Primers specific for *Micromonospora* (see Section 6.2.5)
- Polymerase chain reaction (PCR) reagents (polymerase, buffer, dNTPs, etc.)
- PCR thermocycler
- PCR tubes and pipettes (0.2–10 μL and 10–200 μL)
- Nuclease-, protein-, and biomaterial-free ultrapure water, TBA buffer, and agarose
- Agarose gel electrophoresis equipment, etc.

6.2.4 PCR Primers for Screening Rare Actinomycetes-Specific Genes

Polyketide synthases (PKS) I and II, non-ribosomal peptide synthetases (NRPS), and lanthipeptides are large multi-domain enzyme complexes that synthesize many biologically active compounds, including antibiotics, anticancer agents, and immunosuppressants. Lanthipeptide (lanthionine biosynthesis cyclase LanC) and other lanthipeptides are another class of compounds produced by actinomycetes, which have potent antimicrobial and anticancer activities. PCR-based methods can be used to screen for the presence of these biosynthetic genes in actinomycetes.

6.3 METHODS

6.3.1 Cultivation and Identification of Actinomycetes Species

1. **Sample collection**: Actinomycetes are often found in samples of soil, compost, and other parts of the environment. A sample was taken using a sterile spoon or spatula and placed in a sterile container.
2. **Dilution and plating**: The sample was diluted by adding a small amount to a liquid growth medium, and then, the diluted sample was placed on a solid growth medium.
3. **Incubation**: The plates were placed in an incubator at 25–30°C for 1–4 weeks. Actinomycetes took a long time to grow (up to 30 days).
4. **Choosing colonies**: After incubation, colonies that looked different (e.g., fuzzy or powdery) were chosen for further testing (Suarez and Hardisson, 1985). Most colonies of actinomycetes were dry and looked like powder, but there were some exceptions.
5. **Subculturing**: The chosen colonies were moved to new plates, and this was repeated until pure cultures were obtained.
6. **Characterization**: Once pure cultures of actinomycetes were made, they were described in a number of ways, including by looking at their shape, doing biochemical tests, and using molecular methods like DNA sequencing.
7. Actinomycetes were kept for a long time by freezing them in glycerol or drying them on sterile filter paper.
8. Extraction of the actinomycete's DNA: Once a pure culture was made, the actinomycete's DNA was taken out. There are many DNA extraction kits for this purpose, and the choice of kit will depend on the needs of the project.
9. Amplification of 16S rRNA gene: The 16S rRNA was amplified using PCR with primers that are specific to the PKS-I, PKS-II, NRPS, and LanC genes. The whole PKS-I, PKS-II, NRPS, and LanC genes were present in the PCR product, which was then sequenced using Sanger's DNA sequencing method.

6.3.2 PCR Conditions for the Identification of *Micromonospora*-Specific Genes

- A 25-μL reaction mix was prepared in a PCR tube by adding master mix prepared from dNTPs, Taq buffer+$MgCl_2$, primers (specific to genes), Taq DNA polymerase, and H_2O as per Figure 6.1b.
- PCR conditions were set as per Table 6.1.
- PCR conditions were different for each gene found in *Micromonospora* sp., such as PKS-I, PKS-II, NRPS, and lanthipeptides (LanC).

Steps and genes	Time				Cycles
	PKS-I gene (Tomaseto et al., 2020)	PKS-II gene (Sun et al., 2015; Tomaseto et al., 2020)	NRPS (Tomaseto et al., 2020)	Lanthipeptides (LanC gene)	
Initial denaturation	At 95°C 5 minutes	At 95°C 5 minutes	At 95°C 5 minutes	At 95°C 5 minutes	30–35 cycles
Denaturation	At 95°C 30 seconds	At 95°C 35 seconds	At 95°C 35 seconds	At 95°C 30 seconds	
Annealing	At 58°C 30 seconds	At 55°C 40 seconds	At 55°C 40 seconds	At 58°C 30 seconds	
Extension	At 72°C 4 minutes	At 72°C 1 minute	At 72°C 1 minute	At 72°C 2 minutes	
Extension	At 72°C 10 minutes	At 72°C 10 minutes	At 72°C 10 minutes	At 72°C 10 minutes	

TABLE 6.1
Primer List for Screening Rare Actinomycetes Using Specific Genes

Gene	Forward Primer	Reverse Primer	PCR Product Size
PKS-I (Tomaseto et al., 2020)	K1 F: 5′-TSAAGTCSAACATCGGBCA-3′	M6 R: 5′-CGCAGGTT SCSGTACCAGTA-3′	1100 bp
PKS-II (Sun et al., 2015)	IIPF6: 5′-TSGCSTGCTTCGAYGCSATC-3′	IIPR6: 5′-TGGAANCCGCCGAAB CCGCT-3′	600–700 bp
NRPS (Tomaseto et al., 2020)	A3 F: 5′-GCSTACSYSATSTACACSTCSGG-3′	A7 R: 5′-SASGTCVCCSGTSCGGTAS-3′	700 bp
LanC (author's own work)	MALanC-a: MALanC-a_F-5′-Lancgatcaccctcaccaga caa-3′	MALanC-a: MALanC-a_R-5′-ccaggtaggtcatcacg tcg-3′	171 bp
	MALanC-b: MALanC-b_F-5′-ctctaccacctgcacgcc-3′	MALanC-b: MALanC-b_R-5′-gaaagtctgggtgtggt gtg-3′	151 bp
	MALanC-c: MALanC-c_F-5′-cacgatcaccctcaccaga-3′	MALanC-c: MALanC-c_R-5′-taggtcatcacgtcgcgg -3′	168 bp
	MALanC-d: MALanC-d_F-5′-ctctatgcgactggaccgac-3′	MALanC-d: MALanC-d_R-5′-agcctcttctgcccgttg-3′	197 bp

- After completion of PCR, the amplified gene was viewed by performing agarose gel electrophoresis (Sambrook and Russell, 2006) and identified using Sanger's DNA sequencing method (Sanger et al., 1977).

6.4 CONCLUSION

The biosynthetic genes of the genera *Micromonospora* are particularly important because they are a major source of bioactive compounds, including antibiotics, immunosuppressants, and anticancer agents. Hence, the protocol for PKS-I, PKS-II, NRPS, and LanC genes discussed in this chapter will help to enumerate and study new and rare actinomycetes species like *Micromonospora* sp.

ACKNOWLEDGMENTS

BNR acknowledges part of funds received for research from Yenepoya (Deemed to be University) (Project No.: YU/SeedGrant/104–2021).

CONFLICT OF INTEREST

The authors have no conflict of interest to declare that is relevant to the content of this protocol chapter.

DATA AVAILABILITY

The data used to support the findings of this protocol is available within this chapter.

REFERENCES

Boyanova, L., Hadzhiyski, P., Gergova, R. and Markovska, R. 2023. Evolution of *Helicobacter pylori* resistance to antibiotics: a topic of increasing concern. *Antibiotics*. 12: 332.

De La Hoz-Romo, M.C., Díaz, L. and Villamil, L. 2022. Marine Actinobacteria a new source of antibacterial metabolites to treat acne vulgaris disease: a systematic literature review. *Antibiotics*. 11(7): 965.

De Simeis, D. and Serra, S. 2021. Actinomycetes: a never-ending source of bioactive compounds – an overview on antibiotics production. *Antibiotics*. 10(5): 483.

Hayakawa, M. and Nonomura, H. 1987. Humic acid-vitamin agar, a new medium for the selective isolation of soil actinomycetes. *Journal of Fermentation Technology*. 65(5): 501–509.

Helmy, Y.A., Taha-Abdelaziz, K., Hawwas, H.A.E.H., Ghosh, S., AlKafaas, S.S., Moawad, M.M.M., Saied, E.M., Kassem, I.I. and Mawad, A.M.M. 2023. Antimicrobial resistance and recent alternatives to antibiotics for the control of bacterial pathogens with an emphasis on foodborne pathogens. *Antibiotics*. 12: 274.

Hifnawy, M.S., Fouda, M.M., Sayed, A.M., Mohammed, R., Hassan, H.M., AbouZid, S.F., Rateb, M.E., Keller, A., Adamek, M., Ziemert, N. and Abdelmohsen, U.R. 2020. The genus *Micromonospora* as a model microorganism for bioactive natural product discovery. *RSC Advances*. 10(35): 20939–20959.

Ibnouf, E.O., Aldawsari, M.F. and Ali Waggiallah, H. 2022. Isolation and extraction of some compounds that act as antimicrobials from actinomycetes. *Saudi Journal of Biological Sciences*. 29(8): 103352.

Li, L., Zhu, H.R., Xu, Q.H., Lin, H.W. and Lu, Y.H. 2019. *Micromonospora craniellae* sp. nov., isolated from a marine sponge, and reclassification of *Jishengella endophytica* as *Micromonospora endophytica* comb. nov. *International Journal of Systematic and Evolutionary Microbiology*. 69(3): 715–720.

Lo Grasso, L., Chillura-Martino, D. and Alduina, R. 2016. Production of antibacterial compounds from actinomycetes. *Actinobacteria – Basics and Biotechnological Applications*. doi: 10.5772/61525.

Mast, Y. and Stegmann, E. 2019. Actinomycetes: The antibiotics producers. *Antibiotics*. 8(3): 105.

Mohammadipanah, F. and Dehhaghi, M. 2017. Classification and taxonomy of Actinobacteria. In: Wink, J., Mohammadipanah, F. and Hamedi, J. (eds.). *Biology and Biotechnology of Actinobacteria*. Springer, Cham. doi: 10.1007/978-3-319-60339-1_4.

Rahman, M.M., Alam Tumpa, M.A., Zehravi, M., Sarker, M.T., Yamin, M., Islam, M.R., Harun-Or-Rashid, M., Ahmed, M., Ramproshad, S., Mondal, B., Dey, A., Damiri, F., Berrada, M., Rahman, M.H. and Cavalu, S. 2022. An overview of antimicrobial stewardship optimization: the use of antibiotics in humans and animals to prevent resistance. *Antibiotics*. 11(5): 667.

Sambrook, J. and Russell, D.W. 2006. Agarose gel electrophoresis. *Cold Spring Harbor Protocols*. doi: 10.1101/pdb.prot4020.

Sanders, E.R. 2012. Aseptic laboratory techniques: plating methods. *Journal of Visualized Experiments: JoVE*. (63): e3064.

Sanger, F., Nicklen, S. and Coulson, A.R. 1977. DNA sequencing with chain-terminating inhibitors. *Proceedings of the National Academy of Sciences of the United States of America*. 74(12): 5463–5467.

Seong, C.N., Choi, J.H. and Baik, K.S. 2001. An improved selective isolation of rare actinomycetes from forest soil. *The Journal of Microbiology*. 39(1): 17–23.

Suarez, J. and Hardisson C. 1985. Morphological characteristics of colony development in *Micromonospora chalcea*. *Journal of Bacteriology*. 162: 1342–1344.

Sun, W., Zhang, F., He, L., Karthik, L. and Li. Z. 2015 Actinomycetes from the South China Sea sponges: isolation, diversity, and potential for aromatic polyketides discovery. *Frontiers in Microbiology*. 6: 1048.

Tomaseto, A.A., Alpiste, M.C., de Castro Nassar, A.F. and Destéfano, S.A.L. 2020. Antibacterial activity of phytopathogenic Streptomyces strains against bacteria associated to clinical diseases. *Plant Pathology. Arquivos do Instituto Biológico*. 87: 1–7, e0142020. doi: 10.1590/1808-1657000142020.

Ventola, C.L. 2015. The antibiotic resistance crisis: Part 1: Causes and threats. *P&T*. 40(4): 277–283.

WHO. 2021. Antimicrobial Resistance: Key Facts, World Health Organisation, 17 November 2021. https://www.who.int/news-room/fact-sheets/detail/antimicrobial-resistance, accessed on 17th February 2023.

7 Isolation and Screening of Enzymes from Actinobacteria

Apsara S. Babu and K.V. Bhaskara Rao

7.1 INTRODUCTION

Enzymes are known as biocatalysts as they augment the metabolic rate of a reaction by modulating the free energy in our body [1]. According to reports, the global market for enzymes increases 6%–7% annually which can reach up to billions in future [2]. Among various sources of enzymes covering from unicellular to multicellular organisms, microorganisms are mostly preferred. This is because of their high yield, broad biochemical diversity, rapid growth on various media, and ease of genetic manipulation. In addition to this, bacterial enzymes are more stable and safe when compared to plant- and animal-derived enzymes [3].

Actinobacteria, being the transcendent class of microorganisms, are well characterized and identified for their metabolic adaptability. They are well-known producers of various industrial enzymes having potential applications in biotechnology and biomedical fields. Hence, actinobacteria are continuously studied and employed for the production of various enzymes like protease, amylase, cellulase, chitinase, lipase, pectinase, and so forth [4]. This chapter focuses on the isolation and screening methods of three main enzymes such as proteases, amylases, and lipases. A detailed description of these enzymes and their industrial applications are enlisted in Table 7.1.

7.2 MATERIALS

7.2.1 Isolation Media and Chemicals

The isolation media and chemicals are skim milk casein agar, yeast extract-malt extract agar (YEME) (Table 7.2), starch casein agar, rhodamine-olive oil agar media, starch iodine agar, cycloheximide (25 mg/mL), dinitrosalicylic acid (DNS), and p-NPL (p-nitrophenyl laureate).

7.2.2 Laboratory Equipment

The devices used in the laboratory were petri plates, Erlenmeyer flask, micropipettes, conical flask, and water bath.

7.3 METHODOLOGY

7.3.1 Sample Collection

Soil/sediment samples from various sites are collected in sterile plastic bags and transported to the laboratory. The samples are stored at 4°C for further requirement.

7.3.2 Isolation of Actinobacteria

1. One g of soil/sediment sample was taken and diluted in a serial manner up to 10^{-7} with distilled water.

 DOI: 10.1201/9781003398400-7

TABLE 7.1
Various Enzymes Produced by Actinobacterial Strains and Their Importance and Applications

Enzymes	Importance	Industrial Applications	Enzyme-Producing Actinobacterial Strains	References
Protease	Proteases, also known as peptidyl-peptide hydrolases, are protein-digesting enzymes that hydrolyze long-chain polypeptides into shorter chains and amino acids [5]. A wide range of physiological functions are carried out by proteases starting from the cellular level of the organ to the organism level [6]. Due to their unique catalytic activity and specificity, they have several applications in industries and pharmaceutical sectors.	Pharmaceutical, detergent, leather, photographic, baking, and brewing industries.	*Streptomyces pactum*, *Streptomyces exfoliatus*, *Streptomyces thermoviolaceus*, and *Nocardiopsis* sp.	[4,6]
Lipase	Lipases are a family of triglyceride ester hydrolytic enzymes that mediate the breakdown of triglycerides into free fatty acids and glycerol [7,8]. They are widely distributed in various sources like bacteria, fungi, plants, and animals. However, lipases isolated from bacteria are found to be more stable and economical when used on an industrial scale [9].	Detergents, pharmaceutical, food, medicine, cosmetics, paper industry, and synthesis of biodegradable polymers.	*Streptomyces exfoliates* and *Nocardiopsis alba*	[4,9,10]
Amylase	α-Amylases are one of the most significant enzymes used in industrial and biotechnological studies. They hydrolyze α-1,4 glycosidic linkages of starch polymer into dextrins and gluco-oligosaccharides [3]. Majority of the α-amylases are metalloproteins where they need calcium ions for their structural rigidity, activity, and stability [7]. They are widely present in plants, animals, and microorganisms.	Baking, pharmaceutical, textile, detergent, paper, and distilling industries.	*Streptomyces erumpens* and *Thermobifida fusca*	[3,4]

TABLE 7.2
Media Components of YEME Agar

Components	Amount (g/L)
Malt extract	10
Yeast extract	4
Glucose	4
Agar	20
pH ±7.3	

2. Using the spread plate technique, 0.1 mL of the following dilutions (10^{-5}, 10^{-6}, and 10^{-7}) was plated on various agar plates supplemented with cycloheximide (antifungal agent).
3. The agar plates were incubated at 28°C for 6–14 days for actinobacterial isolates [11].

7.3.3 Preliminary Screening of Actinobacterial Isolates for Enzyme Activities

7.3.3.1 Screening of Protease-Producing Actinobacteria

1. The isolates were spot inoculated on to skim milk casein agar plates.
2. The plates were incubated at 36°C ± 1°C for 5 days.
3. The plates were monitored for clear zone around the colony [12,13].

7.3.3.2 Screening of Amylase Producing Actinobacteria

1. The isolates were spot inoculated on to starch agar plates.
2. The agar plates were incubated at 28°C for 5 days.
3. Following incubation, the agar plates were flooded with Lugol's iodine solution (1%) and left for 5 minutes [14].
4. The plates were monitored for clear zone around the colony [13,15].

7.3.3.3 Screening of Lipase-Producing Actinobacteria

1. The isolates were spot inoculated on to rhodamine-olive oil agar plates.
2. The agar plates were incubated at 28°C for 5 days.
3. The plates were monitored for orange fluorescent halos around the colony under UV light at 366 nm [15].

7.3.4 Secondary Screening Assays for Enzyme Production

The following fermentation media are used for the production of enzymes like protease, amylase, and lipase (Table 7.3).

1. A loop full of culture was inoculated into 50 mL of production media in an Erlenmeyer flask.
2. The flasks were incubated at 28°C for 96 hours on a shaker incubator at 160 rpm.
3. Following incubation, the fermented broth was centrifuged at 8,000 rpm for 15 minutes at 4°C.
4. The supernatant was collected to perform various assays [13].

7.3.4.1 Assay for Determining Protease Activity

1. To a clean test tube, 0.4 mL of casein solution (2% (w/v) dissolved in 50 mM Tris HCl buffer, pH 7.2) and 0.1 mL of actinobacterial crude extract were added.
2. The test tubes were incubated at 65°C for 15 minutes in a water bath.
3. Following incubation, the reaction was terminated by adding 1 mL of 10% trichloroacetic acid to the reaction mixture.
4. The mixture was centrifuged at 5000 rpm for 10 minutes.
5. The supernatant was collected, and the absorbance was measured at 280 nm using a UV–Vis spectrometer [6].

7.3.4.2 Assay for Determining Amylase Activity

1. To a clean test tube, actinobacterial crude extract (1 mL) and 1 mL of 1% starch were added.
2. The test tubes were incubated at 37°C for 15 minutes in a water bath.

TABLE 7.3
Production Media Components for Protease, Amylase, and Lipase Enzymes

Enzyme Media	Composition/liter	References
Protease	2 g of soluble starch, 0.5 g of peptone, 0.5 g of glucose, 0.5 g of yeast extract, 0.5 g of casein, 0.2 g of soybean meal, 0.6 g of $(NH_4)_2SO_4$, 0.8 g of $CaCO_3$, and 0.5 g of NaCl (pH 7.0)	[13]
Amylase	10 g of soluble starch, 2 g of yeast extract, 5 g of peptone, 0.5 g of $MgSO_4$, 0.5 g of KH_2PO_4, 1.5 g of NaCl, and 0.15 g of $CaCl_2$ (pH 7.0)	[13]
Lipase	17 g of pancreatic casein, 3 g of enzymatic digest soybean, 5 g of NaCl, 2.5 g of K_2HPO_4, 2.5 g of glucose, and 1% olive oil (pH 7.5)	[16]

3. Following incubation, the reaction was stopped by adding 2 mL of dinitrosalicylic acid (DNS) reagent and kept in a water bath for 5 minutes.
4. Subsequently, 1 mL of sodium potassium tartrate was added, and the absorbance was measured at 570 nm [3].

7.3.4.3 Assay for Determining Lipase Activity

1. In a clean test tube, 2 µL of p-NPL, 6 µL 0.05 M phosphate buffer, and 2 µL of actinobacterial crude extract were added.
2. The tubes were incubated at 30°C for 10 minutes.
3. The reaction was stopped by adding 30 µL of absolute ethanol, and the absorbance was measured at 380 nm using a UV–Vis spectrophotometer [17].

REFERENCES

1. Singh, K., Bose, H., Richa, K., Karthik, L., Gaurav, K. and Bhaskara, K.V. 2012. "Isolation and characterization of protease producing marine eubacteria." *Journal of Agricultural Technology* 8(5): 1633–1649.
2. Furini, G., Berger, J.S., Campos, J.A., Sand, S.T. and Germani, J.C. 2018. Production of lipolytic enzymes by bacteria isolated from biological effluent treatment systems. *Anais da Academia Brasileira de Ciências* 90: 2955–2965.
3. Rao, K.V., Bose, H., Richa, K., Singh, K., Karthik, L. and Kumar, G. 2013. "RSM mediated optimization of amylase production from marine *Bacillus* sp. VITRKHB." *Research Journal of Pharmaceutical, Biological and Chemical Sciences* 4(4): 523–536.
4. Mukhtar, S., Zaheer, A., Aiysha, D., Abdulla Malik, D.K. and Mehnaz, S. 2017. "Actinomycetes: a source of industrially important enzymes." *Journal of Proteomics & Bioinformatics* 10(12): 316–319.
5. Sayali, D., Karthik, L., Kumar, G. and Bhaskara Rao, K.V. 2013. "Isolation, purification and applications of enzymes from *Bacillus subtilis*." *Research Journal of Pharmaceutical, Biological and Chemical Sciences* 4(3): 1196–1204.
6. Fernandez, R., Ramachandran, R. and Nallakumar, K. 2018. "Screening and characterization of protease producing marine actinobacteria Streptomyces pactum RA71 isolated from Pulicat Lake, Chennai, Tamil Nadu, India." *Asian Journal of Microbiology, Biotechnology, and Environmental Sciences* 20: 618–625.
7. Antony, K. and Santhi, N. 2021. "A review on applications of actinomycetes."*Journal of Critical Reviews* 8(4): 2394–5125.
8. Pizarro, I.F., de Souza, H.F., dos Santos Ferreira, J., Maldonado, R.R., Kamimura, E.S. and Aguiar-Oliveira, E. 2022. "Production of lipase by Actinobacteria." *Methods in Actinobacteriology* 1: 505–512.
9. Susilowati, A.R.I., Sari, S.L.A., Setyaningsih, R., Handarwati, H., Pangastuti, A. and Purwoko, T. 2018. "Lipolytic bacteria isolated from Indonesian sticky rice cake wajik and jenang experiencing with rancidity." *Biodiversitas Journal of Biological Diversity* 19(1): 351–356.
10. Ma, Q., Sun, X., Gong, S. and Zhang, J. 2010. "Screening and identification of a highly lipolytic bacterial strain from barbecue sites in Hainan and characterization of its lipase." *Annals of Microbiology* 60(3): 429–437.
11. Babu, A.S. and Bhaskara Rao, K.V. 2022. "Isolation of Actinobacteria from seaweeds." *Methods in Actinobacteriology* 1: 83–88.
12. Karthik, L., Gaurav, K., Kumar, K. Siva, Thangaradjou, T. and BhaskaraRao, K.V. 2011. "Isolation and characterization of protease producing actinobacteria from marine crab." *Elixir Bio Technology* 39: 5098–5102.
13. Haritha, R., Siva Kumar, K., Jagan Mohan, Y.S.Y.V. and Ramana, T. 2010. "Amylolytic and proteolytic Actinobacteria isolated from marine sediments of Bay of Bengal." *International Journal of Microbiology Research* 1(2): 37–44.
14. Selvam, K., Vishnupriya, B. and Subhash Chandra Bose, V. 2011. "Screening and quantification of marine actinomycetes producing industrial enzymes amylase, cellulase and lipase from south coast of India." *International Journal of Pharma Biology* 2: 1481–1487.
15. Hamedi, J., Kafshnouchi, M. and Ranjbaran, M. 2019. "A study on actinobacterial diversity of Hampoeil cave and screening of their biological activities." *Saudi Journal of Biological Sciences* 26(7): 1587–1595.

16. Sagar, K., Bashir, Y., Phukan, M.M. and Konwar, B.K. 2013. "Isolation of lipolytic bacteria from waste contaminated soil: a study with regard to process optimization for lipase." *International Journal of Scientific & Technology Research* 2(10): 214–218.
17. Ilesanmi, Oluwaseye I., Abiodun E. Adekunle, Japhel A. Omolaiye, Efemena M. Olorode, and Adebayo L. Ogunkanmi. "Isolation, optimization and molecular characterization of lipase producing bacteria from contaminated soil." *Scientific African* 8 (2020): e00279.

8 RSM-Based Media Optimization for Enzyme Production by Streptomyces

Pavithra M., Mohamed Hasan M., and Sai Ramesh A.

8.1 INTRODUCTION

Response surface methodology or modeling (RSM), also known as response surface analysis, is a method of analysis for optimizing two or more quantitative factors that are available. The dependent variables are called responses in the response surface method, whereas the independent variables which are also called factors are referred to as predictor variables. RSM is particularly effective for optimizing variables/factors more realistically than merely the statistical significance test at a certain point. This RSM uses the method of steepest ascent and descent. Because higher-order effects are frequently insignificant, a second-order regression model is used to approximate the response surfaces.

$$y = \beta_0 + \beta_1 x_1 + \cdots \beta_k x_k + \cdots \beta_{11}^2 x_1^2 + \cdots \beta_{kk}^2 x_k^2 + \beta_{12} x_1 x_2 + \cdots \beta_{k-1,k} x_{k-1} x_k$$

The equation represents a second-order regression model (sometimes known as the complete quadratic) for k factors [1].

Types of design:

- Central composite design (CCD)
- Box–Behnken design (BBD)

8.2 THE CENTRAL COMPOSITE DESIGN

The CCD is the most often used approach for response surface design [2]. The CCD is a two-level full factorial or fractional factorial design with extra center and axial points (also known as star points). The axial points are positioned in the middle of each level of the other components, while the center point is inserted in the center. The parameter, α value is determined by the calculation possibilities and the desired precision of the surface response estimation. The location of the points determines the quality of the estimation. The setting of α value and the number of trials in the domain's center have an effect on estimation accuracy. The axial (star) points of the 2^2 design are positioned on the face of the square as in Figure 8.1a. As a result, the design in Figure 8.1b becomes a 3^3 factorial for which a complete quadratic model (second-order regression) for the response surface may be fitted. The lack-of-fit might also be evaluated with the inclusion of numerous center points. α represents the distance between the center and the axial points.

DOI: 10.1201/9781003398400-8

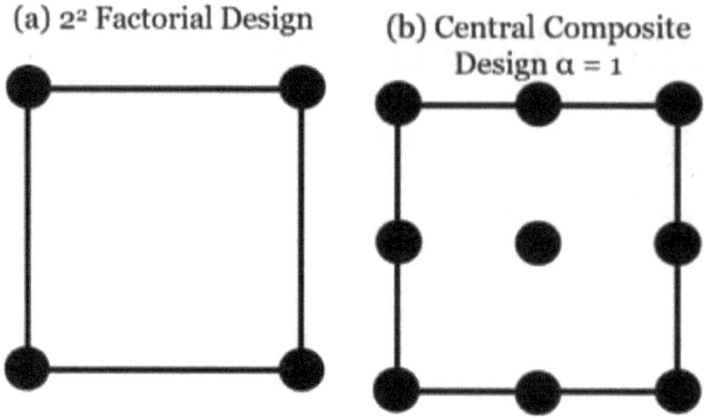

FIGURE 8.1 (a and b) 2^2 Factorial design and central composite design.

8.3 RESPONSE SURFACE MODELING

8.3.1 Major Steps

The optimization of process variables in RSM consists of seven distinct processes.
The processes are as follows:

1. Selection of response
2. Identification of variables
3. Developing an experimental design for contaminant removal
4. Analyzing the regression
5. Forming a quadratic polynomial
6. Developing counter plot/surface plot
7. Improvement of optimum operating conditions by analysis.

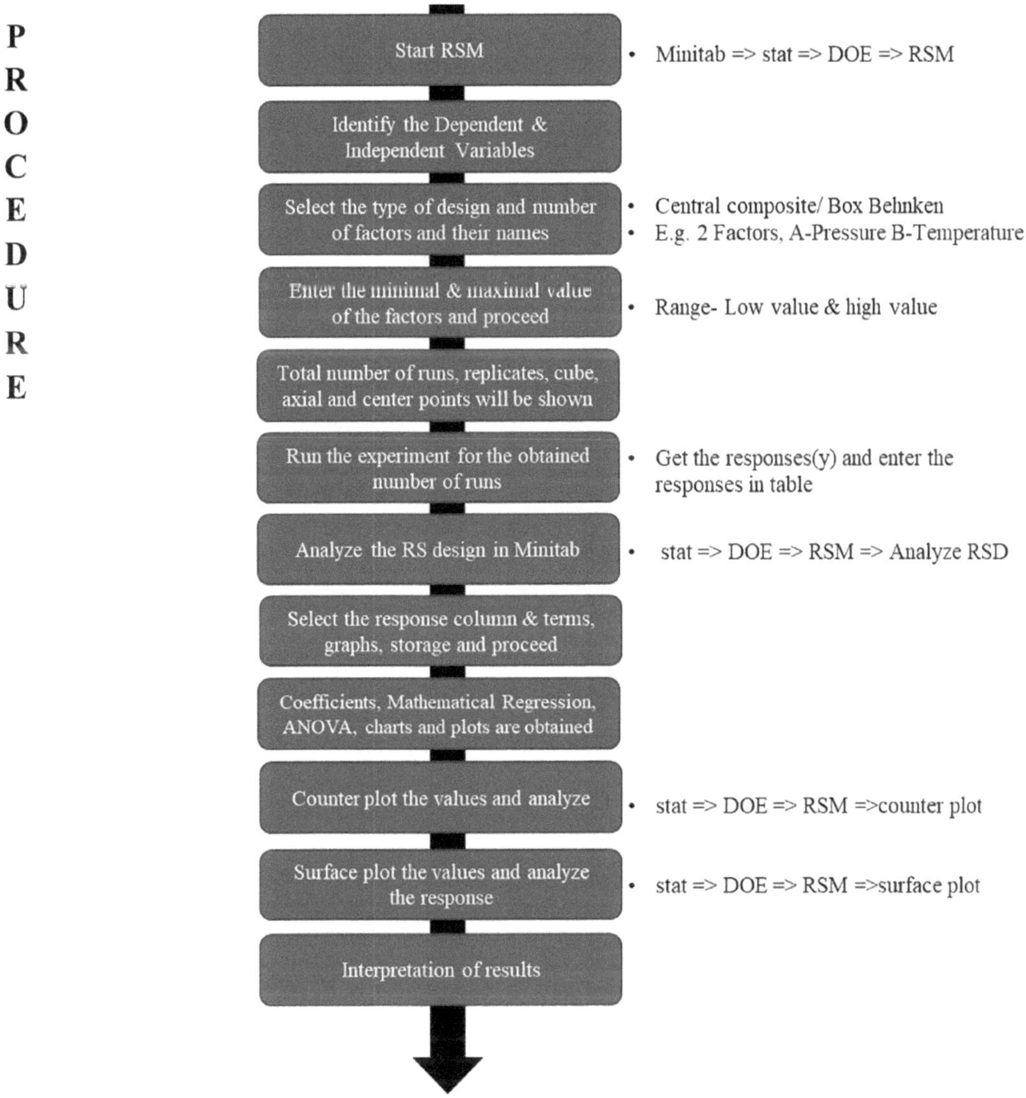

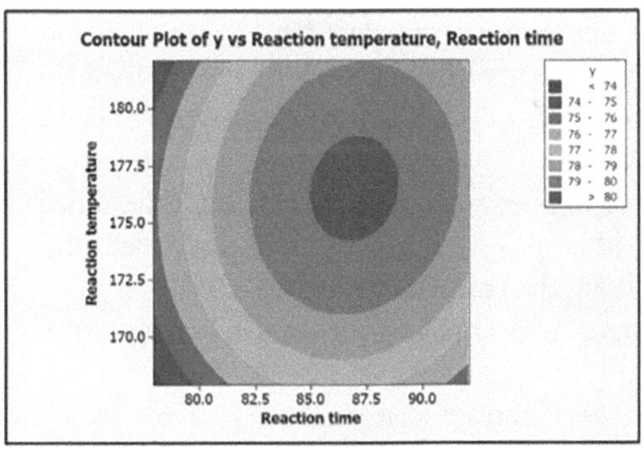

COUNTER PLOT

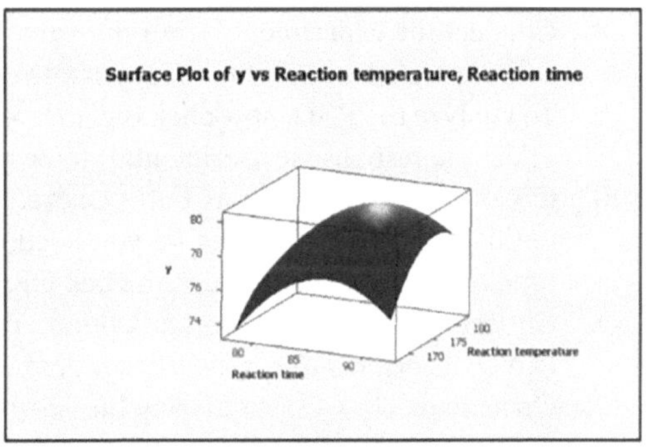

SURFACE PLOT

8.4 PROCEDURE

For example, we take α-amylase production from *Streptomyces*.

- α-Amylase has been found in several microorganisms like bacteria [1], actinobacteria, and fungi [3]. α-Amylases are abundantly scattered in *Streptomyces* species [4] and are employed in sectors such as food, drinks, and fermentation [5]. The production of α-amylase in synthetic medium is economically unfeasible. To lower the costs, these synthetic media must be replaced with more economically accessible products.
- As a result, optimizing the fermentation conditions of the medium, particularly physical and chemical parameters [6], is essential for the progress of any fermentation process.
- Traditional optimization approaches are "one-factor-at-a-time" strategies. Statistical optimization is an alternative strategy that allows for the quick screening of multiple factors and factor interactions, as each component's contribution is highlighted. RSM, a collection of mathematical and statistical tools, is becoming popular as a powerful approach for optimizing conditions for the manufacturing of industrially important goods such as chemicals and enzymes.
- This RSM procedure will help to find out the optimum concentration/value [7] for the selected factors for the cultivation of *Streptomyces* for the production of the enzyme.
  1. Before creating an RSM model, choose the dependent variables (enzyme activity), the response, and independent variables or factors (rice bran, wheat bran, NaCl, $MgSO_4$, and incubation period, in days)

     Minitab version 14 *is used in this chapter to optimize the media conditions for the production of the enzyme (α-amylase).*
  2. Open the Minitab application, from the toolbar click on *stat ⇒ DOE ⇒ Response surface ⇒* Create *Response surface design.*
  3. Choose the *design type* (CCD or BBD) and the *number of continuous factors (here we choose 5 as we have a total of five independent variables)* for your analysis.
  4. Choose the appropriate *number of runs and blocks from designs (here we select full designs with 32 runs)* and then continue with other options.
  5. Type *the name* (rice bran, wheat bran, NaCl, $MgSO_4$, and incubation period, in days) and the *range of the factor.* Example: Low value to high value for the concentration (g/100 mL) selected factors.
  6. Choose how to summarize the printed results by selecting the options for the summary table, based on your own considerations for summarizing the results in the next step.
  7. It will create a table with *stdorder* and *runorders*, after which we can conduct the experiment linearly using the randomized values in an order.
  8. Conduct the experiment for all the values from *the low stdorder* (1 to *n*) and enter *the response (enzyme activity)* in a separate column on the Minitab worksheet (Figure 8.2).
  9. To analyze the RSD, now click on *Stats > DOE > Response Surface > Analyze RSD* and select the response (experimental) to be analyzed.
  10. Choose *the terms* (such as Full Quadratic) or the necessary terms, check all the other options, and then continue (if you need residual plots, click on graphs and select the type of plot you need, and then click OK) to get the result as in Figure 8.3.
  11. Choose the result and Storage option; a popup will appear with the necessary terms (*ANOVA, fits, residuals, coefficient, etc.,*).
  12. Once done, click OK to display the results for the chosen terminologies.
  13. CCD, analysis of variance (ANOVA) table for the responses (from where we can confirm the significant factors from the *P*-value), error, *R*-sq, mathematical regression equation for *y*, and other terminologies will be shown in printed results as in Figure 8.4.

↓	C1 StdOrder	C2 RunOrder	C3 PtType	C4 Blocks	C5 Rice bran	C6 Wheat bran	C7 NaCl	C8 MgSO4	C9 Incubation period	C10 Activity
1	14	1	1	1	8.0	4.0	5	5	7.0	27.1639
2	17	2	-1	1	0.5	5.5	3	3	8.0	29.0039
3	4	3	1	1	8.0	7.0	1	1	10.0	16.0979
4	23	4	-1	1	5.5	5.5	3	-1	8.5	14.9186
5	15	5	1	1	3.0	7.0	5	5	7.0	20.1361
6	11	6	1	1	3.0	7.0	1	5	10.0	18.5032
7	28	7	0	1	5.5	5.5	3	3	8.5	20.1446
8	31	8	0	1	5.5	5.5	3	3	8.5	20.1446
9	12	9	1	1	8.0	7.0	1	5	7.0	24.2063
10	5	10	1	1	3.0	4.0	5	1	7.0	24.5626
11	32	11	0	1	5.5	5.5	3	3	8.5	20.1446
12	7	12	1	1	3.0	7.0	5	1	10.0	11.9040
13	9	13	1	1	3.0	4.0	1	5	7.0	9.6993
14	21	14	-1	1	5.5	5.5	-1	3	8.5	25.4030
15	29	15	0	1	5.5	5.5	3	3	8.5	20.1446
16	27	16	0	1	5.5	5.5	3	3	8.5	20.1446
17	19	17	-1	1	5.5	2.5	3	3	8.5	18.7424
18	8	18	1	1	8.0	7.0	5	1	7.0	19.4548
19	2	19	1	1	8.0	4.0	1	1	7.0	19.8990
20	24	20	-1	1	5.5	5.5	3	7	8.5	13.0934
21	26	21	-1	1	5.5	5.5	3	3	11.5	31.8188
22	16	22	1	1	8.0	7.0	5	5	10.0	23.7329
23	6	23	1	1	8.0	4.0	5	1	10.0	32.3797
24	30	24	0	1	5.5	5.5	3	3	8.5	20.1446
25	20	25	-1	1	5.5	8.5	3	3	8.5	10.7497
26	13	26	1	1	3.0	4.0	5	5	10.0	28.1243
27	10	27	1	1	8.0	4.0	1	5	10.0	33.4282
28	3	28	1	1	3.0	7.0	1	1	7.0	38.6242
29	1	29	1	1	3.0	4.0	1	1	10.0	29.3703
30	25	30	-1	1	5.5	5.5	3	3	5.5	29.3695
31	18	31	-1	1	10.5	5.5	3	3	8.5	31.9418
32	22	32	-1	1	5.5	5.5	7	3	8.5	24.8112

FIGURE 8.2 Minitab home page sheet with experimental values.

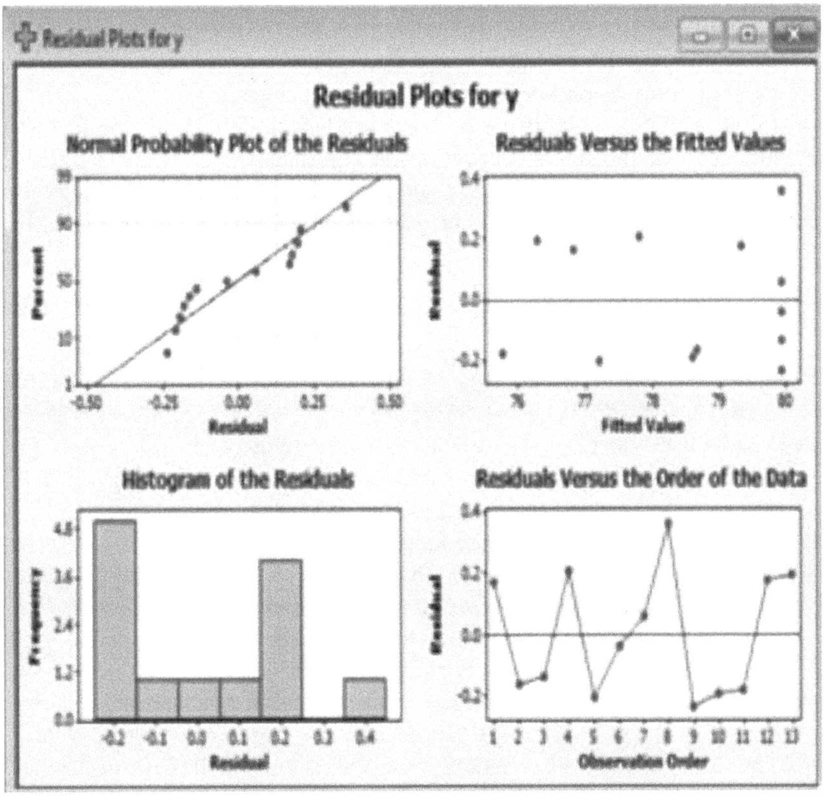

FIGURE 8.3 The residuals vs fitted plot reveals no clear trend, indicating that the residuals are unpredictable. As a result, the residuals are deemed homogeneous (constant). The plot of residuals vs observation order reveals no obvious trend. As a result, under this model, there is no violation of the uncorrelated variance.

```
S = 0.0002806   R-Sq = 100.0%   R-Sq(adj) = 100.0%


Analysis of Variance for Activity

Source          DF   Seq SS    Adj SS    Adj MS            F      P
Regression      20  1549.70  1549.698    77.485  9.83877E+08  0.000
  Linear         5   126.26   133.369    26.674  3.38695E+08  0.000
  Square         5   589.26   574.946   114.989  1.46009E+09  0.000
  Interaction   10   834.18   834.175    83.418  1.05921E+09  0.000
Residual Error  11     0.00     0.000     0.000
  Lack-of-Fit    6     0.00     0.000     0.000      2164.95  0.000
  Pure Error     5     0.00     0.000     0.000
Total           31  1549.70
```

FIGURE 8.4 The analysis of variance (ANOVA) table is used to determine statistical significance. The *p*-value of the model, which is (0.000), is less than the level of significance (0.05). As a conclusion, the null hypothesis that there is no relationship between the dependent and independent variables is rejected. As a result, the entire quadratic model of independent factors has a significant impact on the dependent variable.

- The model summary output table is used to execute the practical significance test. The corrected R^2 value is found to be 98%, suggesting that the model parameters can explain variation in the dependent variable, the response, extremely well. As a result, the model is useful. R^2 and the modified R^2 were used to represent the quality of fit of the second-order polynomial model equation.
- In the ANOVA table, if the p-values are less than 0.05, all the factors are significant, but the inter-reaction is not significant. Lower the p-values greater the significance. Based on the p-values, the coefficients are most significant for all effects. The R^2 should be greater than 95% or it must be near 90%–95% (Figure 8.5).
- OMP: Place Figure 8.5 Here>
- Regression equation in uncoded units [8]

$R1 = 31.0842 - 6.93603X1 + 27.341X2 + 0.133839X3 - 9.84804X4 - 12.0380X5 + 0.394714X1^2 - 0.599826X2^2 + 0.310161X3^2 - 0.383658X4^2 + 1.16107X5^2 - 0.446501X1X2 + 0.257124X1X3 + 0.608727X1X4 + 0.333921X1X5 - 0.875783X2X3 + 0.172782X2X4 - 2.06003X2X5 + 0.453285X3X4 - 0.00308X3X5 + 0.736909X4X5$

- When we substitute the randomized values of the independent factors in this equation with the uncoded coefficient [9], we will get the nearest value of the response that we entered in the activity column doing experimentally.
- For example, when we take $X1 = 8$, $X2 = 4$, $X3 = 5$, $X4 = 5$, and $X5 = 7$, the value we get is *27.164* which is nearest to the experimental value, where $X1$, $X2$, $X3$, $X4$, and $X5$ are the independent variables (rice bran, wheat bran, NaCl, $MgSO_4$, and incubation period), respectively.

14. Analyze the ANOVA table and then *counter/surface plot* the values for optimization.
    Minitab ⇒ stat ⇒ DOE ⇒ Response surface ⇒ counter plot/surface plot
    - The fitted polynomial equations were then shown as three-dimensional surface plots to illustrate the link between the responses and the experimental values of each variable used in this study. By improving the level of each variable, the point optimization technique was applied to maximize the reaction (Figure 8.6).
15. To improve optimization, the obtained counter plot and surface plots are analyzed.

Term	Coef
Constant	31.0842
Rice bran	-6.93603
Wheat bran	27.3410
NaCl	0.133839
MgSO4	-9.84804
Incubation period	-12.0380
Rice bran*Rice bran	0.394714
Wheat bran*Wheat bran	-0.599826
NaCl*NaCl	0.310161
MgSO4*MgSO4	-0.383658
Incubation period*Incubation period	1.16107
Rice bran*Wheat bran	-0.446501
Rice bran*NaCl	0.257124
Rice bran*MgSO4	0.608727
Rice bran*Incubation period	0.333921
Wheat bran*NaCl	-0.875783
Wheat bran*MgSO4	0.172782
Wheat bran*Incubation period	-2.06003
NaCl*MgSO4	0.453285
NaCl*Incubation period	-0.00306704
MgSO4*Incubation period	0.736909

FIGURE 8.5 Terms and coefficients.

8.5 FINDINGS AND INTERPRETATIONS

8.5.1 Counter Plot

The main reason we use response surfaces is that we prefer to see a range of values that would be more comfortable than a single number. For example, in optimization with RSM for response vs temperature and time, 176°g is generally considered to be the most comfortable temperature; we see temperatures between 175 and 177 degrees because they look nice, giving good yields. If we set the temperature between these two, however, there will not be a significant change in the yield, so that is what these surfaces tell us instead of just one number (Figure 8.7).

8.5.2 Surface Plot

For more optimal optimization, the surface plot also gives conclusions with a range of values where the process will be more efficient and where it is less efficient (Figures 8.8–8.10).

For example, in optimization with RSM for response vs temperature and time, here, even at the low temperature with high time, the response yield is less, whereas at the middle range, *the yield is at a higher peak than other range of factors*, showing the optimal range of effectiveness (Figure 8.11).

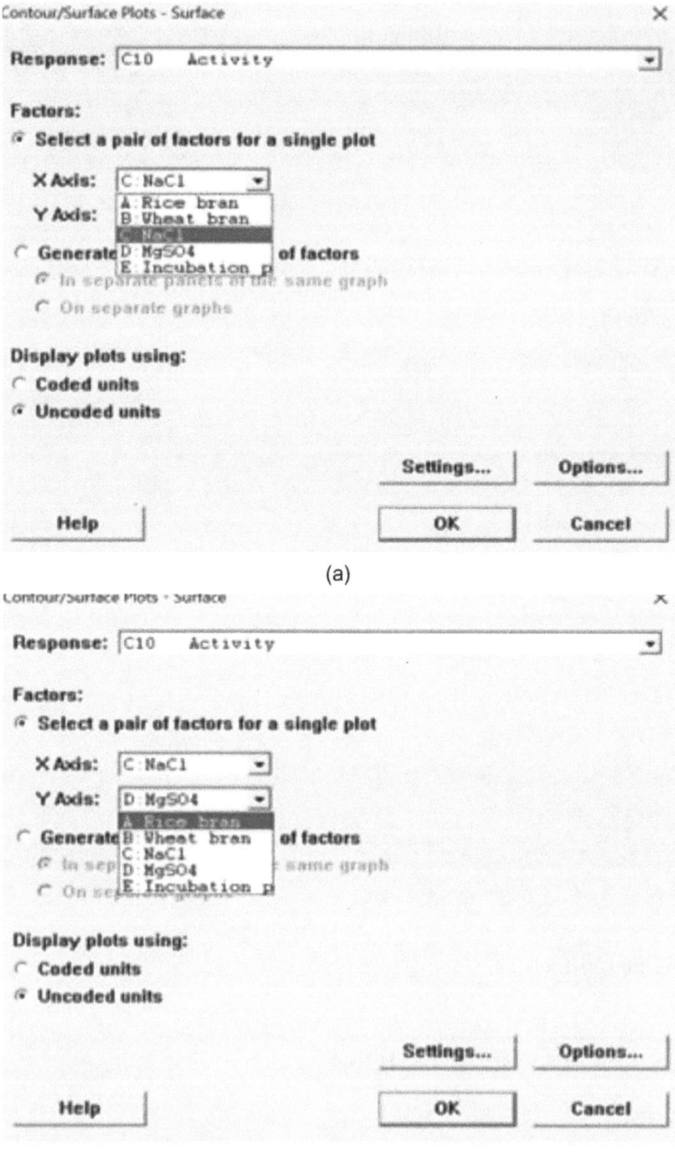

FIGURE 8.6 (a and b) Surface plot/counter plot the factors by selecting different *X*-axis and *Y*-axis for the response (activity).

8.5.3 Response Optimization

Response optimization gives the final optimal solution from the range of values entered for the analysis.

Minitab ⇒ stat ⇒ DOE ⇒ Response Surface ⇒ Response optimizer (Figure 8.12).

Here, the value in red shows the optimum values which gives more yield.

The amylase yield increased with these optimal values of the five parameters, and therefore, the media is optimized with the optimal values obtained above, for getting better yield.

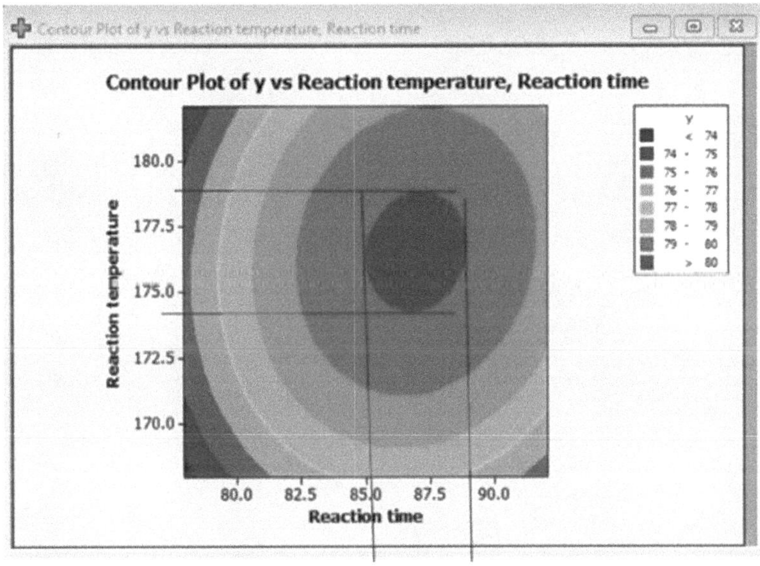

FIGURE 8.7 Example plot showing optimal range of temperature and time.

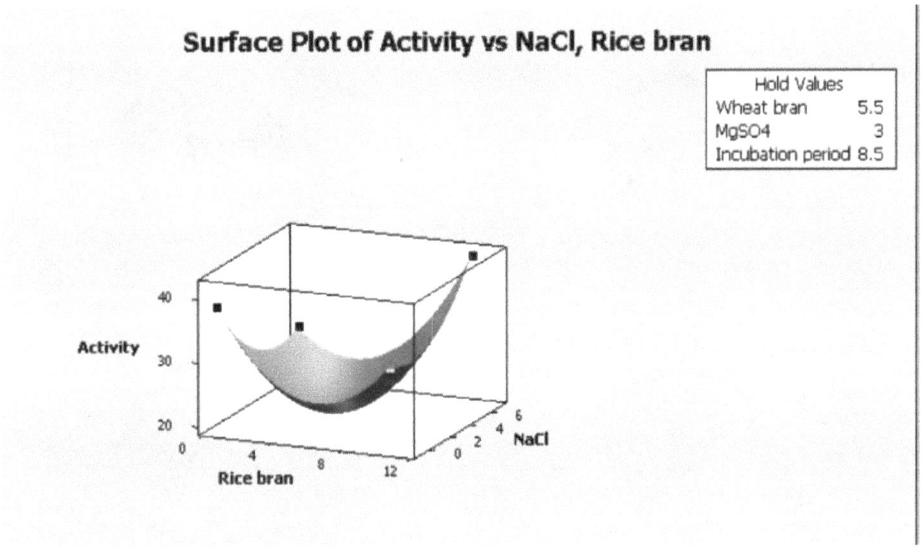

FIGURE 8.8 Enzyme activity vs NaCl, rice bran.

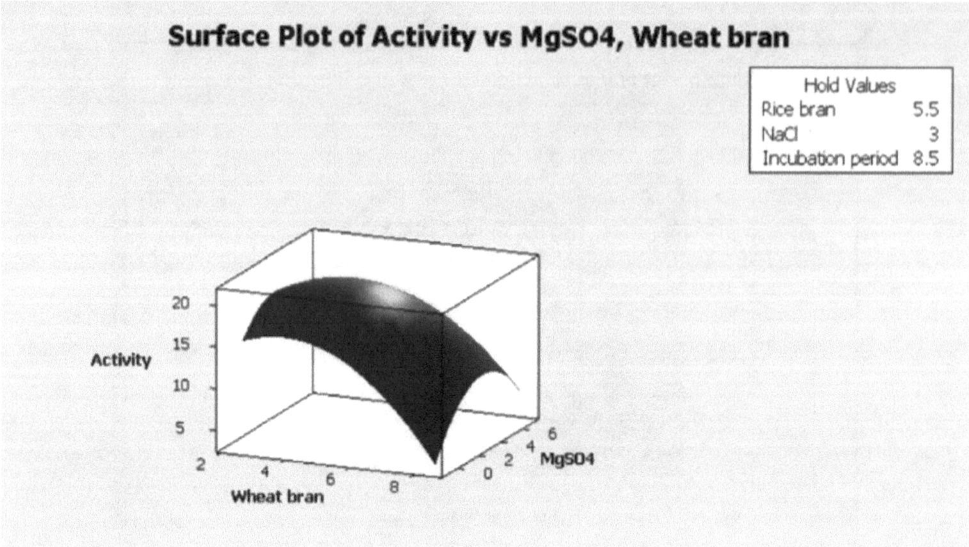

FIGURE 8.9 Enzyme vs $MgSO_4$, wheat bran.

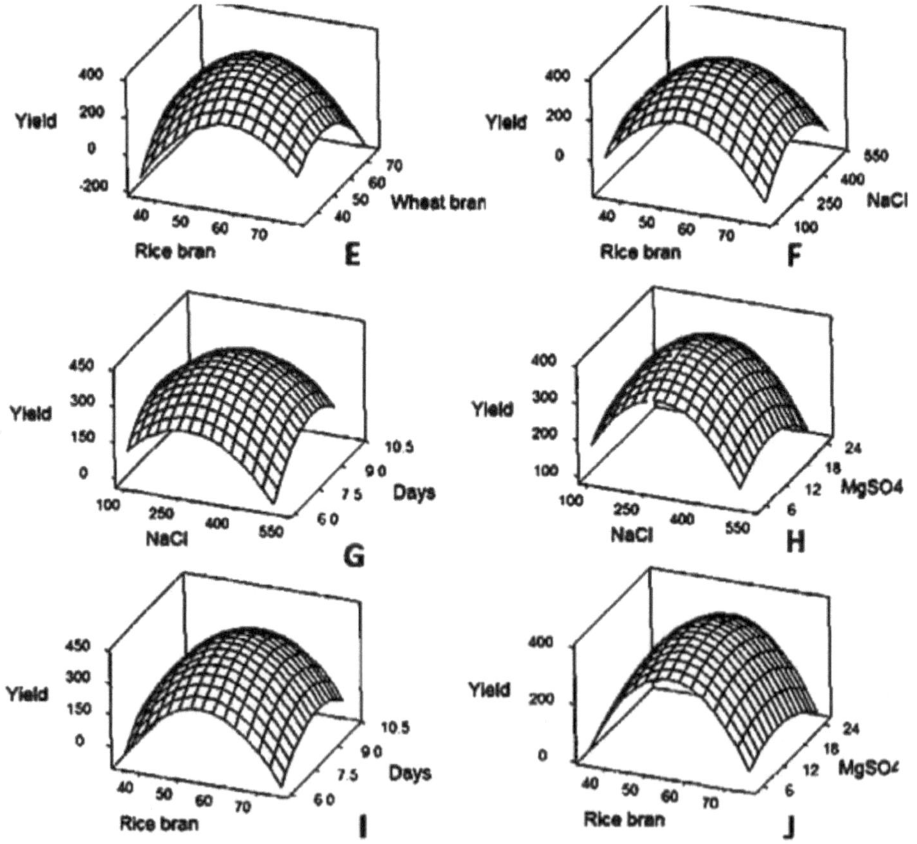

FIGURE 8.10 Plot all the possible factors on *X*-axis and *Y*-axis like plotted above and determine the optimum values by response optimizer.

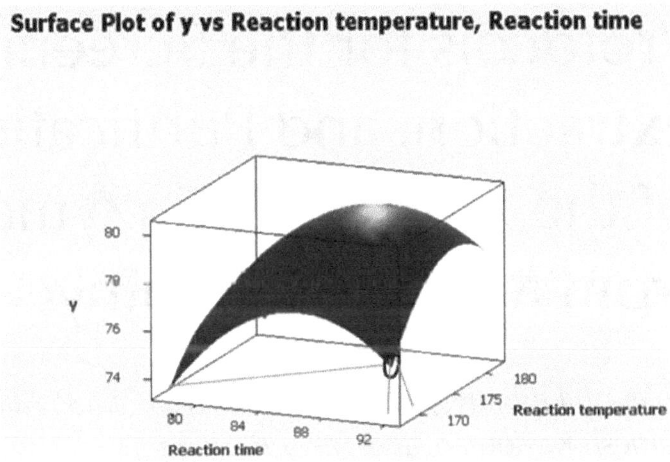

FIGURE 8.11 Surface plot mapping corresponding factor's effect on dependent variable (y).

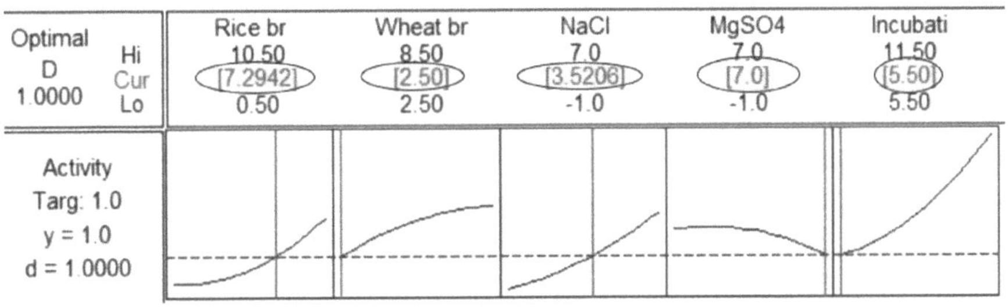

FIGURE 8.12 Optimized values by response optimizer.

REFERENCES

1. Kannan, S., Valliappan, K., Gunasekaran, V. S., Thirunavukarasu, T., and Lakshmanan, K. 2012. Response surface methodology for the optimization of α-amylase production by Streptomyces sp. ML12 using agricultural byproducts. *Biologia* 67:32–40.
2. Khurana, S., Kapoor, M., Gupta, S., and Kuhad, R. C. 2007. Statistical optimization of alkaline xylanase production from *Streptomyces violaceoruber* under submerged fermentation using response surface methodology. *Indian Journal of Microbiology* 47:144–152.
3. Ramachandran, S., Patel, A. K., and Nampoothiri, K. M., et al. 2004. α-Amylase from a fungal culture grown on oil cakes and its properties. *Brazilian Archives of Biology and Technology* 47:309–317.
4. Sourav, K., Maudy, K. S, Sivakumar, K., and Kannan, L. 2006. Isolation and characterization of extra-cellular enzymes producing Actinomycetes from the alimentary canal of estuarine fishes. *Asian Journal of Microbiology, Biotechnology & Environmental Sciences* 8:811–815.
5. Hiroshi, H., Kazuaki, I., Yasuhiro, H., et al. 2001. Novel α-amylase that is highly resistant to chelating reagents and chemical oxidants from the alkaliphilic *Bacillus* isolate KSM-K38. *Applied and Environment Microbiology* 67:1744–1750.
6. Puri, S., Beg, Q. K., and Gupta, R. 2002. Optimization of alkaline protease production from Bacillus sp. by response surface methodology. *Current Microbiology* 44:286–290.
7. Ashwini, K. 2016. Use of response surface methodology for optimization of basal nutrients in production of amylase by Streptomyces gancidicusasd_KT852565. *International Journal of Pharma and Bio Sciences* 7:314–323.
8. Adinarayana, K., and Ellaiah, P. 2002.Response surface optimization of the critical medium components for the production of alkaline protease by a newly isolated *Bacillus* sp. *Journal of Pharmacy & Pharmaceutical Sciences* 5:272–278.
9. Tanyildizi, M. S., Elibol, M., and Özer, D. 2006. Optimization of growth medium for the production of α-amylase from *Bacillus amyloliquefaciens* using response surface methodology. *Journal of Chemical Technology and Biotechnology* 81:618–622.

9 Protocols for the Screening, Extraction, and Purification of the Hydrolytic Enzymes from Marine Actinomycetes

Ankita J. Dobariya, Hina N. Radadiya, Gira P. Mankad, Ramesh K. Kothari, and Satya P. Singh

9.1 INTRODUCTION

Enzymes are useful in varied applications related to food, textiles, cosmetics, dairy, bakery, detergent, paper, leather and pharmaceuticals, and bioremediation [1,2]. Enzymes particularly from extremophiles have greater potential in biotechnology and industries. Among the widely used enzymes, amylases, proteases, lipases, cellulases, and esterases occupy pivotal positions [3–7]. Halophilic enzymes are active and stable at high salinity, low water activity, and inorganic solvents due to certain unique properties, such as hydrophobic core and large numbers of acidic amino acids such as aspartate and glutamate in comparison to the non-halophilic proteins [1,8]. These characteristics prevent the precipitation of the protein at high salt content [2,9].

The marine actinomycetes, such as *Nocardiopsis* and *Streptomyces* species, optimally grow at high salt concentrations and alkaline pH [10,11]. Marine *Streptomyces* sp. produces thermostable amylases that catalyze and maintain stability under alkaline pH. The enzymes from halophiles, such as *Halobacillus* sp., *Halomonas meridiana*, *Chromohalobacter* sp., *and Haloarcula hispanica*, are also able to function with surfactants and detergents [12].

Some microorganisms produce both extracellular and intracellular enzymes. The screening of the microorganisms for their enzyme-producing ability is carried out at two levels: primary and secondary [13–15]. The primary screening is based on the qualitative assessment of the enzyme production, usually performed to screen a large number of organisms. The primary screening is carried out on solid media containing the appropriate substrate of the enzyme [16–18]. After the primary assessment, secondary screening is carried out for the quantitative estimation of the particular enzyme. A detailed description of the primary screening for amylase, protease, and lipase is provided in this chapter [19–21].

After screening, the next step is the extraction from the liquid medium (submerged fermentation) or solid-state fermentation followed by its purification and characterization of the enzyme. The production of enzymes in actinomycetes is in higher quantity; however, it takes a longer time to grow them and optimally produce the enzyme [22].

After the completion of growth, there are varying strategies for the extraction of enzymes depending on whether they are intracellular or extracellular. In the case of intracellular enzymes, the cells need to be ruptured for extraction. This chapter describes the extraction process for extracellular enzymes only. There are several techniques, such as centrifugation, homogenization, precipitation, flocculation, and filtration, to prepare a cell-free extract [23–26]. With respect to the extraction of enzymes from actinomycetes, filtration and centrifugation are the most commonly employed methods. However, mycelia can also be easily removed by filtration in a combination of centrifugation to remove cell biomass and other impurities. After extraction, it requires a series of steps that include

 DOI: 10.1201/9781003398400-9

TABLE 9.1
Summary of the Purification of the Enzymes from Halophiles

Sr. No.	Name of Enzyme	The Matrix Used in the Purification Process	Microbial Strain	References
1.	β-Amylase and serine protease	Q-Sepharose (IEC) and Sephacryl S-200 (SEC)	*Salimicrobium halophilum*	[41]
2.	Serine protease	Sephacryl S-200 (SEC)	*Halogeometricum borinquense*	[42]
3.	Lipase	Q-Sepharose and SP Sepharose (IEC)	*Bacillus atrophaeus*	[21]
4.	Serine proteinase	Bacitracin-Sepharose 4 Fast Flow (HIC)	*Halobacillus* sp.	[43]
5.	Protease	phenyl Sepharose 6B (HIC) and Sephadex G-100 (SEC)	*Chromohalobacter* sp.	[44]
6.	α-Amylase	DEAE cellulose (IEC) and Sephadex G-75 (HIC)	*Marinobacter* sp.	[45]
7.	Lipase	Mono-Q-Sepharose (IEC)	*Haloferax mediterranei*	[46]
8.	α-Amylase	Q-Sepharose (IEC) and Sephacryl S-200 (SEC)	*Nesterenkonia* sp.	[57]
9.	Lipase	Sephadex G-75 (SEC) and DEAE-Sepharose Fast Flow (IEC)	*Chromohalobacter* sp.	[47]
10.	Lipase	Sephadex G-75 (SEC)	*Halobacillus* sp.	[48]
11.	α-Amylase	Phenyl Sepharose 6FF (HIC)	*Bacillus amyloliquifaciens*	[26]
12.	Protease	Phenyl Sepharose 6FF (HIC)	*Oceanobacillus iheyensis Haloalkaliphilic bacterium*	[52]
13.	Protease	Phenyl Sepharose 6FF (HIC)	*Brachystrepto spora xinjiangensis*	[55]
14.	Protease	DEAE cellulose (IEC)	*Nocardiopsis dassonvillei*	[32]
15.	Protease	Phenyl Sepharose 6FF (HIC)	*Oceanobacillus* sp.	[37]
16.	Protease	DEAE cellulose (IEC)	*Streptomyces clavuligerus*	[36]
17.	Protease	Phenyl Sepharose 6FF (HIC)	*Bacillus pseudofirmus*	[56]
18.	α-Amylase	Biogel P 60 (SEC)	*Bacillus licheniform*	[25]

SEC, size exclusion chromatography; IEC, ion exchange chromatography; HIC, hydrophobic interaction chromatography.

conventional and newly developed strategies to improve the yield of the purified protein and the efficacy and cost-effectiveness of the process [27–29]. Various column chromatographic techniques used for protein purification are based on ion exchange, hydrophobic interaction, gel filtration, and ligand-based affinity chromatography [30–32]. After the selection of a suitable chromatographic technique, the conditions for the binding and elution of the desired molecules should be worked out. Other factors such as the flow rate of the mobile phase, the volume of the sample, the physical properties of the matrix, and the chemical nature of the mobile phase are highly significant and thus need to be optimized. In recent years, the strategies and procedures for the effective purification of enzymes from the haloalkaliphilic actinomycetes have been reported (Table 9.1) [33–37].

9.2 MATERIALS AND METHODS FOR THE SCREENING OF AMYLASE, PROTEASE, AND LIPASE

9.2.1 Culture Activation of Actinomycetes Strain

For inoculation, the active culture is important for the assessment of enzyme production in the plate assay method. For the preparation of the young culture, a loopful of the culture is inoculated into the growth medium. For the actinomycetes, it usually takes a minimum of 48 hours to get sufficient cell density as the seed culture. The medium contains appropriate salt concentration with an alkaline pH. The optimum growth conditions, such as temperature, salinity, pH, and shaking, are provided [25,28,38,39].

9.2.2 Screening for Amylase

For the assessment of their ability to produce amylases, the microorganisms are grown on starch agar plates, with certain modifications for the marine actinomycetes [4,17,18,26,40,41]. The actinomycetes of marine habitats require alkaline pH and high salt concentrations for their growth and enzyme production. Therefore, usually, the growth medium contains 1% starch, 0.5% peptone, 0.5% yeast extract, 5% NaCl, and 5% agar. The pH of the medium is maintained at 8–10, for which Na_2CO_3 is used [42]. The young seed culture of the selected strain is inoculated in the form of spots on the solid growth medium. After 4–5 days of growth of actinomycetes at 28°C, amylase production is detected by flooding with the iodine solution (1% potassium iodide and 0.5% iodine). The appearance of a clear zone of hydrolysis against a dark blue background is indicative of amylase catalysis [17,25,43].

9.2.3 Screening for Proteases

The protease activity is usually detected on the gelatin/casein agar plate [44]. For the detection of the protease activity after the growth of the organism, the plates are flooded with Frazier's reagent (20 mL HCl and 15 g mercuric chloride in 100 mL distilled water). The appearance of a clear zone of hydrolysis around the colonies on the gelatin agar plate against the white hazy background is indicative of the protease action. The haloalkaliphilic actinomycetes in general are grown at 28°C for 4–5 days in a medium that contains 0.4% peptone, 0.1% yeast extract, 1.5% gelatin, 5% NaCl, and 5% agar and alkaline pH 8–9 for the screening and detection of proteases [36,37,45].

9.2.4 Screening of Lipase

Sensitive and specific plate assay methods are used for the screening of the lipase-producing microorganisms: for instance, Rhodamine B agar plate, spirit blue agar plate, and Tween-80 agar plate assays. Rhodamine B dye emits orange fluorescence under UV light; while for the spirit blue agar, a pH indicator dye is incorporated in the medium; and for the Tween-80 agar plate, $CaCl_2$ is added with Tween-80 [46,47].

Rhodamine B agar plate: Screening of the lipase-producing strains is carried out on 3% olive oil as a lipase substrate, with 0.4% sodium chloride and 2.8% nutrient agar, at pH 7. After the autoclave, 1% of Rhodamine B solution (0.01% w/v) is incorporated in the above medium [48,49]. Furthermore, a loopful of 48-hour grown culture is spotted onto the Rhodamine B agar plate and incubated at 37°C for 4 days. The detection of the lipase enzyme is observed under UV light at 350 nm.

Tween-80 agar plate: For the preparation of Tween-80 agar, 0.25% peptone, 0.3% yeast extract, 2% agar powder, 0.01% $CaCl_2$, and 1% Tween-80 are added. The pH of the medium was adjusted to 7.0. The other details are as described earlier [50].

Spirit blue agar plate: Spirit blue agar medium contained 1% casein enzymic hydrolysate, 0.5% yeast extract, 0.01% spirit blue, 1.7% agar, and 1% olive, at pH 6.8. For inoculation, 100 µL of the seed culture is added into the well and incubated at 37°C for 4 days. The lipase activity is periodically monitored at 24 hours intervals [51,52].

9.2.5 Measurement of the Zone of Hydrolysis and Expression of the Enzyme Activities

The detection and analysis of the enzyme production by the plate assays is primarily determined by the measurement of the zone of hydrolysis. However, to create more precise data correlating the growth and enzyme production, a ratio of the zone of hydrolysis to the colony diameter is calculated. This ratio (diameter of the zone of clearance (mm)/colony diameter (mm)) represents the enzyme production about the growth of the organisms [17,25,26,52].

9.3 THE EXTRACTION AND PURIFICATION OF THE ENZYMES

The extraction and purification of the hydrolytic enzymes is a crucial part of the characterization and applications. During the production of the enzymes, many other proteins/enzymes and metabolites are produced; therefore, the purification of the targeted enzyme is essential. In general, as an initial step of the purification of the enzymes, precipitation either by salt or solvent is required [7,17].

9.3.1 Materials

Crude enzyme solution, ammonium sulfate, organic solvents (acetone, methanol, and ethanol), magnetic stirrer, and appropriate buffer with optimum salt concentration are the required materials [25,27].

9.3.2 Method

9.3.2.1 Precipitation by Salt/Organic Solvents

The appropriate volume of the crude preparation of the enzyme is kept in a beaker or flask and stirred with a magnetic stirrer under a low temperature. This follows the gradual addition of ammonium sulfate in the crude enzyme preparation. Ammonium sulfate is added to achieve the desired level of saturation, dissolved till homogeneity, and incubated at 4°C overnight. The precipitates are then collected by centrifugation at 10,000 rpm for 10 minutes at 4°C. The precipitate pellets are suspended in inappropriate volumes of the buffer containing salt (NaCl). The enzymes from halophilic microorganisms get denatured in the absence of optimum salt concentration. Similarly, as an alternative for protein precipitation, the organic solvent is used instead of ammonium sulfate [47].

9.3.3 Size Exclusion Chromatography (SEC)

Size exclusion chromatography (SEC) also known as molecular sieve chromatography, gel filtration chromatography, or gel permeation chromatography depends on the molecular size of the molecules to be separated. It is widely used for the separation of macromolecules, such as proteins [25,40].

The basic steps involved in SEC are provided in the following flow chart (Figure 9.1).

9.3.3 Hydrophobic Interaction Chromatography (HIC)

Hydrophobic interaction chromatography is based on the hydrophobicity of the molecules, which are separated on account of their differential surface hydrophobicity (Figure 9.2).

9.3.4 Ion Exchange Chromatography

Ion exchange chromatography (IEC) is commonly used for purification due to (a) high resolution, (b) high protein binding capacity, (c) versatility, (d) simple principle, (e) easy and specific elution, and (f) easy performance.

The basic principle in IEC is the competition between the ions of interest and another ion of the opposite charge on the exchangers. This interaction depends on their net charge. IEC relies on the interaction of the charged molecules in the mobile phase (buffer and sample) with the oppositely

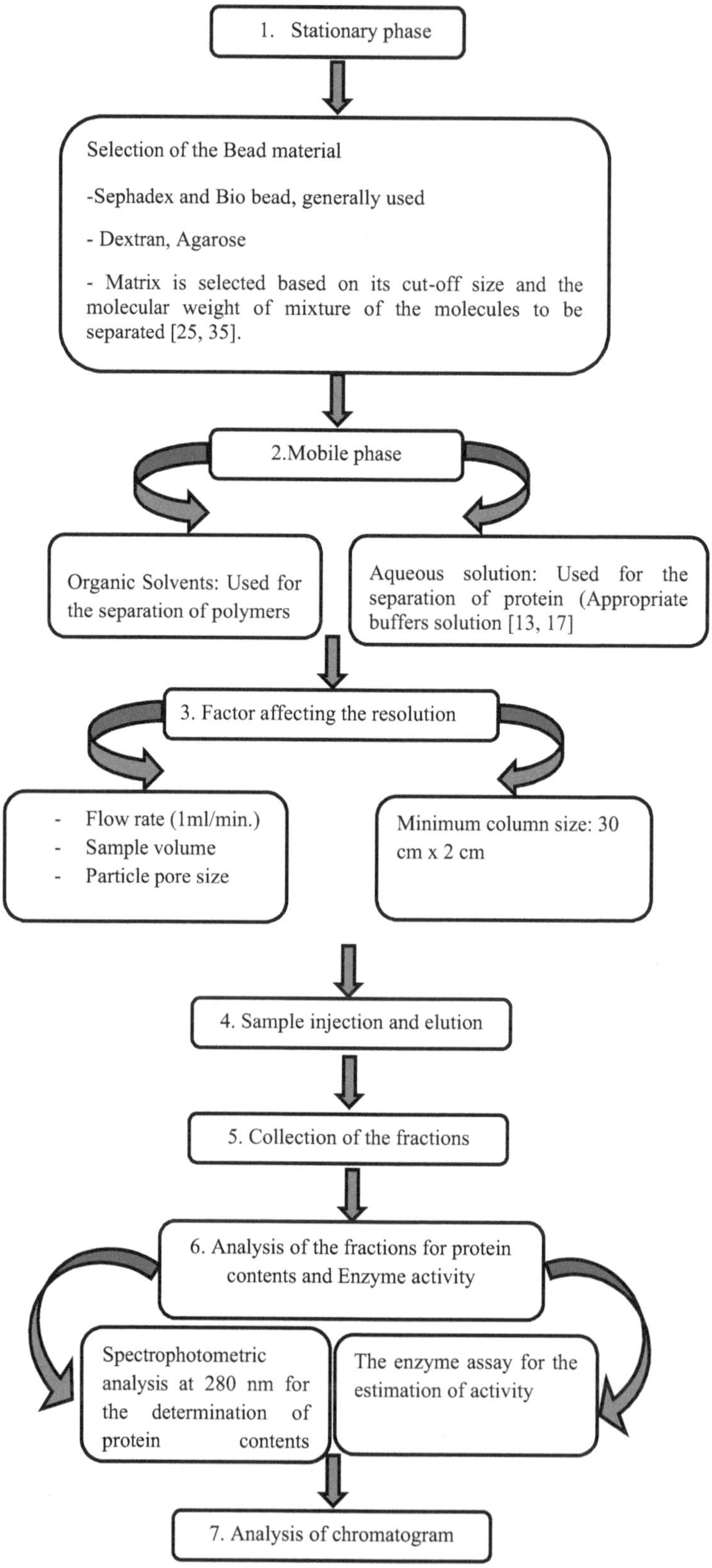

FIGURE 9.1 Basic information on size exclusion chromatography.

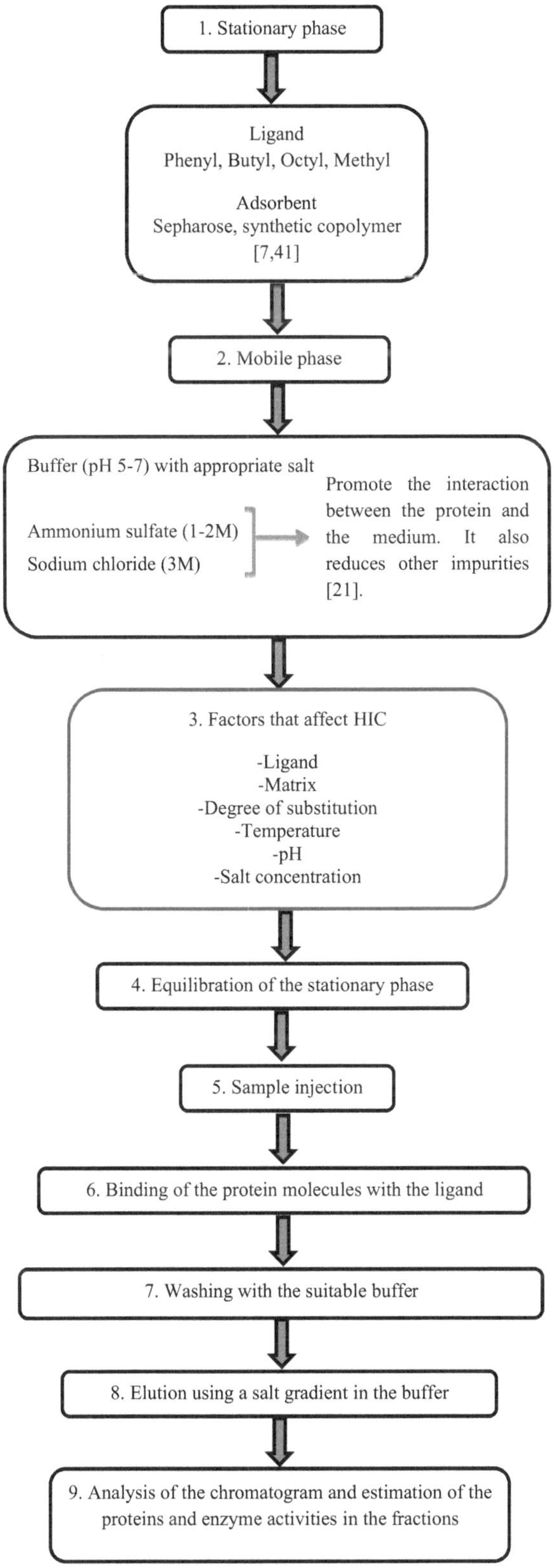

FIGURE 9.2 Basic information of hydrophobic interaction chromatography.

charged groups of the stationary phase. All aqueous solutions for use in IEC should be prepared with distilled or deionized water (Figure 9.3) [53,54].

9.3.5 Sodium Dodecyl Sulfate–Polyacrylamide Gel Electrophoresis (SDS-PAGE) and Native PAGE

9.3.5.1 Materials

The required materials are as follows.

1. Electrophoresis assembly
2. Stock acrylamide solution: 30 g acrylamide and 0.8 g bisacrylamide in a final volume of 100 mL with distilled water and stored at 4°C temperature. Wearing gloves while weighing and handling the acrylamide powder is recommended.
3. Separating gel buffer: 1.8 M Tris–HCl pH 8.8
4. Stacking gel buffer 0.6 M: Tris–HCl pH 6.8
5. 10% ammonium persulfate (freshly prepared)
6. N,N,N,N-tetramethylethylenediamine (TEMED)
7. Sample buffer: 5 mL stacking gel buffer, 0.5 g sodium dodecyl sulfate (SDS), 1mL glycerol, 0.25 mL β-mercaptoethanol, and 0.025 g bromophenol blue, adjusted to a final volume of 50 mL with distilled water.
8. 10% SDS
9. Running buffer: 6 g Tris, 28.8 g glycine, and 1 g SDS, dissolved in a final volume of 1,000 mL with distilled water
10. Staining solution: 0.1 g Coomassie brilliant blue R250 dissolved in 50 mL methanol and distilled water, followed by the addition of 10 mL glacial acetic acid. The ratio of methanol, water, and glacial acetic acid was 50:40:10. The solution was filtered on Whatman no. 1 filter paper.
11. Destaining solution: 10 mL methanol and 7 mL of glacial acetic acid are added to a final volume of 100 mL with distilled water.
12. Appropriate protein markers.

9.3.5.2 Method

1. After setting the electrophoresis assembly, the separating gel is prepared: 2 mL separation gel buffer at pH 8.8, 3.86 mL distilled water, 4.16 mL acrylamide bisacrylamide stock solution, 0.1 mL of 10% SDS, 50 μL of 10% APS, and 14 μL of TEMED. TEMED is added last when the gel mixture is ready for pouring into the gel cassette.
2. The gel solution is carefully mixed and poured into a gel cassette preferably using a pipette, until it reaches the position of the comb.
3. Stacking gel buffer: 1 mL stacking gel buffer, 1.35 mL acrylamide bisacrylamide stock solution, 7.5 mL distilled water, 0.1 mL of 10% SDS, 50 μL of 10% APS, and 14 μL TEMED.
4. Before pouring the stacking gel mixture, the extra water is removed from the gel cassette and then the stacking gel mixture is poured followed by the insertion of the comb. It is then left for polymerization.
5. Sample preparation: To 8 μL of the protein sample, 2 μL of sample buffer is added followed by heat treatment at 95°C for 5 minutes in a water bath.
6. After the polymerization of the gel, the comb is carefully removed. The bubbles, if any, are also removed from the well. The gel cassette is transferred into the buffer tank containing the running buffer avoiding bubble formation.
7. The protein samples of 8–10 μL are loaded into the well. An appropriate volume of the protein marker as per the manufacturer's instruction is also loaded.
8. Electrophoresis is carried out at 60 V.

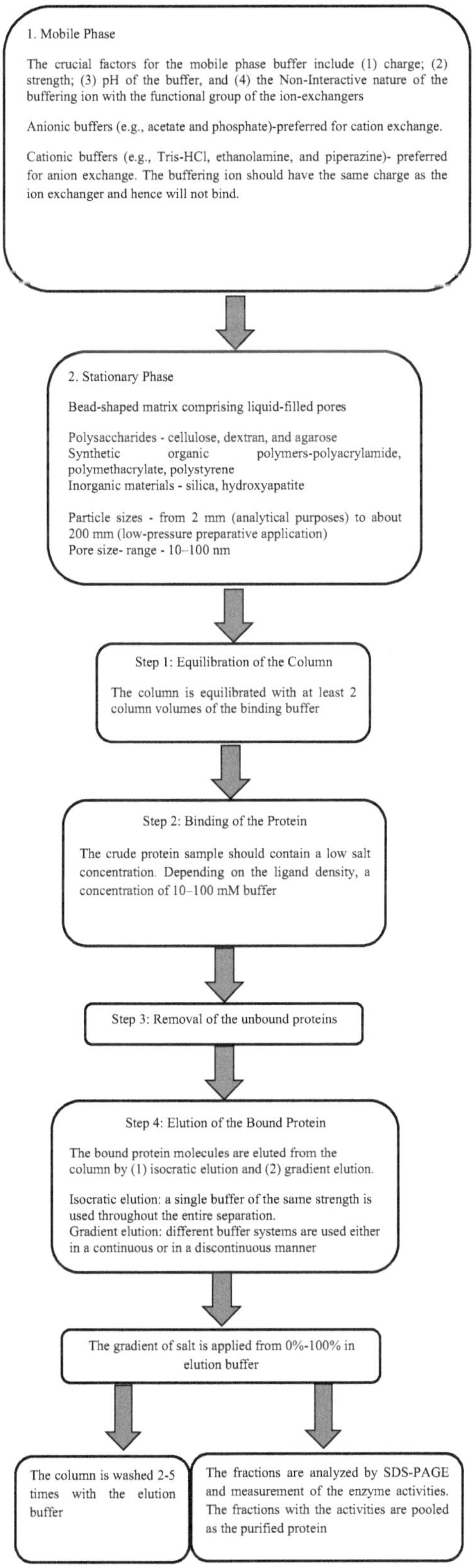

FIGURE 9.3 Basic information on ion exchange chromatography.

9. After judging the appropriate distance of running the samples into the gel, the electrophoresis is terminated and the gel is carefully transferred into the staining solution for overnight, preferably under slow shaking.
10. The gel is then transferred into the destaining solution for the visualization of the protein bands [25,33,36,41,57].

9.3.6 Zymography/Activity Staining

Zymography is a type of polyacrylamide gel electrophoresis (PAGE) also known as native PAGE or non-denaturing PAGE for the detection of enzyme activity. This method is particularly used to detect the functionality of the native proteins/enzymes. If SDS-PAGE is performed, a renaturation procedure would be required to detect activity staining [25]. For activity staining, the specific substrate-containing buffer is included in the gel, and after electrophoresis, the enzyme activity and substrate hydrolysis are detected by putting the gel into a detection solution. For instance, to detect the amylase activity, the starch solution is used as a substrate, while for the protease and lipase, gelatin and triton X-100 are used, respectively [26,45].

9.4 CONCLUSION

This chapter describes the screening, extraction, and purification of extracellular enzymes from the haloalkaliphilic actinomycetes of the saline habitats. Recent studies have revealed that the Actinobacteria from saline habitats prominently produce hydrolytic enzymes such as amylases, proteases, and lipases. Their enzymes possess unique properties and thus can function under harsh conditions. Therefore, the extraction and purification of such enzymes are significant to facilitate their wider applications.

ACKNOWLEDGMENTS

AD and HR acknowledge the award of the ShoDH Fellowships from the Government of Gujarat. The award of the DBT-Multi-Intuitional Project, MoES Net Working Project, UGC-BSR Faculty Fellowship, and International Travel Fellowships from DST-SERB, CSIR, DBT, and UGC to SPS are duly acknowledged. The infrastructural facilities created under the UGC-CAS and DST-FIST Program and by the Saurashtra University are duly acknowledged. RKK gratefully acknowledges the award of the International Travel Fellowships from DST, CSIR, CICS-INSA, Sfam International, and UGC. The contributions of Dr. Dalip Singh Rathore and Dr. Mahejbin Sheikh in the MoES Project on marine actinomycetes as research fellows deserve mention.

REFERENCES

1. Karan, R., Capes, M. D., & Das Sarma, S. (2012). Function and biotechnology of extremophilic enzymes in low water activity. *Aquatic Biosystems*, *8*(1), 1–15.
2. Mehta, V. J., Thumar, J. T., & Singh, S. P. (2006). Production of alkaline protease from an alkaliphilic actinomycete. *Bioresource Technology*, *97*(14), 1650–1654.
3. Van Den Burg, B. (2003). Extremophiles as a source for novel enzymes. *Current Opinion in Microbiology*, *6*(3), 213–218.
4. Elleuche, S., Schröder, C., Sahm, K., & Antranikian, G. (2014). Extremozymes-biocatalysts with unique properties from extremophilic microorganisms. *Current Opinion in Biotechnology*, *29*, 116–123.
5. Schiraldi, C., Acone, M., Giuliano, M., Di Lernia, I., Maresca, C., Cartenì, M., & De Rosa, M. (2001). Innovative fermentation strategies for the production of extremophilic enzymes. *Extremophiles*, *5*(3), 193–198.
6. Bhatt, H. B., & Singh, S. P. (2020). Cloning, expression, and structural elucidation of a biotechnologically potential alkaline serine protease from a newly isolated haloalkaliphilic Bacillus lehensis JO-26. *Frontiers in Microbiology*, *11*, 941.

7. Thakrar, F. J., & Singh, S. P. (2019). Catalytic, thermodynamic and structural properties of an immobilized and highly thermostable alkaline protease from a haloalkaliphilic actinobacteria, *Nocardiopsis alba* TATA-5. *Bioresource Technology, 278*, 150–158.
8. Coronado, M. J., Vargas, C., Hofemeister, J., Ventosa, A., & Nieto, J. J. (2000). Production and biochemical characterization of an α-amylase from the moderate halophile *Halomonas meridiana. FEMS Microbiology Letters, 183*(1), 67–71.
9. Fukushima, T., Mizuki, T., Echigo, A., Inoue, A., & Usami, R. (2005). Organic solvent tolerance of halophilic α-amylase from a *Haloarchaeon, Haloarcula* sp. strain S-1. *Extremophiles, 9*(1), 85–89.
10. Vasavada, S. H., Thumar, J. T., & Singh, S. P. (2006). Secretion of a potent antibiotic by salt-tolerant and alkaliphilic actinomycete *Streptomyces sannanensis* strain RJT-1. *Current Science, 91*(10) 1393–1397.
11. Singh, S. P., Raval, V. H., Purohit, M. K., Thumar, J. T., Gohel, S. D., Pandey, S., & Rawal, C. M. (2012). Haloalkaliphilic bacteria and actinobacteria from the saline habitats: new opportunities for biocatalysis and bioremediation. In *Microorganisms in Environmental Management* (pp. 415–429). Ed. Satyanarayan, T., Springer, Dordrecht.
12. Rathore, D. S., Sheikh, M. A., Gohel, S. D., & Singh, S. P. (2021). Genetic and phenotypic heterogeneity of the *Nocardiopsis alba* strains of seawater. *Current Microbiology, 78*(4), 1377–1387.
13. Kalyani, G., & Rajesh, E. M. (2018). Production and purification of amylase from *Bacillus subtilis* isolated from soil. *International Journal of Engineering and Management Research (IJEMR), 8*(3), 246–254.
14. Siroosi, M., Borujeni, F. B., Amoozegar, M. A., Babavalian, H., & Hassanshahian, M. (2021). Halophilic amylase production and purification from *Haloarcula* sp. strain D61. *Biointerface Research in Applied Chemistry, 11*(1), 7382–7392.
15. Veerapagu, M., Narayanan, A. S., Ponmurugan, K., & Jeya, K. R. (2013). Screening selection identification production and optimization of bacterial lipase from oil spilled soil. *Asian Journal of Pharmceutical and Clinical Research, 6*(3), 62–67.
16. Steele, D. B., & Stowers, M. D. (1991). Techniques for selection of industrially important microorganisms. *Annual Review of Microbiology, 45*(1), 89–106.
17. Shukla, R. J., & Singh, S. P. (2015). Characteristics and thermodynamics of α-amylase from thermophilicactinobacterium, *Laceyellasacchari* TSI-2. *Process Biochemistry, 50*(12), 2128–2136.
18. Rathore, D. S., & Singh, S. P. (2021). Kinetics of growth and co-production of amylase and protease in novel marine actinomycete, *Streptomyces lopnurensis* KaM5. *Folia Microbiologica, 66*(3), 303–316.
19. Samad, M. Y. A., Razak, C. N. A., Salleh, A. B., Yunus, W. Z. W., Ampon, K., & Basri, M. (1989). A plate assay for primary screening of lipase activity. *Journal of Microbiological Methods, 9*(1), 51–56.
20. Suganthi, C., Mageswari, A., Karthikeyan, S., Anbalagan, M., Sivakumar, A., &Gothandam, K. M. (2013). Screening and optimization of protease production from a halotolerant *Bacillus licheniformis* isolated from saltern sediments. *Journal of Genetic Engineering and Biotechnology, 11*(1), 47–52.
21. Ameri, A., Shakibaie, M., Faramarzi, M. A., Ameri, A., Amirpour-Rostami, S., Rahimi, H. R., & Forootanfar, H. (2017). Thermoalkalophilic lipase from an extremely halophilic bacterial strain *Bacillus atrophaeus* FSHM2: purification, biochemical characterization and application. *Biocatalysis and Biotransformation, 35*(3), 151–160.
22. Amoozegar, M. A., Malekzadeh, F., & Malik, K. A. (2003). Production of amylase by newly isolated moderate halophile, *Halobacillus* sp. strain MA-2. *Journal of Microbiological Methods, 52*(3), 353–359.
23. Zhi, W., Song, J., Ouyang, F., & Bi, J. (2005). Application of response surface methodology to the modeling of α-amylase purification by aqueous two-phase systems. *Journal of Biotechnology, 118*(2), 157–165.
24. Hmidet, N., Bayoudh, A., Berrin, J. G., Kanoun, S., Juge, N., & Nasri, M. (2008). Purification and biochemical characterization of a novel α-amylase from *Bacillus licheniformis* NH1: cloning, nucleotide sequence and expression of amyN gene in *Escherichia coli. Process Biochemistry, 43*(5), 499–510.
25. Shukla, R. J., & Singh, S. P. (2015). Production optimization, purification and characterization of α-amylase from thermophilic *Bacillus licheniformis* TSI-14. *Starch-Stärke, 67*(7–8), 629–639.
26. Kikani, B. A., & Singh, S. P. (2011). Single step purification and characterization of a thermostable and calcium independent α-amylase from *Bacillus amyloliquifaciens* TSWK1-1 isolated from TulsiShyam hot spring reservoir, Gujarat (India). *International Journal of Biological Macromolecules, 48*(4), 676–681.
27. Shafiei, M., Ziaee, A. A., & Amoozegar, M. A. (2012). Purification and characterization of a halophilic α-amylase with increased activity in the presence of organic solvents from the moderately halophilic *Nesterenkonia* sp. strain F. *Extremophiles, 16*(4), 627–635.

28. Thumar, J. T., & Singh, S. P. (2007). Secretion of an alkaline protease from a salt-tolerant and alkaliphilic, *Streptomyces clavuligerus* strain MIT-1. *Brazilian Journal of Microbiology, 38*, 766–772.
29. Gohel, S. D., & Singh, S. P. (2012). Purification strategies, characteristics and thermodynamic analysis of a highly thermostable alkaline protease from a salt-tolerant alkaliphilic actinomycete, *Nocardiopsis alba* OK-5. *Journal of Chromatography B, 889*, 61–68.
30. Bharathi, D., & Rajalakshmi, G. (2019). Microbial lipases: an overview of screening, production and purification. *Biocatalysis and Agricultural Biotechnology, 22*, 101368.
31. Dodia, M. S., Rawal, C. M., Bhimani, H. G., Joshi, R. H., Khare, S. K., & Singh, S. P. (2008). Purification and stability characteristics of an alkaline serine protease from a newly isolated Haloalkaliphilic bacterium sp. AH-6. *Journal of Industrial Microbiology and Biotechnology, 35*(2), 121–131.
32. Sharma, A. K., Kikani, B. A., & Singh, S. P. (2020). Biochemical, thermodynamic and structural characteristics of a biotechnologically compatible alkaline protease from a haloalkaliphilic, *Nocardiopsis dassonvillei* OK-18. *International Journal of Biological Macromolecules, 153*, 680–696.
33. Castro-Ochoa, L. D., Rodríguez-Gómez, C., Valerio-Alfaro, G., & Ros, R. O. (2005). Screening, purification and characterization of the thermoalkalophilic lipase produced by *Bacillus thermoleovorans* CCR11. *Enzyme and Microbial Technology, 37*(6), 648–654.
34. Mukherjee, S. (2019). Isolation and purification of industrial enzymes: advances in enzyme technology. In *Advances in Enzyme Technology* (pp. 41–70). Ed. Singh, R. S., Elsevier, Netherlands.
35. Böing, A. N., Van Der Pol, E., Grootemaat, A. E., Coumans, F. A., Sturk, A., & Nieuwland, R. (2014). Single-step isolation of extracellular vesicles by size-exclusion chromatography. *Journal of Extracellular Vesicles, 3*(1), 23430.
36. Thumar, J., & Singh, S. P. (2007). Two-step purification of a highly thermostable alkaline protease from salt-tolerant alkaliphilic *Streptomyces clavuligerus* strain Mit-1. *Journal of Chromatography B, 854*(1–2), 198–203.
37. Pandey, S., Rakholiya, K. D., Raval, V. H., & Singh, S. P. (2012). Catalysis and stability of an alkaline protease from a haloalkaliphilic bacterium under non-aqueous conditions as a function of pH, salt and temperature. *Journal of Bioscience and Bioengineering, 114*(3), 251–256.
38. Pérez-Pomares, F., Bautista, V., Ferrer, J., Pire, C., Marhuenda-Egea, F. C., & Bonete, M. J. (2003). α-Amylase activity from the halophilic archaeon *Haloferaxmediterranei*. *Extremophiles, 7*(4), 299–306.
39. MatpanBekler, F., Güven, K., & GülGüven, R. (2021). Purification and characterization of novel α-amylase from Anoxybacillusayderensis FMB1. *Biocatalysis and Biotransformation, 39*(4), 322–332.
40. DamodaraRao, M., Purnima, A., Ramesh, D. V., & Ayyanna, C. (2002). Purification of α-amylase from Bacillus licheniformis by chromatofocusing and gel filtration chromatography. *World Journal of Microbiology and Biotechnology, 18*(6), 547–550.
41. Li, X., & Yu, H. Y. (2012). Purification and characterization of novel organic-solvent-tolerant β-amylase and serine protease from a newly isolated *Salimicrobiumhalophilum strain* LY20. *FEMS Microbiology Letters, 329*(2), 204–211.
42. Vidyasagar, M., Prakash, S., Litchfield, C., & Sreeramulu, K. (2006). Purification and characterization of a thermostable, haloalkaliphilic extracellular serine protease from the extreme halophilic archaeon *Halogeometricumborinquense* strain TSS101. *Archaea, 2*(1), 51–57.
43. Namwong, S., Hiraga, K., Takada, K., Tsunemi, M., Tanasupawat, S. and Oda, K., 2006. A halophilic serine proteinase from *Halobacillus* sp. SR5-3 isolated from fish sauce: purification and characterization. *Bioscience, Biotechnology, and Biochemistry, 70*(6), 1395–1401.
44. Vidyasagar, M., Prakash, S., Mahajan, V., Shouche, Y. S., & Sreeramulu, K. (2009). Purification and characterization of an extreme halothermophilic protease from a halophilic bacterium *Chromohalobacter* sp. TVSP101. *Brazilian Journal of Microbiology, 40*, 12–19.
45. Kumar, S., & Khare, S. K. (2012). Purification and characterization of maltooligosaccharide-forming α-amylase from moderately halophilic *Marinobacter* sp. EMB8. *Bioresource Technology, 116*, 247–251.
46. Akmoussi-Toumi, S., Khemili-Talbi, S., Ferioune, I., & Kebbouche-Gana, S. (2018). Purification and characterization of an organic solvent-tolerant and detergent-stable lipase from *Haloferaxmediterranei* CNCMM 50101. *International Journal of Biological Macromolecules, 116*, 817–830.
47. Ai, L., Huang, Y., & Wang, C. (2018). Purification and characterization of halophilic lipase of *Chromohalobacter* sp. from ancient salt well. *Journal of Basic Microbiology, 58*(8), 647–657.
48. Esakkiraj, P., Prabakaran, G., Maruthiah, T., Immanuel, G., & Palavesam, A. (2016). Purification and characterization of halophilic alkaline lipase from *Halobacillus* sp. *Proceedings of the National Academy of Sciences, India Section B: Biological Sciences, 86*(2), 309–314.

49. Bharathi, D., Rajalakshmi, G., & Komathi, S. (2019). Optimization and production of lipase enzyme from bacterial strains isolated from petrol spilled soil. *Journal of King Saud University-Science, 31*(4), 898–901.
50. Batubara, I., Mitsunaga, T., & Ohashi, H. (2009). Screening antiacne potency of Indonesian medicinal plants: antibacterial, lipase inhibition, and antioxidant activities. *Journal of Wood Science 55*(3), 230–235.
51. Kumar, D., Kumar, L., Nagar, S., Raina, C., Parshad, R., & Gupta, V. K. (2012). Screening, isolation and production of lipase/esterase producing *Bacillus* sp. strain DVL2 and its potential evaluation in esterification and resolution reactions. *Archives of Applied Science Research, 4*(4), 1763–1770.
52. Purohit, M. K., & Singh, S. P. (2011). Comparative analysis of enzymatic stability and amino acid sequences of thermostable alkaline proteases from two haloalkaliphilic bacteria isolated from Coastal region of Gujarat, India. *International Journal of Biological Macromolecules, 49*(1), 103–112.
53. Cummins, P. M., Rochfort, K. D., & O'Connor, B. F. (2017). Ion-exchange chromatography: basic principles and application. In *Protein Chromatography* (pp. 209–223). Humana Press, New York.
54. Acikara, Ö. B. (2013). Ion exchange chromatography and its applications. *Column Chromatography, 10*, 55744.
55. Gohel, S. D., & Singh, S. P. (2013). Characteristics and thermodynamics of a thermostable protease from a salt-tolerant alkaliphilic actinomycete. *International Journal of Biological Macromolecules, 56*, 20–27.
56. Raval, V. H., Pillai, S., Rawal, C. M., & Singh, S. P. (2014). Biochemical and structural characterization of a detergent-stable serine alkaline protease from seawater haloalkaliphilic bacteria. *Process Biochemistry, 49*(6), 955–962.
57. Shafiei, M., Ziaee, A. A., & Amoozegar, M. A. (2010). Purification and biochemical characterization of a novel SDS and surfactant stable, raw starch digesting, and halophilic α-amylase from a moderately halophilic bacterium, *Nesterenkonia* sp. strain F. *Process Biochemistry, 45*(5), 694–699.

10 Marine Actinomycetes' Hydrolytic Enzymes
A Structural Characterization

A. Sai Ramesh, R. Rajamathanghi, and Lavanya Balaji

10.1 INTRODUCTION

A wide variety of distinct and generally uncharacterized actinomycetes can be found in marine locations. Actinomycin D (*Streptomyces parvulus*), Rakicidin D (*Streptomyces* sp. MWW064), Mitomycin C (*Streptomyces caespitosus*), and Salinosporamide A (*Streptomyces tropica*) are a few examples of potential marine organisms that may produce effective compounds [1].

Among the most prevalent subgroups of Gram-positive bacteria, actinomycetes play a significant role in the microbial community's ability to recycle nutrients in natural substrates and are used in the enzyme industry. Many different actinomycetes' genera have been observed to produce a wide range of enzymes and their products are used in the biotechnological and biomedical industries. They are constantly used in the synthesis of xylanase, protease, cellulose, chitinase, and amylase, among other enzymes [2]. Commercially significant enzymes are produced in large quantities by *Streptomyces* species from various sources. *Thermobifida* sp., *Thermomonospora* sp., *Thermoactinomyces* sp., *Actinomadura* sp., *Nocardiopsis* sp., *and Microbispora* sp. are some of the other organisms. Industrial enzymes such as cellulase, xylanase, amylase, pectinase, protease, and chitinase are reportedly produced by these species [2].

Industrial extracellular enzymes are isolated and characterized in order to conduct a range of application investigations. Historically, chromatographic methods have been used in conjunction with salt or solvent precipitation to achieve partial purification. Enzyme structural characterization is carried out using the resulting pure enzymes [3].

10.2 MALDI TOF ANALYSES

10.2.1 Materials Required

1. One Touch Spot Picker (1.5 mm)
2. Silver destaining solution: Potassium ferricyanide (2 mg/mL); sodium thiosulphate (0.2 mg/mL)
3. 50 mM NH_4 HCO_3 – Aqueous buffer
4. 50 mM NH_4 HCO_3/acetonitrile 1:1 – Organic buffer
5. 10 mM Dithiothreitol (DTT) – in aqueous buffer
6. 50 mM Iodoacetamide solution
7. Trypsin – for PMF (peptide mass fingerprinting)
8. 10% Trifluoroacetic acid (TFA)
9. MALDI sample plate

DOI: 10.1201/9781003398400-10

10.2.2 Procedure

Destaining the gel spot or band

Protein bands or spots cut from Poly Acrylamide Gel Electrophoresis gels will be used for protein identification.

1. Gel piece removal: The desired band should be carefully selected using a picker or sterile scalpel, and the stained gel should be placed on a clean disposable Petri dish in a lightbox.
2. The excised band or spot should be transferred to a microfuge tube and should be washed with aqueous buffer once (5 minutes) and organic buffer twice (5 minutes).
3. For silver-stained spots, it has to be destained using silver destaining solution after 30 minutes; the gel will continue to be a pale yellow colour and need to be twice cleansed with water for 5 minutes and then should be washed with aqueous and organic buffer.
4. The sample must be alkylated at the cysteine to prevent reoxidation – reduction/alkylation provides increased susceptibility to tryptic digestion. The gel spots should be incubated in DTT solution for an hour at 50°C. DTT should be discarded and the spots must be incubated in iodoacetamide for an hour in the dark at room temperature.
5. After discarding the supernatant, the gel spots were washed twice with organic buffer and aqueous buffer for 5 minutes.
6. The gel spots were dried. After which, an aliquot of trypsin with aqueous buffer is added and mixed. The gel spots were rehydrated in an enzyme solution for 10 minutes.
7. Add an equal volume of aqueous buffer and then incubate at 37°C for the entire night. After adding a tenth volume of 10% TFA, vortex for 5 minutes.
8. The acquired supernatant is utilized in the MALDI TOF-MS analysis.

10.2.3 MALDI TOF-MS – Mass Determination of Peptides

The use of a high-quality matrix will yield the best outcomes. For database searches to be accurate, MALDI TOF spectra need to be calibrated. Peptides can be analysed independently from the experimental samples in an external calibration, or they can be mixed with the experimental samples in an internal calibration.

1. Coat the MALDI plate with 0.5 µL of the matrix solution and then let it dry.
2. Fill the dried matrix spot with 2 µL of protein digest and then let it air dry.
3. After 30 seconds, remove the added 8 µL 0.1% TFA from the sample spot until the sample has desalted. (Perform this twice.)
4. To prepare the sample spot for internal calibration, add the calibration mixture.

10.2.4 Peptide Masses Measuring

For optimal signal quality, the laser strength should be adjusted to the lowest possible level. The mass can range from 600 to 3,500, and there can be between 100 and 200 shots per shot. The laser strength must be adjusted until a signal that is good enough to prevent noise is received.

As a general rule, globular structures with hydrophobic cores require amino acid lengths of 50. Topology and secondary structure are used to categorize the globular unit as all α, all β, $\alpha+\beta$, or α/β.

10.2.5 Data Analysis and Identification of Protein

A protein identification method called peptide mass fingerprinting produces tiny peptides by breaking down protein using endoprotease. By using MS analysis, the precise mass of the peptides is determined. The spectral analysis produces the peak list. The software will identify the best match between the peak list and the theoretical peptide peak list produced by the insilico digestion of the

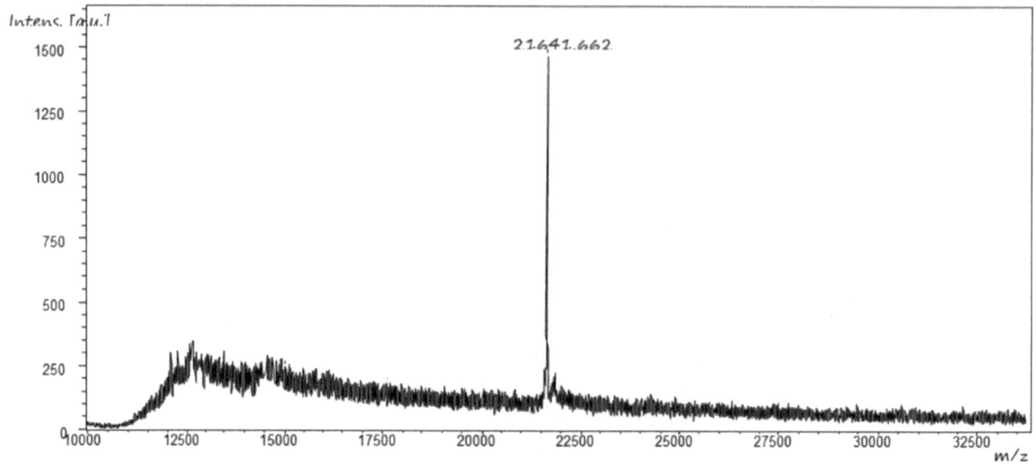

FIGURE 10.1 The representative peak of MALDI TOF analysis; lipase enzyme from halotolerant organism [4].

database proteins. Search engines like Mascot, Peptident, Profound, and MS-Fit are frequently utilized. The option to present the data in the list of masses will be available in the data analysis software. Parameters such as taxonomy, sample origin, digestion enzyme, mis-cleavages, modifications, peptide tolerance, mass values, and mono isotopic measurements (for reflectron mode) should be considered (Figure 10.2).

10.2.6 Trouble Shooting

The following list outlines some typical causes for which a database search might not yield a significant match:

1. Mass spectral noise in the information – Should optimize the peak detection threshold.
2. Sample purity: Two to three proteins can yield high-quality PMF data; however, the score for each distinct match is decreased when peptide masses from various proteins are present.
3. Peptide tolerance should not be too low or too high.
4. A significant source of protein contamination is laboratory dust; it should be avoided from samples, buffers and reagents. The utilization of sterile cabinets for workflow is preferable.
5. Auto digestion of trypsin can be avoided by preferring modified trypsin.
6. SwissProt is a relatively compact, minimally redundant, highly annotated database. Larger databases with several entries for numerous proteins include UniProt, NCBInr, and MSDB.

```
  1 MSTVSPSTDG ARSRPRPLGR FRRRALAALV GLAVPFAGMV GLAAPTQAAE
 51 AAADPSASYT RTQDWGSGFE GKWTVKNTGT APLSGWTLEW DFSAGTKVTS
101 AWDADVTNNG DHWTAKNKSW AGSLAPGASV SFGFNGTGPG TPSGCKLNGA
151 SCDGGSVPGD TPPTAPGTPT ASDLTKNSVK LSWKAATDDK GVKNYDVLRD
201 GAKVATVTAT TFTDQNLAPG TDYSYSVQAR DTADQTGPVS APVKVTTPGD
251 GTGPGPGPRE KINLGYFTEW GVYGRNYHVK NLVTSGSAEK ITHINYSFGN
301 VQGGKCTIGD SFAAYDKAYT AAESVDGVAD TWDQPLRGNF NQLRKLKAKY
351 PHIKVLWSFG GWTWSGGFTD AVKNPAAFAK SCHDLVEDPR WADVFDGIDL
401 DWEYPNACGL SCDSSGPAAL TNMVQAMRAQ FGTDLVTAAI TADASSGGKL
451 DAADYAGAAQ YFDWYNVMTY DFFGAWDKTG PTAPHSALNS YSGIPKAEFH
501 SAAAIAKLKA KGVPASKLLL GIGFYGRGWT GVTQDAPGGT ATGPATGTYE
551 AGIEDYKVLK NTCPATGTVG GTAYAKCGSN WWSYDTPATI KTKMTWAKDQ
601 GLGGAFFWEF SGDTAGGELV AAMDSGLR
```

FIGURE 10.2 Extracellular chitinase C of 630 amino acid residues [5].

7. By carefully describing the taxonomy, fewer database entries need to be searched, which decreases search durations and threshold scores for matches while simultaneously raising the likelihood that any protein identification will be made.
8. Entering the protein's molecular weight can be dangerous because the search will only look at database results in the area surrounding the value you specify; this can usually be left blank [6].

10.3 CIRCULAR DICHROISM

CD spectroscopy is the most important spectroscopic technique for quickly determining the structure of proteins in a solution. In this chapter, a summary of the methodology for the structural analyses of actinomycetes' enzymes is provided.

10.3.1 Materials Required

1. Purified protein sample – (0.8–1 OD is preferable)
2. Suprasil quartz cuvettes
3. 10 mM Sodium phosphate buffer
4. CD spectroscopy

10.3.2 Procedure

1. Check the absorbance of the buffer in a UV spectrophotometer to avoid noise.
2. Check the absorbance of protein in a UV spectrophotometer for sufficient protein concentration.
3. Centrifuge the sample and the buffers after preparation if the absorbance is high.
4. To assess the quality of the CD cell, scan the empty cell.
5. Run the reference sample.
6. Place the sample in the CD for spectral analysis (Caution – Baseline drift has to be monitored).
7. Adjust the bandwidth for spectral resolution.
8. Set three to five scans per sample.

10.4 DATA ACQUISITION

10.4.1 Protein Structure

The peptide chain's form is significantly influenced by the dihedral backbone angles φ and ψ. The protein's φ and ψ numerical values are subject to two limitations:

1. Sterical restrictions caused by side chains.
2. Creating hydrogen bonds between the hydrogen atoms of the alpha-amino groups and the carbonyl groups of peptides results in specific stabilization. The distinctive configuration of backbone angles and hydrogen bonds, also known as structural motifs, is a feature of the secondary structure found in particular protein secondary structural element combinations (Table 10.1).

10.5 SPECTRAL ANALYSIS OF PROTEINS

Tryptophan, tyrosine, phenylalanine, and the aromatic side chains of these three compounds all function as UV chromophores. Of these, the peptide bond is the weakest chromophore, but its

TABLE 10.1
Overview of Important Secondary Structure Elements in Proteins and Typical Values for φ and ψ [7]

Element	φ	Ψ	Comment
Right-handed α-helix	−62	−41	The most prevalent secondary structural element, accounting for 30%–35% of all structures that are known
β-sheet, parallel	−119	−113	Common sheet structures can arise between distant units in a main sequence or between chains
β-sheet, anti-parallel	−139	135	
Polyproline II helix (PPII)	−75	145	Significant amounts of PP2 structure are present in both folded proteins' unordered regions and unfolded proteins
3_{10}-helix	−71	−18	Rely on where you are within the helical framework
π-helix	−57	−70	Found in 15% of protein structures that are known to exist
B-turn, type 1	−64, −90	−27, −7	One of the turn-based constructions. Angles in the turn between the first and second and the second and third amino acids are denoted by values

abundance causes absorption in the far-UV region, which is below 250 nm. The α-helix's intensity can reach three times that of the β strands' spectrum. The absorption at 192 nm is the positive peak and 208 and 222 nm is the minima, which make the α-helix to recognize easily. The β-strand exhibits reduced intensity in the far-UV region, while the aromatic groups absorb light in the 250–320 nm near- and far-UV ranges. At close to the UV area, the disulphide bond between cysteine residues absorbs. Therefore, the protein concentration is determined by measuring the absorption in the near UV. Changes in the tertiary or quaternary structure of the protein, along with modifications to the microenvironment surrounding these side chains, can have an impact on its near-UV CD spectrum. Thus, the near- and far-UV spectra provide information on the physical make-up of proteins in solution and allow one to track conformational changes in response to various environmental factors, including temperature, ionic strength, and pH (Figures 10.3–10.4).

A lowest wavelength of 280 nm and a highest wavelength of 198 nm are shown in the far-UV portion of the CD spectrum of the halotolerant lipase from the VITL8 strain (Figure 10.5).

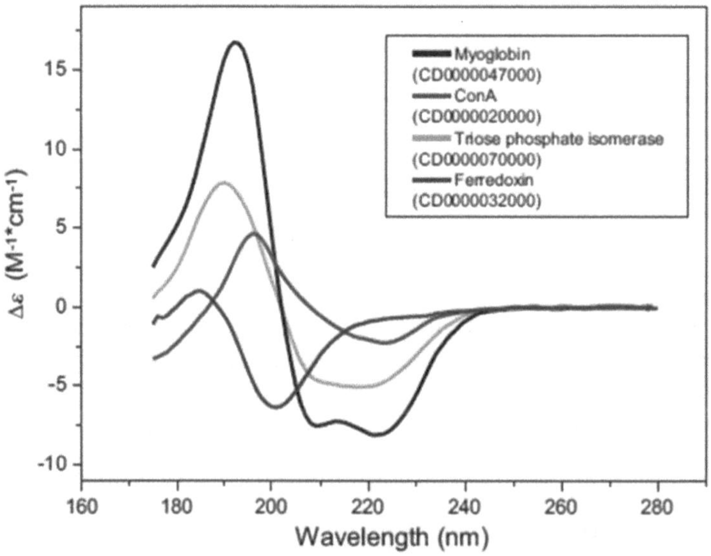

FIGURE 10.3 Far-UV CD spectra of proteins with high α-helix, β-strand, α/β protein, and high irregular structural content are examples. CD information retrieved from the online far-UV CD spectrum database [8].

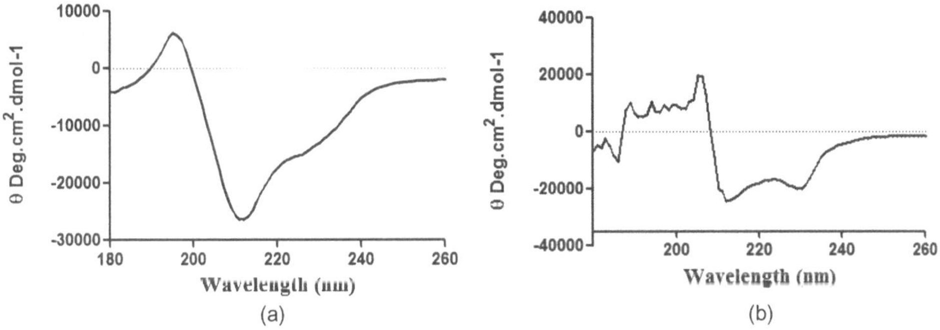

FIGURE 10.4 Far-UV CD spectra beta-pleated protease. (a) The spectra of the native protein, (b) showing denatured protein, where the beta-pleated sheets are disrupted.

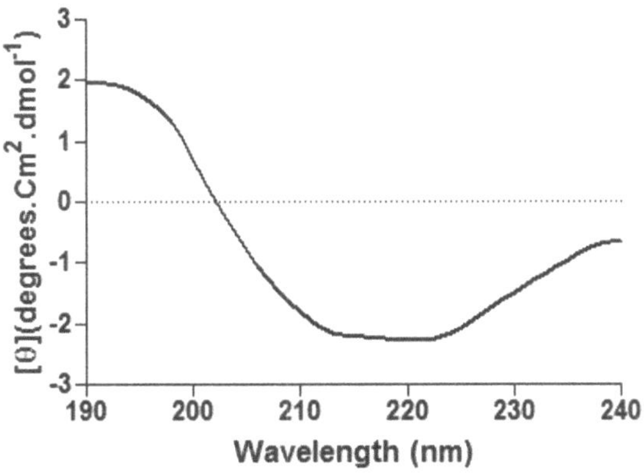

FIGURE 10.5 Halotolerant lipase's CD spectrum in far UV [4].

At 220 nm, the mean residue ellipticity was calculated to be –2 deg cm^2/d mol. A distinctive minimum may be seen in the spectra at 208 and 222 nm. This clearly demonstrates that the α-helix is the main structural feature of the enzyme. The following calculated values corresponded to the secondary structure's contents: The enzyme contains 28.5% β-turns and 38.5% α-helix [4].

CD spectroscopy is used at various solvent and temperature combinations to examine the alterations in the secondary structure of the alkaline protease from *Nocardiopsis alba* TATA-13 and *N. alba* OM-4 (Figure 10.6). The spectra of *N. alba* TATA-13 protease show a clear reverse trend with rising temperatures, although the contents of α-helices and β-sheets have grown in the *N. alba* OM-4 protease. When solvents were present, the α-helix contents of both proteases increased along with a decrease in β-sheets [9, 10].

10.5.1 Troubleshooting

1. Subtraction of the background spectrum is mandatory since the sample cell and buffer also contribute to absorbance
2. The protein to be investigated and the fitting strategy employed should be carefully employed. Model proteins are highly purified and extensively characterized, reducing the amount of uncertainty brought on by concentration mistakes.

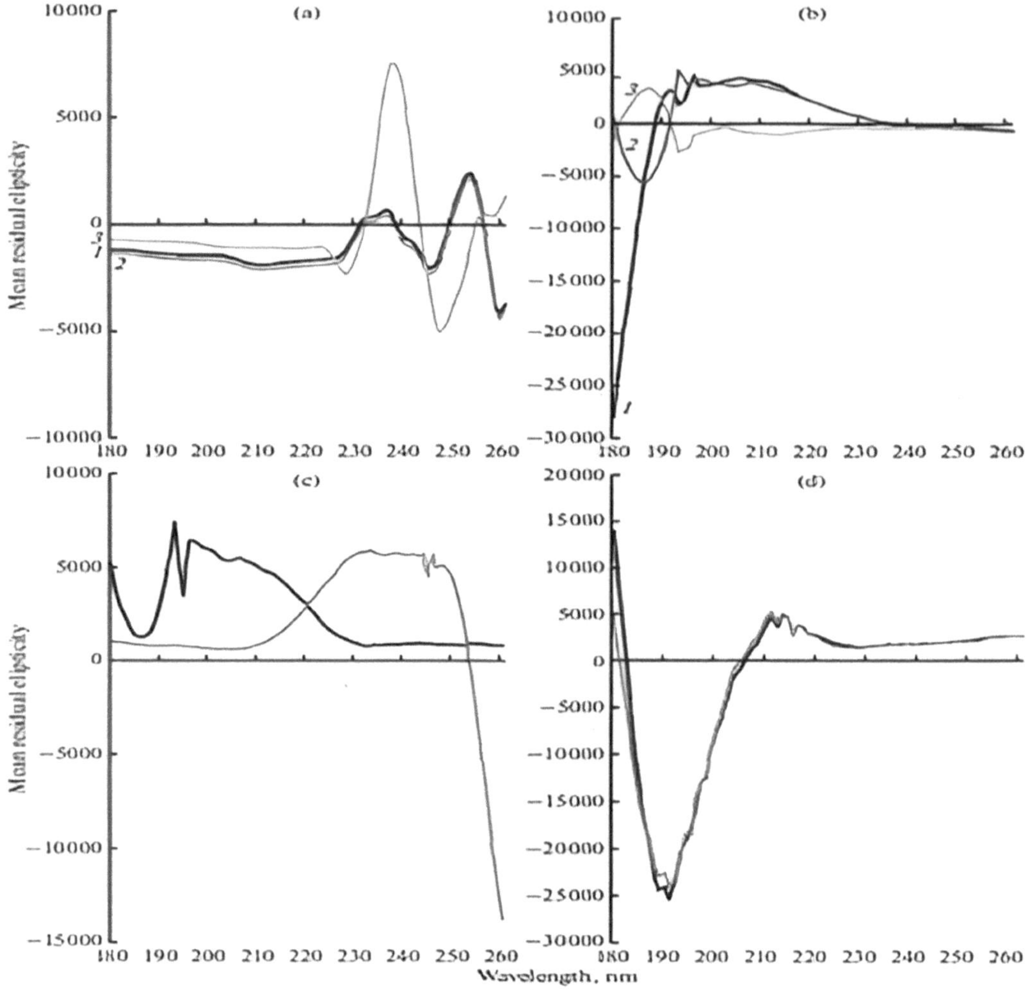

FIGURE 10.6 CD spectral analysis of alkaline protease from haloalkaliphilic *Actinobacteria Nocardiopsis alba* TATA-13 and *Nocardiopsis alba* OM-4. The spectral analysis reveals the structural change at different temperatures and with n-decane and benzene showing the effect of temperature and organic solvent on the conformational change of the enzyme [10].

REFERENCES

1. Ngamcharungchit, C., Chaimusik, N., Panbangred, W., Euanorasetr, J., and Intra, B. 2023 Bioactive metabolites from terrestrial and marine Actinomycetes. *Molecules*, *28*, 5915.
2. Prakash, D., Nawani, N., Prakash, M., Bodas, M., Mandal, A., Khetmalas, M., and Kapadnis, B. 2013. Actinomycetes: a repertory of green catalysts with a potential revenue resource. *BioMed Research International*, 2013, 1–8.
3. Thomas, L., Joseph, A., Arumugam, M., and Pandey, A. 2013. Production, purification, characterization and over-expression of xylanases from actinomycetes. *Indian Journal of Experimental Biology. 5,* 875–884.
4. Balaji, L., and Jayaraman, G. 2014. Metal ion activated lipase from halotolerant Bacillus sp. VITL8 displays broader operational range. *International Journal of Biological Macromolecules*, *67*, 380–386.
5. Nagpure, A., and Gupta, R. K. 2013. Purification and characterization of an extracellular chitinase from antagonistic *Streptomyces violaceusniger. Journal of Basic Microbiology*, *53*(5), 429–439.
6. Webster, J., and Oxley, D. 2012. Protein identification by MALDI-TOF mass spectrometry. *Chemical Genomics and Proteomics: Reviews and Protocols*, 227–240.

7. Hoffmann, S. V., Fano, M., and van de Weert, M. 2016. Circular dichroism spectroscopy for structural characterization of protein. *Analytical Techniques in the Pharmaceutical Sciences*, Chapter 6, 223–251.
8. Whitmore, L., Woollett, B., Miles, A. J., Klose, D. P., Janes, R. W., and Wallace, B. A. 2011. PCDDB: the protein circular dichroism data bank, a repository for circular dichroism spectral and metadata. *Nucleic Acids Research*, *39* (suppl_1), D480–D486.
9. Roberge, M., Lewis, R. N. A. H., Shareck, F., Morosoli, R., Kluepfel, D., Dupont, C., and McElhaney, R. N. 2002. Differential scanning calorimetric, circular dichroism, and fourier transform infrared spectroscopic characterization of the thermal unfolding of xylanase A from *Streptomyces lividans. Proteins: Structure, Function, and Bioinformatics, 50*(2), 341–354.
10. Thakrar, F. J., Kikani, B. A., Sharma, A. K., and Singh, S. P. 2018. Stability of alkaline proteases from haloalkaliphilic actinobacteria probed by circular dichroism spectroscopy. *Applied Biochemistry and Microbiology*, *54*, 591–602.

11 Isolation of Beta-Glucanase Encoding Genes from Actinobacterial Genus *Streptomyces* spp. and its Overexpression in *Escherichia coli* Heterologous System

Lekshmi K. Edison and N.S. Pradeep

11.1 INTRODUCTION

Industrial enzymes are currently sold as commodity goods, which necessitates continuous production and affordable manufacturing costs in order to remain competitive in the world markets. Heterologous expression is regarded as a key step in building modern industrial biotechnology's cost-effective, innovative production methods for recombinant enzymes. Prokaryotic expression systems are a low-cost method of producing recombinant enzymes that can produce large quantities of an enzyme from high cell density cultures. Because of its high growth rate, low maintenance requirements, low cost of fermentation, and understandable genetics, *Escherichia coli* is the organism of choice for heterologous protein production. However, it is widely acknowledged that species-specific changes in codon use due to genetic code degeneracy significantly reduce protein expression levels in *E. coli*-like heterologous systems. The genome of *Streptomyces* contains an increased amount of GC, which is greater than 70%, and various compositional biases in codon usage may dramatically reduce the amount of protein expression in heterologous hosts. In addition to this, a large amount of codon use variability was discovered among the various *Streptomyces* genomes (Wu et al., 2005). Because of this, the expression yield is low when *E. coli* is used as the host system for protein production. This is due to inconsistencies in the adaption of the codon. Earlier studies concluded that a multiparametric approach to optimizing codons, taking into consideration factors like GC content, codon quality, ribosome entrance site, secondary structure formation, and so on, could be the best way to achieve high levels of overexpression in heterologous systems (Edison et al., 2020; Xiong et al., 2006; Teng et al., 2007). The optimization of the coding sequence of the target protein in accordance with the codon bias of the expression host results in a considerable increase in the amount of protein produced, often a 10- to 50-fold increase on average (Edison et al., 2022; Sinclair and Choy, 2002; Kanekiyo et al., 2005). This protocol explains the codon optimization of beta-glucanases gene sequences from *Streptomyces* strains according to the codon bias of *E. coli* for the successful expression in the *E. coli* heterologous expression system.

DOI: 10.1201/9781003398400-11

11.2 MATERIALS AND METHODS

11.2.1 Designing Primers for Beta-Glucanase Genes

1. Retrieve beta-glucanases gene sequences from the phylogenetic neighboring species of respective *Streptomyces* spp. from the NCBI (National Center for Biotechnology Information) database.
2. Align these sequences using Clustal Omega (https://www.ebi.ac.uk/Tools/msa/clustalo/) and pick the primers from conserved regions of the aligned sequences.
3. Check the oligo quality parameters using Oligo Analysis Tool, Eurofins Genomics (https://eurofinsgenomics.eu/en/ecom/tools/oligo-analysis/).
4. Send the primer sequences for synthesis of designed oligos to any commercial oligo-synthesizing companies

11.2.2 PCR Amplification and SequencingAmplify the Beta-Glucanase Genes from *Streptomyces* spp. Strains Using the Designed Primers

1. Prepare 50 μL of the PCR (Polymerase chain reaction) mixture by adding 20–50 ng template DNA, 0.5–1 μM of forward primer (10 μM stock), and 0.5–1 μM of reverse primer (10 μM stock), 200 μM dNTPs (10 mM stock), 1–1.25 units of Taq DNA polymerase, and 1× final concentration of Taq Reaction Buffer (10× stock).
2. Carry out the PCR reaction in a temperature profile of an initial denaturation at 95°C for 2 minutes followed by 35 cycles of denaturation at 95°C for 1 minutes, annealing at 66°C for 30 seconds, extension at 72°C for 1.30 seconds, and a final extension at 72°C for 5 minutes.
3. Visualize and document the PCR products on 1%–2% of ethidium bromide-agarose gel.
4. Purify the PCR-amplified product using Wizard® SV Gel and PCR clean-up system (Promega Corporation, Wisconsin, USA) according to the manufacturer's instructions.

11.2.3 Sequencing of PCR-Amplified Products

1. Sequence the PCR product based on Sanger's dideoxy chain termination method.
2. Check the sequences using Sequence scanner software v5.2 (Applied Biosystems) and assembled using BIOEDIT v7.2.5.

11.2.4 Codon Optimization and Gene Synthesis

1. Optimize the coding regions of isolated beta-glucanases genes using the GeneArt™ GeneOptimizer® expert software from Life Technologies, Invitrogen, in order to maximize their expression in *E. coli* expression system. (The GeneOptimizer® expert software uses an approach of sliding window algorithm with multiparametric optimization of DNA sequences for codon optimization.) The optimization parameters include codon usage, GC content, ribosomal entry sites, and RNA instability motifs.
2. The optimized genes synthesize as oligonucleotides and then assemble by a batch-polymerase extension reaction of overlapping nucleotides.
3. The assembled fragments clone into a pMA-T vector (Geneart™, Germany) and transform to *E. coli* K12 OmniMAX™ 2 T1R cells.
4. For verification of sequences, purify the plasmid, amplify, and sequence using optimized beta-glucanases gene-specific primers.

11.2.5 Construction of *E. coli* Expression Vector Beta-Glucanase Genes

1. The Champion™ pET101 directional TOPO® Expression vector is one of the best choices for the expression of beta-glucanase genes.
2. For cloning in Champion™ pET101 directional TOPO® Expression vector, design the forward primer with 5′CACC overhang sequence, which base pairs with GTGG overhang sequence present in the pET101/D-TOPO® vector. Design the reverse primers without complementary stop codons, so that it allows to hybridize just downstream of the stop codon. Primers should synthesize without 5′ phosphates to allow the directional blunt-end ligation to the pET101/D-TOPO® vector.
3. Amplify the optimized beta-glucanases genes from pMA-T construct using Phusion™ Flash High-Fidelity PCR Master Mix (Thermo Fisher Scientific, Inc., USA) for obtaining blunt-end PCR product.
4. Prepare 50 µL of PCR reaction mixture by adding 20–50 ng template DNA, 0.5–1 µM of forward primer (10 µM stock), 0.5–1 µM of reverse primer, and 1× concentration of Phusion™ Flash High-Fidelity PCR Master Mix.
5. Carry out the PCR reaction in a temperature profile of an initial denaturation at 98°C for 10 seconds followed by 30 cycles of denaturation at 98°C for 1 seconds, extension at 72°C for 15 seconds, and a final extension at 72°C for 1 minute.
6. Clone the purified PCR products into pET101/D-TOPO® expression vector in a 2:1 molar ratio of PCR product: vector. Set up cloning reaction in the total volume of 6 µL by adding 2 µL of 20 ng of fresh PCR product, 1 µL of salt solution, 1 µL of pET101/D-TOPO® vector, and the remaining volume of molecular biology-grade water.
7. Gently mix the reaction components and incubate at room temperature (22°C–23°C) for 10 minutes. Then, keep it on ice.

11.2.6 Transformation of BL21 Star™ (DE3) One Shot® *E. coli* Expression Strains with Beta-Glucanase pET101 Expression Constructs

1. Transform the fresh competent *E. coli* BL21 Star™ (DE3) One Shot® cells with the expression constructs.
2. Mix 3 µL of each cloning reaction product with *E. coli* BL21 Star™ (DE3) One Shot® expression competent cells and incubate on ice for 30 minutes.
3. The cells were exposed to heat shock at 42°C for 30 seconds and immediately placed in ice.
4. Add 250 µL of S.O.C (Super Optimal broth with Catabolite repression) medium and incubate at 37°C at 200 rpm for 1 hour.
5. Spread 200 µL of transformation mixture on LB (Luria-Bertani) agar plates supplemented with 100 µg/mL ampicillin and incubate at 37°C overnight.
6. Add the entire transformation reaction mixture to 10 mL of LB broth supplemented with 50 µg/mL carbenicillin and incubate at 37°C overnight.

11.2.7 Induction of Protein Expressions in BL21 Star™ (DE3) One Shot® *E. coli* Cells

1. Inoculate 500 µL of LB-grown overnight cultures into 500 mL of terrific broth (TB) culture and incubate at 37°C till an $O.D_{600}$ reaches 0.5–0.8.
2. The culture is induce with o.1-1mM IPTG and the plates are incubate for 3 hours at 37°C.

3. Harvest the cells by centrifugation at 5,000 rpm and wash with PBS (phosphate buffered saline) buffer.

   Then, resuspend the cells in lysis buffer (50 mM sodium phosphate, 500 mM NaCl, 10 mM imidazole, 10% glycerol, pH 8.0) with 1 mg/mL lysozyme and 1% (v/v) protease inhibitor cocktail and incubate on ice for 30 minutes.
4. 10 μg/mL RNase A and 5 μg/mL DNase were added to the lysate and incubated on ice for 10–15 minutes.
5. Collect the cell lysate by centrifugation at 10,000*g* for 10 minutes at 4°C. Confirm the heterologous expression by SDS-PAGE and western blot.

REFERENCES

Edison, L.K., Dan, V.M., Reji, S., & Pradeep, N.S. (2020). A strategic production improvement of *Streptomyces* Beta glucanase enzymes with the aid of codon optimization and heterologous expression. *Biosciences Biotechnology Research Asia,* 17(3), 587–599.

Edison, L.K., Dan, V.M., & Pradeep, N.S. (2022). Application of codon optimization and recombinant expression of beta-glucanases. In: Pradeep, N., & Edison, L. K. (eds.) *Microbial Beta Glucanases.* Interdisciplinary Biotechnological Advances. Springer, Singapore, pp. 157–170.

Kanekiyo, M., Matsuo, K., Hamatake, M., Hamano, T., Ohsu, T., Matsumoto, S., ... Honda, M. (2005). Mycobacterial codon optimization enhances antigen expression and virus-specific immune responses in recombinant *Mycobacterium bovis* bacille Calmette-Guerin expressing human immunodeficiency virus type 1 Gag. *Journal of Virology*, 79(14), 8716–8723.

Sinclair, G. & Choy, F.Y.M. (2002). Synonymous codon usage bias and the expression of human glucocerebrosidase in the methylotrophic yeast *Pichia pastoris. Protein Expression and Purification,* 26, 96–105.

Teng, D., Fan, Y., Yang, Y.L., Tian, Z.G., Luo, J., & Wang, J.H. (2007). Codon optimization of *Bacillus licheniformis* β-1,3-1,4- glucanase gene and its expression in Pichia pastoris. *Applied Microbiology and Biotechnology*, 74(5), 1074–1083.

Wu, G., Culley, D.E., & Zhang, W. (2005). Predicted highly expressed genes in the genomes of *Streptomyces coelicolor* and *Streptomyces avermitilis* and the implications for their metabolism. *Microbiology*, 151, 2175–2187.

Xiong, A.S., Yao, Q.H., Peng, R.H., Zhang, Z., Xu, F., Liu, J.G., Han, P.L., & Chen, J. M. (2006). High level expression of a synthetic gene encoding *Peniophora lycii* phytase in methylotrophic yeast *Pichia pastoris. Applied Microbiology and Biotechnology*, 72(5), 1039–1047.

12 Isolation and Screening of Enzyme Inhibitors from Marine Actinobacteria

Apsara S. Babu and K.V. Bhaskara Rao

12.1 INTRODUCTION

Enzyme inhibitors are considered to be effective tools not only for the study of enzyme structures and mechanisms but also have major applications in industry, medicine, biotechnology, and agriculture [1,2]. In medicine, enzyme inhibitors are probably used as curative agents for the treatment of several diseases, including bacterial, viral, fungal, and parasitic infections. In addition to that, they also portray an important role in the treatment of cancer, immunological, cardiovascular, and neurodegenerative diseases as well as carbohydrate-dependent diseases such as diabetes, obesity, and arteriosclerosis [1].

Various enzyme inhibitors have seemed to be isolated from microorganisms. Among them, marine actinobacterial enzyme inhibitors pose significant and interesting features. The harsh marine environmental conditions enable them to produce unique bioactive compounds of clinical, pharmaceutical, and nutraceutical value [3]. Enzyme inhibitors from actinobacteria synthesize low molecular weight compounds compared to plant and animal enzyme inhibitors [4]. They are usually found in the culture filtrate of microorganisms and are products of macromolecular hydrolysis. According to the reports, the first low molecular weight enzyme inhibitor was isolated from *Streptomyces* species. Over and above 60 enzyme inhibitors have been identified. Some of them are antipain, plasmin, trypsin, leupeptin, chymotrypsin, and cathepsin B [5]. This chapter mainly discusses on the isolation and screening methods of three main enzyme inhibitors (protease inhibitors, α-glucosidase inhibitors, and lipase inhibitors) from marine actinobacteria.

12.1.1 Actinobacterial Enzyme Inhibitors

Several actinobacterial enzyme inhibitors have been identified. Some of them are detailed in the following sections.

12.1.1.1 Protease Inhibitors (PIs)

Protease inhibitors are a pervasive class of enzymes which regulate physiological as well as pathological functions in living beings. The majority of the PIs are small proteins that inactivate target proteases in the infectious process of human diseases such as cancer, arthritis, pancreatitis, muscular dystrophy, malaria, thrombosis, leishmaniasis, and AIDS [6]. In addition to this, they also act as defense molecules in intracellular and extracellular environments [7]. Depending on the type of protease they inhibit, PIs are classified as serine, cysteine, aspartic, and metalloprotease inhibitors. Most of the proteases commonly present in pathogens play critical roles like evasion of host immune defenses, procuring of nutrients for development and proliferation, and assisting tissue damage during infection [3]. Therefore, targeting these proteases can be used as future drug targets. Some of the protease inhibitors isolated from actinobacteria include leupeptin, pepstatin, phosphoramidon, etc. [8] (Figures 12.1–12.3).

DOI: 10.1201/9781003398400-12

12.1.1.2 Lipase Inhibitors

The key lipolytic enzyme named pancreatic lipase secreted by the pancreas plays a major role in the digestion of triacylglycerols. The mechanism of action of pancreatic lipase inhibition is broadly studied to arbitrate the potency of anti-obesity agents [1]. The main function of lipase inhibitors is to minimize fat absorption, thereby suppressing the activity of pancreatic lipase [9]. Some of the lipase inhibitors isolated from *Streptomyces* sp. include lipstatin, panclicins, and ebelactone B [10].

12.1.1.3 Alpha-Glucosidase Inhibitors

Alpha-glucosidase inhibitors are considered important as they are used to treat carbohydrate dependent diseases like type 2 diabetes and obesity. They work by suppressing the absorption of glucose and thereby mitigate high post-prandial blood glucose levels in diabetic patients [11]. In addition to that, they are found to be effective in treating a wide range of diseases like cancer, AIDS, and lysosomal storage disorders. Kumar et al. (2018) identified an α-glucosidase inhibitor from marine actinobacteria *Streptomyces coelicoflavus* SRBVIT13. The identified compound can be an ideal candidate for regulating postprandial hyperglycemic effects [12].

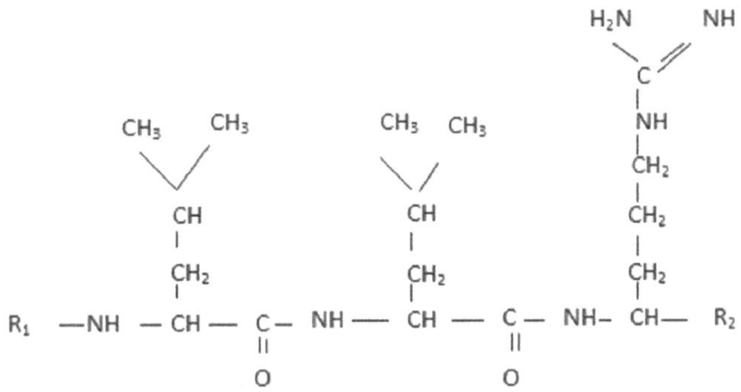

FIGURE 12.1 Structure of leupeptin. (Source: Kamarudheen et al. (2018).)

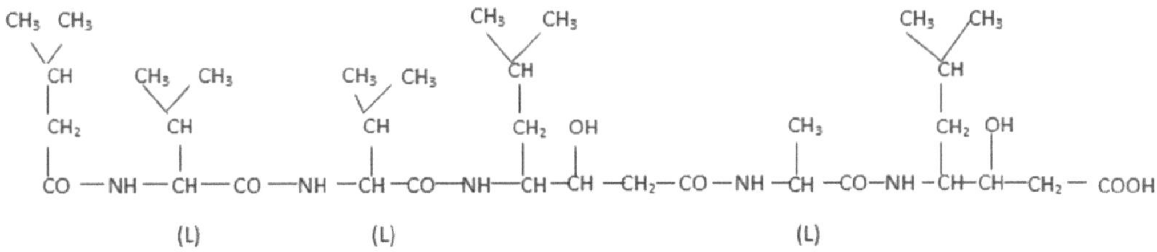

FIGURE 12.2 Structure of pepstatin. (Source: Kamarudheen et al. (2018).)

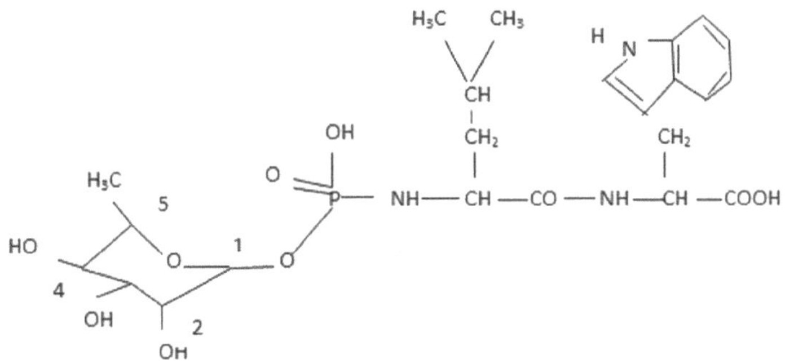

FIGURE 12.3 Structure of phosphoramidon. (Source: Kamarudheen et al. (2018).)

12.2 MATERIALS

12.2.1 Isolation Media and Chemicals

Actinomycetes isolation agar, starch casein agar, oatmeal agar (ISP3), potassium tellurite agar medium (Table 12.1), glycerol asparagine agar base (ISP5), starch casein nitrate agar (Table 12.2), soluble starch media (Table 12.3), nalidixic acid (25 mg/mL), cycloheximide (25 mg/mL), N-α-ben zoyl-DL-arginine-p-nitroanilide (BAPNA), p-nitrophenyl butyrate (P-NPB) and p-nitrophenyl-α-D -glucopyranoside (PNPG) were used.

12.2.2 Laboratory Equipment

Petri plates, Erlenmeyer flask, micropipettes, conical flask, and rotary shaker incubator were used.

TABLE 12.1
Media Components of Potassium Tellurite Agar

Components	Amount (g/L)
Potassium tellurite	0.1
Peptone	5.0
Yeast extract	2.5
pH ± 6.4	

TABLE 12.2
Media Components of Starch Casein Nitrate Agar

Components	Amount (g/L)
Soluble starch	10
Casein	0.3
K_2HPO_4	2
KNO_3	2
NaCl	2
$CaCO_3$	0.02
$MgSO_4{\cdot}7H_2O$	0.05
$FeSO_4{\cdot}7H_2O$	0.01
Agar	17

TABLE 12.3
Media Components of Soluble Starch Media

Components	Amount (g/L)
Soluble starch	20
Yeast extract	2
Glucose	10
Calcium carbonate	3
1 mL of trace elements (0.5% $FeSO_4$, $CuSO_4$, $ZnSO_4$ and $MnCl_2$).	

12.3 METHODOLOGY

12.3.1 Sample Collection and Processing

Marine sediment samples (from backwaters, estuaries, mangroves) are collected in sterile plastic bags and transported to the research laboratory. The collected samples are then pre-treated using various methods.

12.3.2 Isolation of Actinobacterial Isolates

1. Take 1 g of pre-treated sediment sample and sequentially dilute up to 10^{-7} dilution with seawater.
2. Plate 100 μL of 10^{-5}, 10^{-6} and 10^{-7} dilutions on various agar plates supplemented with nalidixic acid (antibacterial agent) and cycloheximide (antifungal agent).
3. Incubate the agar plates at 28°C for 6–14 days for actinobacterial isolates and 7–21 days for *Streptomyces* sp. [13].

12.3.3 Fermentation of Isolated Actinobacterial Cultures

1. Inoculate a loop full of culture into 50 mL of SS broth (pH 7.5) in a 250-mL conical flask containing 50% seawater and 50% distilled water.
2. Incubate the broth at 28°C on a rotary shaker incubator at 180 rpm for 2–3 days.
3. Transfer the seed inocula (10%) into a 1-L Erlenmeyer flask containing 200 mL production media.
4. Incubate the inoculated seed cultures at 28°C in a shaker incubator at 200 rpm for 72 hours.
5. Following incubation, centrifuge the fermented broth at 8,000 rpm for 10 minutes at 4°C [3,14].
6. Collect the supernatant to perform various assays.

12.3.4 Assay for Determining Enzyme Inhibitor Activity

12.3.4.1 Protease Inhibitor Activity

1. To a sterile test tube, add 0.5 mL of marine actinobacterial crude extract, 1 mg/mL trypsin solution (0.5 mL), and 0.1 mL of 0.05 M Tris–HCl (pH 8.0) and mix well.
2. Incubate the tubes at 37°C for 15 minutes in a shaker incubator.
3. After incubation, add 1 mL of 0.8 mM BAPNA to the reaction mixture and mix well. Incubate the tubes at 37°C for 15 minutes in a shaker incubator.
4. Stop the reaction by adding 5% TCA to the reaction mixture. Let the tubes stand for 20 minutes at room temperature [3,14,15].
5. Filter the solution using Whatman no. 1 filter paper and measure the absorbance at 410 nm using a UV–Vis spectrometer.
6. Calculate the percentage inhibition of actinobacterial crude extract using the formula.

$$\text{Inhibiting activity}(\%) = \frac{\text{Absorbance of control} - \text{absorbance of test}}{\text{Absorbance of control}} \times 100$$

12.3.4.2 Lipase Inhibitor Activity

1. Add 25 μL of P-NPB (dissolved in 10 mL of acetonitrile), 50 μL of pancreatic lipase (dissolved in 0.1 M phosphate saline buffer at pH 7.2, 0.15 M NaCl, and 0.5% Triton-X-100), 25 μL of actinobacterial crude extract, and 100 μL of buffer phosphate saline to a 96-well microtiter plate.

2. Homogenize and incubate the reaction mixture at 37°C for 20 minutes [9].
3. Record the absorbance at 400 nm using a microtiter plate reader.
4. Calculate the percentage inhibition using the formula.

$$\text{Percentage inhibition} = \frac{\text{Absorbance of control} - \text{absorbance of test}}{\text{Absorbance of control}} \times 100$$

12.3.4.3 Alpha-Glucosidase Inhibitor Activity

1. Add 50 µL of yeast alpha-glucosidase (1 U/mL in 0.1 M potassium phosphate buffer, pH 7.0) and 20 µL of varying concentrations of actinobacterial crude extract (20, 40, 60, 80, and 100 µg/mL) to a 96-well microtiter plate. Pre-incubate the test sample mixture for 5 minutes at 37°C [16].
2. Then, add 20 µL of 1 mM PNPG and incubate at 37°C for another 20 minutes [12].
3. After incubation, add 50 µL of 0.1 M Na_2CO_3 solution to terminate the reaction.
4. Record the absorbance at 405 nm using a microtiter plate reader (Bio-TEK, USA).
5. Calculate the percentage of inhibition using the formula.

$$\%\,\text{inhibition} = \frac{\text{Absorbance of control} - \text{absorbance of test}}{\text{Absorbance of control}} \times 100$$

REFERENCES

1. Manivasagan, P., Venkatesan, J., Sivakumar, K., and Kim, S.K. "Actinobacterial enzyme inhibitors: a review." *Critical Reviews in Microbiology* 41, no. 2 (2015): 261–272.
2. Sivakumar, K., Sahu, M.K., Thangaradjou, T., and Kannan, L. "Research on marine actinobacteria in India." *Indian Journal of Microbiology* 47 (2007): 186–196.
3. Sreedharan, V., and Rao, K.V.B. "Efficacy of protease inhibitor from marine Streptomyces sp. VITBVK2 against Leishmania donovani: An in vitro study." *Experimental Parasitology* 174 (2017): 45–51.
4. Raja, S., Ganesan, S., Sivakumar, K., and Thangaradjou, T. "Screening of marine actinobacteria for amylase enzymes inhibitors." *Indian Journal of Microbiology* 50 (2010): 233–237.
5. Bano, N., Siddiqui, S., and Amir, M. "House of industrially important bioactive metabolites: A review on actinobacteria." *Indian Journal of Biotechnology* 18 (2019): 293–304.
6. Bijina, B., Chellappan, S., Krishna, J.G., Basheer, S.M., Elyas, K.K., Bahkali, A.H., and Chandrasekaran, M. "Protease inhibitor from *Moringa oleifera* with potential for use as therapeutic drug and as seafood preservative." *Saudi Journal of Biological Sciences* 18, no. 3 (2011): 273–281.
7. Shintre, N.A., Tamhane, V.A., Baig, U.I., Pund, A.S., Patwardhan, R.B., and Deshpande, N.M. "Diversity of culturable actinobacteria producing protease inhibitors isolated from the intertidal zones of Maharashtra, India." *Current Microbiology* 77 (2020): 3555–3564.
8. Kamarudheen, N., and Rao, B. "An overview of protease inhibitors from Actinobacteria." *Research Journal of Biotechnology* 13 (2018): 1.
9. Winanda, W., Batubara, I., and Lestari, Y. "Pancreatic lipase inhibitory activity of endophytic Actinobacteria from *Rhododendron* spp." *Biosaintifika: Journal of Biology & Biology Education* 13, no. 2 (2021): 178–184.
10. Kilaru, N.B. "Production of novel lipase inhibitor from Streptomyces sp." PhD diss., University of Mysore (2005).
11. Sathish Kumar, S.R., and Bhaskara Rao, K.V. "Postprandial anti-hyperglycemic activity of marine *Streptomyces coelicoflavus* SRBVIT13 mediated gold nanoparticles in streptozotocin induced diabetic male albino Wister rats." *IET Nanobiotechnology* 10, no. 5 (2016): 308–314.
12. Kumar, S.S., and Rao, K.B. "Efficacy of alpha glucosidase inhibitor from marine Actinobacterium in the control of postprandial hyperglycaemia in Streptozotocin (STZ) induced diabetic male albino Wister rats." *Iranian Journal of Pharmaceutical Research: IJPR* 17, no. 1 (2018): 202.
13. Babu, A.S., and Bhaskara Rao, K.V. "Isolation of Actinobacteria from seaweeds." *Methods in Actinobacteriology* (2022): 83–88.

14. Karthik, L., Kumar, G., Keswani, T., Bhattacharyya, A., Chandar, S.S., and Bhaskara Rao, K.V. "Protease inhibitors from marine actinobacteria as a potential source for antimalarial compound." *PloS One* 9, no. 3 (2014): e90972.
15. Kamarudheen, N., Khaparde, A., Gopal, S., and Bhaskara Rao, K.V. "Unraveling a natural protease inhibitor from marine Streptomyces griseoincarnatus HK12 active against Chikungunya virus." *Microbiological Research* 252 (2021): 126858.
16. Siddharth, S. "Isolation and characterization of bioactive compounds with antibacterial, antioxidant and enzyme inhibitory activities from marine-derived rare actinobacteria, Nocardiopsis sp. SCA21." *Microbial Pathogenesis* 137 (2019): 103775.

13 Screening of Amylase and Lipase Enzyme Inhibitors from Actinomycetota

Ramzan Ahmed, Noor Afsha, Shiela Chetri, and Sirikanjana Thongmee

13.1 INTRODUCTION

Actinomycetota are a broad category of filamentous, aerobic, and Gram-positive actinobacteria. They are recognised as one of the most abundant sources of antibiotics and other bioactive components. In fact, this group of microorganisms has been responsible for the discovery of 80% of naturally produced antibiotics (Khanna et al., 2011), particularly Actinomycetota belonging to the genus *Streptomyces*, which is also well recognised for producing important antibiotic, anticancer, and antihelminthic compounds (Thakur et al., 2009; Zothanpuia et al., 2017; Zengler et al., 2005). Recent research has revealed that secondary metabolites from the *Nocardiopsis* and *Streptomyces* species have antioxidant and anti-diabetic activities (Sharma and Gupta, 2008). Despite their enormous potential, few studies have reported actinomycetes' enzyme inhibitory activity (Nakashima et al., 2009; Chang and Tseng, 2006).

Enzyme inhibitors have attracted a lot of interest as important tools for understanding enzyme structures and reaction mechanisms, as well as for prospective applications in medicine and agriculture (Bode and Huber, 1992; Terashita et al., 1980). It can be used to diagnose and treat pancreatitis and inflammation (Delston et al., 2003). Amylase inhibitors have been proposed for use in pharmaceutical or agricultural applications, such as the treatment of diabetes, pre-diabetes, adiposity, obesity, hyperlipidaemia, gastric ulcers, gastritis, duodenal ulcers, and caries in humans, as well as the control of insects in agriculture (Wehmeir and Piepserberg, 2004). Amylase inhibitors available today are mostly chemically produced compounds that may have negative impacts on human health (Lin et al. 2022). Amylase inhibitors are also called "starch blockers" because they stop enzymes from breaking down starch, like amylase, and slow down the rate at which starch is turned into sugar. α-Amylase and α-glucosidase are enzymes that convert carbohydrates into soluble monosaccharides (Stuart and Gulve, 2004). The inhibition of these enzymes reduces the rise in blood glucose levels by delaying the uptake of ingested carbs. As a result, these substances provide excellent therapeutic targets for treating diseases like diabetes mellitus and other diseases.

Lipases play a crucial role in the onset of obesity, which leads to cardiovascular disease and type I diabetes. According to Franson and Rossner (2000), lipase inhibitors attach to gut lipases and prevent the breakdown of dilatory triglycerides into fatty acids and monoglycerides (Yun, 2010). As a result, obesity is decreased since the absorption of lipids is reduced. As a byproduct of their origin from the breakdown of macromolecules, the enzyme inhibitors of microbial origin have a significantly lower molecular weight than those of animal and plant origin. Recognising the significance of amylase and lipase enzyme inhibitors, particularly those originating from Actinomycetota, the purpose of this chapter was to focus on the screening of amylase and lipase enzyme inhibitors from Actinomycetota, which could lead to new enzyme inhibitors that can be used for many different things.

 DOI: 10.1201/9781003398400-13

13.2 MATERIAL AND METHODS

13.2.1 Screening of A-*Amylase* Enzyme Inhibitors by Plate Methods (Imada and Simidu, 1988)

13.2.1.1 Requirements

Starch, amylase (500 μg/mL in 50 mM Tris–HCl solution, buffer pH 7.0), ethanol, iodine solution, and agar medium (compositions – 0.6% polypeptone, 1.0% soluble starch, 0.1% bacto-yeast extract, and 1.5% bacto-agar in full-strength saltwater, pH 7) are required for the screening of alpha-amylase enzyme inhibitors by plate methods.

13.2.1.2 Procedure

1. The starch agar medium was prepared, and the pH was brought to 7 before sterilisation.
2. The medium was sterilised at 121°C for 15 minutes and then cooled down, poured into the petri dishes, and left to solidify.
3. The plate was inoculated with a fresh 24-hour bacterial culture by stabbing and then incubated at 25°C for 5 days.
4. Incubation was followed by the addition of ethanol, which killed the colonies on the plate.
5. After that, the plates were sprayed with an amylase solution (500 g/mL in a 50 mM Tris–HCl buffer with a pH of 7.0), incubated for 12 hours at 37°C, and then flooded with an iodine solution.
6. A purple colour inhibitory zone (halo), which served as an indicator of the development of the inhibitor, was produced by the inhibitor-producing strain around the colony (Figure 13.1).

13.2.1.3 Expected Results

Colonies that develop an inhibitory zone of a purple colour are thought to be those of a strain that developed an alpha-amylase inhibitor (Figure 13.1).

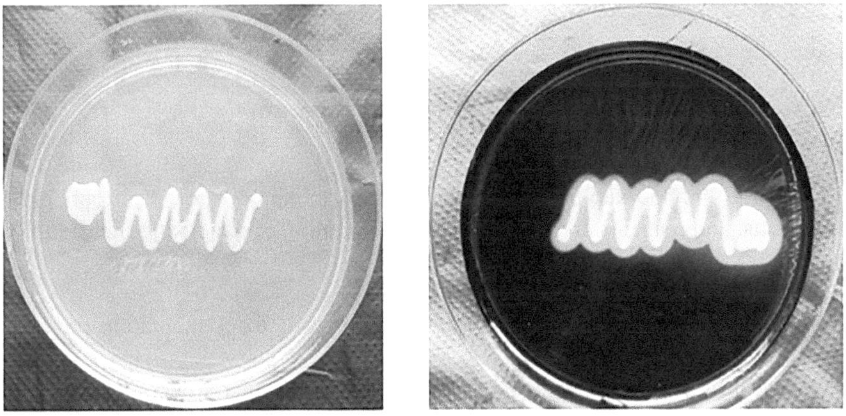

Before — After adding iodine solution

FIGURE 13.1 Inhibitory zone of Actinomycetota. (Imada and Simidu, 1988; Ubalua, 2014 open access.)

13.2.2 Screening of α-*Amylase and Lipase* Enzyme Inhibitors by Spectrophotometry Methods

13.2.2.1 Preparation of Crude Extract from Actinomycetota

In brief, the selected actinomycetes were grown in a shaker incubator at 30°C and 120 rpm for 7 days in Soybean Casein Digest Medium with 1% dextrose added. After proper incubation, the broths were centrifuged at 4 C for 30 min at a rate of 10,000 rpm. The supernatants were examined for the existence of a component with biological activity and utilised as a raw extract of metabolites from the particular Actinomycetota.

13.2.2.2 α-*Amylase*

The α-amylase inhibitor assay mixture consisted of 0.5 mL of raw extracts from Actinomycetota bacteria and 0.5 mL of Na_3PO_4 buffer (0.02 M, pH 6.9) with 0.5 mg/mL of α-amylase solution. After a pre-incubation period of 10 min at 37°C, the reaction was started by adding up to 0.5 mL of 1% starch (prepared in phosphate buffer of 0.02 M, at a pH of 6.9). The entire reaction mixture was incubated for 10 min at 37°C, and 1 mL of dinitrosalicylic acid reagent was injected to stop the reactions. The tubes were immersed in boiling water for 5 min before being brought to room temperature. The enzyme inhibition rate was estimated by measuring the quantity of glucose expressed at 540 nm with and without raw products from bacteria (Pratibha et al., 2013). The degree of inhibition was estimated using the standard α-amylase inhibitor acarbose as follows:

Amylase inhibition (%) $= X - Y/X \times 100$

where X is the absorbance of the control sample at 540 nm and Y is the absorbance when extract is present.

13.2.2.3 Lipase

To test the raw extracts' ability to block lipase activity, we mixed 1 mL of each extract with 3 mL of an olive oil emulsion comprising 5% olive oil, 10% gum Arabic, and 1 mL of an *Aspergillus niger* lipase solution containing 1.56 mg/mL. This mixture was then incubated at 37°C for 30 min. A phosphate buffer from the reaction mixture was used to make the control instead of the crude extract. Titration of the reaction mixture against 0.05 M NaOH with phenolphthalein (0.1% ethanol) as a pH indicator allowed for the determination of the free fatty acids (Selvam et al., 2010). The following formula was used to calculate the extracts' lipase inhibition activity (%):

Lipase inhibition (%) $= X - Y/X \times 100$

where X represents the enzyme activity under control and Y represents the activity in the presence of extract.

13.2.3 Screening of Pancreatic Lipase Inhibitors (Chandwad and Gutte, 2019)

13.2.3.1 Requirements

$C_{24}H_{39}NaO_4$, Na_3PO_4 monobasic, porcine pancreatic lipase, isopropanol, p-nitrophenyl palmitate, media components, and reagents are required for the screening of pancreatic lipase inhibitors.

13.2.3.2 Metabolite Extraction and Fermentation

1. Actinomycetota were isolated and maintained at 2°C–8°C for fermentation and metabolite production.
2. In a conical flask, 1 mL suspension was added to a 25-mL medium containing 1.5% (w/v) soya bean meal, 1% glycerol, 2% dextrose, and bacteria culture taken from a slant with 5 mL of 2% peptone broth and incubated at 30°C for 48 h at 200 rpm in a shaker incubator.

3. The ingredients in the growth medium were 1% glycerol, 0.5% yeast extract, 2.5% (w/v) soya bean meal, and 2% dextrose. A 10%-grown culture was used to inoculate the growth media in a flask for 8 days at 28°C, with a duplicate flask included.
4. After fermenting for a week, the metabolites were extracted with the same amount of isopropyl alcohol and ethyl acetate to get totally concentrated crude material solvents by rotavapor. The obtained raw materials assisted in the determination of lipase inhibitory action and were stored between 2°C and 8°C. To prepare a stock solution for the enzymatic assay, the crude extract was dissolved in DMSO.

13.2.3.3 Primary Screening

1. To identify PL inhibitors, actinomycetes were screened utilising the procedures described in this chapter based on the appearance of p-nitrophenol's yellow colour. This screening process took less time and did not necessitate a lengthy procedure; it just required a few chemicals.
2. The test solution was mixed with 0.1 mL of acetonitrile (containing 10 mg of extract) in 1 mL of enzyme and 1 mL of acetonitrile, the negative control tube contained 1 mL of acetonitrile and 1 mL of enzyme (20% w/v), and both solutions were incubated for 1 h. In preincubated control and test tubes, 1 mL of p-nitrophenol palmitate reagents (20 mg/mL acetonitrile) was mixed, and the volume was raised to 4 mL. The reaction tubes were kept in an incubator at 37°C for half an hour. At the beginning and end of the incubation, each extract's colour changes were noted and compared to the control. Expression of no enzyme inhibition was indicated by a dark yellow colour, moderate enzyme inhibition was indicated by a fair yellow colour, and extreme enzyme inhibition was indicated by a colourless solution.

13.2.3.4 Secondary Screening

1. Raw extracts were analysed for lipase inhibition using a spectrophotometric experiment. The experiment was conducted by observing the substrate hydrolysis, or p-nitrophenol, at 400 nm. The substrate, p-nitrophenyl palmitate, was hydrolysed by lipase in the assay's basic mechanism to produce p-nitrophenol (yellow colour).
2. When an inhibitor was present, the enzyme did not react with the substrate and appeared as a colourless solution.

Preparation of enzyme

As a model enzyme, porcine pancreatic lipase was used; it was resuspended in reaction buffer at a concentration of 10 mg/mL.

Preparation of buffer

Hundred millimolar Na_3PO_4 monobasic, 150 mM NaCl, and 0.5 mL Triton X-100 were combined to create 100 mL of reaction buffer in distilled water, and the pH of the mixture was raised to 7.4.

13.2.3.5 Assay Protocol

In this test, we measured the rate of substrate hydrolysis or p-nitrophenol formation at 400 nm. The test mixture (0.5 mL) contained 225 μL of reaction buffer and 50 mM p-nitrophenyl butyrate (from stock) (pH 7.4). The process was started by adding 0.1 mL of prepared enzyme. The inhibitor-containing fraction was solubilised in a minimal amount of dimethyl sulphoxide (10 mg/mL), and 25 μL was added to the reaction mixture containing dimethyl sulphoxide as the control agent. The temperature at which a reaction begins was 37°C, and it took 3 minutes. The percentage of inhibition was calculated in comparison with a solvent control. When the enzyme reaction time had elapsed for 3 minutes, the percentage ratio of the enzyme's activity in the presence of inhibitors to its activity in the absence of inhibitors was used to compute relative activity. The percentage of inhibition was estimated using the following formula:

$$\text{Inhibition}(\%) = \frac{\Delta A_{540\text{nm}}(\text{Uninhibited test}) - \Delta A_{540nm}(\text{Inhibited test})}{\Delta A_{540\text{nm}}(\text{Uninhibited test})} \times 100$$

ACKNOWLEDGEMENTS

Dr. Ramzan Ahmed acknowledges the Ministry of Higher Education, Science, Research and Innovation and the Thailand Science Research and Innovation Project, under Kasetsart University Reinventing University Program 2022 for Post-Doctoral Fellowship (Ref. No. 6501.0207/8434, dated 26 August 2022).

CONFLICT OF INTEREST

The authors declare no conflicts of interest related to the content of this protocol chapter.

DATA AVAILABILITY

This chapter includes all the data created or evaluated during this assessment.

REFERENCES

Bode W, Huber R. 1992. Natural protein proteinase inhibitors and their interaction with proteinases. *European Journal of Biochemistry*, 204: 433–451. DOI:10.1111/j.1432-1033.1992.tb16654.x.

Chandwad SC, Gutte SL. 2019. Screening of Actinomycetes for lipase inhibitors production. *International Journal of Pharmacy and Biological Sciences*, 9(3): 277–28. DOI:10.21276/ijpbs.2019.9.3.39.

Chang TS, Tseng M. 2006. Preliminary screening of soil actinomycetes for anti-tyrosinase activity. *Journal of Marine Science and Technology*, 14(3): 190–193. DOI:10.51400/2709-6998.2073.

Delston RB, Kothary MH, Shangraw KA, Tall BD. 2003. Isolation and characterization of a zinc-containing metalloprotease expressed by *Vibrio tubiashii*. *Canadian Journal of Microbiology*, 49: 525. DOI:10.1139/w03-067.

Franson K, Rossner S. 2000. Fat intake and food choices during weight reduction with diet, behavioural modification and a lipase inhibitor. *Journal of Internal Medicine*, 247: 607–614. DOI:10.1046/j.1365-2796.2000.t01-1-00666.x.

Imada C, Simidu U. 1988. Isolation and characterization of an α-amylase inhibitor producing actinomycete from marine environment. *Nippon Suisan Gakkaishi*, 54(10): 1839–1845. DOI:10.2331/suisan.54.1839.

Khanna M, Solanki R, Lal RN. 2011. Selective isolation of rare actinomycetes producing novel antimicrobial compounds. *International Journal of Advanced Biotechnology and Research*, 2(3): 357–375.

Nakashima T, Anzai K, Kuwahara N, et al. 2009. Physicochemical characters of a tyrosinase inhibitor produced by Streptomyces roseolilacinus NBRC 12815. *Biological and Pharmaceutical Bulletin*, 32: 832–836. DOI:10.1248/bpb.32.832.

Pratibha S, Thenmozhi M, Kannabiran K. 2013. Screening of glycolytic enzyme inhibitory activity of *Streptomyces* isolates from brine spring and marine sediments of India. *International Journal of Pharmaceutical Research and Review*, 2: 5–11.

Selvam K, Visnupuria B, Bose SC. 2010. Screening and quantification of marine actinomycetes producing industrial enzymes amylase, cellulase and lipase from south coast of India. *International Journal of Pharmaceutical and Biological Archive*, 2: 1481–1485.

Sharma SK, Gupta VK. 2008. *In vitro* antioxidant studies of *Ficus racemosa* linn root. *Pharmacognosy Magazine*, 13: 70–74.

Stuart AR, Gulve EA. 2004. Chemistry and biochemistry of type 2 diabetes. *Chemical Reviews*, 104: 1255–1282. DOI:10.1021/cr0204653.

Terashita T, Kono M, Murao S. 1980. Promoting effect of S-PI on fruiting of *Lentinus edodes*. *Transactions of the Mycological Society of Japan*, 21: 137–140.

Thakur D, Bora TC, Bordoloi GN, Mazumdar S. 2009. Influence of nutrition and culturing conditions for optimum growth and antimicrobial metabolite production by Streptomyces sp. 201. *Journal of Mycology Medicine*, 19: 161–167. DOI:10.1016/j.mycmed.2009.04.001.

Ubalua AO. 2014. The use of corn starch for growth and production of α-amylase from *Bacillus subtilis*. *Journal of Microbiology Research*, 4(2): 153–160. DOI:10.5923/j.microbiology.20140404.01.

Wehmeir UF, Piepserberg W. 2004. Biotechnology and molecular biology of the α-glucosidase inhibitor acarbose. *Applied Microbiology and Biotechnology*, 63: 613–625. DOI:10.1007/s00253-003-1477-2.

Yun W. 2010. Possible anti-obesity therapeutics from nature: A review. *Phytochemistry*, 71: 1625–1641. DOI:10.1016/j.phytochem.2010.07.011.

Zengler K, Paradkar A, Keller M. 2005. New methods to access microbial diversity for small molecule discovery. In: Zhang L, Demain AL, eds. *Natural Products: Drug Discovery and Therapeutic Medicine*. Totowa, NJ: Humana Press, 275–293. DOI:10.1007/ 978-1-59259-976-9_12.

Zothanpuia, Passari AK, Chandra P, et al. 2017. Production of potent antimicrobial compounds from Streptomyces cyaneofuscatus associated with fresh water sediment. *Frontiers in Microbiology*, 8. DOI:10.3389/fmicb.2017.00068.

Lin Q, Qiu C, Li X, Sang S, et al. 2022. The inhibitory mechanism of amylase inhibitors and research progress in nanoparticle-based inhibitors. *Critical Reviews in Food Science and Nutrition*, 63(33), 12126–12135. DOI:10.1080/10408398.2022.2098687

14 Screening and Characterization of Carotenoid Pigment from the Actinobacteria

Jayachandra S. Yaradoddi, Merja H. Kontro, Chetan D. M., and Gangadhar Megeri

14.1 INTRODUCTION

Till today, more than 10,500 types of bioactive molecules, including antibiotics, enzymes, enzyme inhibitors, single-cell proteins and pigments, are recovered from Actinomycetes; these organisms carry more diverse secondary metabolites compared to other microbes. Among these metabolites, pigments are known to be more versatile compounds [1]. The current trend, majorly based on the extraction, purification, characterization and applications of these pigments, is of higher demand [2].

The main influential factor in the exponential progression of bacterial pigment research is mainly their immense industrial applications [3]. One of the unique characteristic features of these pigments is that they are environmentally friendly, easily degradable through biological processes and non-hazardous and have greater stability compared to plant-derived and synthetic pigments. Pigments, namely anthraquinone, carotenoids, zeaxanthin, melanin and lycopene, are generally derived from microbes. By considering the health risks associated with using synthetic dyes, in recent times, there has been a tremendous growth in the natural pigments-based metabolites, which are of greater demand [4].

There has been a continuous effort in the economical and industrial scale-up of the unique metabolites. Microbial-derived pigments are well known for their pharmaceutical value, and they are also used as coloring agents in various industrial applications [5]. A recent report has revealed that enormous work is happening around the globe on biological producers of pigments in search of better substitute substances for synthetic dyes. There has been a significant increase in the product value proposition of naturally produced pigments over synthetic colors. Date back to 2011, the global sales of natural pigments were estimated to be $600 million and attained about 29% growth from 2007, which is approximately 7% higher. As per the Markets and Markets report published in January 2023, the natural color market has grown to USD 5.5 billion in 2021 and is forecasted to reach 6.7 billion USD by 2026. The pigments obtained using microbial sources could be the best choice instead of pigments that are recovered from plant and animal sources. Due to their safety aspects, natural pigments are the default choice over synthetic colorants. The uniqueness of the natural pigments lies in their cost-effective methods, including the ease of extraction and purification processes. Actinomycetes have indicated significant demand considering their potential to produce pigments.

As carotenoids are significant natural antioxidants that reduce the production of free radicals and, as a result, lessen skin photodamage, they have caught the attention of the cosmetic industry [6]. Moreover, they exhibit a promising anti-UV property, which encourages the creation of

DOI: 10.1201/9781003398400-14

formulations for photoprotective cosmetics. Contrarily, it aims to cut down on the amount of chemical filters or replace them with sunscreens made of natural ingredients [7].

Since they have potent antioxidant effects, β-carotene, lycopene and astaxanthin are some of the most extensively discussed carotenoids and are utilized as photoprotectants [8]. After consumption, β-carotene accumulates in the skin and is later converted by the body into vitamin A, which aids in the production of melanin [9]. It is also known as a precursor to provitamin A and is a potent O_2 inhibitor and free radical scavenger.

The phylum Actinobacteria contains the majority of the microorganisms in this environment's soil microbiota, and these species are crucial to the habitat's ecological balance [10]. This group is representative in terms of richness and diversity and may grow in harsh environments, such as soils with low moisture content that are common in semi-arid regions [11]. They produce a variety of colored pigments, such as carotenoids, that define the group and aid in its resistance to the environment. Moreover, they create vital secondary metabolites and extracellular enzymes [6].

The presence of several yeasts in the microbiome of these harsh Brazilian habitats also stands out as actinobacteria. According to Da Silva Andrade et al. [12], it is possible to produce β-carotene under cost-effective circumstances from the yeast *Rhodotorula glutinis* that was isolated from the semi-arid Pernambucano soil. Moreover, this carotenoid has been shown to effectively limit the proliferation of tumor cells in cases of breast cancer and promyelocytic leukemia.

These microbial pigments carry the advantages of biodegradability, lower toxicity and higher stability compared to plants and synthetic colorants. Microbial pigments can include anthraquinones, carotenoids, zeaxanthin, melanin and lycopene. The microbial pigments, besides biomedical applications, can also be employed in agricultural, environmental and industrial applications [6–11]. Compared to the bacteria of halophilic origin [12,13], actinobacterial strains have more potential in enzyme production.

14.2 MATERIALS AND METHODS

14.2.1 Isolation of Actinobacteria

Instead of isolating the actinobacteria, the *Rhodococcus* sp. strain (NCIM 5126) was procured from the collection center and was derived from the oil-polluted coastal region of Mumbai.

This microbial strain was allowed to grow on starch casein agar (SCA) media for further maintenance.

The composition of the SCA is included in Table 14.1.

TABLE 14.1
Media composition of SCA

Ingredients	g/L
K_2HPO_4	2
Starch	10
NaCl	2
KNO_3	2
$MgSO_4.7H_2O$	0.05
Casein	0.3
$FeSO_4.7H_2O$	0.01
$CaCO_3$	0.02
Agar	20
pH	7.2

14.2.2 Morphological Characters

To understand the phenotypic features of the isolates, Gram staining was carried out to identify the Gram-positive and Gram-negative nature of the cultures.

Size, mycelium, colony morphologies and pigment production were observed for each actinobacterial culture after growing them on the nutrient agar for about 20 days.

The motility test of each actinobacterial isolate was carried out using the conventional hanging drop technique with the help of an 18-hour fresh culture. Individual isolates were allowed to grow on nutrient broth through repeated subcultures and shake-flask studies.

14.2.3 Pigment Production

Actinobacterial culture broth was prepared using the Luria-Bertani modified (LBM) medium, which consists of peptone 2.0 g/L, starch 10 g/L and yeast extract 4.0 g/L, with a pH of 7.2, with the help of freshly prepared selected isolates grown on the starch nitrate agar medium.

Furthermore, the broth medium was allowed to incubate in shake condition (200 rpm) at about 30°C temperature for 7 days.

14.2.3.1 Pigment Extraction

Once the formation of orange color is observed in culture media, it is understood that the strain *Rhodococcus* produced carotenoid pigments [7]. Based on our previous reports, the extraction process is given as follows:

1. The solvent ethyl acetate was utilized in a significant quantity as an extracting medium for the pigments, and the extracted pigments were kept under the rotary evaporator.
2. The sample containing ethyl acetate and pigment extract was poured into the round bottom flask. Vortexing helped to slowly evacuate the system.
3. Then, the sample was heated with a warm water bath; the highest amount of solvent was collected in the collecting flask.
4. Furthermore, the crude extract of orange pigment was obtained in the round bottom flask.

14.2.3.2 Thin-Layer Chromatography (TLC)

1. Initially, the crude extract of the pigment was loaded on the prepared TLC plate.
2. A standard developing solution was poured into the chromatography chamber that was already saturated in the mobile phase.
3. Later, the TLC plate was inserted within the chamber.
4. Furthermore, the TLC plate was dried using a hot air oven and the pigment/bio-colorant was visualized under normal light [14].

14.2.3.3 One-Factor-at-a-Time (OFAT) Analysis

1. It is one of the significant monothetic methods generally involved in the experimental design process.
2. It tests effectively one factor at a time instead of multiple factor analysis simultaneously.
3. It provides extensive characteristic features of the pigments.

14.2.4 Identification Methods for the Extracted Pigments

14.2.4.1 Fourier Transform Infrared (FTIR) Analysis

The functional groups and the chemical bonds in the fully dried pigment were recognized by using FTIR (Shimadzu FTIR-8400S, Japan).

During the analysis, about 1 mg of tested material and 300 mg of potassium bromide (KBr) were inserted into the disks.

Then, infrared spectra measurement allowed to read between 400 and 4,000 cm^{-1} [15].

14.2.4.2 Raman Spectroscopy

For analyzing the Raman spectra, the Raman Senterra instrument with a provision of multi-wavelength potential was operated at 785 nm range.

A power supply of 50 mW with a broad range of 400–4,000 cm^{-1} was used to determine the effects of excitation wavelengths on the pigment spectrum.

Laser irradiation was performed on an object under an optical microscope, and the scattered light source from the samples was collected by the optics, passed through a pinhole and a monochromator and detected with a charge-coupled device (CCD) [15].

14.2.4.3 Gas Chromatography-Mass Spectrometry (GC-MS) Analysis

1. A capillary column of phenyl methyl siloxane 5% with a dimension of 30.0 m × 250 μm × 0.25 μm was used, and helium was mainly used as a carrier gas at 1.0 mL/min.
2. Programming of column T was done at 90°C temperature for 1 min, followed by an increment of temperature of 8°C/min to 205°C, next 5°C/min to 240°C and then 8°C/min to 300°C.
3. Simultaneously, the MS instrument ran at 70 eV. The compounds were determined by comparing the mass spectral data of the samples with those of standard compounds from the National Institute of Standard and Technology (NIST) spectral library [15].

14.2.4.4 Pigment Elemental Composition

Elemental composition analysis for the pigments was done using the energy-dispersive X-ray (EDX) instrument to determine the elemental composition of the investigated pigments.

14.3 EFFECTS OF VARIOUS SOURCES ON YIELD

14.3.1 Carbon Source

Optimal production of the pigment was achieved by various sources of carbon such as maltose, dextrose, starch, sucrose and fructose by constantly maintaining the media components with varying carbon sources.

The pigment production medium consisted of fructose/dextrose/sucrose/maltose with a concentration of 1 g/mL, whereas the LB broth of 1.3 g/100 mL was prepared in Erlenmeyer's flask of 100 mL volume of medium and sterilized.

The 24-hour culture was used to inoculate the flask containing medium with varied carbon sources and incubate the flasks on the orbital shakers at 120 rpm for 20 days for the pigment production. The culture biomass was filtered and oven-dried prior to weighing.

14.3.2 Nitrogen Source

The effect of nitrogen sources was evaluated using casein, peptone, ammonium nitrate and ammonium sulfate. The pigment production medium was kept constant with only varying the nitrogen sources.

The typical production medium consisted of casein/peptone/ammonium nitrate/ammonium sulfate of 1 g/mlL, and by adding this to the nutrient broth of 1.3 g/100 mL medium, it was sterilized.

The 24-hour-grown culture was inoculated and incubated at 120 rpm for 20 days. The culture biomass was filtered and oven-dried prior to weighing.

14.3.3 Effect of Temperature and pH

For testing the temperature and pH range for the optimal production of the actinobacterial pigment, the same inoculated culture medium was kept under different temperatures such as 30°C, 35°C, 40°C, 45°C and 50°C, and for pH range determination, various ranges of pH of the culture medium such as 4, 5, 6, 7 and 8 were used.

Furthermore, actinobacterial cultures grown under various temperatures and pH conditions were kept under a shaker for incubation. The culture biomass was filtered and oven-dried prior to weighing.

14.4 POTENTIAL ANTIMICROBIAL ACTIVITIES

The potential antimicrobial activity of the extracted pigment was tested against various pathogenic microbes with the aid of the disk diffusion technique.

The 24-hour-grown swab pathogenic microbial culture was inoculated on the agar plates and further loaded with liquified extract (100 μL) on a 6-mm sterile disk (pigmented coated disk) on the medium surface.

Then, the prepared plates were incubated at 40°C for 24 hours. Potential antimicrobial activity was determined by measuring the zone of inhibition around the pigmented coated disk.

REFERENCES

[1] Mohan Reddy, K., Siva Deepthi, S., Parameshwar, A.B., Jayachandra, S.Y., and Sulochana, M.B., 2015. Thermo and alkali tolerant exo-inulinase produced by *Streptomyces* sp. isolated from unexplored terrestrial habitat. *International Journal of Current Research and Academic Review*, 3(10), 354–363.

[2] Neethu, K., Thahiya, N., and Kokati Venkata Bhaskara, R., 2019. Biosynthesis, characterization and antagonistic applications of extracellular melanin pigment from marine Nocardiopsis sps. *Indian Journal of Pharmaceutical Education and Research*, 53(2), S112–S120.

[3] Venil Kulandaisamy, C., and Lakshmanaperumalsamy, P., 2009. An insightful overview on microbial pigment, prodigiosin. *Electronic Journal of Biology*, 5, 49–61.

[4] Pandey, N., Jain, R., Pandey, A., and Tamta, S., 2018. Optimisation and characterization of the orange pigment produced by a cold adapted strain of Penicillium sp. (GBPI_P155) isolated from mountain. Mycology, 9(2), 81–92.

[5] Hardeep Tuli, S., Chaudhary, P., Beniwal, V., and Sharma Anil, K., 2014. Microbial pigments as natural color sources: current trends and future perspectives. *Association of Food Scientists & Technologists (India)*, 52(8), 4669–4678.

[6] Stahl, W., and Sies, H., 2012. β-Carotene and other carotenoids in protection from sunlight. *The American Journal of Clinical Nutrition*, 96, 1179S–1184S. https://doi.org/10.3945/ajcn.112.034819

[7] Tosato, M.G., Orallo, D.E., Fangio, M.F., Diz, V., Dicelio, L.E., and Churio, M.S., 2016. Nanomaterials and Natural Products for UV-Photoprotection. In: Grumezescu, A. M., Ed., *Surface Chemistry of Nanobiomaterials*, Elsevier, Amsterdam, pp. 359–392. https://doi.org/10.1016/B978-0-323-42861-3.00012-1

[8] Silva, V., Martins, C.M., and Silveira, S.C., 2015. Atividade celulolítica de actinobactérias de região semiárida do Ceará. *Enciclopédia Biosfera*, 11, 2026–2036. https://doi.org/10.18677/Enciclopedia_Biosfera_2015_016

[9] Amsaveni, R., Sureshkumar, M., Vivekanandhan, G., Bhuvaneshwari, V., Kalaiselvi, M., Padmalochana, K., and Preethikaharshini, J., 2015. Screening and isolation of pigment producing Actinomycetes from soil samples. *International Journal of Biosciences*, 2, 24–28.

[10] Da Silva, I.L., Coel, L.C.B.B., and Da Silva, L.A.D.O., 2015. Biotechnological potential of the Brazilian Caatinga Biome. *Advances in Research*, 5, 1–17.

[11] Canganella, F., and Wiegel, J., 2011. Extremophiles: from abyssal to terrestrial ecosystems and possibly beyond. *Naturwissenschaften*, 98, 253–279. https://doi.org/10.1007/s00114-011-0775-2

[12] Da Silva Andrade, R.F., Lima, R.A., Ribeaux, D.R., Waleska, H., Araújo, C., Franco, L.O., Pessoa-Júnior, A., and Campos-Takaki, G.M., 2016. Production of β-carotene by a newly isolated Rhodotorula glutinis UCP1555 strain and cytotoxic effect evaluation. *Journal of Chemistry*, 10, 212–220. https://doi.org/10.17265/1934-7375/2016.05.003

[13] Kazi, Z., Hungund, B.S., Dessai, D., Banapurmath, N.R., and Yaradoddi, J.S., 2022. Potential biomedical implications of bioactive pigment and silver nanoparticles produced by the actinobacteria *Rhodococcus* sp. NCIM 5126. *Research Journal of Biotechnology*, 17(5), 160–169.
[14] Kazi, Z., Hungund, B.S., Yaradoddi, J.S., et al., 2022. Production, characterization, and antimicrobial activity of pigment from *Streptomyces* species. *Journal of Nanomaterials*, 2022, Article ID 3962301, 8 pages. https://doi.org/10.1155/2022/3962301
[15] Sharma, P., and Thakur, D., 2020. Antimicrobial biosynthetic potential and diversity of culturable soil actinobacteria from forest ecosystems of Northeast India. *Scientific Reports*, 10, 4104. https://doi.org/10.1038/s41598-020-60968-6

15 Screening of Biosurfactant Activity of Actinobacteria

R. C. Jaysree

15.1 INTRODUCTION

Surfactants synthesized from the microbial community are generally called biosurfactants. They are usually amphiphilic in nature, sharing both hydrophilic and hydrophobic moieties produced extracellularly or cell bounded such as lipopolysaccharides [1]. Actinobacteria have shown the characteristics of both bacteria and fungi. They are commonly present in soil, water (i.e. marine and freshwater ecosystems) and air. Actinobacteria have been playing an essential role in the decomposition of complex mixtures of dead plants and animals. The bioactive compounds isolated from actinobacteria have contributed to the field of medicine as antimicrobial and anti-cancer agents, as well as contributed to various biotechnological processes such as pigment production, enzyme production, bioherbicide development and nanoparticle synthesis [2]. They also have been identified to produce a lipopeptide biosurfactant, i.e. daptomycin, which is used for the treatment of Gram-positive bacteria and is marketed by Cubist Pharmaceuticals as Cubicin [2].

Biosurfactants have been found to have vast applications across various industries, including environmental remediation, agriculture, therapeutics, biotechnology, food processing, cosmetics and petroleum. It is widely used in different fields due to its diverse properties like emulsification, surface activity, anti-adhesive, antimicrobial, antioxidant, stabilizing, wetting and dispersing properties [3]. In this chapter, isolation of actinobacteria from different sources and qualitative screening methods have been discussed to evaluate the presence of biosurfactants produced by actinobacteria.

15.2 SCREENING FOR BIOSURFACTANT PRODUCTION

15.2.1 Isolation of Actinobacteria from Different Environmental Sources

Different environmental sources include soil, water and air. Water samples are of two types, i.e. freshwater and marine water sources (35 ppt salinity). For the isolation of biosurfactant-producing actinobacteria, samples can be collected from areas where oil spillages usually occur, e.g. petrol bunks, waterlogging places, agricultural lands and wastewater treatment facilities.

15.2.1.1 Soil Collection and Isolation

- Soil was collected from the rhizosphere region from a depth of 2 cm to 5 cm in a sterile zip-lock bag.
- The zip-lock bag was transported to the laboratory after analysing the pH and temperature of the soil.
- One gram of soil was weighed, and it was serially diluted for the isolation of actinobacteria from the soil.
- Nutrient mediums like actinomycete isolation agar and starch casein agar were used for the isolation of actinobacteria.
- Solid agar medium was prepared and autoclaved at 121°C for 20 minutes at 15 lbs pressure. Oil was incorporated after autoclaving, i.e. 2 mL of oil in each 100 mL of medium.

DOI: 10.1201/9781003398400-15

- The oil was added to the medium before pouring it into the Petri plates, when the temperature of the medium had decreased to 45°C.
- A total of 0.1 mL of serially diluted sample was taken from each 10^{-5}, 10^{-6} and 10^{-7} dilutions, and the sample was spread plated in the oil-incorporated solid agar medium and kept for 48–72 h in an incubator at 37°C.

15.2.1.2 Water Collection and Isolation

- Water was collected in sterile vials or screw-capped glass tubes.
- Before being transported to the laboratory, the pH and temperature of the water were analysed.
- One millilitre of water sample was used for the isolation of actinobacteria and serially diluted.
- The last three steps mentioned in Section 2.1.1 were repeated.

15.2.1.3 Enrichment Method

- One gram of soil or 1 mL of water was directly added to the nutrient broth medium.
- For the isolation of biosurfactant-producing actinobacteria, oil was also added as a nutrient source to the nutrient medium in the proportion of 2 mL for 100 mL of nutrient broth.
- This was incubated in a rotary shaker at 121 rpm for 48 hours at 37°C.
- The sample was serially diluted and inoculated in a solid medium already incorporated with oil for 48–72 hours.

15.2.1.4 Result

Whitish or grey powdery colonies formed after 48 hours in clusters and gave out the odour of soil that was wet after a rainfall.

15.2.2 Isolation of Biosurfactant from Actinobacteria

- Isolated pure culture colonies of actinobacteria grown in a Petri plate or agar slant were taken.
- They were inoculated in a liquid broth medium and kept for growth for 48 hours in a rotary shaker at 121 rpm, in an aerobic condition.
- For the isolation of biosurfactant, the liquid broth was taken in 15-mL centrifuge tubes and centrifuged at 10,000 rpm for 10 minutes.
- The supernatant was collected for further evaluation and quantification of the biosurfactant.

15.2.3 Oil Displacement

- A Petri plate was taken and kept on a dark background, so that the process could be visualized perfectly by the naked eye.
- Twenty millilitres of distilled water was added to the Petri plate.
- Four drops of mineral oil were added to the surface of distilled water making a thin film of oil. Since the oil density was less, it floated on the water.
- Ten microlitres of cell-free supernatant of actinobacteria was added to the layer [4].
- For negative control, deionized water was used.
- For positive control, sodium dodecyl sulphate (SDS) or cetyltrimethylammonium bromide (CTAB) was used.

15.2.3.1 Preparation of Positive Control: 0.5 g of SDS/CTAB Mixed in 10 mL of Water

- Ten microlitres of SDS/CTAB was added to the thin film of oil to determine the positive control.
- Displacement of oil was evaluated as positive and measured in centimetres [4].

15.2.3.2 Inference

If the supernatant has good biosurfactant activity, the values will be nearly >25 cm. The measurement can be done using a ruler. Usually, the readings are taken in diameter.

15.2.4 Drop Collapse Test

- Two microlitres of oil in 96-well microtitre plates was added and equilibrated for 30 minutes at 37°C.
- Five microlitres of cell-free broth was added on top of the oil.
- After 1 minute, the shape of the drop was observed.

15.2.4.1 Inference

If the culture supernatant collapses or breaks the oil drop, it indicates a positive result, and if the drop remains in beaded form, it is considered negative, which can be compared with distilled water as a control [5].

15.2.5 Haemolytic Activity

- Blood agar plates were prepared by adding agar plates with sheep blood.
- Blood was added to the medium before pouring it into the Petri plates, when the temperature of the medium had decreased to 45°C.
- Cultures of actinobacteria were streaked onto the agar plates and incubated for 24 hours at 37°C.

15.2.5.1 Inference

If there is a presence of a zone of clearance around the colony, then it is considered positive, and if no zone of clearance is visualized, then it is considered negative. The concentration of biosurfactant produced is directly proportional to the zone of clearance as explained by Mulligan [6].

15.2.6 Emulsification Test

Strains of actinobacteria were analysed for emulsification, according to the experimental method followed by Das [7].

- Five hundred microlitres of the supernatant was taken in a glass test tube.
- Thousand microlitres of oil was added to the glass test tube.
- The mixture was vortexed for 2 minutes left at room temperature and observed for the stability of the emulsion for 24 hours.
- The relative emulsion volume ($E_{24\%}$) was calculated as follows:

E_{24} (%) = Height of the emulsion (cm)/Total height of the liquid (cm) × 100

15.2.6.1 Inference

If the supernatant has good biosurfactant activity, the values will be >50%. The height of the emulsion can be measured using a ruler.

15.2.7 Cell Hydrophobicity Test

The cell hydrophobicity test is also called microbial adherence to hydrocarbon as given by Rosenberg [8].

- In this test, actinobacteria were cultured in a liquid culture broth for 24–48 hours.
- In the log phase, cells of actinobacteria were collected and concentrated by centrifuging at 12,000 rpm for 5 minutes.
- Phosphate buffer was prepared at pH 7.2.
- The actinobacteria cells were washed with phosphate buffer two times and measured using a calorimeter at an absorbance of 600 nm so that the initial OD value was equal to 1.00 OD, which was denoted as *Ao*.
- About 0.2 mL of oil was added to the cell suspension and vortexed for 2 minutes. Mineral oil or engine oil was used for the experiment.
- The cell suspension was incubated for 24 hours at room temperature.
- After the period of incubation, the OD of the liquid phase was measured (A_1).
- The degree of hydrophobicity was measured using the following formula:

$$[1 - (Ao/A_1)] \times 100$$

- Cells in phosphate buffer were used as a negative control.

15.2.7.1 Inference

The cell hydrophobicity should be >0.3 for actinobacteria having biosurfactant activity. The cells having biosurfactant activity would be adhered to the hydrocarbon source.

15.2.8 Surface Tension Measurement

The surface tension of an unknown liquid was measured using an instrument called a stalagmometer [9]. The surface tension of liquids changed with a slight difference in temperature.

- The temperature of the water was determined with the help of a thermometer, since for different temperatures, the surface tension changed. For example, at 27.75°C, the water surface tension was 71.59 mN/m (standard value). The mass of distilled water was calculated by using a stalagmometer.

15.2.8.1 Mass Calculation by Stalagmometer

Depending on the temperature and surface tension, the size of the drop of the liquid changes. With regard to this, the volume and density also change. Since by definition you know, surface tension is the force exerted on the surface of a liquid.

- An empty glass beaker (m_1) was weighed.
- Water taken in the stalagmometer till the mark.
- One drop of water was released from the stalagmometer, and the final mass on the weighing balance (m_2) of the glass beaker was measured. The mass of water was calculated as follows:

$$= m_2 - m_1$$

- The mass of chemical surfactant (SDS or CTAB) or soap solution and cell-free broth was measured by following the above method.
- The surface tension of the chemical surfactant and supernatant was determined by substituting in the below formula.
- The surface tension was calculated as follows:

$$\sigma \text{ (unknown)} = \sigma \text{ } (H_2O) \times m \text{ (unknown)}/m \text{ } (H_2O)$$

where σ denotes the surface tension of a liquid and m is the mass of the liquid.

15.2.8.2 Inference

If the value of the surface tension of the supernatant is <45 mN/m, the biosurfactant produced by the actinobacteria has good capability to reduce the surface tension.

15.2.9 Tributyrin Test

This test was performed to determine the lipid-degrading capability of actinobacteria [10].

- Tributyrin base agar or nutrient agar was prepared and autoclaved, and tributyrin oil was added to the medium before pouring it into the Petri plates when the temperature of the medium had decreased to 45 C.
- Actinobacteria cultures were streaked, inoculated onto plates of solid agar medium and incubated for 24 h at 37 C.

15.2.9.1 Inference

- A clear zone present around the colonies indicates positive, and no zone production indicates negative.

15.2.10 Stability

Biosurfactants show stability at varied pH levels, temperatures and salt concentrations.

15.2.10.1 pH

- Five hundred microlitres of cell-free broth was taken in each test tube such that each test tube had different pH, named as 1–14, respectively, with respect to different pH.
- The pH was adjusted by using 0.5 M HCL or 0.5 N NaOH.
- Thousand microlitres of oil was added to each test tube and vortexed for 2 minutes.
- The test tubes were incubated for 24 hours.
- After 24 hours of incubation, the emulsification index (E_{24}) was determined using the formula discussed in Section 2.6.

15.2.10.1.1 Inference

If the emulsification index is >60%, then biosurfactant has good stability in all the pH conditions.

15.2.10.2 Temperature

- Five hundred microlitres of supernatant was taken in each of the five test tubes and named with a particular temperature with respect to the temperature in which it has to be maintained, i.e. 10°C, 20°C, 37°C, 70°C and 100°C.
- The test tubes were incubated for 30 minutes at the particular temperature.
- After incubation, 1000 μL of oil was added to each test tube and vortexed for 2 minutes.
- The test tubes were incubated for 24 hours, and the emulsification index (E_{24}) was determined using the formula discussed in Section 2.6.

15.2.10.2.1 Inference

If the emulsification index is >60%, then biosurfactant has good stability in all the temperature conditions.

15.2.10.3 NaCl Concentration

- Five hundred microlitres of cell-free broth was taken in each of the ten test tubes and named with the particular percentage of NaCl concentration, i.e. 1%–10%.
- NaCl was added to each test tube according to the given concentration.
- Thousand microlitres of oil was added and vortexed for 2 min.
- After 24 hours of incubation, the emulsification index (E_{24}) was determined using the formula discussed in Section 2.6.

15.2.10.3.1 Inference

If the emulsification index is >60%, then biosurfactant has good stability in varied NaCl concentrations.

15.2.11 Thin-Layer Chromatography

The cell-free broth of actinobacteria was used to detect the presence of biosurfactants by performing thin-layer chromatography. Silica gel ready-made sheets were used for this process.

- The cell-free broth was added to the silica sheets with the help of capillary tubes.
- The solvent chloroform:methanol (2:1) was prepared, which acted as a mobile phase for the test.
- When the solvent front reached the other end, with the help of forceps, the TLC sheet was removed from the solvent.
- The identification of an unknown compound was performed by calculating the retention factor.
- The retention factor value (R_f) was calculated as the distance travelled by the compound (in centimetres) divided by the distance travelled by the solvent front (in cm) [11].
- The identification of the compound was done by using iodine vapours. Iodine vapours were used to detect biosurfactant-containing carbohydrate groups. The identification of the compound by iodine vapours was observed as brown spots on the TLC paper.
- Ninhydrin spray was also used for the identification of protein-based biosurfactants. The blue spots made by ninhydrin were used for the identification of the compound.

15.3 CONCLUSION

Diverse applications of biosurfactants in the fields of medicine, environment, cosmetics and pharmaceutical companies have led to the identification of good strains of microorganisms capable of producing biosurfactants. Oil displacement, emulsification index, surface tension and stability tests are different qualitative tests that can be used for screening biosurfactants produced by actinobacteria. The different procedures that have been discussed will be significantly helpful for the screening and evaluation of biosurfactant-producing actinobacteria isolated from different contaminated or uncontaminated environmental sources.

REFERENCES

1. Henrichsen, J. (1972). Bacterial surface translocation: a survey and a classification. *Bacteriological Reviews*, 36(4):478–503.
2. Ranjani, A., Dhanasekaran, D., & Gopinath, P.M. (2016). An introduction to Actinobacteria. *Actinobacteria – Basics and Biotechnological Applications*. Dharumadurai Dhanasekaran and Yi Jiang (Editors) 1–5. InTech.
3. Jaysree, R.C., & Rajendran, N. (2022). A review on biosurfactants. *International Journal of Research in Pharmaceutical Sciences*, 13(4):359–366.

4. Morikawa, M., Hirata, Y., & Imanaka, T. (2000). A study on the structure–function relationship of lipopeptide biosurfactants. *Biochim Biophys Acta*, 1488:211–218.
5. Bodour, A.A., & Maier, R.M. (1998). Application of a modified drop collapse technique for biosurfactant quantification and screening of biosurfactant producing microorganisms. *Journal of Microbiological Methods*, 32(3):273–280.
6. Mulligan, C.N., Cooper, D.G., & Neufeld, R.J. (1984). Selection of microbes producing biosurfactants in media without hydrocarbons. *Journal of Fermentation Technology*, 62:311–314.
7. Das, M., Das, S.K., & Mukherjee, R.K. (1998). Surface active properties of the culture filtrates of a *Micrococcus* species grown on n-alkenes and sugars. *Bioresource Technology*, 63(3):231–235.
8. Rosenberg, M., Gutnick, D., & Rosenberg, E. (1980). Adherence of bacteria to hydrocarbons: a simple method for measuring cell-surface hydrophobicity. *FEMS Microbiology Letters*, 9(1):29–33.
9. Ramasami, P., & Rughooputh, S. (2005). Accurate determination of surface tension using a Traube Stalagmometer and a low cost experimental design.
10. Carrillo, P.G., Mardaraz, C., Pitta-Alvarez, S.J., & Giulietti, A.M. (1996). Isolation and selection of biosurfactant-producing bacteria. *World Journal of Microbiology and Biotechnology*, 12:82–84.
11. Bare, K.J., & Read, H. (1987). Fast atom bombardment MS to identify materials separated on HPTLC plates. *Analyst*, 7/2:433–436.

16 Traditional Screening of Actinomycetes for Antibacterial, Antifungal, and Antiviral Properties

Sai Ramesh A., Mohamed Hasan M., Kiran Sharma B., and Pavithra M.

16.1 INTRODUCTION

Actinomycetes are a group of Gram-positive filamentous bacteria that resemble fungi and are found in a range of environments, including soil. They can even be found in severe environments because they are so prevalent in nature. A class of typical soil microorganisms commonly referred to as "thread or ray bacteria" is referred to as "actinomycetes," a non-taxonomic term. A range of natural compounds are known to be produced by microorganisms that live in harsh environments, and they are also known to have special systems for coping with challenging circumstances. Actinomycetes have a slower growth rate than other organisms, which makes them more difficult to grow. They are known for their ability to break down more challenging organic substances like chitin, a complex polysaccharide found in the exoskeletons of insects and other creatures. However, a range of elements, including soil type, location, agriculture, and organic matter, have a significant impact on the diversity and richness of actinomycetes found in any given soil [1]. They represent a distinct evolutionary line of organisms that comprises branching filaments, pleomorphic and coccoid forms, and significant chemical and physical varieties. These bacteria are very significant to researchers, pharmaceutical firms, and agricultural businesses since they have been demonstrated to be excellent repositories of medicinal antibiotics.

16.2 PRELIMINARY CROSS-STREAK TESTING OF ACTINOMYCETES FOR ANTIBACTERIAL ACTIVITY

More than 70% of the surface of our world is covered by water, and water is where life on Earth first emerged. Experts believe that biological variety is higher in certain marine environments, such as coral reefs and the deep-sea floor than in tropical rainforests. Because maritime environmental circumstances differ greatly from those on land, marine actinomycetes have distinct properties from their terrestrial counterparts and are capable of producing a variety of bioactive substances.

16.2.1 PROCEDURE

1. The strains were injected into nutrient agar plates in the middle by a single streak, as shown in Figure 16.1. The plates were incubated at 30°C for 3 days.
2. Ten different strains of bacteria were used as test organisms: *Bacillus subtilis* ATCC 11774, *Staphylococcus aureus* ATCC 1026, *Proteus mirabilis* ATCC 12453, *Klebsiella pneumoniae* ATCC 10031, *Escherichia coli* ATCC 11229, *Pseudomonas aeruginosa* ATCC

DOI: 10.1201/9781003398400-16

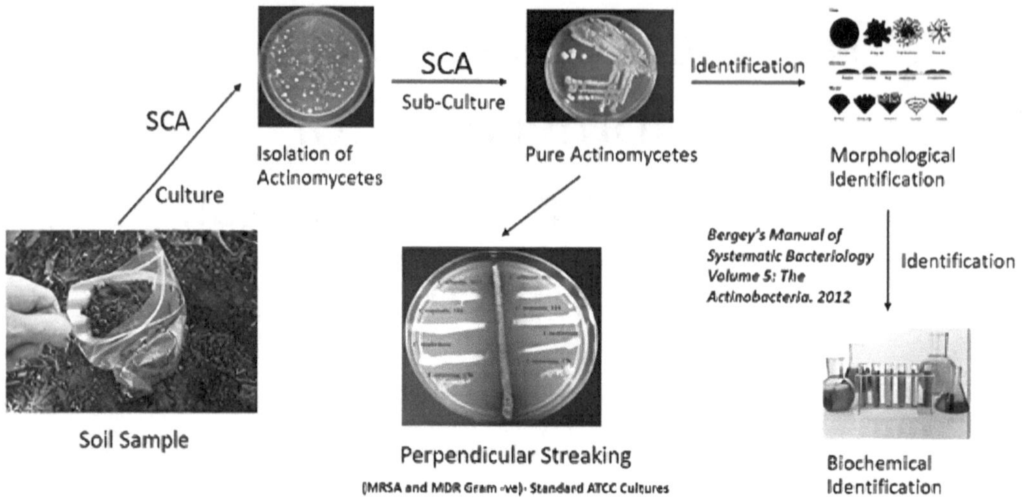

FIGURE 16.1 Expected results of isolation of actinomycetes.

10145, *Vibrio fischeri* ATCC 700601, *Proteus vulgaris* ATCC 49132, *Salmonella typhi* ATCC 700931, and *Vibrio harveyi* ATCC 1116

3. The test organisms were added to the agar [2] substrate at 90° to the antagonist. The plates were incubated at 37°C for 24 hours.
4. The diameter of the inhibition zones allowed us to observe the microbial inhibitions.
5. In order to identify them microscopically and macroscopically by Gram's staining, two of the best negative Actinomycetes isolates were picked from all of the others. These isolates were then employed for further investigation.

16.2.2 Extraction of Antimicrobial Compounds and Screening for Antibacterial Activity

16.2.3 Materials Required

1. Nutrient broth
2. Whatman filter paper
3. Screw-capped bottle
4. Orbit shaker
5. Centrifuge
6. Filtration solvents
7. Storage plates
8. Mueller-Hinton agar
9. Nutrient agar plates
10. Test bacteria
11. Sterile distilled water
12. Sterile pipettes
13. Sterile agar plate puncher
14. Antibacterial compound or extract
15. Sterile paper discs
16. Test tubes
17. Sterile saline solution

16.2.4 Production of Active Components

- For the synthesis of active compounds, the shake-flask technique was used with any of the following media: Luria-Bertani agar medium, glucose-yeast extract-malt agar medium, and inorganic salt agar medium.
- Fermentation was carried out in a broth medium. The broth medium was precisely weighed, dissolved in 100 mL of pure water before being autoclaved for 15 minutes at 121°C (15 pounds), and then allowed to cool.
- A 10% culture of the selected strain was inoculated into the medium.
- The inoculated broth was shaken at 200 rpm at 30°C for a duration of 14 days to maintain the fermentation.

16.2.5 Recovery of the Active Components

- The broth was centrifuged at 10,000 rpm for 10 minutes at 4°C in order to isolate the mycelial biomass following a 14-day incubation period.
- Filtration using Whatman filter paper was used to collect the supernatant.

16.2.6 Extraction of Active Components

- The recovered supernatant was combined with an equal amount of solvent (methanol, ethyl acetate, and acetone).
- The solvent-supernatant combination is then homogenized in a homogenizer by agitation for 45 minutes.
- To eliminate residues of fermentation broth, the solvent was isolated from the broth using a separating funnel and centrifuged at 5,000 rpm for 15 minutes (Hotam 2013).
- The extracts were dried in a water bath at 87°C until a dark brown gum-like substance was produced.
- The raw antibiotic was gathered and dehumidified in a 40°C oven.
- Purification was performed on the obtained residue.

16.2.7 Antimicrobial Activity of Crude Extracts and Minimal Inhibitory Concentration (MIC) Determination

- Antimicrobial activity of the crude extract was carried out using well diffusion for the exact determination of zones of inhibition [3,4].

16.2.8 Procedure

1. Nutrient agar plates were prepared and allowed to cool and solidify.
2. The plates with a bacterial suspension were inoculated using a sterile swab, and the bacteria were spread over the surface of the agar and incubated at 30°C to 37°C for 24 hours.
3. A stock solution of the crude extract was prepared in sterile distilled water.
4. A series of dilutions of the stock solution were prepared in sterile saline solution in test tubes, such as 1:2, 1:4, and 1:8.
5. Using a sterile agar plate puncher, wells were created in the agar surface of the plates.
6. Ten to twenty microlitres of each dilution of the extract was applied into the wells, and a positive control was added in a well (e.g. streptomycin).
7. The plates were incubated at the appropriate temperature for the bacteria to be tested.
8. After incubation, the zones of inhibition (clear areas around the well) were measured in millimetres using a ruler (Figure 16.2).

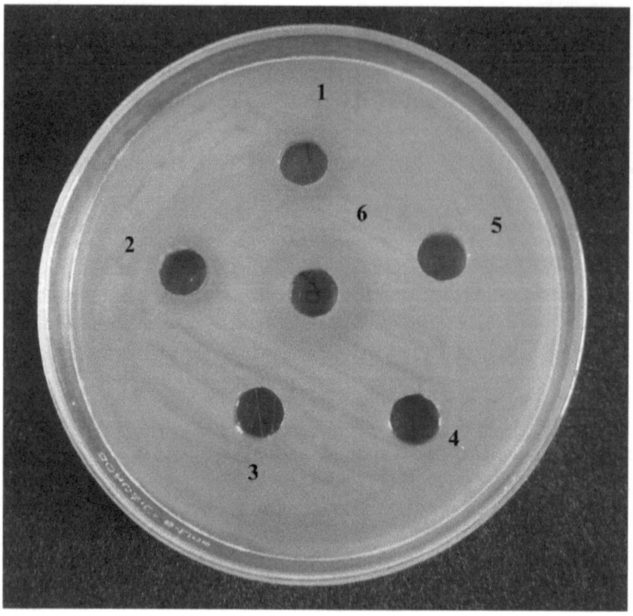

FIGURE 16.2 Antibacterial activity of the crude extracts determined by measuring the zone of inhibition.

9. The minimum inhibitory concentration (MIC) of the extract was determined as the lowest dilution that produced an inhibition zone, or the test tube serial dilution method was used to determine the MIC.

16.2.9 MIC Determination Assay with Test Tube Serial Dilution

16.2.9.1 Materials Required

- Nutrient broth or Mueller-Hinton broth
- Test bacteria
- Sterile distilled water
- Sterile pipettes
- Antibacterial compound or extract
- Test tubes
- Sterile saline solution

16.2.9.2 Procedure

1. Nutrient broth or Mueller-Hinton broth was prepared and distributed into the test tubes.
2. A stock solution of crude extract was prepared in sterile distilled water.
3. A series of dilutions of the stock solution were prepared in sterile saline solution in test tubes, such as 1:2, 1:4, and 1:8.
4. A bacterial suspension was prepared and diluted to a specific concentration that would allow the bacteria to grow in the broth but not create turbidity.
5. One millilitre of the bacterial suspension was added to the test tubes containing the broth and the extract at various concentrations.
6. Control test tubes containing only broth and bacteria and test tubes containing only broth and the extract were included.
7. The test tubes were incubated at the appropriate temperature for the bacteria to be tested.
8. After incubation, the test tubes were incubated for bacterial growth. The MIC was the lowest concentration of the extract at which no visible growth of bacteria was observed.

16.3 SCREENING FOR ANTIFUNGAL PROPERTIES

Antifungal well diffusion and MIC determination assays using test tube serial dilution are commonly used methods for evaluating the antifungal activity of compounds or extracts. The following protocol outlines the steps for conducting these assays:

16.3.1 Antifungal Activity of Crude Extracts and Minimal Inhibitory Concentration (MIC) Determination

16.3.1.1 Materials Required

- Potato dextrose agar plates
- Test fungi
- Sterile distilled water
- Sterile pipettes
- Sterile agar plate puncher
- Antifungal compound or extract
- Sterile paper discs
- Test tubes
- Sterile saline solution

16.3.1.2 Procedure

1. Potato dextrose agar plates were prepared and allowed to cool and solidify.
2. The plates with a fungal suspension (e.g. *Candida albicans* and *Aspergillus fumigatus)* were inoculated using a sterile swab, and the fungi were spread over the surface of the agar and incubated around 30°C to 37°C for 24–48 hours until the fungal colonies had grown.
3. A stock solution of the extract was prepared in sterile distilled water.
4. A series of dilutions of the stock solution were prepared in sterile saline solution in test tubes, such as 1:2, 1:4, and 1:8.
5. Using a sterile agar plate puncher, wells were created in the agar surface of the plates.
6. Ten to twenty microlitres of each dilution of the antifungal compound or extract was applied to sterile paper discs and placed into the wells.
7. The plates were incubated at the appropriate temperature for the fungi to be tested.
8. After incubation, the zones of inhibition (clear areas around the well) were measured in millimetres using a ruler (Figure 16.3).
9. The minimum inhibitory concentration (MIC) of the antifungal compound or extract was determined as the lowest dilution that produced an inhibition zone.

16.3.2 MIC Determination Assay with Test Tube Serial Dilution

16.3.2.1 Materials Required

- Sabouraud broth
- Test fungi
- Sterile distilled water
- Sterile pipettes
- Antifungal compound or extract
- Test tubes
- Sterile saline solution

16.3.2.2 Procedure

1. Sabouraud broth was prepared and distributed into the test tubes.
2. A stock solution of the crude extract was prepared in sterile distilled water.

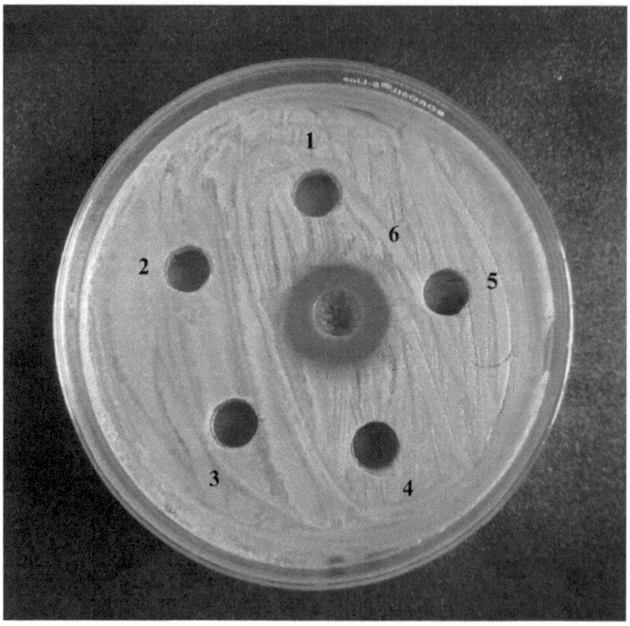

FIGURE 16.3 Antifungal activity of the crude extracts determined by measuring the zone of inhibition.

3. A series of dilutions of the stock solution were prepared in sterile saline solution in test tubes, such as 1:2, 1:4, and 1:8.
4. A fungal suspension (*Candida albicans*) was prepared and diluted to a specific concentration that would allow the fungi to grow in the broth but not create turbidity.
5. One millilitre of the fungal suspension was added to the test tubes containing the broth and the antifungal compound or extract at various concentrations.
6. Control test tubes containing only broth and fungi and test tubes containing only broth and antifungal compound or extract were included.
7. The test tubes were incubated at the appropriate temperature for the fungi to be tested.
8. After incubation for 24 hours, the test tubes were examined for fungal growth. The MIC was the lowest concentration of the antifungal compound or extract at which no visible growth of fungi was observed [8].

16.4 SCREENING FOR ANTIVIRAL PROPERTIES

16.4.1 Antiviral Activity of the Active Compounds

- The herpes simplex virus type 1 (HSV-1) strain was grown in Vero cells [7].
- All cells were cultivated and developed in Minimum Essential Medium supplemented with 10% fetal bovine serum [7].
- The experiment was carried out in microtitre plates with 96 wells and 30,000 cells each well to determine the antiviral activity of the active components (Van *et al.*, 2011) [5].
- After that, the cell culture was kept in an atmosphere with 5% CO_2 humidity and 37°C for 24 hours.
- Following a 24-hour incubation period, 100 and 50 g dilutions of extracts were applied [6].
- The cells were then cultured for an additional hour.
- Following this period, viral dilutions (50 µL) were administered in triplicates.
- For an additional 96 hours, the plates were incubated.

- As a control, monolayers of cells treated just with MEM were employed.
- Viral titres were measured using tissue culture (TCID50) at 50% infective dosages.
- Antiviral activity was calculated using the difference in virus titre between untreated and treated infected control cultures.
- The isolated strain had adequate antiviral efficacy against herpes simplex virus type 1.

REFERENCES

1. Agadagba, S.K. (2014). Isolation of Actinomycetes from soil. *Journal of Microbiology Research*, 4(3): 136–140.
2. Hsu, S.C., & Lockwood, J.L. (1975). Powdered chitin agar as a selective medium for enumeration of Actinomycetes in water and soil. *Applied Microbiology*, 29(3): 422–426.
3. Hotam, S., Chaudhar, Jayprakash Yadav, Anju R. Shrivastava, Smriti Singh, Anil K. Singh, and Natrajan Gopalan. 2013. Antibacterial activity of actinomycetes isolated from different soil samples of Sheopur (A city of central India). *Journal of Advanced Pharmaceutical Technology & Research*, 4(2): 118–23.
4. Ganesan, P., Reegan, A.D., David, R.H.A., Gandhi, M.R., Paulraj, M.G., Al-Dhabi, N.A., & Ignacimuthu, S. (2017). Antimicrobial activity of some actinomycetes from Western Ghats of Tamil Nadu, India. *Alexandria Journal of Medicine*, 53(2): 101–110.
5. Van Wezel, G.P. and McDowall, K.J., 2011. The regulation of the secondary metabolism of Streptomyces: new links and experimental advances. *Natural Product Reports*, 28(7): 1311–1333.
6. Collee, J.G., Fraser, A.G., Marmion, B.P., Simmons, A. (Eds.), Mackie, T.J., & McCartney, J.E. (1996). *Practical Medical Microbiology*. Churchill Livingstone, New York.
7. Berezin, V., Abdukhakimova, D., Trenozhnikova, L., et al. (2019). Antiviral activities of extremophilic actinomycetes extracts from Kazakhstan's unique ecosystems against influenza viruses and paramyxoviruses. *Virology Journal*, 16(150): 1–16.
8. Wiegand, I., Hilpert, K., & Hancock, R.E.W. (2008). Agar and broth dilution methods to determine the minimal inhibitory concentration (MIC) of antimicrobial substances. *Nature Protocols*, 3(2): 163–175.

17 Isolation and Screening of Actinobacteria with Antibacterial Activity and Its Production through Shake Flask Culture Method

Durairaj Thirumurugan and Ramasamy Vijayakumar

17.1 INTRODUCTION

Actinobacteria is a large, prominent subgroup among bacteria. They are Gram-positive bacteria with a high concentration of guanine (G) and cytosine (C) bases, which make up 55%–70% of their DNA [1]. They are found all over terrestrial and aquatic habitats. Over 95% of all actinobacteria strains recovered from soil are streptomycetes [2,3]. Actinobacteria resemble fungi in their filamentous structure, branching pattern, and conidia generation. Therefore, they also go by the name "ray fungi" [4]. Actinobacteria produce branching mycelium of two types: aerial and substrate mycelium [5]. Actinobacteria produce a diverse range of bioactive metabolites that are used in different industries and for human health. Over 22,000 biologically active secondary metabolites are produced by microorganisms, of which about half are produced by actinobacteria. Among metabolites, the antibiotics, enzymes, immunosuppressive, and antitumor agents are notable ones. With at least 350 genera recognized as of this writing, actinobacteria are incredibly diverse. To harvest *Streptomyces* for their maximum application, this chapter describes a procedure for the isolation and cultivation of actinobacteria isolated from ecologically inhospitable environment and the production of antibacterials from such bacteria which will serve as a lead for new antimicrobial drugs.

17.2 ISOLATION OF ACTINOBACTERIA

Over the past few decades, a vast number of actinobacteria have been isolated and screened from different sources, and they are responsible for 70%–80% of the commercial secondary metabolites that are currently sold. There are three general targets for obtaining actinobacteria from the environment: (a) isolation of actinobacteria from special environment, (b) isolation of selective (known) actinobacteria, and (c) isolation of unknown actinobacteria. This target can only be achieved with isolation media that promotes the growth of actinobacteria and inhibits the growth of other organisms. In general, selective media with special components should be used that inhibit other bacteria and fungi. In addition to this, the extrinsic factors (temperature, pH, salinity, pressure, etc.) are more important.

17.2.1 Isolation of Thermophilic Actinobacteria

Thermophilic bacteria are isolated from diverse environments such as compost piles, coal refuse, and hot-water tanks which are very difficult to isolate.

The below compositions are made for 1,000 mL distilled water.

DOI: 10.1201/9781003398400-17

17.2.1.1 Media Used

17.2.1.1.1 Starch-Agar Medium

Soluble starch – 2 g, peptone – 5 g, beef extract – 3 g, agar – 20 g.

17.2.1.1.2 Modified Czapek Dox's (MCD) Agar Medium

Sucrose 20 g, $NaNO_3$ 2.0 g, K_2HPO_4 1.0 g, $MgSO_4.7H_2O$ 0.5 g, KCl 0.5 g, $FeSO_4.7H_20$ 0.01 g, *vitamin mixture 3.7 mg, agar–agar 25 g, pH 7.2.

17.2.1.1.3 Oatmeal Agar Medium

Oatmeal 20 g, *vitamins (mixture) 3.7 mg, * salt solution 1.0 mL, agar–agar 20 g, pH 7.2.

17.2.1.1.4 Glycerol-Asparagine Agar Medium

L-asparagine 1.0 g, glycerol 10.0 g, K_2HPO_4 1.0 g, *vitamin mixture 3.7 mg, *trace salt 1.0 mL, agar–agar 20 g, pH 7.3.

*Composition of salt solutions (trace salt): $FeSO_4 \cdot 7H_2O$ 0.1 g, $MnCl_2$ 0.1 g, $ZnSO_4 \cdot 7H_2O$ 0.1 g.

Prepare the above media with 1.0 mg of penicillin and 50 mg of $K_2Cr_2O_7$ to avoid other bacterial and fungal contaminants.

Dilute 1 g of well-mixed sample (10-fold) up to 10^{-9} separately.

Inoculate the plates with the above selective medium with 0.1 mL of diluents from 10^{-2} to 10^{-3} dilutions and incubate at 55°C for 7–15 days.

17.2.2 Isolation of Halophilic Actinobacteria

17.2.2.1 Media Used

17.2.2.1.1 Modified Starch Casein Agar

Soluble starch 10.0 g, casein 0.3 g, KNO_3 2 g, $CaCO_3$ 0.02 g, $FeSO_4$ 10 mg, *salt mixture, agar–agar 25 g, pH 7.2±7.4.

17.2.2.1.2 Glycerol Aspargine Agar

L-Asparagine 1.0 g, glycerol 10 g, K_2HPO_4 1.0 g,

*vitamin mixture 3.7 mg, *trace salt 1 mL, agar–agar 20 g, pH 7.2±7.4.

17.2.2.1.3 Humic Acid-Vitamin Agar Medium

Na_2HPO_4 0.05%; KCl 0.17%, $MgSO_4 \cdot 7H_2O$ 0.005%, $FeSO_4 \cdot 7H_2O$ 0.001 g, $CaCO_3$ 0.002 g, pH 7.2, humic acid 0.1 g, agar 25 g, pH 7.5.

17.2.2.1.4 Horikoshi Medium

Glucose 10 g, yeast extract 5 g, peptone 5 g, K_2HPO_4 1 g, $MgSO_4.7H_2O$ 0.2 g, salt mixture* 1 mL, agar 25, pH 7.2±7.5.

Supplement the media with the antibiotic namely nalidixic acid (40 g) to inhibit Gram-negative bacteria.

Dilute 1 g of well-mixed sample (10-fold) up to 10^{-9} separately.

Inoculate the plates with the above selective medium with 0.1 mL of diluents from 10^{-2} to 10^{-3} dilutions and incubate at 28°C±2°C for 7–15 days.

17.2.3 Isolation of Acidophilic Actinobacteria

17.2.3.1 Culture Media

17.2.3.1.1 Starch Casein Agar

Soluble starch 10.0 g, casein 0.3 g, KNO_3 2.0 g, $CaCO_3$ 0.02 g, $FeSO_4$ 10.0 mg, *salt mixture, agar–agar 25.0 g, pH 4.0±4.5.

17.2.3.1.2 Glycerol Asparagine Medium

(Same as the above)+salt mixtures*, agar 25 g, pH 7.2–7.4.

17.2.3.1.3 Soil-Extract Medium

Soil extracts, meat extracts 3.0 g, peptone 5.0 g, salt mixtures*, agar 25 g, pH 4.0±4.5.

Soil-extract composition: Heat 400 g of soil at 120°C for 1 hour, filter, and make it to volume of 1 L.

17.2.3.1.4 Oatmeal Agar

Oatmeal 20.0 g, *vitamin mixture 3.7 mg, *trace salts 1.0 mL, agar–agar 20 g, pH 4.0±4.5.

Supplement the above media with the antibiotics, namely, cycloheximide 50 mg+nystatin 50 mg+nalidixic acid 20 mg.

Dilute 1 g of well-mixed sample (10-fold) up to 10^{-9} separately.

Inoculate the plates with the above selective medium with 0.1 mL of diluents from 10^{-2} to 10^{-3} dilutions and incubate at 28±2°C for 7–15 days.

17.3 SCREENING OF ACTINOBACTERIA FOR ANTIBACTERIAL ACTIVITY

There are many more secondary metabolites and applications that are yet to be discovered, giving actinobacteria the potential to lead to the identification of novel chemicals that are active against microbes, which opens up intriguing new opportunities for biotechnology and biomedical research. In general, two basic screening techniques were used to check the antibacterial activity of actinobacteria.

17.3.1 Primary Screening

Primary screening is a qualitative method used to evaluate the potentiality of axenic culture.

1. Make a single streak of a pure culture of actinobacteria at the center of modified-nutrient agar.
2. Incubate the plates at 28°C±2°C for 4 days.
3. Following incubation, make a single cross streak (90° angle) of test organism (pathogenic microbes) against the actinobacterial streak and incubate at 37°C for 1–2 days.
4. Following incubation, the diameter of the inhibition zone is to be measured in millimeter.
5. Select the potential isolate for secondary screening.

17.3.2 Secondary Screening

The potential actinobacteria with antibacterial activity were selected for further (secondary) screening.

1. Inoculate the potential actinobacteria in starch-casein broth and incubate in a rotary shaker at 28°C±2°C for 7 days (physical parameters and concentration of nutrients supporting maximum antibacterials production can be confirmed by culturing actinobacteria in media with different concentration of nutrients like C and N).
2. Then, filter the fermented media and centrifuge at 4,000 rpm for 30 minutes.
3. Separate the supernatant by sterile Whatman no. 1 filter and Millipore filter.
4. Extract the antibacterial compounds from cell-free extracts with equal (1:1) volume of different solvents, namely methanol, ethyl acetate, petroleum ether, ethyl alcohol, and chloroform.
5. Collect and concentrate the supernatant by using a rotary evaporator at 50°C. Cool and store the concentrated sample at 4°C for further experiments.
6. Transfer a loopful of the test organism in to 5 mL of nutrient broth and keep it for 4 hours at 37°C.
7. After incubation, compare the turbidity of the test organism with McFarland standards 0.5.
8. Spread the lawn test culture over the Muller–Hinton agar plates and make a 6-mm diameter well using sterile cork borer.
9. Further, fill 100 µL of actinobacterial extract in all the wells and incubate at 37°C for 24 hours.
10. After inhibition, measure the zone around each well in diameter and compare with the control.
11. The presence and the diameter of inhibitory zone around the well indicate antibacterial activity while no zone indicates the absence of activity.

17.4 FINDINGS AND RECOMMENDATIONS

The procedure for the screening of the antibacterial activity of actinobacteria is explained in this protocol. The results of the preliminary and secondary screening of isolates of actinobacteria are given in Figures 17.1 and 17.2. After screening several isolates of actinobacteria for their antibacterial production at the primary level, the isolates with noticeable activity alone have to be evaluated

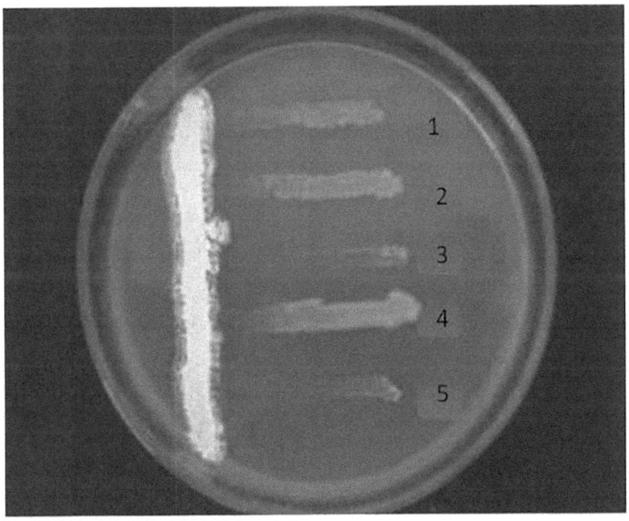

FIGURE 17.1 Antibacterial activity of *Streptomyces albogriseolus* ECR64 in primary screening. (1) *Vibrio cholerae* (2) *Vibrio parahaemolyticus*, (3) *Vibrio alginolyticus*, (4) *Aeromonas hydrophila*, and (5) *Pseudomonas fluorescens*.

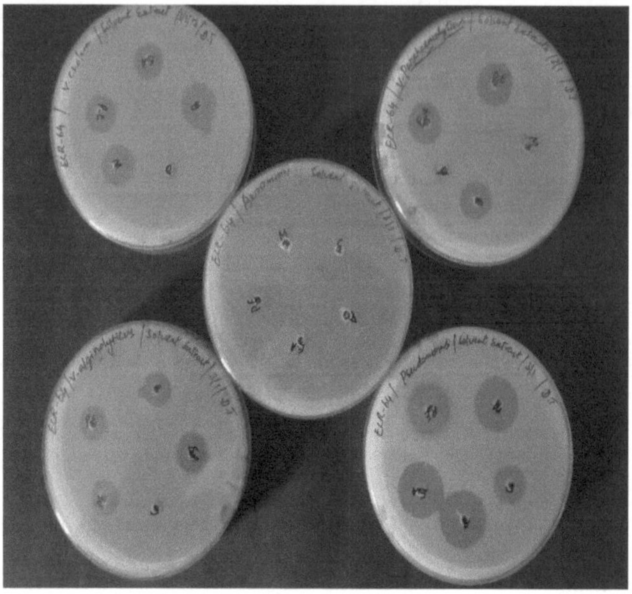

FIGURE 17.2 Antibacterial efficacy of *Streptomyces albogriseolus* ECR64 metabolites. M, methanol; EA, ethyl acetate; PE, petroleum ether; EA, ethyl alcohol; C, chloroform.

further for enriching the production of antibacterials. The selection of solvents is the crucial process in the production of antibacterials. Antibacterial efficacies of actinobacteria are maximum in ethyl acetate extract than the other solvents. Maximum production of antibacterial compounds is observed in starch casein medium with pH between 7 and 8, temperature at 30°C, starch as a carbon source, and KNO_3 as a nitrogen source, during the 9th day of incubation (Table 17.1).

TABLE 17.1
Optimization of Physico-Chemical Parameters for Higher Antibacterial Compound Production by *Streptomyces albogriseolus* ECR64

Name of the Parameter	Zone of Inhibition (mm)			
	Bacillus subtilis	*Staphylococcus aureus*	*Escherichia coli*	*Proteus vulgaris*
		pH		
5	4	–	–	6
6	11	5	5	10
7	20	19	10	19
8	18	9	10	8
9	12	6	7	3
		Temperature (°C)		
20	6	5	4	6
30	12	11	17	20
40	10	10	12	9
50	3	4	6	–
		Salinity (%)		
1	4	5	3	–
2	10	7	7	6

(Continued)

TABLE 17.1 (*Continued*)
Optimization of Physico-Chemical Parameters for Higher Antibacterial Compound Production by *Streptomyces albogriseolus* ECR64

Name of the Parameter	Zone of Inhibition (mm)			
	Bacillus subtilis	*Staphylococcus aureus*	*Escherichia coli*	*Proteus vulgaris*
4	16	13	14	9
8	14	10	9	7
16	6	6		
		Incubation Period (days)		
3	–	4	6	–
6	8	10	10	6
9	16	16	14	14
12	16	15	15	13
15	16	15	14	9
		Carbon Sources (10 g/L)		
Dextrose	7	6	9	3
Maltose	7	10	6	8
Mannitol	7	9	9	13
Starch	10	14	14	19
Sucrose	3	7	7	9
		Nitrogen Sources (2 g/L)		
Alanine	9	10	10	8
Glycine	10	12	6	9
Phenylalanine	6	7	7	7
Tyrosine	7	13	9	11
KNO_3	11	15	13	16

17.5 CONCLUSION

Isolation of actinobacteria from the various types of samples in the different ecological area using more than one cultivation media will bring out the diversity of actinobacteria with multiple antibacterials. Screening methods with optimized physico-chemical parameters, solvent selection, extraction, separation, and purification of the antibacterials are vital processes for the new drug discovery. Discovery of new antimicrobial drugs is very essential in the current scenario of the pharmaceutical world to combat emerging microbial diseases.

ACKNOWLEDGEMENTS

The authors are thankful to Bharathidasan University, India, for providing the necessary facilities.

CONFLICT OF INTEREST

The authors have no conflict of interest to declare that is relevant to the content of this protocol chapter.

DATA AVAILABILITY

The data is part of Ph.D. Thesis of the author's student. The hard and soft copy of the thesis is available in the Library, Bharathidasan University, India.

REFERENCES

1. Ventura, M., Canchaya, C., Tauch, A., Chandra, G., Fitzgerald, G.F., Chater, K., and van Sinderen, D. 2007. Genomics of actinobacteria: tracing the evolutionary history of an ancient phylum. *Microbiology and Molecular Biology Review* 71: 495–548.
2. Barka, E.A., Vatsa, P., Sanchez, L., Gaveau-Vaillant, N., Jacquard, C., Meier-Kolthoff, J.P., Klenk, H.P., Clément, C., Ouhdouch, Y., and van Wezel, G.P. 2015. Taxonomy, physiology, and natural products of actinobacteria. *Microbiology and Molecular Biology Review* 8: 1–43.
3. Takahashi, Y., and Nakashima, T. 2018. Actinomycetes, an inexhaustible source of naturally occurring antibiotics. *Antibiotics* 7: 45.
4. Wang, Y., Zhang, Z.S., Ruar, T.S., Wang, Y.M., and Ali, S.M. 1999. Investigation of actinomycetes diversity in the tropical rainforests of Singapore. *Journal of Industrial Microbiology and Biotechnology* 23: 178–187.
5. Chaudhary, H.S., Yadav, J., Shrivastava, A.R., Singh, S., Singh, A.K., and Gopalan, N. 2013. Antibacterial activity of actinomycetes isolated from different soil samples of Sheopur (A city of central India). *Journal of Advanced Pharmaceutical Technology and Research* 4: 118–123.

18 Screening of Antimycobacterial Property of Actinobacteria

Muniyan Rajiniraja

18.1 INTRODUCTION

Actinomycetes are Gram-positive bacteria mostly present in soil organisms which have characteristics common to bacteria and fungi and yet possess sufficient distinctive features to delimit them into a distinct category. Actinomycetes population is considered as one of the major group of soil population (Kuster, 1968), even being isolated from a diverse range of marine samples, including sediments obtained from deep-sea (Walker et al., 1975), which have antibacterial, antimycobacterial properties, etc.

Tuberculosis (TB) is one of the ancient and most prevalent diseases caused by *Mycobacterium tuberculosis* and infects about one-third of the world's population. It is a Gram-positive, rod-shaped and acid-fast bacteria (2–4 μm in length and 0.2–0.5 μm in width), one of the most prolific infectious agents affecting humans worldwide. It is an obligate aerobe and requires a host typically for growth and reproduction. TB is spread from person to person through the air when people with lung TB cough, sneeze or spit to propel the TB germs into the air. A person needs to inhale only a few of these germs to become infected. Since 1993, the World Health Organization (WHO) has declared TB a global emergency (WHO, 2022). Bioactive compounds from actinobacteria offer hope to fulfil these needs and have been used for curing various diseases, including TB.

M. tuberculosis infection begins with LTBI form (latent TB infection) in which it does not cause any symptoms (asymptomatic) since the host immune system controls the infection. Nearly 5%–10% of individuals are not able to control the initial infection and will develop **primary tuberculosis**. On the other hand, a large group of the remaining 90%–95% of individuals is left with LTBI. *M. tuberculosis* can remain dormant in macrophages and other cells for decades among this large group of people. The dormant bacteria would become an active form in later stage along with those 5%–10% LTB infections, as a result of various factors that can lead to reduced alertness of the immune system. Active tuberculosis among this group is referred to as **reactivation** tuberculosis.

In this chapter, we focused on *in vitro* qualitative and quantitative inhibition studies against tuberculosis strains especially for the laboratory strains such as *Mycobacterium tuberculosis* (H37Ra) and *Mycobacterium smegmatis.*

18.2 MATERIALS AND METHODS

18.2.1 Preparation of Actinobacterial Extract

1. The isolates of actinobacteria are inoculated into ISP2 medium in a shaker incubator maintaining 120 rpm for 7 days at 28°C and centrifuged at 13,000 rpm for 10 minutes at 4°C to remove cell debris.
2. The resultant supernatant is mixed thoroughly with ethyl acetate in the ratio of 1:1 (v/v) and left overnight.

DOI: 10.1201/9781003398400-18

3. The solvent phase is separated from the aqueous phase and concentrated in the rotary evaporator at 40°C under reduced pressure.
4. Thus, the obtained extract is allowed to dry and mixed with 100% dimethyl sulfoxide (DMSO) to prepare stock concentration of the extract in mg/mL (Jann Eldy et al., 2021).

18.2.2 Antimycobacterial Study

18.2.2.1 Bacterial Strains

1. The bacterial strains *Mycobacterium tuberculosis* H37Ra (*Mtb*) (MTCC 300) and *Mycobacterium smegmatis* (MTCC 6) were purchased from MTCC Chandigarh, India.
2. The strains were sub-cultured periodically as per the specifications and were maintained as slants and glycerol stocks.
3. Seventy-two- and forty-eight-hour-old pure cultures of *Mtb* were prepared for use during the study whenever required.

18.2.3 Culture Medium

18.2.3.1 Solid Culture

1. *M. tuberculosis* H37Ra and *M. smegmatis* were sub-cultured by suspending 6.2 g of Loewenstein–Jensen (LJ) medium base (Fluka-63237) in 100 mL water containing 2 mL glycerol (Sigma–Aldrich) and autoclaved (15 lbs for 15 min).
2. The egg emulsion (50 mL) was mixed with the base to obtain a uniform mixture.
3. Slants were prepared by dispersing medium (10 mL) in sterile screw-capped bottles and allowed to coagulate by heating in water bath at 85°C–90°C for 45 minutes (NCIM).
4. After solidification, *M. tuberculosis* and *M. smegmatis* were streaked and incubated at 37C for 4 weeks.

18.2.3.2 Liquid Culture

1. The pure culture was isolated from the slant (LJ medium) and inoculated into 5 mL Middlebrook 7H9 broth base supplemented with 10% (v/v) ADC (Accumix™) and 0.2% (v/v) glycerol (Sigma–Aldrich).
2. The inoculated media was maintained at 37°C for 72 hours on a shaker incubator until an OD_{600} of 0.6–0.8 was reached (equivalent to 3×10^7 cfu/mL). The culture was used for antimycobacterial assay (Molina-Salinas, 2007; Middlebrook and Cohn, 1958).

18.2.4 Qualitative Assessment of Antimycobacterial Activity

18.2.4.1 Agar Well Diffusion Method

1. The antimycobacterial activity of the extracts was assessed by the agar well diffusion assay. Agar medium was prepared by dissolving 2.35 g of Middlebrook 7H9 broth base in 450 mL of distilled water along with 2 mL of glycerol and 2% agar (2 g/100 mL medium); the mixture was autoclaved at 121°C (15 lbs pressure) for 10 minutes.
2. The agar medium was allowed to cool to room temperature and 25 mL of the medium was dispensed into sterile Petri plates.
3. The medium was allowed to solidify (Vladimir and Juan Gabriel, 2010). The plates were pre-seeded with *M. tuberculosis.*
4. About 50 μL (2 mg/mL) of the sample was loaded in each well (10 mm dia). The plates were incubated for 72 hours at 37°C.

5. The diameter of the zone of inhibition was measured (Sun and Zhang, 1999; Parish and Stroker, 1998). Rifampicin (2 µg; contains 1 mg/mL) served as a positive control, whereas DMSO was used as a negative control.
6. All assays were carried out in triplicates. The average data obtained from triplicate experiments were calculated and tabulated along with its standard deviation (mean ± SD).

18.2.4.2 Contact Bioautography

1. The extracts were resolved using thin layer chromatography, using suitable developing solvents. The plates were exposed to UV light (365 or 254 nm).
2. The TLC plate was flipped and placed onto the pre-inoculated agar layer and left for approx. 45 min to enable diffusion of the separated compounds into the agar plates (Meyers and Smith, 1964; Narasimhachari and Ramachandran, 1967).
3. Subsequently, the TLC sheet was removed and the plate was incubated for 72 hours at 37°C.
4. The zone of clearance on the agar surface at particular spots indicated the presence of the compound with antimycobacterial activity.

18.2.5 Quantitative Assessment of Antimycobacterial Activity

18.2.5.1 Determination of Minimum Inhibitory Concentration (MIC)

1. Minimum inhibitory concentration (MIC) was determined using microplate alamar blue assay (MABA) and resazurin microtitre assay (REMA) in a 96-well micro titter plates.
2. Liquid culture was prepared and served as uniform in equal quantity (0.5 mL each) to measure the MIC values; various concentrations (range of 0.1–50 mg/mL) of the extracts were prepared. The plates were incubated at 37°C for 7 days.
3. After this, 8:5 alamar blue (MABA assay) or 0.6 mM resazurin dye (REMA assay) and 20% Tween 80 (2 mL of alamar blue for MABA assay or 0.6 mM resazurin dye for REMA assay) and 1.25 mL of 20% Tween 80 were added. The total volume of the organism in each well was 180 µL.
4. The micro titre plates were again incubated at 37°C for 24 hours and then fluorescence was measured at 590 nm.
5. MIC is defined as the lowest concentration of the extract that could prevent the colour of alamar blue or resazurin dye to change from blue to pink (visual determination).
6. This indicates that the concentration of compound is sufficient for inhibiting 90% growth of the organism, as inferred from the values of the zone of clearance (Prescott, 1999; Collins & Franzblau, 1997).
7. The MIC determined can be statistically calculated using the formula:

$$\text{MIC (90\% inhibition)} = 0.1 * (\text{control} - \text{sample/control})$$

18.3 RESULTS

18.3.1 Acid-fast Staining

1. Acid-fast staining was done to confirm the presence of purity of culture. The Ziehl Neelsen staining technique used to stain *M. tuberculosis* and *M. smegmatis* was performed and the slides were observed under 100× magnification.
2. It was inferred that the presence of dark pink-coloured rod-shaped clusters of organism that resulted from the primary stain, i.e., carbol fuchsin, indicated the presence of *M. tuberculosis* and *M. smegmatis*.
3. The morphological representation of the organism is given in Figure 18.1.

18.3.2 Qualitative Assay

18.3.2.1 Agar Well Diffusion Method

Antimycobacterial disc diffusion assay was carried out for petroleum ether extract of chromatographic fractions (F1 to F6). From the results, it should be concluded that among all the fractions, only F3 and F4 had antimycobacterial activity (Figure 18.2). The positive control used was

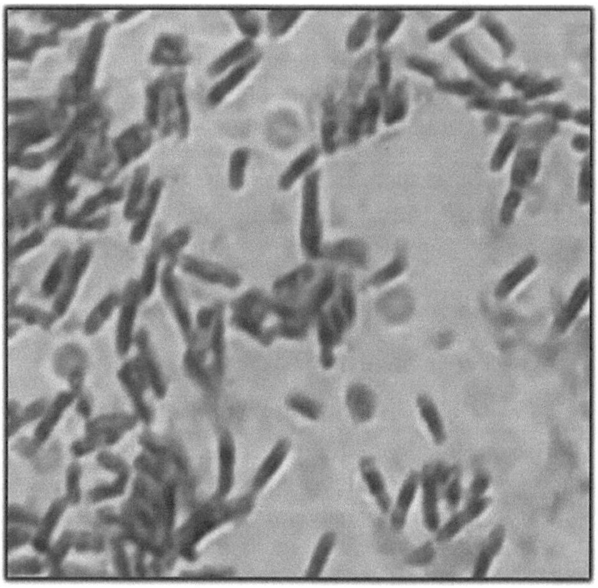

FIGURE 18.1 *M. smegmatis* (MTCC 06) acid-fast staining showing pink rod-shaped organism under 100× magnification.

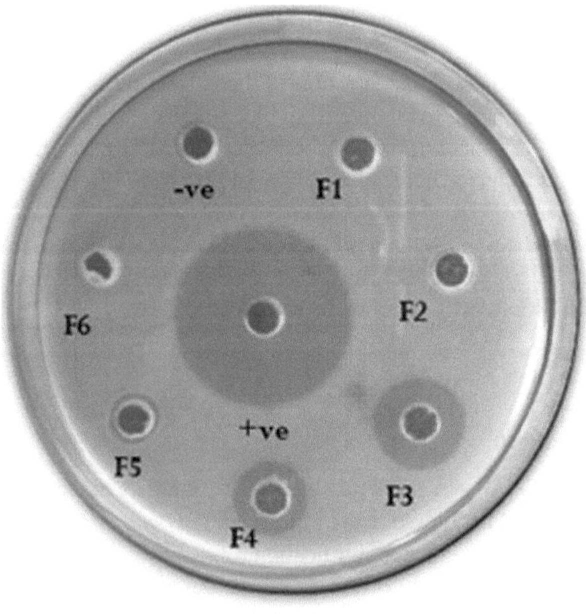

FIGURE 18.2 Antimycobacterial assay showing zone of inhibition of chromatographic fractions of petroleum ether extract. Volume of samples loaded = 20 μL (1 mg/mL), +ve control = Rifampicin (2 μL), −ve control = DMSO.

TABLE 18.1
Zone of Inhibition of Aqueous Extracts of Selected Actinobacteria Strains

S No.	Labelled Fraction	Zone of Clearance	Remark
1	F1	−ve	Rejected
2	F2	−ve	Rejected
3	**F3**	**+ve**	**Selected**
4	**F4**	**+ve**	**Selected**
5	F5	−ve	Rejected
6	F6	−ve	Rejected

rifampicin (2 μL) in concentration 1 mg/mL and DMSO was used as the negative control. The results are represented in Table 18.1.

18.3.2.2 Contact Autobiography

Petroleum ether extract of the actinobacteria extract was separated initially using TLC technique with a mixture of solvents, petroleum ether:chloroform in the set ratio of 4:1. The plate was allowed to stand in a closed chamber until the solvent travels ~6 cm on the TLC sheet. Then, the sheet is dried and visualized under long and short UV light to mark the sample spots. It is then inverted in pre-inoculated agar plate. The TLC plate is removed after ~45 minutes and the plate is incubated at 37°C for 72 hours. After incubation, spots that showed antimycobacterial activity developed a clear zone around the area of diffusion. The result for the petroleum ether extract of the sample is given in Figure 18.3.

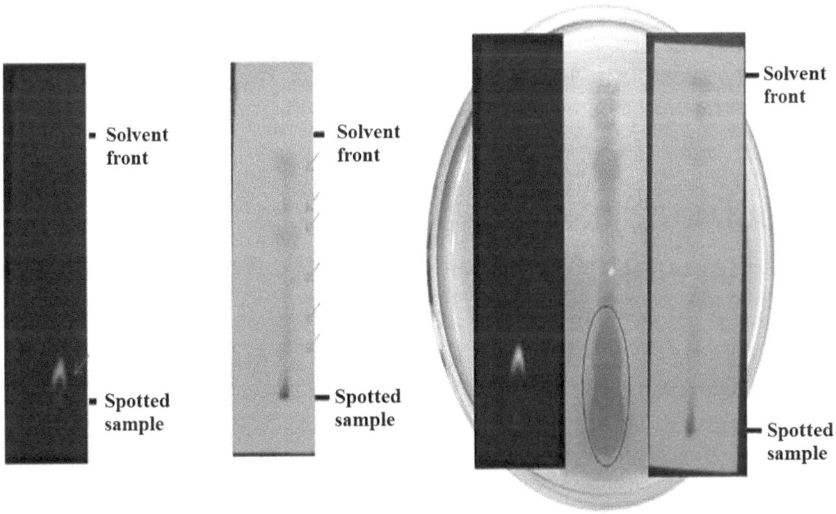

FIGURE 18.3 TLC sheet (under UV visible and short UV) showing sample spots and zone of clearance for separated sample spot observed on agar plate.

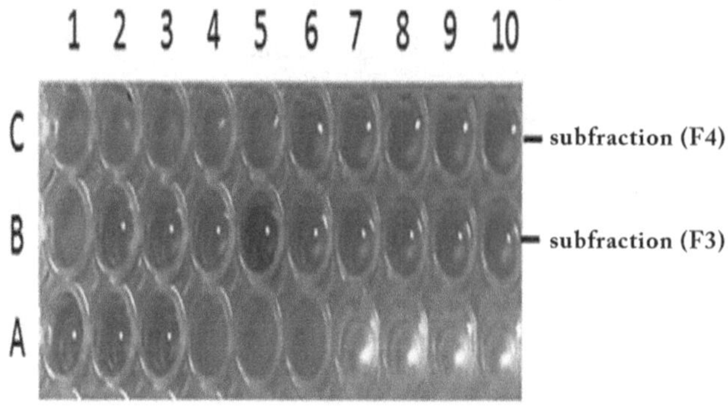

FIGURE 18.4 A1, A2 and A3 are +ve control (Rifampicin (5 μg)); A4, A5 and A6 are –ve control (DMSO (20 μL)); A7, A8, A9 and A10 are blank; B1 = 2 μg; B2 = 4 μg; B3 = 6 μg; B4 = 8 μg; B5 = 10 μg; B6 = 12 μg; B7 = 14 μg; B8 = 16 μg; B9 = 18 μg; B10 = 20 μg of subfraction (F3); C1 = 2 μg; C2 = 4 μg; C3 = 6 μg; C4 = 8 μg; C5 = 10 μg; C6 = 12 μg; C7 = 14 μg; C8 = 16 μg; C9 = 18 μg; C10 = 20 μg of subfraction (F4). This image was taken 10 minutes after adding alamar blue (10 μL). Pink wells represent the absence of inhibition, whereas blue wells indicate the inhibition of *M. tuberculosis* growth.

18.3.3 Quantitative Assessment of Antimycobacterial Activity by Broth Dilution Method

18.3.3.1 Microplate Alamar Blue (MABA Assay)

The broth dilution method was carried out on 96-well plate. The intensity of colour change from pink to blue upon addition of alamar blue dye (MABA assay) after 7 days of incubation in 37°C was measured at 590 nm. The colour change and intensity indicate the ability of the organism to reduce the dye. The oxidized form of the dye is blue which is reduced by viable cells into pink colour. Here, the wells A1, A2 and A3 are +ve control (Rifampicin (5 μg)). Wells A4, A5 and A6 are –ve control (DMSO (20 μL)) and A7, A8, A9 and A10 were kept blank. The wells B1 to B10 contained sample subfraction 1 in concentration 2–20 μg in double dilution method. B1 contained 2 μg, and B10 had 20 μg. Similarly, wells C1 to C10 had sample subtraction 2. The results revealed that the fraction 1 at various concentrations had an MIC of **22.3** μg/mL and fraction 2 had an MIC of **66.7** μg/mL when measured at 590 nm (Figure 18.4).

ACKNOWLEDGEMENT

The authors would like to thank Vellore Institute of Technology (VIT), Vellore, for providing infrastructure facilities and supporting this research work.

REFERENCES

Collins, L.A., and Franzblau, S.G. (1997). Microplate alamar blue assay versus BACTEC 460 system for high-throughput screening of compounds against *Mycobacterium tuberculosis* and *Mycobacterium avium*. *Antimicrobial Agents Chemotherapy*, 41: 1004–1009.

Daquioag, J.E.L., and Penuliar, G.M. (2021). Isolation of Actinomycetes with cellulolytic and antimicrobial activities from soils collected from an urban green space in the Philippines. *International Journal of Microbiology*, 14: 2021. https://doi.org/10.1155/2021/6699430.

Kuster, E. (1968). Taxonomy of soil actinomycetes and related organisms. In: Gray, S., Parkinson, T., editors. *Ecology of Soil Bacteria*. Liverpool University Press, Liverpool, 322–336.

Meyers, E., and Smith, D. (1964). Bioautography of antibiotic spread-layer chromatograms. *Journal of Chromatography*, 14: 129–132.

Middlebrook, G., and Cohn, M.L. (1958). Bacteriology of tuberculosis: laboratory methods. *American Journal of Public Health*, 48: 844–853.
Molina-Salinas, G.M., Pérez-López, A., Becerril-Montes, P., Salazar-Aranda, R., Said-Fernández, S., and Waksman de Torres, N. (2007). Evaluation of the flora of Northern Mexico for in vitro antimicrobial and antituberculosis activity. *Journal of Ethnopharmacology*, 109: 435–441.
Narasimhachari, S., and Ramachandran. (1967). A simple bioautographic technique for identifying biologically active material on thin-layer chromatograms. *Journal of Chromatography*, 27: 494.
Parish, T., and Stroker, N.G. (1998). *Mycobacteria Protocols: Methods in Molecular Biology*. Humana Press, Totowa. NJ.
Prescott, M.L., Harley, J., and Donald, P. (1999).Antimicrob chemother. In: *Microbiology*, 2nd edition. Brown Publisher, Cincinnati, OH, p. 325.
Vladimir, V.K., and Juan Gabriel, B.S. (2010). Antimycobacterial susceptibility testing methods for natural products research. *Brazilian Journal of Microbiology*, 41: 270–277.
Walker, J.D., and Colwell, R.R.(1975). Factors affecting enumeration and isolation of actinomycetes from Chesapeake Bay and Southeastern Atlantic Ocean sediments. *Marine Biology*, 30: 193–201.
World Health Organization. (2022). Global Tuberculosis Report. Available at https://www.who.int/teams/global-tuberculosis-programme/tb-reports.

19 Screening Actinomycetes for Novel Antimycobacterial and Anticancer Molecules

Balaji Muralikrishnan and Vipin Mohan Dan

19.1 INTRODUCTION

Streptomycin was the first antibiotic isolated from an actinomycete that went on to become a wonder drug during the Second World War to treat tuberculosis. From there on, the golden era of antibiotic discovery from soil bacteria began and thousands of newer active principles were isolated which were mostly either antimicrobial or antineoplastic in nature. Both of these properties are high priority even now. Numerous compounds have been discovered from this group of microbes as well as from other natural sources (fungi, Bacillus and plants). However, the necessity of novel ones is increasing to deal with drug resistance. This is because of erroneous use of antibiotics during the post-colonial era that had led to mutations that favoured the survival of pathogens even after being exposed to a previous known susceptible concentration. Thus, 'drug resistance' emerged, and it is slowly percolating and seen in most of the deadly pathogens such as *Mycobacterium tuberculosis*. In addition, India is home to 24% of all TB infections in the world and many of these cases could turn into drug-resistant state if the old regime of drugs is not supplemented or replaced by newer effective ones. This drug resistance phenomenon is not restricted to microbes but also found in tumorigenic and cancerous cells. Therefore, the issue in hand is larger, and if left unhandled, we might push the human race into another pandemic.

Novel bioactive molecules are the key and to discover them, it is important to explore untapped resources to isolate newer strains of actinomycetes. One such resource is the biodiversity hotspot, the Western Ghats. It is a home for rich biodiversity and its soil should harbour novel strains of actinomycetes which in turn might possess novel cryptic biosynthetic gene clusters in their genome. When these clusters are rightly activated (effective fermentation method), the secrets behind the DNA sequence can be identified, and they could produce extraordinary bioactive molecules with completely fresh mechanism of action which would sustain the growing situation.

Screening for basic antimycobacterial activity can be performed by cross-streak method in which the isolated actinomycete strains are grown as a straight line (each) bisecting the Petri dish. After it matures (exuberantly grown), the test strain (*Mycobacterium smegmatis* – surrogate model organism of *M. tuberculosis*) is streaked perpendicular to it and allowed to grow further. If the actinomycete produces an antimycobacterial molecule, then it would inhibit the growth of *M. smegmatis*. On the contrary, to identify anticancer agents, activity-guided fractionation is inevitable, and therefore, we have to test the crude extract initially to confirm its cytotoxicity. Most of these antibiotics produced from actinomycetes have been secreted, and hence, the source of the novel molecules is the spent fermented medium of the actinomycetes. Identification of the right antibiotic production medium is the first step and that can be achieved through trial and error with different carbon sources and innovative culturing methods. (*Sometimes, carbon sources are not all required and a full nitrogen meal plus phosphorus would do the trick.*) Therefore, it is always recommended to identify the growth of the isolate in different growth media and identify the moderate to exuberant growth providing one. Next, choose the growth medium that supports moderate growth as the antibiotic production medium and use it for fermentation. In shake flask fermentation,

 DOI: 10.1201/9781003398400-19

it is essential to disintegrate the clumps formed by actinomycetes in the medium. (*Therefore, it is recommended to place a steel coil inside the flask and let it function as a multi-baffled structure to disintegrate the forming clumps.*) Later, activity-guided fractionation helps to identify the active fraction in the crude extract. The active fraction can be chemically identified by analytical techniques such as mass spectrometry, NMR and artificial intelligence-guided algorithms.

19.2 MATERIALS AND METHODS

Strains required: *Streptomyces* sp., OA161, *Streptomyces* sp., OA293, *Mycobacterium smegmatis*, *M. tuberculosis* H37Rv. **Cell lines**: L6, SW480, THP-1, PBMCs.

Media required: ISP-2 medium, ISP-4 medium, RPMI 1640, FBS (fetal bovine serum) ethanol, Middle Brook 7H9 medium, Tween 80.

Chemicals and others: Agarose, Silica 60 F254 silica TLC plates, C18 reverse phase column, HPLC grade solvents (ethyl acetate, methanol, chloroform, acetone, hexane), mass spectrometry graded solvents (acetonitrile, methanol and water), deuterated acetone, resazurin, MTT, TE buffer, DNTPs, primers, Taq polymerase and buffers, RNase and isopropanol.

19.3 SCREENING FOR ANTIMYCOBACTERIAL ACTIVITY AND ANTICANCER ACTIVITY

19.3.1 Cross-Streak Method

1. Prepare a spore suspension of *Streptomyces* sp., OA161 in ISP-2 medium.
2. On an ISP-2 solid medium (in a petri plate), using a sterile inoculation loop, streak the spore suspension bisecting the plate in the centre and allow it to grow for 72 hours.
3. Meanwhile, a log phase culture of *M. smegmatis* has to be prepared. For this, inoculate *M. smegmatis* on LB medium with Tween 80 (0.025%) and incubate for 48 hours which would be the ideal log phase culture. (*Like any other microorganism, Mycobacterium does not grow homogenously in a liquid rather they form aggregates. Therefore, Tween 80, a surfactant is added to the growth medium to ensure homogenous growth of* M. smegmatis.)
4. Now, pick up the ISP-2 *Streptomyces* grown plate, and using a sterile inoculation loop, streak the log phase grown *M. smegmatis* perpendicular to the grown *Streptomyces.*
5. Incubate the plate at 37°C for 24 hours and check for the growth of *M. smegmatis.* A zone of clearance shows that an antimycobacterial molecule is produced by the *Streptomyces* sp.

(*This is the first evidence of antimycobacterial-producing Streptomyces in a screening program. However, in order to confirm its activity, the crude extract from the Streptomyces spent medium should be tested against M. tuberculosis using a redox dye – resazurin. This dye in the presence of live bacteria gets reduced to form a pink-coloured resorufin while if the crude extract kills the bacteria, the resazurin dye remains blue.*) (Figures 19.1 and 19.2).

19.3.2 Fermentation and Preparation of Organic Extract

1. ISP-2 medium was identified as the antibiotic-producing medium of *Streptomyces* sp., OA161/293. Therefore, inoculate the isolate in ISP-2 medium and incubate for 6 days at 28°C and 250 rpm.
2. Lyophilize the spent medium to a powder form. Divide the lyophilized culture into six parts and extract each part with 1 mL of water, ethyl acetate, methanol, acetone, hexane and chloroform, respectively. (*All the organic solvents used here cover a spectrum of polarity considering the fact that at least in one of these, the active principle gets segregated out to ease the process of its isolation and purification.*)

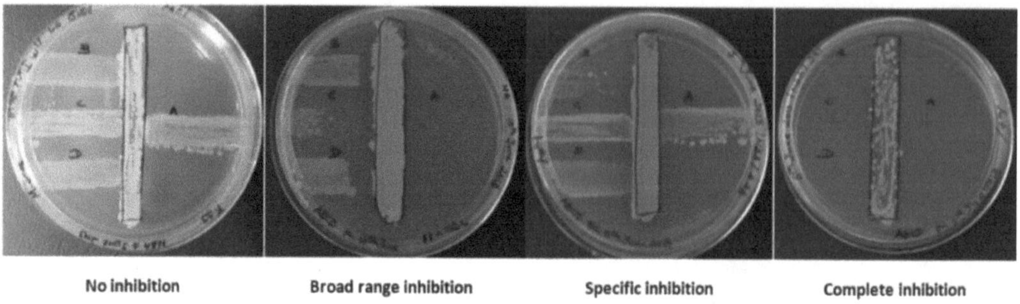

FIGURE 19.1 Cross-streak method.

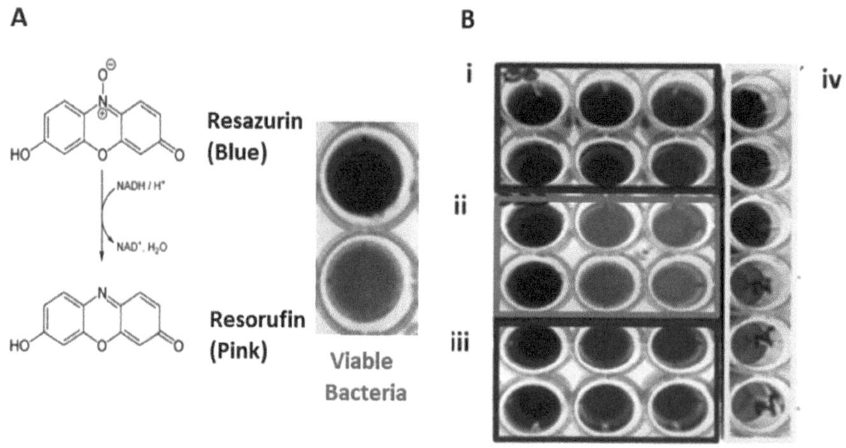

FIGURE 19.2 Principle of resazurin microtitre assay.

3. Except the aqueous extract, spin down the other organic extracts and concentrate them to suspend in DMSO. (*While preparing these extracts, the concentrations of them are noted. These are utilized for further testing on M. tuberculosis using REMA for antimycobacterial activity as well as for anticancer activity using MTT assay on cancer cell lines.*)

19.3.3 Resazurin Microtitre Assay (REMA)

1. Perform this assay in a 96-well plate. (*It is important to chart out the plate first to avoid confusion during serial dilution and keep the right controls.*)
2. The wells in the periphery of the 96-well plate are filled with sterile water (*This is to avoid the medium from drying out as the plate is incubated at 37°C for 7 days*).
3. Prepare Middle Brook 7H9 medium with Tween 80 (0.025%) as per the manufacturer's protocol for growing *M. tuberculosis* H37Rv. Using a Bio-safety level-3 chamber, serially dilute the extracts in Middle Brook 7H9 medium followed by adding 1.5×10^6 bacteria (McFarland Standard 1) to each well.
4. Keep rifampicin at 1 µg/mL (in DMF) as positive control; no drug as negative control and a separate control well with the drug and no bacteria. (*As this entire assay works on the redox property of the dye, it is essential to be sure that the active principle does not reduce the dye to provide a false negative.*)
5. Incubate the plates in no light (covered with aluminium foil) at 37°C for 7 days.

6. On day 8, prepare a 0.025% resazurin dye and filter sterilize it. To each well of the 96-well plate, add 20 µL of the prepared resazurin dye and check the negative control to turn pink by observing the plate every 3 hours.
7. When a complete turnover to a bright pink appears, photograph the plate against the white light. A gradation from blue to pink shows the minimum inhibitory concentration of the active principle from each of the extracts prepared (Figure 19.3).

19.3.4 MTT Assay

This assay is based on a dye MTT – a tetrazolium dye which forms purple crystals in live cells but not in dead cells. Later, these crystals can be solubilized using isopropanol and read calorimetrically.

1. Animal cells (L6, SW480 and THP-1) are maintained in RPMI medium with 10% fetal bovine serum in T-75 tissue culture flask at 37°C, 5% CO_2.
2. Using hemocytometer, count 20,000 and seed it in each of the well of the 96-well plate. All the cell lines used here are adhering cells and thus in 24 hours of incubation, the cells get attached to the surface of the wells.
3. Now, remove the spent media with FBS and replace it with fresh RPMI medium without FBS but with serially diluted active principles or the extracts prepared in Section 3.2.
4. Incubate for another 24 hours and add 20 µL of the prepared MTT dye (5 mg/mL) to each of the 96 wells with cells seeded. Allow it for 4 hours to develop the purple crystals inside the cells and after that solubilize with 200 µL of isopropanol.
5. Read the plate at 540 nm using a plate reader. Values of untreated cells are considered 100% and compared to it the active principle treated cells are also analysed and the percentage of survival is noted. (*80% death is considered to be the effective initial anticancer activity which should not be found in control cells such as L6; otherwise, the observed killing would be simply cytotoxicity without any specific activity*) (Figure 19.4).

19.3.5 Thin Layer Chromatography and Bioautography

1. Using Silica 60 F254-coated aluminium sheets (7.5 cm long) as a stationary phase, load the crude extract (ethyl acetate extract in this case) and allow it to dry.
2. Set up a TLC running chamber with mobile phase (hexane:acetone::1:1) and allow it to saturate for 40 minutes prior to run.
3. Open the top lid of the saturated chamber and place the silica stationary phase loaded with crude extract in the mobile phase such that the mobile phase contacts the plate just below the sample loaded. (*This ensures capillary action and no loss of compounds into the mobile phase.*)

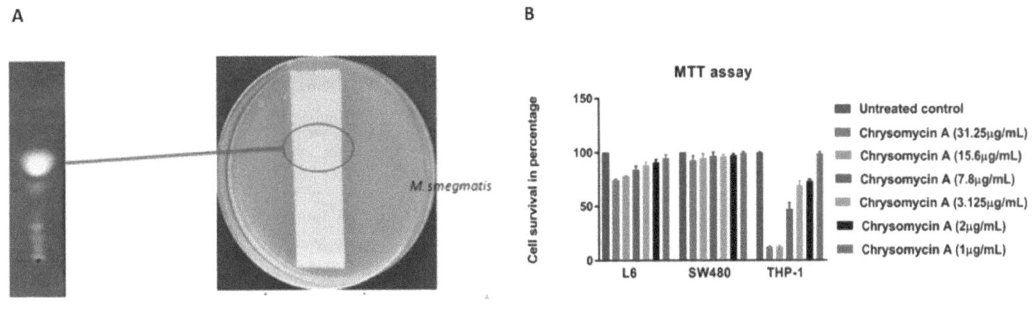

FIGURE 19.3 A sample of resazurin microtitre assay. ('i' and 'iii' represent active principles having antimycobacterial activity, 'ii' represents no activity and 'iv' has positive and negative controls in triplicates.)

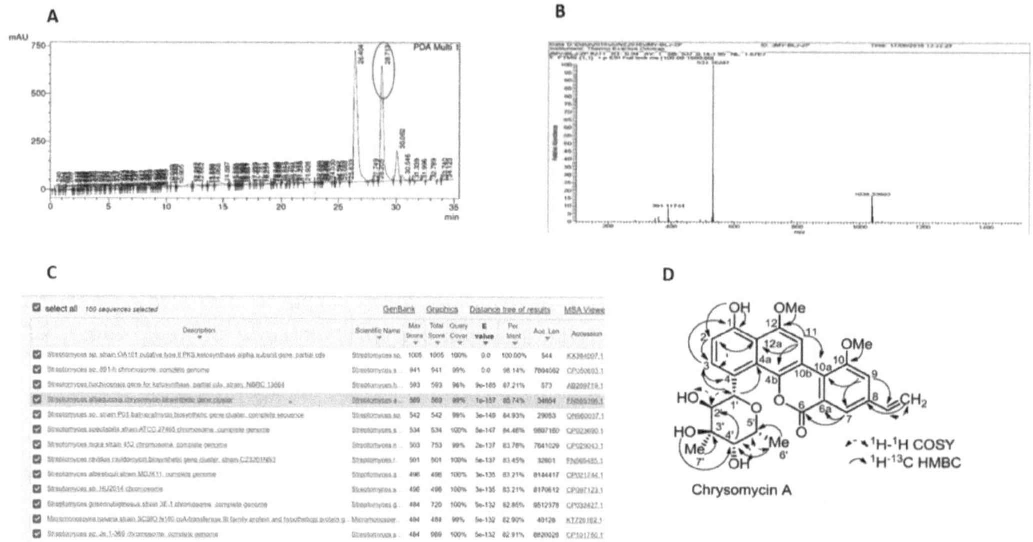

FIGURE 19.4 A sample of MTT assay using chrysomycin A as the active principle.

4. Allow the TLC plate to develop for 35 minutes or until 1 cm from the top.
5. Shine short and long wavelength UV light to detect the separated secondary metabolites on the plate.
6. Dry the developed TLC plate to get rid of the organic solvents and place the thin strip of the TLC plate inside a sterile Petri plate.
7. Meanwhile, prepare Mueller–Hinton agar (0.6% agar), and before it solidifies, add 1 mL of 0.2 OD *Mycobacterium smegmatis* to the soft agar, mix it and pour it gently over the TLC plate placed inside the Petri dish.
8. Incubate this Petri dish at 37°C for 24 hours and check for the zone of inhibition. Now, match the fraction that has the antimycobacterial activity with their Rf values to identify the antimicrobial fraction.

(*In the case of finding the active fraction in a crude extract with anticancer activity, each fraction identified after shining the UV light is scraped, re-extracted, concentrated and re-dissolved in DMSO to test it on cell lines using MTT assay.*)

19.3.6 HPLC, MALDI-TOF and NMR Analysis

1. Once the active fraction is identified through bioautography or by activity-guided identification (*in the case of anticancer fraction*), that particular fraction is scraped off the TLC plate and re-extracted with ethyl acetate or the same organic solvent used to prepare the crude extract.
2. Evaporate the ethyl acetate to re-dissolve the extract in acetonitrile (*this is performed because acetonitrile is compatible with the C18 reverse phase column in HPLC*).
3. Ten microliters of this acetonitrile extract is used as the injection liquid in HPLC. A water-methanol gradient is used as the mobile phase. Specifically, start the run at 0 minute with 35% methanol and gradually increase the concentration of methanol to 80% at 35 minutes; flow rate at 1 mL/min (*this is a standardized method for isolating this particular compound – chrysomycin A*).
4. Wash the column with 80% methanol for 5 minutes, followed by 100% methanol for another 5 minutes, before loading the next sample.

5. Once the individual peaks are observed in the chromatogram (using either a PDA detector or a UV detector), note down the eluting time and use it to elute the purified separated compounds in the subsequent rounds of HPLC run.

   (No shoulder peaks are allowed, and if they are found, identify the time at which the shoulder is found. Then, set a longer method of gradual increase in solvent concentration which would remove the shoulder peaks to give individual peaks. Individual peaks often have single isolated compounds in purified form; however, there can be more than one compound in a single peak elution.)
6. Place the eluted aqueous liquid at –80°C to freeze it overnight and lyophilize it to a dry powder the next day.
7. Dissolve the lyophilized powder using methanol (mass spectrometry grade) at 1 mg/mL and use it for mass spectrometry analysis.
8. Briefly, mix appropriate substrate (ion donors) with the sample and fire it to go for a flight, and the time of flight is calculated by the software to represent compounds by its m/z. *(The mother peak is identified and the other noises in the background are nullified to get a chromatogram of the mass spectrometry.)*
9. For NMR analysis, mix the sample with deuterated acetone and record its spectrum (^{1}H and ^{14}C NMR) using 500 and 125 MHz on a Bruker DRX 500 NMR instrument.
10. Note down the chemical shifts in parts per million (ppm) and decoupling constants in Hz. *(Based on these values and insights from the identification of the type of biosynthetic gene cluster present in the Streptomyces isolate, a common chemical backbone is assumed and the structure of the molecule can be deduced.)*

19.3.7 Identification of Biosynthetic Gene Clusters

19.3.7.1 Genomic DNA Isolation:

Streptomyces and other actinomycetes have relatively thicker cell walls and it would be different to break them using conventional techniques. Therefore, it is advised to either freeze fracture the cell walls or break the cell walls using zirconium beads.

1. A 48–hour-old liquid culture should serve as a starting material for the isolation of DNA. Therefore, for this, a single colony from the pure culture has to be inoculated into 20 mL of ISP-2 medium and incubated at 200 rpm for 48 hours prior to the DNA isolation.
2. On the day of isolation, the culture should be transferred and centrifuged at 5,000 rpm for 10 minutes. Discard the supernatant, and using a sterile loop, transfer the culture to a sterile pestle and motor.
3. Now, add 5–10 mL liquid nitrogen to freeze the cell pellet, and using a pestle, grind it several times until it looses temperature and liquefies. Add 500 µL of 1X PBS (phosphate-buffered saline) and 50 µL of lysozyme (20 mg/mL) to the grind and mix them well using a micropipette.
4. After incubating them at 37°C for 1 hour, add 20 µL of proteinase and 50 µL of 10% SDS and incubate at 55°C for 30 minutes.
5. Later, 600 µL of phenol is added and vortexed. The mixture was centrifuged and only the aqueous layer was transferred and re-extracted with equal volume of chloroform. *These steps can be repeated to achieve pure DNA.*
6. The aqueous layer from the previous step is added with 2.5 volumes of cold absolute ethanol and 10% sodium acetate solution (3M) to precipitate the genomic DNA.
7. Incubation at –20°C improves the DNA yield; even otherwise, handling in room temperature also gives good concentration of DNA (1–2 mg/mL) which is amicable for downstream PCR and sequencing application.

8. After precipitation, add 200 μL of 70% ethanol to wash off impurities by centrifuging for 10,000 rpm for 2 minutes.
9. Discard the ethanol and air dry the pellet for 15 minutes inside the laminar hood.
10. Now, add 50 μL of TE buffer (pH-8) to the dried pellet and dissolve the precipitated DNA *(if it remained insoluble, placing it at 4°C is recommended).*
11. The DNA has to be estimated using a nano-drop and a 260/280 value of 1.8 would ensure its purity.
12. Agarose gel electrophoresis is performed to ensure the isolated product is DNA by running the sample at 80 V for 45 minutes. A thick band just below the loading well is an ideal representation of good genomic DNA.

19.3.8 Identification of the Biosynthetic Gene Cluster in the Actinomycetes Strain

The primers used for the partial amplification of biosynthetic gene clusters are given in Table 19.1.

19.3.8.1 PCR

1. The reaction mix for a 50 μL amplification reaction has been given below.

10× PCR buffer with $MgCl_2$	-10 μL
Template DNA (50–200 ng)	-1 μL,
Forward primer (20 mM)	-1 μL
Reverse primer (20 mM)	-1 μL
dNTP mix (5 mM)	-2 μL,
Taq DNA polymerase (Sigma)	-1 U/μL
Sterile MilliQ water	-34 μL.
Total	-50 μL.

The reaction conditions are: initial denaturation at 95°C for 5 min, followed by 30 cycles of denaturation at 95°C for 30 s, annealing at 54°C–72°C for 30 s and extension at 72°C for 45 s. A final extension was performed at 72°C for 7 min.

The PCR product was purified using GE Healthcare gel band purification kit and sequenced directly using a Taq BigDye Terminator Cycle Sequencing Kit in ABI automated DNA sequencer (Applied Biosystems). The sequences were compared with the help of BLAST tool, and the similar biosynthetic gene cluster is identified to get the backbone of the tentative compound. (*Later, comparing with the NMR data and performing COSY combinations, the final structure of the compound is deduced.*)

TABLE 19.1
List of Primers for Identification of the Biosynthetic Gene Clusters

Primers	Gene Cluster
5′-TSGCSTGCTTGGAYGCSATC-3′ (F) 5′-TGGAANCCGCCGAABCCTCT-3′ (R)	Type II polyketide synthases
5′- GCSTACSYSATSTACACSTCSGG-3′(F) 5′-SASGTCVCCSGTSCGGTAS-3′(R)	Non-ribosomal peptide synthetases
5′-CCGCVCACATCACSGSCCTCGCSGTGAACATGCT-3′(F) 5′-GCAGGCKCCGTCSACSGTGTABCCGCCGCC-3′ (R)	Enediyne-specific polyketide synthases

Abbreviation: Y= C or T, R=A or G, N=A or T or C or G.

19.4 CONCLUSION

Identification of novel molecules against cancer and *Mycobacterium tuberculosis* has to be a continuous process; only then the pipeline of developing new lead molecules would be replenished. The rationale behind this process is that, against all these drugs, the cells/microbes might gain resistance just like what they have developed against the older drugs today, and therefore, keeping the pipeline live with newer drugs is the only way out to evade infection-related morbidity. Developing effective vaccines against these diseases is the primary objective; unfortunately, an effective vaccine in near time is at far reach, and until then, novel drugs are the ones to rely.

ACKNOWLEDGEMENTS

This work was supported by CSIR-OSDD, India, and intramural funds of Rajiv Gandhi Centre Biotechnology, Department of Biotechnology, India.

CONFLICT OF INTEREST

The authors have no conflict of interest to declare that is relevant to the content of this protocol chapter.

DATA AVAILABILITY

The data used to support the findings of this protocol is available upon reasonable request. The nucleotide sequences of some microbial strains obtained in this protocol have been deposited in the GenBank database with accession number (KX KX364007.1).

BIBILOGRAPHY

Aspatwar, Ashok, et al. β-CA-specific inhibitor dithiocarbamate Fc14-584B: a novel antimycobacterial agent with potential to treat drug-resistant tuberculosis. *Journal of Enzyme Inhibition and Medicinal Chemistry* 32.1 (2017): 832–840.

Dan, Vipin Mohan, et al. Molecular networking and whole-genome analysis aid discovery of an angucycline that inactivates mTORC1/C2 and induces programmed cell death. *ACS Chemical Biology* 15.3 (2020): 780–788.

Dan, Vipin Mohan, et al. Streptomyces sp metabolite (s) promotes Bax mediated intrinsic apoptosis and autophagy involving inhibition of mTOR pathway in cervical cancer cell lines. *Scientific Reports* 8.1 (2018): 2810.

Genilloud, Olga. Actinomycetes: still a source of novel antibiotics. *Natural Product Reports* 34.10 (2017): 1203–1232.

He, Haiyin, et al. Mannopeptimycins, novel antibacterial glycopeptides from Streptomyces hygroscopicus, LL-AC98. *Journal of the American Chemical Society* 124.33 (2002): 9729–9736.

Muralikrishnan, Balaji, et al. Anti-microbial activity of chrysomycin A produced by Streptomyces sp. against Mycobacterium tuberculosis. *RSC Advances* 7.58 (2017): 36335–36339.

Muralikrishnan, Balaji, et al. Chrysomycin A inhibits the topoisomerase I of Mycobacterium tuberculosis. *The Journal of Antibiotics* 75.4 (2022): 226–235.

Zhu, Hua, Stephanie K. Sandiford, and Gilles P. van Wezel. Triggers and cues that activate antibiotic production by actinomycetes. *Journal of Industrial Microbiology and Biotechnology* 41.2 (2014): 371–386.

20 *In Vitro* Antioxidant Protocols for Actinobacteria

Charles Lekhya Priya

20.1 INTRODUCTION

Free radicals are collectively called reactive oxygen species (ROS). Free radicals under the category of ROS are superoxide anion (O2•-), reactive hydroxyl (OH•), peroxyl (ROO•), hydroperoxyl (HO2•), alkoxyl (RO•) and peroxyl (RO2•). Free radicals are formed within the body either by enzymatic and non-enzymatic reactions or by endogenous and exogenous sources. Free radicals at a low or moderate level play an important role in normal physiological functions of the human body such as energy production and phagocytosis and also contribute to the functioning of intercellular signalling. However, when they exceed the moderate level, free radicals cause serious effects that initiate cellular damage to the DNA, RNA, proteins and lipids, leading to the emergence of numerous disorders such as atherosclerosis, liver cirrhosis, diabetes, cardiovascular disease, neurological disease, pulmonary disease, rheumatoid arthritis and cancer (Lekhya and Bhaskara Rao, 2014).

Antioxidants are the substances that give protection to the human body against free radical-mediated tissue damage. Cells of the human body are protected from oxidative stress by a wide range of mechanisms, particularly endogenous enzymatic and non-enzymatic antioxidant systems (Shukla et al., 2009). Enzymatic antioxidants include superoxide dismutase (SOD), catalase (CAT), glutathione peroxidase (GPx) and glutathione reductase (GR), and non-enzymatic antioxidants include vitamin E, vitamin C and reduced glutathione (Halliwell, 2011). SOD acts as a first line of defence against ROS by metabolizing superoxide radicals to hydrogen peroxide and water. CAT, GPx and the rest of the peroxidases are a group of enzymes that break down hydrogen peroxide to oxygen and water, thus contributing to the protection of cells from free radical damage (Jos et al., 1999). When there is severe or continuous oxidative stress, endogenous antioxidant defence mechanisms become insufficient to afford protection to the cells against free radicals. Recently, it has also been revealed that hyperglycaemia leads to the glycation of antioxidant enzymes, which leads to the alteration in the structure and function of antioxidant enzymes. As a result, they lose their capacity to do free radical detoxification (Desai et al., 2013). Thus, there is a greater necessity to continuously provide exogenous antioxidants to neutralize oxidative stress within the human body. There are synthetic antioxidants that are readily available like butylated hydroxyanisole, butylated hydroxytoluene and propyl gallate but possess toxic and carcinogenic effects on humans (Sheng et al., 2011). Therefore, there is a renewed interest among researchers to invade natural sources for their use as antioxidants. Natural products are enormous sources of significant and novel compounds with various medicinal properties. Actinobacteria are Gram-positive, filamentous bacteria that are supreme secondary metabolite producers (Riddhi et al., 2017). Recent scientific reports state that the marine environment is a potent source of new antioxidant compounds or biologically active substances (Syed and Rekha, 2019; Rammali et al., 2022). Antioxidant studies help us to analyse the efficacy of actinobacteria as a potential antioxidant.

DOI: 10.1201/9781003398400-20

20.2 SAMPLE PREPARATION

20.2.1 Preparation of Extract

The isolates of actinobacteria were inoculated into ISP2 medium in a shaker incubator maintained at 120 rpm for 7 days at 28°C and centrifuged at 13,000 rpm for 10 minutes at 4°C to remove cell debris. The resultant supernatant was mixed thoroughly with ethyl acetate in the ratio of 1:1 (v/v) and left overnight. The solvent phase was separated from the aqueous phase and concentrated in a rotary evaporator at 40°C under reduced pressure. Thus, the obtained extract was allowed to dry and mixed with 100% DMSO to prepare a stock concentration of the extract in mg/mL (Jann Eldy et al., 2021).

20.3 DETERMINATION OF ANTIOXIDANT ACTIVITY

The antioxidant property of actinobacteria extracts was studied by following the standard protocols. Data was given as mean ± SD ($n = 3$). The IC_{50} value was measured from the graph of the standard using the formula "$y = mx+c$" from the slope of the graph. The lower the IC_{50} value, the stronger the antioxidant capacity of the extract.

a. DPPH radical scavenging activity
b. Superoxide radical scavenging activity
c. Reducing power activity
d. Total antioxidant capacity
e. Metal chelating activity
f. β-Carotene bleaching inhibition activity
g. Lipid peroxidation inhibition activity

20.3.1 DPPH Radical Scavenging Activity

DPPH radical scavenging activity uses stable free radical 1,1-diphenyl-2-picrylhydrazyl (DPPH) and is based on the reduction of methanol solution of DPPH to nonradical form DPPH-H in the presence of a hydrogen-donating antioxidant, thus resulting in the colour change of DPPH solution from purple to yellow, and the absorbance is measured spectrophotometrically.

The extracts of actinobacteria are examined for DPPH free radical scavenging potential. Antioxidants convert 1,1-diphenyl-2-picrylhydrazyl (DPPH) to 1,1-diphenyl-2-picrylhydrazine. The presence of strong antioxidants in the extracts is directly proportional to the decreased absorbance, which is measured at 517 nm.

Free radical scavenging activity of the actinobacteria can be performed by DPPH radical scavenging activity (Brand-Williams et al., 1995).

20.3.1.1 Procedure

- Different concentrations (20, 40, 60, 80 and 100 μg/mL) of the extract were made using distilled water.
- One millilitre of DPPH solution (0.2 mM/mL) was added to 2 mL of each dilution and mixed thoroughly.
- The mixture was incubated in the dark at 20°C for 40 minutes. Absorbance was measured at 517 nm using a UV–visible spectrophotometer (Varian, Inc., CA, USA) with methanol as a blank and ascorbic acid as a positive control.

The percentage DPPH radical scavenging of the isolates was calculated, and IC_{50} values of the extracts were determined according to the following formula:

% DPPH radical scavenging = [(Ac – At) / Ac] × 100
where Ac is the absorbance of the control (DPPH) and At is the absorbance of the tested sample. The free radical scavenging activity of the extract increased in a dose-dependent manner.

20.3.2 Superoxide Radical Scavenging Activity

Superoxide anion is a weak oxidant that leads to the production of hydroxyl radicals, which are strong and hazardous. This method depends on the principle of nitro blue tetrazolium (NBT) reduction by the generation of superoxide radicals, which turn NBT to blue colour in the presence of light. Superoxide radical scavenging activity aims to measure the capability of extracts to inhibit the blue colour of NBT spectrophotometrically at 560 nm.

The superoxide radical scavenging activity of the actinobacterial isolates is evaluated by the procedure of Nishikimi et al. (1972). IC_{50} values of extracts are determined in comparison with the standard ascorbic acid.

20.3.2.1 Procedure

- A reaction mixture was made with 1 mL of nitro blue tetrazolium (NBT) solution (156 mM NBT in phosphate buffer, pH 7.4), 1 mL of nicotinamide adenine dinucleotide (NADH) solution (468 mM NADH in phosphate buffer, pH 7.4) and 1 mL of the extract at different concentrations (125–1,000 µg/mL).
- To this reaction mixture, 100 µL of phenazine methosulphate (PMS) (60 mM PMS in phosphate buffer, pH 7.4) was added and incubated at 25°C for 5 minutes, and the absorbance was measured at 560 nm using a UV–visible spectrophotometer. Ascorbic acid is used as a control.

The percentage superoxide anion generation was calculated, and IC_{50} values of the extracts were measured using the following formula:
% Superoxide radical scavenging activity = [(Ac – At) / Ac] × 100
where Ac is the absorbance of the control and At is the absorbance of the tested sample.

20.3.3 Reducing Power Activity

Reducing power assay method is based on the principle that substances, which have reduction potential, react with potassium ferricyanide (Fe^{3+}) to form potassium ferrocyanide (Fe^{2+}), which then reacts with ferric chloride to form a ferric–ferrous complex that has an absorption maximum at 700 nm.

The higher the absorbance of the reaction mixture, the greater the reducing potential of the extract. The isolates of actinobacteria are evaluated for reducing power activity by ferric ion-reducing power assay (Oyaizu, 1986). IC_{50} values of actinobacterial extracts are calculated.

20.3.3.1 Procedure

- Different dilutions (125, 250, 500 and 1,000 µg/mL) with phosphate buffer (2.5 mL, 0.2 M, pH 6.6) were mixed thoroughly with 2.5 mL of 1% potassium ferricyanide.
- The mixture was incubated at 50°C for 20 minutes. After incubation, 2.5 mL of 10% trichloroacetic acid was added to the mixture and centrifuged at 3,000 rpm for 10 minutes.
- A volume of 2.5 mL of the supernatant was mixed with an equal volume of distilled water and 0.5 mL of 0.1% $FeCl_3$ and incubated at 40°C for 20 minutes.
- Absorbance was measured at 700 nm using a UV–visible spectrophotometer. The higher the absorbance of the reaction mixture, the higher its reducing power capacity.

20.3.4 H_2O_2 Radical Scavenging Activity

Hydrogen peroxide, which occurs naturally at low concentrations in the environment, enters the human body through inhalation and contact with the eyes or skin. H_2O_2 inside the body gets decomposed into oxygen and water, which leads to the generation of hydroxyl radicals. These hydroxyl radicals are capable of initiating lipid peroxidation, resulting in DNA damage.

Hydrogen peroxide scavenging activity of actinobacterial isolates is determined by the method of Ruch et al. (1989).

20.3.4.1 Procedure

- Actinobacterial dilutions were prepared (125, 250, 500 and 1,000 µg/mL).
- Two millilitres of each dilution was mixed with 1.2 mL of hydrogen peroxide solution (40 mM in phosphate buffer, pH 7.4) and incubated for 10 minutes.
- The absorbance of the solution was taken at 230 nm using a UV–visible spectrophotometer against a blank solution containing the plant extract without H_2O_2.
- The percentage scavenging of hydrogen peroxide was calculated using the following formula:

  $$\% \text{ Scavenging of } H_2O_2 = [(Ac - At)/Ac] \times 100$$

  where Ac is the absorbance of the control and At is the absorbance of the sample.
- The hydrogen peroxide scavenging activity was also expressed as ascorbic acid equivalence (µg/mL). IC_{50} values of each isolate's hydrogen peroxide scavenging activity were also estimated (Lekhya and Bhaskara Rao, 2016).

20.3.5 Metal Chelating Activity

Metal ions in excess are capable of generating hydroxyl radicals that have a catalytic effect on lipid peroxidation. In this assay, ferrozine (3-(2-pyridyl)-5,6-diphenyl-1,2,4-triazinep,p′-disulphonic acid) reacts with free ferrous ions Fe (II), resulting in strong absorbance at 562 nm. Metal chelating activity aims to measure the capacity of test extracts to chelate free ferrous ions, thereby reducing absorbance at 562 nm in a dose-dependent manner. Metal chelating activity of the actinobacterial isolates is determined by the method of Dinis et al. (1994).

20.3.5.1 Procedure

- The extracts were diluted to make 125, 250, 500 and 1,000 µg/mL dilutions. To 2 mL of each dilution, 100 µL of 2 mM ferrous chloride and 400 µL of 5 mM ferrozine were added and allowed to react for 10 minutes at room temperature.
- The absorbance was measured at 562 nm using a UV–visible spectrophotometer.
- A mixture of ferrous chloride and ferrozine was used as a control. Each experiment was performed in triplicate at each concentration.
- The percentage inhibition of the ferrous ion was calculated by comparing the results of the test with those of the control using the following formula:

$$\text{Percentage metal chelating activity} = [1 - (At / Ac)] \times 100$$

where At is the absorbance of the test sample and Ac is the absorbance of the control.

Metal chelating activity of the extracts was also expressed as Na2EDTA equivalence (µg/mL). IC_{50} values of the isolates were calculated.

20.3.6 β-Carotene Bleaching Inhibition Activity

β-Carotene is a yellowish-orange-coloured fat-soluble hydrocarbon that on reaction with linoleic acid turns to a light milky colour due to linoleic acid oxidation. If an antioxidant is added to the reaction mixture, it scavenges linoleic acid free radicals, thereby retaining the colour of β-carotene. Thus, the stronger the antioxidant in the extracts, the greater the colour and the higher the absorbance at 470 nm.

β-Carotene bleaching assay is performed by following a standard protocol reported earlier by Kulisic et al. (2004).

20.3.6.1 Procedure

- β-Carotene (2 mg) was mixed with chloroform (1 mL) followed by the addition of linoleic acid (25 μL) and Tween 40 (200 μL).
- The mixture was evaporated at 40°C in a rotary evaporator to remove chloroform.
- Distilled water (50 mL) was immediately added to the above mixture with agitation to form an emulsion.
- The test samples (100 μL) at different concentrations were mixed with linoleic acid emulsion (3 mL) and incubated at 50°C for 120 minutes.
- Experimental control was prepared by the addition of reaction mixture (3 mL) and distilled water (100 μL) and incubated at 50°C for 120 minutes.
- Absorbance of the sample was measured at 470 nm using the test samples at respective concentrations (125, 250, 500 and 1,000 μg/mL) with distilled water as a blank.
- Initial and final OD values of experimental control were taken at 470 nm, the percentage of antilinoleic peroxidation was calculated and IC_{50} values of the extracts were measured using the following formula:

% Antilinoleic acid peroxidation = [(AS120 – AC120/AC0 – AC120)] × 100
where AS120 = absorbance of the test sample after incubation for 120 minutes
AC120 = absorbance of the control after incubation for 120 minutes
AC0 = absorbance of the control before incubation for 120 minutes.

20.3.7 Lipid Peroxidation Inhibition Activity

Lipid peroxidation is a process of free radical chain reaction mechanism causing oxidative degradation of lipids, resulting in cell damage. This assay aims to determine the efficacy of the extracts to act as antioxidants in the inhibition of lipid peroxidation.

The lipid peroxidation inhibition activity is evaluated by using goat liver homogenate, as reported earlier by Mandal and Chatterjee (1980).

20.3.7.1 Procedure

- Actinobacterial isolates at different concentrations (125, 250, 500 and 1,000 μg/mL) were mixed with 2.8 mL of 10% goat liver homogenate and 0.1 mL of 50 mM $FeSO_4$ and incubated for 30 minutes at 37°C. After incubation, 2 mL each of 10% trichloroacetic acid (TCA) and 0.67% thiobarbituric acid (TBA) in acetic acid (50%) was mixed with 1 mL of the above reaction mixture and kept for boiling at 100°C for an hour followed by centrifugation (R-8C Laboratory Centrifuge, Remi, India) at 10,000 rpm for 5 minutes.
- Absorbance of the test sample was taken at 535 nm against the reagent blank.
- The control was without the isolate and $FeSO_4$ and induced without the isolate.
- Antilipid peroxidation percentage was calculated, and IC_{50} values of each extract were measured using the following formula:

% Antilipid peroxidation = [(Ai – As) / (Ai – Ac)] × 100
where Ai = absorbance of Fe^{2+}-induced peroxidation
As = absorbance of the test sample
Ac = absorbance of the control.

REFERENCES

Adedapo, A.A., Jimoh, F.O., Afolayan, A.J., and Masika P.J. 2009. Antioxidant properties of the methanol extract of the leaves and stems of *Celtis Africana. Records of Natural Products*. 3: 23–31.

Brand-Williams, W., Cuvelier, M.E., and Berset, C. 1995. Use of a free radical method to evaluate antioxidant activity. *Food Science and Technology*. 28: 25–30.

Desai, N.K., Bhabhor, P.H., Domadia, D., and Bhatt, J.D. 2013. Role of antioxidant therapy in management of type-2 diabetes mellitus. *National Journal of Integrated Research in Medicine*. 4(2): 1–6.

Dinis, T.C., Maderia, V.M., and Almeida, L.M. 1994. Action of phenolic derivatives (acetaminophen, salicylate and 5-amino salicylate) as inhibitors of membrane lipid peroxidation and as peroxyl radical scavengers. *Archives of Biochemistry and Biophysics*. 315(1): 161–169.

Halliwell, B., and Gutteridge, J.M.C. 2007. *Free Radicals in Biology and Medicine*. 4th edition, Clarendon Press, Oxford.

Jann, E.L.D., and Gil, M.P. 2021. Isolation of actinomycetes with cellulolytic and antimicrobial activities from soils collected from an urban green space in the Philippines. *International Journal of Microbiology*. 2021: 1–14.

Jos, E.M.M., Cristina, P.G., and Ignacio, N.D.C. 1999. Antioxidant enzymes and human diseases. *Clinical Biochemistry*. 32(8): 595–603.

Kulisic, T., Radonic, A., Katalinic, V., and Milos, M. 2004. Use of different methods for testing antioxidative activity of oregano essential oil. *Food Chemistry*. 85(4): 633–640.

Lekhya Priya, C., and Bhaskara Rao, K.V. 2014. Phytochemical composition, antidiabetic and antioxidant activities of *Acalypha indica* Linn (Euphorbiaceae) stem extract: An in vitro study. *Research Journal of Biotechnology*. 9(4): 1–9.

Lekhya Priya, C., and Bhaskara Rao, K.V. 2016. Postprandial antihyperglycemic and antioxidant activities of *Acalypha indica* Linn stem extract: An in vivo study. *Pharmaceutical Magazine*. 12(47): 475–481.

Mandal, T.K., and Chatterjee, S.N. 1980. Ultraviolet- and sunlight-induced lipid peroxidation in liposomal membrane. *Radiation Research*. 83: 290–292.

Nishikimi, M., Rao, A.N., and Yagi, K. 1972. The occurrence of superoxide anion in the reaction of reduced phenazine methosulfate and molecular oxygen. *Biochemical and Biophysical Research Communications*. 46: 849–853.

Riddhi, N. D. Raghawendra Kumar, Avinash Mishra, Kalpana, H.M., and Bhavanath, J. 2017. Antibacterial and antioxidant activities of novel Actinobacteria strain isolated from gulf of Khambhat, Gujarat. *Frontiers in Microbiology*. 8: 1–16.

Ruch, R.J., Cheng, S.J., and Klaunig, J.E. 1989. Prevention of cytotoxicity and inhibition of intracellular communication by antioxidant catechins isolated from Chinese green tea. *The Journal of Carcinogenesis*. 10(6): 1003–1008.

Sheng, Z.W., Hong, W.M., Gao, J.H., Yang, B., Zhang, W.M., Dou, H.T., and Jin Z.Q. 2011. Antioxidant properties of banana flower of two cultivars in China Using 2,2-diphenyl-1-picrylhydrazyl (DPPH,) reducing power, 2,2'-azinobis-(3-ethylbenzthiazoline-6-sulphonate (ABTS) and inhibition of lipid peroxidation assays. *African Journal of Biotechnology*. 10(21): 4470–4477.

Shukla, S., Mehta, A., Bajpai, V.K., and Shukla S. 2009. *In vitro* antioxidant activity and total phenolic content of ethanolic leaf extract of *Stevia rebaudiana* Bert. *Food and Chemical Toxicology*. 47: 338–2343.

Syed, Z.H., and Rekha, G. 2019. Antimicrobial activity of ethyl acetate extract of antibiotic isolated from fresh water actinomycetes. *International Journal of Research Culture Society*. 3(6): 56–61.

Oyaizu, M. 1986. Studies on products of browning reactions: Antioxidative activities of products of browning reaction prepared from glucosamine. *Japan Journal of Nutrition*. 44(6): 307–315.

21 Protocol for Antileishmanial Activity Using Marine Actinobacteria

Veena Sreedharan, Karthik Loganathan, and K.V. Bhaskara Rao

21.1 INTRODUCTION

21.1.1 LEISHMANIASIS

Among all the vector borne diseases leishmaniasis is the most complex and distinct, which is caused by intramacrophage protozoan of the genus Leishmania (Oliveira et al., 2009). This parasite is transmitted when bitten by female sand flies which belong to the genera, Phlebotomus, in the old world, and Lutzomyia, in the new world (Singh, 2016). According to a recent World Health Organization (WHO) report (2020), 1.5–2 million new cases of cutaneous leishmaniasis and 500,000 cases of visceral leishmaniasis are reported each year. Leishmaniasis is endemic in nearly 90 countries worldwide and close to 12 million new cases of leishmaniasis are reported worldwide annually (WHO, 2020). The different types of Leishmaniasis includes visceral leishmaniasis (VL), cutaneous leishmaniasis (CL) and muco-cutaneous leishmaniasis (ML) (Herwaldt, 1999). Among them VL is the most severe type of leishmaniasis, which is also well-known as "kala azar" caused by the *Leishmania donovani* (Davis et al., 2004).

21.1.2 ENZYME COULD BE A POSSIBLE THERAPEUTIC TARGET

Leishmanial pathogenesis is triggered by the protease enzymes present on the surface of Leishmania species which causes a breakdown of cell surface proteins leading to ingress inside host cells (Drag and Salvesen, 2010). Leishmania parasites, like other parasites, need metabolic pathways for existence and virulency (McConville and Naderer, 2011). Several metabolic mechanisms have been recognized as being critical for existence inside the cells and development of diseases based on current genomic and biochemical analyses of Leishmania's metabolic capabilities, and investigation of the virulency of metabolic mutants (Chawla and Madhubala, 2010; Bhargava and Singh, 2012). Some metabolic pathways, such as glycolysis, purine synthesis, glyoxalase, the trypanothione system, and folate biosynthesis, need many Leishmania enzymes. Protein kinases, tropoisomerases, metacaspases, and, most crucially, proteases ensure that surface proteins are broken down and that infection is maintained (Chawla and Madhubala, 2010; Chandra et al 2010; Singh et al., 2012). These enzymes are the only ones that are thought to be good Leishmania targets.

21.1.3 LEISHMANIAL ZINC METALLOPROTEASE AS DRUG TARGETS

The predominant surface glycoprotein on Leishmania promastigotes is zinc metalloprotease (GP63), also known as main surface proteases (MSP) (King et al., 1987; Button and McMaster, 1988; Chaudhuri and Chang, 1988). Metalloproteases are involved in every stage of the parasite's life cycle, from attachment to the sand fly gut to life within the macrophage host. The MSP interacts

 DOI: 10.1201/9781003398400-21

to the CR3 receptor on macrophages and assists in promastigote internalization (Brittingham et al., 1995). Protease inhibitor drug bioproduction with a concentration on leishmanial Metalloprotease will be quite effective in controlling *Leishmania* species pathogenesis.

21.1.4 Enzyme Inhibitors from Marine Actinobacteria

Many enzyme inhibitors are also extracted from terrestrial organism, even if they are structurally identical to marine microbial enzymes (Umezawa, 1972; Hamato et al., 1992). In contrast, the properties of enzyme inhibitors in marine creatures differ significantly from those in terrestrial organisms (Imada, 2004). One of the possible producers of enzyme inhibitors is Streptomyces isolated from the terrestrial environment (Umezawa, 1972). Actinomycin (Katz and Weissbach, 1962), Candicidin (Liras et al., 1977) and Chloramphenicols (Jones and Westlake, 1974) and Streptomycin (Liras et al., 1977) are examples of secondary metabolites extracted marine actinobacteria that act as enzyme inhibitors (Martin and Demain, 1980). The amount of research on enzyme inhibitors from marine microorganisms is limited (Newman and Cragg, 2007; Skropeta et al., 2011). Enzyme inhibitors have long been regarded as important tools in pharmacology and agriculture, owing to their biotechnological potential (Bode and Huber, 1992). (Terashita et al., 1980). The third most important product of marine actinobacteria is enzyme inhibitors (Bode and Huber, 1992). Even though it is used to research enzyme shapes and reaction mechanisms, it was only recently introduced into pharmacology (Bode and Huber, 1992). These discriminating inhibitors can be employed to inactivate target proteases in disorders like high blood pressure, malaria, pancreatitis, emphysema, AIDS, and cancer (Demuth, 1990). Amylase inhibitors are effective tools for regulating carbohydrate-dependent illnesses such as hyperlipemia, obesity diabetes (Puls and Keup, 1973; Bo-Linn et al., 1982). Karthik et al. (2014) extracted a protease inhibitor (peptide) from deep-sea Nicobar sand samples from marine actinobacterial that showed promising action against malarial parasites P. falciparum. The effectiveness of inhibitor for α-glucosidase from sea actinobacteria for the treatment of hyperglycemia in Streptozotocin induced diabetic male rats was recently reported by Sathish and Bhaskar (2018). In light of the scarcity of antileishmanial drug research, the goal of this chapter is to highlight the crucial protocols that are useful for isolation and evaluation antileishmanial activity *in vitro* and *in vivo* using marine actinobacteria.

21.2 MATERIAL AND METHODS

21.2.1 In Vitro Antileishmanial Activity

The antileishmanial activity of protease inhibitor producing isolate marine actinobacteria can be performed using FACS assay

21.2.2 Fluorescence Activated Cell Sorting Assay

In fluorescence activated cell sorting assay J774A.1 macrophages should be maintained in Dulbecco's modified Eagle's medium (DMEM) in a CO_2 (5%) incubator for 8–12 hours. Culture the macrophage in a 12 well Neubauer's chamber slide to a cell density of 1 × 105 cells/well and infect with Red fluorescence protein (RFP) parasites in the ratio of 10:1 (parasite/macrophage) and leave for 12 hours in 5% CO_2 incubator so that they could get internalized properly. To remove un-internalized parasites, after incubation period wash the cells with DMEM medium. Treat the RFP infected parasites cells with different concentrations of actinobacterial extract for 48 hours in CO_2 incubator. Use Miltefosine drug as a positive control with distilled water as a negative control. Scrap the cells in 1X PBS and acquire in FC500 flow cytometer after 24 and 48 hours of treatment. The percentage of parasitic inhibition will be then calculate using Eq. 21.2 based on their Mean fluorescent values and compare with the untreated ones (Plock et al., 2001).

$$PI = 100 - T \times 100/C \tag{3.1}$$

Where, PI is the percentage of inhibition, T is the mean fluorescence intensity of the treated sample and C is the mean fluorescence intensity of the control sample.

21.2.3 Direct Promastigotes Treatment Assay Using Actinobacteria

Grow stationary phase, 74 hours old culture of *Leishmania donovani* (2×10^6/mL promastigotes) on 25% fetal bovine serum supplemented RPMI media. Promastigotes should be then distributed in 24 well culturing plates. Add plates with Actinobacterial extract (100 mg/mL) in triplicate. Use saline and Amphotericin B as negative and positive control. Incubate all cell-plates at 25°C and then analyze using microscope after 24, 48 and 72 hours in a Neubauer's chamber.

21.2.4 In Vivo Antileishmanial Activity

21.2.4.1 Animals and Parasites

Golden hamsters which are also known as *Mesocricetus auratus*were used for in-vitro study. Animals weighing 40–50g were used which were kept in plastic cages with proper air supply, water and fed with commercially available food or ad libitum. Strain 2001 of *Leishmania donovani* parasite were grown in heat inactivated fetal bovine serum supplemented RPMI-1640 media. These parasites were also preserved in golden hamsters through sequential passage from amastigotes to promastigotes.

21.2.5 Infection in Golden Hamsters with Promastigotes

Harvesting of promastigotes were done by centrifugation for 15 minutes at 5000 rpm. The harvested promastigotes were washed thrice in PBS solution. Promastigotes (1×10^7 per 0.1 mL) were resuspended in PBS solution. Finally, these harvested promastigotes were intracardially inoculated into golden hamsters in 0.1 mL PBS solution. After 25–30 days parasitic burden was checked by carrying out the splenic biopsies.

21.2.6 Splenic Amastigotes Infection and Drug Treatment

Autopsy of animals carrying 60 days old infection was done and spleens were removed carefully and aseptically. Spleens were then homogenized and centrifuged for 5 minutes at 1000 rpm at 5°C to filter out the tissue remains. Collected solution was again centrifuged for 15 minutes at 3,000 rpm and pellets were collected. Followed by washing of pellets twice using PB solution. Golden hamsters were the intracardially infected with 1×10^7 of amastigotes in PB solution. After 30 days of infections treatment was started using crude extract. Animals were treated orally with actinobacterial extract. Two different dose actinobacterial extract was used for the study. Positive control used was Miltefosine drug at a concentration of 40 mg/kg. On 48 the 7th days after last dose of the compound was also administered again splenic biopsy was performed. Parasitic burden was checked for all treated groups, untreated control groups and animals treated with standard drugs. All treated groups were then kept for 30 days to check the survival rate in comparison with the untreated controls (Anuradha et al., 2005).

21.3 CONCLUSION

Predominantly, 80% of natural compounds that are isolated from marine actinobacteria showed very good bioactivities. The most shown bioactivity includes antimicrobial, antioxidant, anticancer and anti-HIV activity. Obviously, this is just the beginning of isolation of bioactive compounds from marine actinobacteria. Till now more than 130 natural compounds were isolated which belong to classes like peptides, steroids, isoprenoids etc. Among these natural compounds many compounds are new which belong to novel metabolites. Above mentioned protocols will help you screen the potent marine actinobacteria isolated marine environment that have antileishmanial activity. The findings of such study imply that marine actinobacteria have strong antileishmanial activity and could provide the starting point for the creation of effective medications to treat visceral leishmaniasis.

REFERENCES

Anuradha, D., Singh, N., Sundar, S., and Singh, N. 2005. Refractoriness to the treatment of sodium stibogluconate in Indian kala-azar field isolates persist in in vitro and in vivo experimental models. *Parasitology Research.* 96(4): 216–223.

Bhargava, P., and Singh, R. 2012. Developments in diagnosis and antileishmanial drugs. *Interdisciplinary Perspective of Infectious Diseases.* 12(5): 626–838.

Bode, W., and Huber, R. 1992. Natural protein proteinase inhibitors and their interaction with proteinases. *European Journal of Biochemistry.* 204(2): 433–445.

Bo-Linn, G.W., Santa Ana, C.A., Morawski, S.G., and Fordtran, J.S. 1982. Starch blockers-their effect on calorie absorption from a high-starch meal. *The New England Journal of Medicine.* 307(23): 1413–1416.

Brittingham, A., Morrison, C.J., McMaster, W. R., McGwire, B. S., Chang, K. P. and Mosser, D. M. 1995. Role of the leishmania surface protease GP63 in complement fixation, cell adhesion, and resistance to complement-mediated lysis. *Journal of Immunology.* 155(6): 3102–3111

Button, L. L. and McMaster, W. R. 1988. Molecular cloning of the major surface antigen of Leishmania. *Journal of Experimental Medicine.* 167(2): 724–729.

Chandra, S., Ruhela, D., Deb, A. and Vishwakarma, R. A. 2010. Glycobiology of the Leishmania parasite and emerging targets for antileishmanial drug discovery. *Expert Opinion on Therapeutic Targets.* 14(7):739–75.

Chaudhuri, G. and Chang, K. P. 1988. Acid protease activity of a major surface membrane glycoprotein (GP63) from Leishmania mexicana promastigotes. *Molecular and Biochemical Parasitology.* 27(1): 43–52.

Chawla, B. and Madhubala, R. 2010. Drug targets in Leishmania. *Journal of Parasitological Diseases.* 34(1): 1–13.

Davis, A. J., Murray, H. W. and Handman, E. 2004. Drugs against leishmaniasis: a synergy of technology and partnerships. *Trends in Parasitology.* 20(2): 73–76.

Demuth, H. U. 1990. Recent developments in inhibiting cysteine and serine proteases. *Journal of Enzyme Inhibition and Medicinal Chemistry.* 3(4): 249–278.

Drag, M. and Salvesen, G. S. 2010. Emerging principles in protease-based drug discovery. *Nature Review Drug Discovery.* 9(9): 690–701.

Hamato, N., Takano, R., Kamei-Hayashi, T., Imada, C. and Hara, S. 1992. Leupeptins produced by the marine Alteromonas sp. B-10-31. *Bioscience, Biotechnology, and Biochemistry.* 56(8): 1316–1318.

Herwaldt, B. L. 1999. Leishmaniasis. *Lancet.* 54 (9185): 1191–1199.

Imada, C. 2004. Enzyme inhibitors of marine microbial origin with pharmaceutical importance. *Marine Biotechnology.* 6(3): 193–198.

Jones, A and Westlake, D.W. 1974. Regulation of chloramphenicol synthesis in Streptomyces sp. 3022a. Properties of arylamine synthetase, an enzyme involved in antibiotic biosynthesis. *Canadian Journal of Microbiology.* 20(11): 1599–1611.

Karthik, L., Gaurav, K., Tarun, K., Arindam, B., Sarath, C.S and Bhaskara Rao, K. V. 2014. Protease inhibitors from marine actinobacteria as a potential source for antimalarial compound. *PLoS One.* 9(3): 1–13.

Katz, E and Weissbach, H. 1962. Biosynthesis of the actinomycin chromophore; enzymatic conversion of 4-methyl-3-hydroxyanthranilic acid to actinocin. *Journal of Biological Chemistry.* 237: 882–886.

King, D. L., Chang, Y. D. and Turco, S. J. 1987. Cell surface lipophosphoglycan of Leishmania donovani. *Molecular and Biochemical Parasitology.* 24(1): 47–53.

Liras P., Villanueva J. R. and Martin, J.F. 1977. Sequential expression of macromolecule biosynthesis and candicidin formation in Streptomyces griseus. *Journal of General Microbiology.* 102(2): 269–277.

Martin J. F. and Demain L. A. 1980. Control of antibiotic biosynthesis. *Microbiological Review.* 44(2): 230–251.

McConville, M. J. and Naderer, T. 2011. Metabolic pathways required for the intracellular survival of leishmania. *Annual Review of Microbiology.* 6(1): 543–561.

Newman, D. J. and Cragg, G. M. 2007. Natural products as sources of new drugs over the last 25 years. *Journal of Natural Products.* 70(3): 461–477.

Oliveira, M. and Hassani, K. 2009. Protease inhibitors as prophylaxis against leishmaniasis: new hope from the major surface protease gp63. *Future Medicinal Chemistry.* 2(4): 539–542.

Plock, A., Sokolowska-Kohler, W. and Presber, W. 2001. Application of flow cytometry and microscopical methods to characterize the effects of herbal drugs on Leishmania spp. *Experimental Parasitology.* 97(3): 141–153.

Puls, W. and Keup, U. 1973. Influence of an a-amylase inhibitor (BAY d 7791) on blood glucose, serum insulin and NEFA in starch loading tests in rats, dogs and man. *Diabetologia.* 9(2): 97–101.

Sathish, K. and Bhaskara Rao, K. V. 2018. Efficacy of alpha glucosidase inhibitor from marine actinobacterium in the control of postprandial hyperglycaemia in streptozotocin (STZ) induced diabetic male albino Wister rats. *Iranian Journal of Pharmaceutical Research.* 17(1): 202–214.

Singh, N., Kumar, M and Singh, R. K. 2012. Leishmaniasis: Current status of available drugs and new potential drug targets. *Asian Pacific Journal of Tropical Medicines.* 5(6): 485–497.

Singh, S. P., Reddy, D. C. S., Rai, M and Sundar, S. 2016. Serious underreporting of visceral leishmaniasis through passive case reporting in Bihar, India. *Tropical Medicine and International Health.* 11(6): 899–905.

Skropeta, D., Pastro, N and Zivanovic, A. 2011. Kinase inhibitors from marine sponges. *Marine Drugs.* 9(10): 2131–2154.

Terashita, T., Kono, M and Murao, S. 1980. Promoting effect of S-Pion fruiting of Lentinus edodes. *Transaction of the mycology society of Japan.* 21(4): 137–140.

Umezawa, H. 1972. Enzyme Inhibitor of Microbial Origin, Proceedings of the Sixth International Congress of Pharmacology. Mechanism of Toxicity and Metabolism. 17–31.

WHO. 2020. Leishmaniasis Fact Sheet. WHO Report on Global Surveillance of Epidemic-prone Infectious Diseases - Leishmaniasis. https://www.who.int/news-room/fact-sheets/detail/leishmaniasis

22 Shining a Light on the Potent Antimalarial Activity of Actinomycetes Extract

A Comprehensive In Vitro *and* In Vivo *Evaluation*

A. Vishnu Kirthi, L. Karthik, Tarun Keswani, Anirban Sengupta, and Shilpi Garg

22.1 INTRODUCTION

Malaria is a serious health problem, especially in tropical and subtropical regions, transmitted by infected mosquitoes. It can cause fever, headaches, and even death if not treated. Malaria is a leading cause of illness and death globally, particularly in low-income countries in Africa, Asia, and Latin America. In 2019, around 229 million cases of malaria were reported worldwide, causing an estimated 409,000 deaths. While these numbers have reduced since 2000, when there were 262 million cases and 839,000 deaths, eliminating malaria by 2030, as planned by the World Health Organization, is still a significant challenge (Garcia, 2010; Milner, 2018; Peltola, 1974; Randall and Seidel, 1985; Recht et al., 2017; Weinman, 1972; WHO, 2021, 2022; Moxon et al., 2020; Varo et al., 2020; Walter and John, 2022).

Effective treatment is essential to reduce the burden of the disease and save lives. Several classes of antimalarial drugs are available, including chloroquine and hydroxychloroquine, artemisinin and its derivatives, atovaquone and proguanil, mefloquine, and primaquine.

Chloroquine and hydroxychloroquine have been used for decades to treat uncomplicated malaria, but the emergence of drug-resistant strains and potential retinal toxicity limit their use (Thomé et al., 2013; Coban, 2020). Artemisinin and its derivatives are potent antimalarial drugs used in combination therapies, but their high cost and emerging resistance are limitations. Atovaquone and proguanil are commonly used for prophylaxis but are limited by the emergence of drug-resistant strains (Petersen et al., 2011). Mefloquine has neuropsychiatric side effects and is used less frequently (Hanboonkunupakarn et al., 2019). Primaquine is used for *Plasmodium vivax* and *Plasmodium ovale* malaria but can cause hemolysis in patients with G6PD deficiency (WHO, 2016).

22.2 *IN VITRO* ANTIPLASMODIAL STUDY

Materials

- Dulbecco's modified eagle medium (DMEM)
- Sodium hydroxide (NaOH)
- RPMI-1640 medium
- Fetal bovine serum (FBS)
- Human red blood cells (RBCs)

DOI: 10.1201/9781003398400-22

- Parasites of *Plasmodium falciparum* strain (FCR-3)
- Giemsa stain
- Microplate reader
- Spectrophotometer
- Hemocytometer

22.3 METHODS

Preparation of the Extract

- Cultivate the actinomycetes in a suitable medium (such as ISP-2 or Gause's synthetic agar) at optimal temperature and pH for 7–14 days until the growth is abundant.
- Harvest the biomass of actinomycetes and extract it with an appropriate solvent (such as ethyl acetate, methanol, or chloroform).
- Concentrate the extract by evaporation under reduced pressure, followed by lyophilization to obtain a dry powder.

22.3.1 Preparation of Test Compounds

Selection of Test Compounds

- Choose a set of known antimalarial drugs, such as chloroquine, quinine, or artemisinin, as positive controls.
- Prepare a set of test compounds, such as plant extracts or synthetic compounds, to be evaluated for their antiplasmodial activity.
- Prepare a stock solution of each test compound in dimethyl sulfoxide (DMSO) at a concentration of 10 mg/mL.
- Prepare a working solution of each test compound in RPMI-1640 medium supplemented with 10% fetal bovine serum (FBS) at a concentration of 10 µg/mL.

22.3.2 Preparation of Parasite Cultures

Culture of *Plasmodium falciparum*

- Maintain a culture of *Plasmodium falciparum* in human erythrocytes using standard protocols.
- Synchronize the parasite culture using 5% sorbitol solution to obtain a highly synchronized culture of late-stage parasites.
- Adjust the parasitemia to 1% using RPMI-1640 medium supplemented with 10% FBS.

22.3.3 Evaluation of Antiplasmodial Activity

Microscopy-Based Assay

- Dispense 100 µL of RPMI-1640 medium supplemented with 10% FBS into each well of a 96-well microtiter plate.
- Add 100 µL of the working solution of each test compound to the appropriate wells.
- Add 100 µL of the parasite culture to each well.
- Add 100 µL of the extract solution at various concentrations (ranging from 1 to 100 µg/mL) to the appropriate wells.
- Incubate the plate for 48 hours at 37°C in a 5% CO_2 atmosphere.

- After 48 hours, add 20 µL of a 0.5% glutaraldehyde solution to each well to fix the cells.
- Add 100 µL of Giemsa stain to each well and incubate for 20 minutes at room temperature.
- Wash the wells with distilled water and allow them to dry.
- Count the number of infected erythrocytes and uninfected RBCs.

Preparation of Parasite Cultures

1. Obtain *P. falciparum* strain 3D7 from a culture bank or a collaborator.
2. Maintain *P. falciparum* cultures in complete RPMI 1640 medium supplemented with 10% human serum, 25 mM HEPES, 100 µg/mL gentamicin, and 0.5% Albumax II at 37°C in a humidified atmosphere containing 5% CO_2.
3. Synchronize the parasite cultures using 5% D-sorbitol to obtain parasites at the ring stage.
4. Change the medium daily to maintain healthy cultures and synchronize the parasites.
5. Use Giemsa-stained thin blood smears to monitor parasite growth and synchronize the cultures.

Evaluation of Antiplasmodial Activity

1. Prepare stock solutions of the test compounds and actinomycetes extract in DMSO.
2. Prepare serial dilutions of the test compounds and actinomycete extract in complete RPMI 1640 medium to obtain final concentrations ranging from 0.01 to 100 µg/mL.
3. Add 180 µL of the diluted test compounds and actinomycete extract to each well of a 96-well plate.
4. Add 20 µL of synchronized *P. falciparum* culture (1% parasitemia and 1% hematocrit) to each well of the 96-well plate.
5. Incubate the plate at 37°C in a humidified atmosphere containing 5% CO_2 for 72 hours.
6. After incubation, add 20 µL of Alamar Blue to each well and incubate for an additional 6 hours.
7. Measure the fluorescence using a microplate reader with excitation at 560 nm and emission at 590 nm.
8. Calculate the percentage inhibition of parasite growth using the following formula: % inhibition = (1 − (test fluorescence control/untreated control)) × 100.
9. Calculate the IC_{50} values using non-linear regression analysis.

Parasite Growth Inhibition Assay

- The assay can be performed using the modified schizont maturation inhibition assay.
- Parasite cultures should be synchronized at the ring stage and seeded in 96-well microplates.
- Different concentrations of the actinomycete extract should be added to the wells in triplicate, along with a positive control (artemisinin) and a negative control (media only).
- The plates should be incubated for 48 hours at 37°C in a gas mixture containing 5% CO_2, 5% O_2, and 90% N_2.
- After incubation, the plates should be frozen at –20°C for at least 24 hours.
- The plates should be thawed, and the contents of each well should be transferred to a new plate for staining with 100 µL of 1% Giemsa solution.
- The plates should be incubated for 30 minutes, and the Giemsa solution should be removed.
- The plates should be washed with distilled water and air-dried.
- The number of mature schizonts per 1,000 erythrocytes should be counted under a microscope.
- The percentage of parasite growth inhibition should be calculated using the following formula:

% Growth Inhibition = (1 − [Mean parasitemia of test sample/Mean parasitemia of control]) × 100

Cytotoxicity Assay

- The cytotoxicity of the actinomycete extract can be evaluated using the MTT (3-(4, 5-dimethylthiazolyl-2)-2, 5-diphenyltetrazolium bromide) assay.
- Vero cells should be seeded in 96-well microplates at a density of 5,000 cells per well and incubated for 24 hours.
- Different concentrations of the actinomycete extract should be added to the wells in triplicate, along with a positive control (doxorubicin) and a negative control (media only).
- The plates should be incubated for 48 hours at 37°C in a humidified atmosphere containing 5% CO_2.
- After incubation, the medium should be removed, and 100 μL of MTT solution (0.5 mg/mL in PBS) should be added to each well.
- The plates should be incubated for 4 hours at 37°C in a humidified atmosphere containing 5% CO_2.
- The medium should be removed, and 100 μL of DMSO should be added to each well to solubilize the formazan crystals.
- The absorbance of each well should be measured at 570 nm using a microplate reader.
- The percentage of cell viability should be calculated using the following formula: % Cell Viability = (Mean absorbance of test sample/Mean absorbance of control) × 100

These methods can be used to determine the antiplasmodial activity and cytotoxicity of the actinomycete extract. The results can be used to evaluate the potential of the actinomycete extract as a treatment for malaria and to optimize the concentration of the extract for further studies.

Data Analysis

1. Use GraphPad Prism software for statistical analysis and data visualization.
2. Perform statistical analysis using one-way ANOVA (Analysis of Variance) followed by Dunnett's multiple comparisons test to compare the means of the treated groups with the untreated control group.
3. Determine the significance level as $p < 0.05$.

22.2 *IN VIVO* ANTIPLASMODIAL STUDY USING *PLASMODIUM BERGHEI*

22.2.1 Procedure

Chemical Requirements

1. Chloroquine diphosphate
2. Parasitized erythrocytes
3. Test compounds
4. Normal saline
5. Giemsa stain
6. Phosphate-buffered saline (PBS)

7. Sodium bicarbonate
8. Heparin
9. Trisodium citrate dihydrate
10. Ethanol

Materials Required List

1. Mice infected with *Plasmodium* species
2. Sterile needles and syringes
3. Hemacytometer
4. Microscope slides
5. Cover slips
6. Incubator
7. Centrifuge
8. Chemical fume hood
9. Glassware (beakers, flasks, pipettes, etc.)
10. Sterile filter paper
11. Timer
12. Sterile disposable gloves
13. Tissue culture plates

22.2.2 Things to Remember

When conducting *in vivo* antiplasmodial activity testing, there are several things to remember to ensure accurate and reliable results:

1. Use appropriate animal models and strains that are suitable for the *Plasmodium* species being tested.
2. Ensure that the animals are properly housed and cared for, and that the experiment is conducted in compliance with ethical guidelines and regulations.
3. Standardize the inoculum size and route of infection to ensure that the parasitemia levels are consistent across the experimental groups.
4. Use appropriate controls, including untreated infected mice and uninfected mice, to compare with the treated groups.
5. Choose appropriate time points for parasitemia evaluation to ensure that the test compound is given enough time to exert its effect.
6. Ensure that the test compound is properly dissolved and administered to the animals at the correct dose and route.
7. Use appropriate safety measures, such as wearing gloves and working in a chemical fume hood, when handling potentially hazardous chemicals and reagents.
8. Use appropriate quality control measures to ensure the accuracy and reproducibility of the results.
9. Analyze the data carefully and use appropriate statistical methods to assess the significance of the results.
10. Report the results accurately and transparently, including any limitations or potential sources of error.

By keeping these considerations in mind, researchers can ensure that their *in vivo* antiplasmodial activity testing is rigorous, reliable, and informative.

22.2.3 Selection of the Mice

Selecting the appropriate mice for the in vivo antiplasmodial activity test is an important step in ensuring reliable and reproducible results. Here are the steps for selecting mice for the test:

1. Select the appropriate strain of mice: Different strains of mice may have different susceptibilities to *Plasmodium* infection. Consult the literature to determine which strain is appropriate for the type of *Plasmodium* species being tested.
2. Age and weight of mice: Select mice that are between 6 and 8 weeks old and weigh between 18 and 25 g. This ensures that the mice are mature enough to handle the experimental procedure and have developed a sufficient immune response to the infection.
3. Check for pre-existing conditions: Mice with pre-existing infections, injuries, or other health conditions should be excluded from the test, as these factors can affect the outcome of the experiment.
4. Acclimation: Acclimate the mice to the experimental conditions for at least 7 days prior to the start of the experiment. This allows the mice to adapt to the new environment and reduces stress-induced factors that can interfere with the experimental results.
5. Randomization: Randomly assign the mice to treatment groups to avoid bias in the experimental design. Ensure that the number of mice in each group is sufficient to achieve statistical significance.
6. Infection: Inoculate each mouse with an appropriate dose of *Plasmodium* parasites to induce a malaria infection. This is typically done by injecting infected blood or by infecting mosquitoes and allowing them to feed on the mice.
7. Monitoring: Monitor the mice for signs of illness or distress throughout the experiment. This includes observing for changes in behavior, body weight, and parasite levels in the blood.

By following these steps, you can ensure that the mice selected for the in vivo antiplasmodial activity test are healthy, mature, and suitable for the experimental procedure. This will help to produce reliable and meaningful results that can be used to evaluate the efficacy of potential antimalarial drugs.

Preparation of Parasitized Mice

1. Obtain mice infected with *Plasmodium* species.
2. House the mice in a specific pathogen-free facility with controlled temperature and humidity.
3. Provide the mice with a suitable diet and water *ad libitum*.
4. Monitor the mice for signs of illness, such as lethargy, anemia, and weight loss.
5. Keep the parasitemia levels of the mice between 5%–10% for experimental use.

Administration of Test Compound

1. Dissolve the test compound in normal saline or other appropriate solvent to obtain a suitable concentration.
2. Administer the test compound to the mice via the appropriate route of administration (e.g., oral, intraperitoneal, subcutaneous).
3. Monitor the mice for any signs of toxicity or adverse effects.
4. Collect blood samples at appropriate time points for parasitemia evaluation.

Parasitemia Evaluation

1. Collect blood samples from the infected mice by tail vein puncture.
2. Mix the blood with heparin to prevent clotting.
3. Dilute the blood with PBS and count the number of infected erythrocytes using a hemocytometer.
4. Prepare thin blood smears on microscope slides and fix them with ethanol.
5. Stain the smears with Giemsa stain.
6. Observe the stained smears under a microscope and count the number of infected erythrocytes in a minimum of 1000 erythrocytes.
7. Calculate the percentage parasitemia using the formula: (Number of infected erythrocytes/Total number of erythrocytes) $\times$ 100.

22.2.4 Procedure

Preparation of Test Compounds

1. Selection of Test Compounds: Test compounds are usually selected based on their potential antiplasmodial activity. They can be natural or synthetic compounds and may have been previously tested *in vitro* for their ability to inhibit the growth of malaria parasites.
2. Solubility Assessment: Test compounds are often dissolved in appropriate solvents to prepare stock solutions, which are subsequently diluted to the desired concentrations for testing. However, some compounds may be poorly soluble in common solvents such as dimethyl sulfoxide (DMSO), ethanol, or water. In such cases, alternative solvents or surfactants may be used to improve solubility.
3. Vehicle Selection: The selection of a suitable vehicle for dissolving test compounds is critical. The vehicle should be non-toxic and should not interfere with the pharmacological activity of the compound being tested. Commonly used vehicles include DMSO, ethanol, and water. The maximum concentration of the vehicle in the test solution should be kept below 1% (v/v) to avoid any potential toxic effects.

Preparation of Parasitized Mice

1. Parasite Strain Selection: *Plasmodium berghei* is one of the commonly used malaria parasite strains for *in vivo* antiplasmodial activity testing. Parasites are usually maintained in laboratory animals, such as mice or rats, by serial passage.
2. Mice Selection: Mice used for *in vivo* antiplasmodial activity testing should be healthy and free of any pre-existing infections or diseases. Ideally, mice of similar age, weight, and gender should be selected to reduce variability in the results.
3. Parasitemia Establishment: Parasitemia is established by injecting infected erythrocytes into the mouse. Typically, a 10% parasitemia level is established by intraperitoneal injection of infected erythrocytes.
4. Parasitemia Monitoring: Parasitemia levels are monitored by examining Giemsa-stained thin blood smears under a microscope. Smears are prepared from tail vein blood samples, and parasitemia levels are calculated by counting the number of infected erythrocytes in a minimum of 10,000 erythrocytes.

Administration of Test Compound

1. Dose Calculation: The dose of the test compound is usually calculated based on the body weight of the mouse and is expressed in milligrams per kilogram (mg/kg) of body weight.

2. Route of Administration: The route of administration of the test compound depends on the properties of the compound and its solubility in the vehicle solution. Common routes of administration include intraperitoneal injection, oral gavage, or subcutaneous injection.
3. Dosing Regimen: The dosing regimen may vary depending on the compound being tested and may involve single or multiple doses per day. The duration of treatment may also vary depending on the experimental design.

Parasitemia Evaluation

1. Blood Collection: Blood samples are collected from the mouse by tail vein puncture or retro-orbital bleeding. The use of heparinized syringes is recommended to prevent blood clotting.
2. Preparation of Blood Smears: Blood smears are prepared by placing a small drop of blood on a microscope slide, spreading it into a thin film, and allowing it to air dry. The slides are then fixed with methanol or ethanol and stained with Giemsa.
3. Microscopic Examination: Giemsa-stained blood smears are examined under a microscope using oil immersion. The percentage of parasitized erythrocytes is determined by counting the number of infected erythrocytes in a minimum of 10,000 erythrocytes.

Data Analysis: The parasitemia level is expressed as a percentage and is calculated using the following formula:

Parasitemia (%) = (Number of infected erythrocytes/Total number of erythrocytes) × 100

The percentage of parasitized erythrocytes is recorded for each mouse at each time point. The parasitemia levels are plotted over time to generate a parasitemia curve, which provides a visual representation of the effectiveness of the test compound in reducing the parasitemia levels.

The following parameters can be used to assess the antiplasmodial activity of the test compound:

1. Percentage Inhibition: The percentage inhibition of parasitemia is calculated by comparing the parasitemia levels in the treated group with those in the control group.
   Percentage inhibition = [(Parasitemia in control group − Parasitemia in treated group)/ Parasitemia in control group] × 100
2. Mean Survival Time: Mean survival time is the time at which 50% of the mice in a group have died. This parameter is used to assess the toxicity of the test compound.
3. Chemoprophylactic Activity: Chemoprophylactic activity is the ability of the test compound to prevent the establishment of parasitemia. This can be assessed by administering the test compound before parasite inoculation and monitoring the parasitemia levels over time.
4. Therapeutic Activity: Therapeutic activity is the ability of the test compound to reduce parasitemia levels after parasite inoculation. This can be assessed by administering the test compound after the establishment of parasitemia and monitoring the parasitemia levels over time.

22.2.5 Giemsa Staining

Giemsa staining for the *in vivo* antiplasmodial activity protocol, along with some important precautions:

- *Materials required:*
- Giemsa stain
- Ethanol
- Microscope slides
- Cover slips
- Sterile filter paper

Precautions

- Wear gloves when handling the slides and staining solution.
- Work in a well-ventilated area or under a chemical fume hood to avoid inhalation of the staining solution.
- Ensure that the staining solution is properly diluted according to the manufacturer's instructions to prevent over-staining or under-staining.
- Avoid prolonged exposure of the staining solution to light, as this can degrade the solution.

Steps

1. Prepare thin blood smears on microscope slides using a drop of blood obtained from the infected mice by tail vein puncture. Make sure to spread the blood evenly to create a thin smear.
2. Allow the slides to air dry for at least 30 minutes at room temperature. Do not use a heat source to dry the slides, as this can cause uneven staining.
3. Fix the smears by immersing the slides in 100% ethanol for 1 minute. This step helps to preserve the integrity of the cells and prevent them from washing off during the staining process.
4. Prepare a working solution of Giemsa stain by diluting the stock stain solution with distilled water according to the manufacturer's instructions. The working solution should be freshly prepared for each staining session.
5. Stain the fixed smears with the Giemsa working solution for 20–30 minutes. Make sure that the staining solution covers the entire smear and gently agitate the slide to ensure even staining.
6. Rinse the slides with distilled water and blot them dry with sterile filter paper. Rinse the slides thoroughly to remove any excess stain but do not rub the slide, as this can cause the cells to detach.
7. Mount the slides with cover slips using mounting medium. Apply a small drop of mounting medium to the center of the smear and carefully place the cover slip over it. Press down gently to remove any air bubbles.
8. Observe the stained smears under a microscope using a 100x oil immersion objective lens. Adjust the focus and lighting as needed to visualize the cells and identify any parasites.

Note: The staining time may need to be adjusted based on the quality of the stain, the thickness of the blood smear, and the intensity of the color desired. It is also important to avoid over-staining, which can result in obscured cell morphology, or under-staining, which can make it difficult to identify parasites.

22.3 CONCLUSION

In conclusion, the protocol chapter for the *in vitro* and *in vivo* bioassay of antimalarial activity using actinomycete extract provides a detailed and comprehensive guide for conducting experiments in this field. The protocol outlines the steps required to prepare the actinomycete extract, conduct the in vitro bioassay using the *Plasmodium falciparum* parasite, and carry out the in vivo bioassay using mice infected with *Plasmodium berghei.*

The protocol is designed to ensure accuracy, reproducibility, and validity of the results obtained, with an emphasis on minimizing experimental errors and variability. It includes detailed descriptions of the materials and methods used, such as the preparation of the culture media and the methods for determining the minimum inhibitory concentration (MIC) and the effective dose (ED). Overall, the protocol chapter serves as a valuable resource for researchers interested in conducting studies on the antimalarial activity of actinomycete extracts. By following the detailed steps outlined in this protocol, researchers can obtain reliable and meaningful results that can contribute to the development of new and effective antimalarial drugs.

REFERENCES

Coban, C. 2020. The host targeting effect of chloroquine in malaria. *Current Opinion in Immunology* 66:98–107.

Garcia, L.S. 2010. Malaria. *Clinics in Laboratory Medicine* 30(1):93–129. doi:10.1016/j.cll.2009.10.001.

Hanboonkunupakarn, B., van Der Pluijm, R.W., Hoglund, R., Pukrittayakamee, S., Winterberg, M., Mukaka, M., Waithira, N., Chotivanich, K., Singhasivanon, P., and White, N.J.. 2019. Sequential open-label study of the safety, tolerability, and pharmacokinetic interactions between dihydroartemisinin-piperaquine and mefloquine in healthy Thai adults. *Antimicrobial Agents and Chemotherapy* 63(8):e00060–19.

Milner, D.A., Jr. 2018. Malaria pathogenesis. *Cold Spring Harbor Perspectives in Medicine* 8(1). doi:10.1101/cshperspect.a025569.

Moxon, C.A., Gibbins, M.P., McGuinness, D., Milner, Jr., D.A., and Marti, M. 2020. New Insights into malaria pathogenesis. *Annual Review of Pathology* 15:315–343. doi:10.1146/annurev-pathmechdis-012419-032640.

Peltola, H. 1974. [Malaria]. *Duodecim* 90(16):1085–97.

Petersen, I., Eastman, R., and Lanzer, M. 2011. Drug-resistant malaria: molecular mechanisms and implications for public health. *FEBS Letters* 585(11):1551–1562.

Randall, G., and Seidel, J.S. 1985. Malaria. *Pediatric Clinics of North America* 32(4):893–916. doi: 10.1016/s0031-3955(16)34861-1.

Recht, J., Siqueira, A.M., Monteiro, W.M., Herrera, S.M., Herrera, S., and Lacerda, M.V.G. 2017. Malaria in Brazil, Colombia, Peru and Venezuela: current challenges in malaria control and elimination. *Malaria Journal* 16(1):273. doi:10.1186/s12936-017-1925-6.

Thomé, R., Costa Pinto Lopes, S., Trindade Maranhão Costa, F., and Verinaud, L. 2013. Chloroquine: modes of action of an undervalued drug. *Immunology Letters* 153(1–2):50–57.

Varo, R., Chaccour, C., and Bassat, Q. 2020. Update on malaria. *Medicina Clinica (Barc)* 155(9):395–402. doi: 10.1016/j.medcli.2020.05.010.

Walter, K., and John, C.C. 2022. Malaria. *JAMA* 327(6):597. doi:10.1001/jama.2021.21468.

Weinman, D. 1972. Malaria. *Disease-a-Month* 1–40. doi:10.1016/s0011-5029(72)80005-1.

WHO. 2016. Testing for G6PD deficiency for safe use of primaquine in radical cure of P. vivax and P. ovale: Policy brief. World Health Organization.

WHO. 2021. Malaria's WHO moment. *Nature Biotechnology* 39(11):1322. doi: 10.1038/s41587-021-01129-6.

WHO. 2022. *World Malaria Report 2022*. World Health Organization.

23 Screening of Marine Actinobacteria for Postprandial Antidiabetic Activity In-Vitro and In-Vivo Methods

S.R. Sathish Kumar

23.1 INTRODUCTION

Diabetes mellitus is a chronic metabolic disorder characterized by the elevation of blood glucose level with disturbances of carbohydrate, fat and protein metabolism [1]. Diabetes mellitus is categorized into three types such as type I diabetes, type II diabetes and gestational diabetes. Postprandial hyperglycemia (PPHG) is defined as exaggerated increase of high blood sugar level after a meal. In healthy persons, the pancreas enhances the secretion of insulin to regulate the postprandial blood glucose level [2]. In type II diabetes patients, pancreas fails to secrete insulin or insulin resistance which leads to postprandial hyperglycemia [3]. In postprandial hyperglycemic condition, after food intake, the production of insulin hormone is reduced, and low secretion of glucagon leads to the improper metabolism of glucose in liver and kidney. Hence, there is no sufficient uptake of glucose and subsequently increases the elevation of high PPHG [4,5].

Currently, diabetes mellitus is regulated with the help of several commercial drugs such as biguanides, sulfonylureas, meglitinides and thiazolidinediones. Metformin (biguanides) is one of the common drug which is currently used for treating type II diabetes and helps to control the secretion of hepatic glucose, enhances the insulin sensitivity and glucose absorption by the activation of GLUT (glucose transporters). The modern therapeutic application of antidiabetic includes effective drugs, both oral and injectable forms. α-Glucosidase inhibitor (AGI) is one of the main therapeutic mode for the treatment of type II diabetes. In diabetic patients, AGI helps to delay the metabolism of carbohydrates and regulates the elevation of high postprandial blood glucose level. Acarbose, voglibose and miglitol are some of the commercially available AGI which is currently used to treat type II diabetes. AGI prevents the hypoglycemic condition and also controls the micro- and macrovascular diseases. The most common side effects of AGI are gastrointestinal irritation, flatulence, diarrhea and abdominal discomfort [6].

Marine environment covers almost 70% of the earth surface and contains different biodiversity ecosystems [7,8]. The microbial diversity of marine environment is not completely explored, and organisms present in these environments are extremely rich sources of bioactive compounds. The characteristics of marine actinobacteria are completely different from terrestrial actinobacteria, and therefore, the bioactive compounds from marine actinobacteria are unique in their structure and properties [9–11]. The microbiologically derived enzyme inhibitors are widely used in industries for various applications. However, enzyme inhibitors from marine microorganisms were less studied [12]. The secondary metabolites from actinobacteria may act as an enzyme inhibitor for several diseases [13,14]. Karthik et al. reported that the peptide compound was identified from *Streptomyces* sp LK3 (JF710608) isolated from marine sediments [15]. The purified peptide compound showed

DOI: 10.1201/9781003398400-23

significant inhibition of protease enzyme and showed antiplasmodial activity. So, the screening of antidiabetic compounds from marine actinobacteria will provide a new platform for the development of antidiabetic molecules.

23.2 METHODS

23.2.1 Extraction of Bioactive Compounds from Marine Actinobacteria

1. Potent actinobacterial isolates are inoculated into 50 mL of production media (SS broth) in 100 mL Erlenmeyer flasks separately and incubated at 28°C for 7 days on a rotary shaker (REMI) at 100 rpm.
2. The supernatant is treated with different solvents like petroleum ether, ethyl acetate, n-butanol, chloroform and hexane in the ratio of 1:1 in a separating funnel.
3. After extraction, the collected samples need to be dried using a rotary vacuum evaporator.
4. The dried sample can be stored in a refrigerator and used as test sample for further analysis.

23.2.2 Primary Screening of Actinobacteria for α-Glucosidase Inhibitor Activity and α-Amylase Inhibitor Activity

23.2.2.1 Yeast α-Glucosidase Inhibitor Activity

1. An aliquot of 10 µL yeast α-glucosidase (1 U/mL in 0.1 M potassium phosphate buffer) is added to 50 µL of phosphate buffer (50 mM; pH 6.8) and mixed with 20 µL of varying concentration of actinobacterial crude extract or solvent extract (20, 40, 60, 80 and 100 µg/mL) separately in a 96-well microtiter plate.
2. The test sample mixture needs to be preincubated for 5 minutes at 37°C and 20 µL of 1 mM p-nitrophenyl α-D-glucopyranoside (PNPG) is added to the mixture as substrate.
3. After incubation at 37°C for 30 minutes, 50 µL of 0.1 M sodium carbonate solution was added to stop the reaction.
4. Acarbose drug can be used as a positive control and distilled water as negative control. The enzyme inhibition activity was estimated by recording the absorbance at 405 nm in a microtiter plate reader (Bio-TEK, USA).
5. The percentage of inhibition was calculated as % Inhibition = [(AC − AS)/AC] ×100, where AC and AS are the absorbance of the control and test correspondingly [16].

23.2.2.2 α-Amylase Inhibitor Activity

1. This test is carried out by taking 500 µL of different concentrations of actinobacterial crude extracts (20, 40, 60, 80 and 100 µg/mL) and 500 µL of 0.02 M sodium phosphate buffer (pH 6.9) containing α-amylase solution (0.5 mg/mL) at 25°C for 10 minutes.
2. After pre-incubation, 500 µL of 1% starch solution in 0.02 M sodium phosphate buffer (pH 6.9) is added to each tube at 5-second intervals.
3. The reaction mixtures are then incubated at 25°C for 10 minutes, and the reaction is stopped by adding 1.0 mL of dinitrosalicylic acid (DNS).
4. The test tubes were then incubated in a boiling water bath for 5 minutes and cooled to room temperature.
5. The reaction mixture was diluted with 10 mL of distilled water and absorbance was measured at 540 nm in a UV-vis spectrophotometer (U-2800, Japan). The known porcine pancreatic α-amylase inhibitor acarbose was used a positive control [17].

23.2.2.3 Mammalian α-Glucosidase Inhibitor Activity

1. An aliquot of 10 μL mammalian α-glucosidase (1 U/mL in 0.1 M potassium phosphate buffer) was added to 50 μL of 50 mM phosphate buffer (pH 6.8) and mixed with 20 μL of varying concentration of actinobacterial crude extract or solvent extract (20, 40, 60, 80 and 100 μg/mL) separately in a 96 well microtiter plate.
2. The test sample mixture was preincubated for 5 minutes at 37°C and 20 μL of 1 mM PNPG was added to the mixture as substrate.
3. After incubating at 37°C for 30 minutes, 50 μL of 0.1 M sodium carbonate solution was added to stop the reaction. Acarbose drug can be used as a positive control and distilled water as negative control.
4. The enzyme inhibition activity was estimated by recording the absorbance at 405 nm in a microtiter plate reader (Bio-TEK, USA). The percentage of inhibition was calculated as % Inhibition = [(AC − AS)/AC] × 100, where AC and AS are the absorbance of the control and test, respectivley [18].

23.3 IN-VIVO POSTPRANDIAL HYPERGLYCEMIC ASSAY

23.3.1 Experimental Animals

In this in vivo study, adult male albino Wistar rats were utilized to assess the postprandial antidiabetic activity of marine actinobacterial extracts. All test animals received approval from the Institutional Animal Ethics Committee.

23.3.2 Induction of Diabetes

1. The adult male albino Wistar rats should be subjected to be fasting for a period of 16 hours; after fasting, the animals need to be induced for diabetes using streptozotocin (STZ). STZ is prepared by using fresh citrate buffer (0.1 M, pH 4.5) and administrated intraperitoneally at a concentration of 45 mg/kg body weight of each animal.
2. After 3 days of animal growth, the fasting blood glucose level ranging more than 250 mg/dL is considered as diabetic.

23.3.3 Antihyperglycemic Activity of Marine Actinobacterial Extract in Sucrose-Loaded Normal and Diabetic Rats

1. To access the antihyperglycemic activity of marine actinobacterial extract or solvent extract, a total of 24 normal rats need to be divided into four groups in sucrose-loaded normal rats.
2. Group I was administrated with sucrose (2 g/kg body weight) alone and considered as control. Group II was loaded with sucrose (2 g/kg body weight) and acarbose (5 mg/kg body weight). 3. Group III was co-administrated with sucrose (2 g/kg body weight) and actinobacterial extract (300 mg/kg body weight) and group IV was treated with sucrose (2 g/kg body weight) and double the concentration of actinobacterial extract (600 mg/kg body weight).
4. After sucrose administration in animals, the blood glucose level was monitored on 0th, 30th, 60th and 120th minute of the experiment using a glucometer (Model: One touch Horizon™).
5. The variation in blood glucose level after oral sucrose administration was measured and denoted as delta blood glucose. The same experimental animal model was followed for sucrose-loaded diabetic rats [19].

23.3.4 Antihyperglycemic Activity of Marine Actinobacterial Extract in Glucose-Loaded Normal Rats

1. To access the antihyperglycemic activity of actinobacterial extract by the inhibition of α-glucosidase enzyme, a total of 12 normal rats is used and divided into two groups.
2. The first group is considered as control and loaded with glucose (2 g/kg body weight), while the second group received glucose (2 g/kg body weight) and actinobacterial extract (600 mg/kg body).
3. After treatment, the blood glucose level was measured on 0th, 30th, 60th and 120th minute of the experiment.
4. The change in blood glucose level at the initial stage after glucose treatment was observed and indicated as delta blood glucose [19].

23.3.5 Antihyperglycemic Activity of Marine Actinobacterial Extract in Glucose-Loaded Diabetic Rats

1. To access the antihyperglycemic activity of marine actinobacterial extract by the inhibition α-glucosidase enzyme, a total of 12 diabetic rats were used and divided into two groups.
2. The first group was considered as control and loaded with glucose (2 g/kg body weight), while the second group was treated with glucose (2 g/kg body weight) and marine actinobacterial extract (600 mg/kg body).
3. After treatment, the blood glucose level was measured on 0th, 30th, 60th and 120th minute of the experiment.
4. The change in blood glucose level at the initial stage after glucose treatment was analyzed and denoted as delta blood glucose [19].

23.4 EXPECTED RESULTS

All the isolated actinobacterial colonies need to be screened for the α-glucosidase inhibitor activity along with standard drug molecules as control. In an in-vitro assay, the enzyme inhibition values of actinobacterial extract are needed to compare with the standard drug to understand the efficiency for inhibiting the α-glucosidase enzymes. Similarly, after primary screening, the potent actinobacterial extract or solvent extract need to be studied in an animal model to identify the efficiency for inhibiting α-glucosidase enzyme. In this in-vivo model, the blood glucose level of the animal model treated with actinobacterial extract or solvent extract needs to be compared with the animal group treated with standard drug and control. The animal group treated with actinobacterial extract or solvent extract needs to show a significant reduction in blood glucose level while comparing with control and drug. This change in blood glucose level indicates that the marine actinobacterial isolates have the efficiency in regulating the postprandial hyperglycemia.

REFERENCES

1. Ramachandran A, Snehalatha C, Kapur A, Vijay V, Mohan V, Das AK, Rao, P.V., Yajnik, C.S., Prasanna Kumar, K.M. and Nair, J.D. 2001. Diabetes Epidemiology Study Group in India (DESI). High prevalence of diabetes and impaired glucose tolerance in India: National Urban Diabetes Survey, *Diabetologia*, Vol. 44, pp. 1094–101. DOI: 10.1007/s001250100627
2. Hongcheng Z, Guangxin W, Trust Beta and Jie Dong. 2015. Inhibitory properties of aqueous ethanol extracts of propolis on alpha-glucosidase, *Evid. Based Complement. Altern. Med.*, Vol. 1, pp. 1–7
3. Stumvoll M, Goldstein BJ and Van Haeften T. 2005. Type 2 diabetes: principles of pathogenesis and therapy, *Lancet*, Vol. 365, pp. 1333–1346.

4. Mitrakou A, Kelley D, Mokan M, Veneman, T, Pangburn, T, Reilly J and Gerich J. 1992. Role of reduced suppression of glucose production and diminished early insulin release in impaired glucose tolerance, *N. Engl. J. Med.*, Vol. 326, pp. 22–29.
5. Meyer C, Woerle HJ, Dostou JM, Welle SL and Gerich JE. 2004. Abnormal renal, hepatic, and muscle glucose metabolism following glucose ingestion in type 2 diabetes, *Am. J. Physiol. Endocrinol. Metab.*, Vol. 287, pp. 1049–1056.
6. Heiner L. 2002. Acarbose an update of its therapeutic use in diabetes treatment. *Clin. Drug Investig.*, Vol. 22, pp. 141–156
7. Valli S, Suvathi Sugasini S, Aysha OS, Nirmala P, Vinoth Kumar P and Reena A. 2012. Antimicrobial potential of Actinomycetes species isolated from marine environment, *Asian Pac. J. Trop. Biomed.*, Vol. 2, pp. 469–473.
8. Sathish Kumar SR and Bhaskara Rao KV. 2012. In-vitro antimicrobial activity of marine actinobacteria against multidrug resistance Staphylococcus aureus, *Asian Pac. J. Trop. Biomed.*, Vol. 10, pp. 787–792.
9. Haefner B. 2003. Drugs from the deep: marine natural products as drug candidates, *Drug Discov. Today*, Vol. 8, pp. 536–544.
10. Sathish Kumar SR and Bhaskara Rao KV. 2012. In vitro antagonistic and hemolytic activity of marine Actinobacteria, *J. Adv. Life Sci.*, Vol. 3, pp. 7–12.
11. Kathiresan K, Balagurunathan R and Selvam MM. 2005. Fungicidal activity of marine actinomycetes against phytopathogenic fungi, *Indian J. Biotechnol.*, Vol. 4, pp. 271–276.
12. Imada C. 2004. Enzyme inhibitors of marine microbial origin with pharmaceutical importance, *Mar. Biotechnol.*, Vol. 6, pp. 193–198.
13. Imada C. 2005. Enzyme inhibitors and other bioactive compounds from marine actinomycetes, *Antonie Van Leeuwenhoek*, Vol. 87, pp. 59–63.
14. Martin JF and Demain AL. 1980 Control of antibiotic biosynthesis, *Microbiol. Rev.*, Vol. 44, pp. 230–51.
15. Karthik L, Kumar G, Keswani T, Bhattacharyya A, Reddy BP and Bhaskara Rao KV. 2013. Marine actinobacterial mediated gold nanoparticles synthesis and their antimalarial activity, *Nanomed.: Nanotechnol. Biol. Med.*, Vol. 9, pp. 951–960.
16. Pistia-Brueggeman G and Hollingsworth RI. 2001. A preparation and screening strategy for glycosidase inhibitors, *Tetrahedron*, Vol. 57, pp. 8773–8778.
17. Suthindhiran KR, Jayasri MA and Kannabiran K. 2009. α-Glucosidase and α-amylase inhibitory activity of Micromonospora sp. VITSDK3 (EU551238), *Int. J. Integr. Biol.*, Vol. 6, pp. 115–120.
18. Jo SH, Ka EH, Lee HS, Apostolidis E, Jang HD and Kwon YI. 2009. Comparison of antioxidant potential and rat intestinal α-glucosidases inhibitory activities of quercetin, rutin, and isoquercetin, *Int. J. Appl. Res.*, Vol. 2, pp. 52–60.
19. Mohamed Sham Shihabudeen H, Priscilla DH and Thirumurugan K. 2011. Cinnamon extract inhibits a-glucosidase activity and dampens postprandial glucose excursion in diabetic rats, *Nutr. Metab.*, Vol. 46, pp. 1–11.

24 Exploring the Larvicidal Activity of Actinomycetes Extracts

A. Vishnu Kirthi, L. Karthik, Tarun Keswani, Anirban Sengupta, Shilpi Garg, and Chitrakshi Consul

24.1 INTRODUCTION

Mosquitoes transmit deadly diseases, such as malaria, dengue fever, Zika virus, and Chikungunya virus, and have become resistant to traditional insecticides, requiring more effective control methods. Actinomyces is a genus of bacteria that is known to produce a variety of bioactive compounds with potential therapeutic and insecticidal properties (Kaushik et al. 2022; Subramani and Aalbersberg 2012). Actinomycetes extract has been extensively studied for its larvicidal activity against a range of insect pests (Tanvir, Sajid, and Hasnain 2014; Loganathan et al. 2013; Kirthi and Karthik 2018). Actinomycetes are soil-dwelling bacteria that produce a variety of secondary metabolites with diverse biological activities. Their metabolites have been shown to exhibit potent larvicidal activity against mosquitoes, ticks, and other insect pests. Here are some scientific reasons for the larvicidal activity of Actinomyces extract:

1. **Production of insecticidal metabolites**: Actinomyces has been shown to produce a range of secondary metabolites that exhibit insecticidal activity, such as polyketides, terpenoids, and alkaloids. These compounds can disrupt insect physiology and development, leading to larval mortality (Arasu et al. 2013).
2. **Disruption of insect gut microbiota**: Actinomyces can also colonize the gut of insect larvae and produce antimicrobial compounds that disrupt the gut microbiota, leading to larval death (Sikorowski and Lawrence 1994).
3. **Activation of insect immune response**: Actinomyces can trigger the immune response of insects, leading to the production of antimicrobial peptides that can kill the larvae (Couceiro et al. 2016; Tomilova et al. 2016).
4. **Selective toxicity**: Actinomyces extracts have been shown to exhibit selective toxicity against insect larvae, with little or no toxicity to non-target organisms such as mammals and birds (Ho et al. 2020; Kim et al. 2020).

The larvicidal activity of actinomycetes is mainly attributed to the production of secondary metabolites such as polyketides, alkaloids, and terpenoids. These compounds have been shown to disrupt the development and physiology of insects, leading to mortality. Additionally, actinomycetes extracts can modulate the gut microbiota of insect larvae, which can lead to a reduction in larval survival. Furthermore, actinomycetes extracts have been shown to have a high degree of selectivity, with low toxicity towards non-target organisms such as mammals and birds. Thus, actinomycetes have emerged as a promising source of natural larvicides that can provide a sustainable and eco-friendly alternative to chemical insecticides for controlling insect pests that transmit diseases to humans and animals. Further research is needed to explore the full potential of actinomycetes extracts for larvicidal activity and to develop effective formulations for field application.

DOI: 10.1201/9781003398400-24

24.2 MATERIALS REQUIRED

Actinomycetes cultures, nutrient agar plates, ethanol or methanol, sterile distilled water, centrifuge tubes, rotary evaporator, mosquito larvae, glass beakers, micropipettes, sterile tips, incubator, microscope slides and coverslips, stereomicroscope, analytical balance, and software for statistical analysis like GraphPad.

24.2.1 PROCEDURE

Collection of Actinomycetes cultures: Actinomycetes cultures can be obtained from soil samples, compost, or plant roots. Isolate and purify the cultures before proceeding.

Preparation of Actinomycetes extracts: Grow actinomycetes colonies on nutrient agar plates. Harvest and wash the colonies with sterile distilled water. Transfer the colonies to centrifuge tubes and mix with ethanol or methanol. Centrifuge the mixture at 5,000 rpm for 10 minutes. Collect and filter the supernatant and then concentrate it using a rotary evaporator. Weigh and store the extract in a sterile container.

Preparation of mosquito larvae: Collect mosquito larvae from stagnant water bodies, such as ponds or tanks. Identify the larvae and transfer them to glass beakers containing 50 mL of sterile distilled water. Starve the larvae for 24 hours before the experiment.

Larvicidal activity assay: Test the actinomycetes extracts for their potential larvicidal activity against mosquito larvae using the following steps:

a. Dilute the extracts with sterile distilled water to prepare different concentrations (10, 20, 50, 100, and 200 ppm).
b. Transfer 25 larvae to each glass beaker containing 50 mL of sterile distilled water.
c. Add different concentrations of extracts to each glass beaker and keep a control beaker without any extracts.
d. Incubate the beakers at 27°C ± 1°C for 24 hours.
e. Count the number of dead and alive larvae under a stereomicroscope and calculate the percentage of larval mortality.
f. Determine the LC50 value of each extract using probit analysis.

Data analysis: Calculate the percentage of larval mortality for each concentration of actinomycetes extracts. Plot the data on a graph and determine the concentration required to kill 50% of the larvae (LC50) using probit analysis. Analyze the data using analysis of variance (ANOVA) and use Tukey's honestly significant difference (HSD) test to determine the significance of differences between the means. Consider results statistically significant at $p < 0.05$.

Determination of mode of action: To determine the mode of action of the most effective extract, perform a biochemical analysis. Test the extract for its effect on the acetylcholinesterase (AChE) activity of mosquito larvae. Measure the AChE activity using spectrophotometric analysis.

Confirmation of bioactive compounds: Confirm the presence of bioactive compounds in the extract using chromatographic techniques, such as high-performance liquid chromatography.

24.2.1.1 Data Analysis

The data analysis part of larvicidal activity using actinomycetes extract involves several steps to interpret and draw conclusions from the collected data.

a. The mortality rate of the mosquito larvae at each concentration of the extract was calculated. This is typically expressed as a percentage of the total number of larvae used.
b. Next, the data were analyzed using statistical software such as SPSS or GraphPad Prism. One of the most commonly used statistical methods for analyzing larvicidal data

is probit analysis. Probit analysis is a statistical method that uses a linear regression model to estimate the concentration required to kill a certain percentage of the larvae (LC50 and LC90). The analysis determines the dose-response relationship between the concentration of the extract and the mortality rate of the larvae.

c. Once the LC50 and LC90 values are determined, they can be used to compare the potency of the extract with other larvicides or to evaluate the potential of the extract for mosquito control.
d. In addition to probit analysis, other statistical tests such as ANOVA can be used to determine the significance of the results. ANOVA is used to test whether there is a significant difference between the mortality rates of the larvae at different concentrations of the extract.
e. And the data analysis part of larvicidal activity using actinomycetes extract involves statistical methods such as probit analysis and ANOVA to interpret and draw conclusions from the collected data. These methods can help to evaluate the potential of actinomycetes extract as a source of larvicidal compounds for mosquito control.

24.3 CONCLUSION

Mosquitoes are vectors of several deadly diseases that affect millions of people worldwide, including malaria, dengue fever, Zika virus, and Chikungunya virus. Conventional chemical insecticides have been widely used for controlling mosquito populations, but the development of resistance has rendered these methods ineffective. The search for alternative and sustainable control methods has led to the investigation of natural sources of larvicidal agents, such as actinomycetes. Actinomycetes are a group of soil bacteria known for their ability to produce a wide range of bioactive compounds, including antibiotics and insecticides. This protocol provides a detailed guide for screening actinomycetes extracts for their potential larvicidal activity against mosquito larvae.

REFERENCES

Arasu, Mariadhas Valan, Naif Abdullah Al-Dhabi, Valsalam Saritha, Veeramuthu Duraipandiyan, Chinnasamy Muthukumar, and Sun-Ju Kim. 2013.'Antifeedant, larvicidal and growth inhibitory bioactivities of novel polyketide metabolite isolated from Streptomyces sp. AP-123 against Helicoverpa armigera and Spodoptera litura, *BMC Microbiology*, 13: 1–6.

Couceiro, Joel da Cruz, Wesler Luiz Marcelino, Karina Dias Amaral, Lailla Cristina Gandra, Danival José de Souza, and Terezinha Maria Castro Della Lucia. 2016. Effects of entomopathogenic fungi on the mortality and immune system of the leaf-cutting ant A cromyrmex subterraneus subterraneus, *Entomologia Experimentalis et Applicata*, 161: 152–59.

Ho, Louis K, Martin Daniel-Ivad, Swathi P Jeedigunta, Jing Li, Konstantin G Iliadi, Gabrielle L Boulianne, Thomas R Hurd, Craig A Smibert, and Justin R Nodwell. 2020. Chemical entrapment and killing of insects by bacteria, *Nature Communications*, 11: 4608.

Kaushik, Pallavi, Prachi Jain, Sneha Keelka, and Mukesh Kumar Sharma. 2022. Pharmacological properties of bacterial bioactive molecules, *Therapeutic Implications of Natural Bioactive Compounds*, 3: 262–75.

Kim, Jong Hoon, Jae Young Choi, Dong Hwan Park, Dong-Jin Park, Min Gu Park, So Young Kim, Yoon Jung Ju, Jun Young Kim, Minghui Wang, and Chang-Jin Kim. 2020. Isolation and characterization of the insect growth regulatory substances from actinomycetes, *Comparative Biochemistry and Physiology Part C: Toxicology & Pharmacology*, 228: 108651.

Kirthi, Arivarasan Vishnu, and Loganathan Karthik. 2018. Inhibition of Mosquito Vectors of Malaria and Filariasis Using Marine Microorganisms and Their Associated Compounds. In: Brij Kishore Tyagi, Dharumadurai Dhanasekaran (Eds.), *Microbial Control of Vector-Borne Diseases* (CRC Press), pp. 27–36.

Loganathan, Karthik, Gaurav Kumar, Arivarasan Vishnu Kirthi, Kokati Venkata Bhaskara Rao, and Abdul Abdul Rahuman. 2013. Entomopathogenic marine actinomycetes as potential and low-cost biocontrol agents against bloodsucking arthropods, *Parasitology Research*, 112: 3951–59.

Sikorowski, PP, and AM Lawrence. 1994. Microbial contamination and insect rearing, *American Entomologist*, 40: 240–53.
Subramani, Ramesh, and William Aalbersberg. 2012. Marine actinomycetes: an ongoing source of novel bioactive metabolites, *Microbiological Research*, 167: 571–80.
Tanvir, Rabia, Imran Sajid, and Shahida Hasnain. 2014. Larvicidal potential of Asteraceae family endophytic actinomycetes against Culex quinquefasciatus mosquito larvae, *Natural Product Research*, 28: 2048–52.
Tomilova, Oksana G, Vadim Yu Kryukov, Bahytzhan A Duisembekov, Olga N Yaroslavtseva, Maksim V Tyurin, Natalia A Kryukova, Valery Skorokhod, Ivan M Dubovskiy, and Viktor V Glupov. 2016. Immune-physiological aspects of synergy between avermectins and the entomopathogenic fungus Metarhizium robertsii in Colorado potato beetle larvae, *Journal of Invertebrate Pathology*, 140: 8–15.

25 Evaluation of Actinobacteria's Antibiofilm Properties Using Qualitative and Quantitative Methods

Nisha, Gaurav Pant, and Gaurav Kumar

25.1 INTRODUCTION

A biofilm is a complex microbiome structure that adheres to a surface and comprises a range of bacterial colonies or a single type of cell in a group. Biofilms can also contain a single type of cell in a group. These cells encase themselves in a gooey extracellular matrix composed of extracellular polymeric materials (EPS), which are often made up of microbial extracellular DNA, polysaccharides, and proteins. The formation of bacterial biofilms is a significant challenge in a wide variety of industries, including those dealing with food and medicines, water and petroleum, amongst others. A wide variety of medical equipment, such as catheters, heart valves, prosthesis, surgical pins, and so on are all susceptible to the formation of bacterial biofilms, which can result in infections. In addition, bacteria that form biofilms are responsible for about one million cases of catheter-associated urinary tract infections (CAUTI) each year as well as around 80% of all microbial infections that occur in the body. Because the microbes that are present in the biofilms are particularly good at resisting a wide variety of antibiotics, there is an urgent need for the development of novel alternative methods and substances (Balagurunathan et al., 2015; Sharma et al., 2019).

It is commonly recognised that bioactive compounds produced by actinomycete bacteria are used in the pharmaceutical industry, as well as the agricultural industry, and the food industry. It has been demonstrated that this particular strain of bacteria possesses the ability to produce active antibiofilm chemicals. For instance, *Streptomyces albus* and *Streptomyces akiyosheinsis* exhibited antibiofilm action against *Staphylococcus aureus* and *Vibrio harveyi* (Miller and Waturangi, 2022). In addition, the chemicals that are produced by actinomycetes strains have shown an inhibitory and destructive effect against the formation of biofilm in aquatic pathogens such as *V. harveyi*, *Aeromonas hydrophila*, and *Streptococcus agalactiae* (Raissa et al., 2020).

Owing to the antibiofilm potential and wide acceptability of Actinobacteria and its products in various industries, they can be excessively explored for the antibiofilm activity against a variety of organisms. In light of this, the purpose of this chapter is to cover a variety of qualitative and quantitative approaches that have been utilised in the course of research on the antibiofilm activity of Actinobacteria.

25.2 MATERIALS AND METHODS

25.3 BIOFILM-FORMING TEST MICROORGANISMS

Some gram-positive and gram-negative bacteria capable of forming biofilms, i.e., biofilm-forming *Pseudomonas aeruginosa, Pseudomonas fluorescens, Escherichia coli, Vibrio cholerae, Staphylococcus epidermidis, Staphylococcus aureus*, and any other can be used as the test organisms.

DOI: 10.1201/9781003398400-25

25.4 PREREQUISITES

A. Standard biofilm generating assays, such as the Congo red agar (CAR) method and the microtiter plate test (MtP), should be utilised in order to verify whether or not the organisms being tested are capable of producing biofilm (Christensen et al., 1985; Christensen, 1989).
B. The agar well diffusion assay and the lowest inhibitory concentration (MIC) assay are the two tests that should be used to evaluate the antimicrobial activity of an actinobacteria extract or substance. As later, antibiofilm treatments will be carried out with consideration to MIC (Selvameenal et al., 2009; CLSI, 2015).

25.5 SCREENING OF BIOFILM INHIBITION PROPERTY OF ACTINOBACTERIA

Microtiter plate assay (Raissa et al., 202)

- Inoculate the nutrient broth with the organisms to be tested and then incubate it at 37 degrees Celsius for 24 hours.
- In the 96-well polystyrene plates, simultaneously add 200μl of test culture suspension and actinobacterial extract or substance.
- To create a positive control, use the bacterial suspension, and to create a negative control, use nutritional broth.
- Incubate the microtiter plate at 37°C for 24 hours.
- After the incubation period has ended, the planktonic cells and media on the plates should be discarded.
- After giving the adhered biofilm cells a double washing with sterile normal saline, allow them to dry off naturally.
- After that, stain the cell with crystal violet at a concentration of 0.2% (w/v) for 2 minutes.
- Throw away the crystal violet and repeat the process of rinsing the cells twice with sterile normal saline before drying them. This will remove the stain from the cells.
- Add 200 μL of decolouriser (1N HCL) and resuspend to dissolve the stained cells.
- Transfer the decoloriser into fresh plates containing 96 wells at the same time.
- Using a microplate reader set to 600 nm, one may determine the absorbance of each individual well.

Tube assay: (James and Kumar, 2022)

- Inoculate the test tubes with the bacterial cultures after preparing two test tubes of nutrient broth with a pH of 7 and then sterilising the test tubes.
- The first tube is labelled as the control, and the second tube is enriched with Actinobacteria extract and other Actinobacteria-related chemicals and labelled as the test.
- In a shaking incubator set to 120 revolutions per minute, incubate both the tubes at 37°C for 48 hours.
- As soon as the incubation period is through, the medium from the test tubes is removed.
- Clean the inside walls of both tubes with sterile normal saline, which will remove any unattached particles, and then allow the tubes to air dry after the process is complete.
- After that, put 5 mL of crystal violet stain with a weight-to-volume ratio of 0.2% in each test tube and let it sit for 1 minute.
- Crystal violet stain should be drained from the tubes and then they should be washed once more with regular saline to eliminate any remaining stain.
- Observe the results and take the picture for the same.

Inhibition of biofilm on various substrates (Kaur 2021)

- Pick two or three distinct kinds of base materials, such as glass, metal, and plastic.
- In order to eliminate any possibility of contamination, all of the plastic, glass, and metal surfaces should be cut into a rectangular shape and then autoclaved.
- Take 100 mL of nutrient broth and divide it between two conical flasks with a capacity of 250 mL each. In one flask, add the compounds that were extracted from the Actinobacteria at a concentration equal to half of their minimal inhibitory concentration (MIC), while the other flask serves as the control.
- The fresh bacterial cultures should be inoculated into both of the flasks at this point.
- A glass substrate is positioned at a slanted angle inside each flask before they are placed in an incubator at 37°C for a period of 48 hours. Carry out the previous steps with the other surfaces and cultures.
- After incubation, recover all of the base materials, let them to air dry, and then stain them using the simple staining approach (0.2% (w/v) crystal violet, followed by washing and air drying).
- After that, take images of the same thing on a white backdrop in order to visualise the formation of biofilm.

25.6 INHIBITION OF VARIOUS STAGES OF BIOFILM

Inhibition of early biofilm formation (Ferreira et al., 2017)

- Take the flat bottom microtiter plate and place 200 µL of the freshly created bacterial culture in each of the six wells.
- In the first three wells, add Actinobacteria compound/extract at half of MIC concentration while the rest three wells will remain untreated and used as control.
- In order to achieve initial adhesion, incubate the plates at 37°C for 2 hours.
- After the incubation period is complete, remove any planktonic cells from the wells by washing them with sterile normal saline.
- After that, pour fresh nutrient broth into each well and place the plates in an incubator at 37°C for 48 hours.
- For the purpose of quantification, wash the well twice with sterile normal saline, and then stain it for 2 minutes with 0.2% (w/v) crystal violet.
- Again, wash the wells twice with sterile normal saline and then add 200 µL of decoloriser to each well (1N HCL).
- With a microtiter plate reader, determine the absorbance of the microtiter plate at a wavelength of 600 nm.

Inhibition of microcolony (Ferreira et al., 2017)

- Take the flat bottom microtiter plate and place 200 µL of the freshly created bacterial culture in each of the six wells.
- In order to achieve initial adhesion, the microtiter plate should be incubated at 37°C for 2 hours.
- After cleaning the wells with normal saline, treat them for 1 minute with a Actinobacteria compound/extract at their MIC.
- Wash the wells with sterile normal saline and then add 200 µL of nutritional broth to each well.
- After that, incubate the plates at 37°C for 48 hours.
- Quantification of the biofilm is carried out in the same manner as was explained in the prior section.

Inhibition of mature preformed biofilm (Ferreira et al., 2017)

- Take the flat bottom microtiter plate and place 200 μL of the freshly created bacterial culture in each of the six wells.
- In order to achieve initial adhesion, the microtiter plate should be incubated at 37°C for 2 hours.
- After cleaning the wells with sterile normal saline, fill them with freshly prepared nutrient broth.
- Further incubate the plates at 37°C for a period of 48 hours.
- After that, the medium is withdrawn from the wells, and it is washed in sterile normal saline in order to get rid of any loose cells.
- Pour a solution containing 100 μL of Actinobacteria compound/extract into each well at MIC and then incubate the plates for 1 minute.
- The solution is withdrawn by aspiration after the incubation step.
- After washing the plate with sterile normal saline, fill the wells with nutritious broth that has been freshly made.
- Do the entire procedure once again after a break of 12 hours for the next 48 hours.
- To perform quantification, follow the technique outlined in the prior section once more.

25.7 OBSERVATIONS AND CALCULATIONS

25.8 TISSUE CULTURE PLATE ASSAY

The tissue culture plate assay is a standard method for determining the degree to which a sample has showed antibiofilm activity. At the conclusion of the process, the optical density (OD) of both the treated and the untreated wells is measured, and the results are utilised to quantify the biofilm inhibition. It is determined to be an antibiofilm agent if the OD values of the control samples are higher than those of the samples that were treated (Table 25.1).

TABLE 25.1
Sample Observations

Treatment	Sample Code	OD
Untreated well	UT	1.2
Treated well	Actinobacteria sample 1	0.6
Treated well	Actinobacteria sample 2	0.4

25.9 FORMULA

Biofilm inhibition (%)=

[1–(OD_{600} of cells treated with sample/OD_{600} of the untreated control)×100]

25.10 CALCULATIONS

Actinobacteria sample 1 biofilm inhibition (%)= [1–(0.6/1.2)×100] = 50%
Actinobacteria sample 2 biofilm inhibition (%)= [1–(0.4/1.2)×100] = 66.6%

25.11 TUBE ASSAY

The tube method is a qualitative assay for determining whether or not the substances have an antibiofilm characteristic. During the experiment, a control tube that has not been treated is contrasted with a tube that has been treated using visible observations. The presence of a visible film lining the walls of the control tube, in contrast to significantly less or no film lining in the treated tube, is evidence that the sample possesses antibiofilm properties (Figure 25.1).

25.12 INHABITATION OF BIOFILMS ON VARIOUS SUBSTRATES

According to the published research, the production of biofilm can differ across different surfaces; hence, it is essential to investigate the antibiofilm property across various surfaces. This approach, like the tube assay, is a qualitative assay for determining whether or not the samples have an antibiofilm property; however, it is performed on a variety of substrates, such as glass, plastic, metal, and so on. During the test, a control surface that has not been treated is contrasted with a surface that has been treated using visible observations. The presence of visible film lining the surface of the control, as opposed to significantly less or no film lining on the surface of the treated area, is evidence that the sample possesses antibiofilm properties (Figure 25.2).

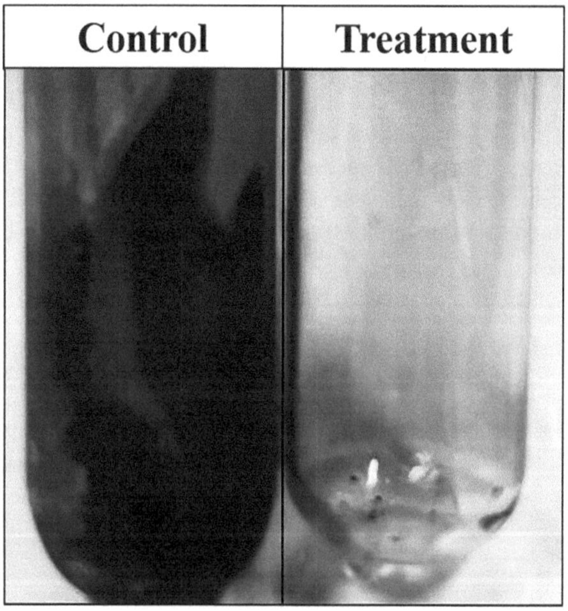

FIGURE 25.1 Antibiofilm activity of Actinobacteria extract/compounds in test tube assay.

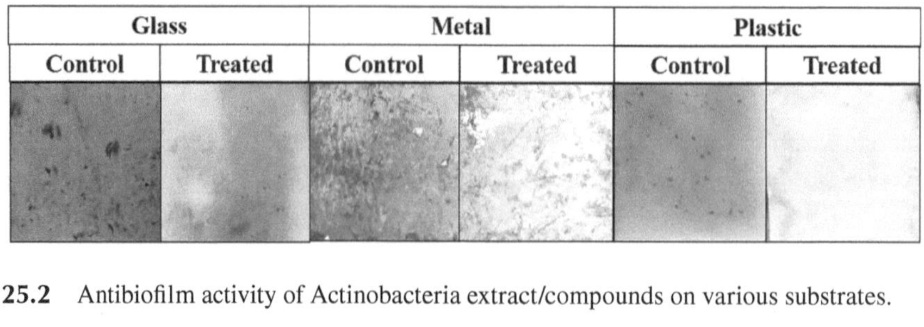

FIGURE 25.2 Antibiofilm activity of Actinobacteria extract/compounds on various substrates.

25.12.1 Inhabitation of Various Stages of Biofilms

The process of biofilm creation is comprised of a sequence of processes, the first of which is the attachment of microorganisms to a surface. This attachment results in the establishment of a micro-colony, which in turn leads to the formation of early biofilm and mature biofilm. Therefore, to establish a sample as antibiofilm in nature, screening of the sample for the various stages of biofilm development is of utmost importance. The observations and the analysis of the results are carried out in a manner that is analogous to that of a tissue culture plate test (Table 25.2).

TABLE 25.2
Sample Observations

Various Stages of Biofilms	Treatment	OD
Inhibition of early biofilm formation	Untreated well	0.9
	Treated well	0.5
Inhibition of microcolony biofilm	Untreated well	1.1
	Treated well	0.7
Inhibition of mature preformed biofilm	Untreated well	1.4
	Treated well	0.8

25.13 FORMULA

Inhabitation of various stages of biofilms (%)=

$[1-(OD_{600}$ of cells treated with sample/OD_{600} of the untreated control)×100]

25.14 CALCULATIONS

Inhibition of early biofilm formation (%)= [1−(0.5/0.9)×100] = 54%
Inhibition of microcolony biofilm (%)= [1−(0.7/1.1)×100] =62%
Inhibition of mature preformed biofilm (%)= [1−(0.8/1.4)×100] = 56%

25.15 CONCLUSION

The ability of bacteria to create biofilms allows them to survive even in environments that are hostile to their growth, such as those with high levels of toxic chemicals, antimicrobial compounds, pH, high salt, pressure, and temperature. These biofilms are to blame for a variety of issues that arise in the fields of medicine, petrochemicals, agriculture, food, and water, as well as maritime transportation. At this time, the only physical and chemical methods of removing biofilm that are available either require complex instrumentation or frequently result in an excessive amount of water being wasted and the discharge of toxic effluents into the environment, or they are less effective. In recent years, it has come to light that Actinobacteria exhibit potent antibiofilm activity against a wide range of microbial biofilms. As a result, this chapter has described several in vitro methods for the qualitative and quantitative methods of determining the level of antibiofilm activity exhibited by Actinobacteria. Qualitative methods provide simple approaches for the antibiofilm activity; however, they do not quantify the degree of activity. Each method has its own set of pros and weaknesses; quantitative methods, on the other hand, involve a modest amount of complexity and also contribute to the determination of the level of antibiofilm.

ACKNOWLEDGEMENTS

The authors would like to express their gratitude to Lovely Professional University in Phagwara, Punjab, India, for supplying them with the resources necessary to conduct research in the applicable topic and produce this chapter.

CONFLICT OF INTEREST

The authors have no conflict of interest to declare that is relevant to the content of this protocol chapter.

REFERENCES

Balagurunathan, R., Sangeetha, M., & Shanmugasundaram, T. (2015). Anti-biofilm compounds from Actinobacteria. *ENVIS Newsletter*, *13*(3), 2–5.

Christensen, B. E. (1989). The role of extracellular polysaccharides in biofilms. *Journal of Biotechnology*, *10*(3–4), 181–202.

Christensen, G. D., Simpson, W. A., Younger, J. J., Baddour, L. M., Barrett, F. F., Melton, D. M., & Beachey, E. H. (1985). Adherence of coagulase-negative staphylococci to plastic tissue culture plates: a quantitative model for the adherence of staphylococci to medical devices. *Journal of Clinical Microbiology*, *22*(6), 996–1006.

Clinical and Laboratory Standards Institute. (2015). M07–A10. *Methods for dilution antimicrobial susceptibility tests for bacteria that grow aerobically; approved standard.*

Ferreira, G. L. S., Rosalen, P. L., Peixoto, L. R., Pérez, A. L. A. D. L., Carlo, F. G. D. C., Castellano, L. R. C.,... & Castro, R. D. D. (2017). Antibiofilm activity and mechanism of action of the disinfectant chloramine T on Candida spp., and its toxicity against human cells. *Molecules, 22*(9), 1527.

James, C., & Kumar, G. (2022). Biosynthesis of gold nanoparticles and inhibition of various stages of bacterial biofilms formed by drug-resistant *Aeromonas hydrophila, Escherichia coli, and Klebsiella pneumoniae. Archives of Microbiology*, *204*(12), 719.

Kaur, T., Putatunda, C., Vyas, A., Kumar, G. (2021) Zinc oxide nanoparticles inhibit bacterial biofilm formation via altering cell membrane permeability. *Preparative Biochemistry & Biotechnology, 51*, 309–319.

Miller, T., & Waturangi, D. E. (2022). Antibiofilm properties of bioactive compounds from Actinomycetes against foodborne and fish pathogens. *Scientific Reports*, *12*(1), 18614.

Raissa, G., Waturangi, D. E., & Wahjuningrum, D. (2020). Screening of antibiofilm and anti-quorum sensing activty of Actinomycetes isolates extracts against aquaculture pathogenic bacteria. *BMC Microbiology*, *20*, 1–10.

Selvameenal, L., Radhakrishnan, M., & Balagurunathan, R. (2009). Antibiotic pigment from desert soil actinomycetes; biological activity, purification and chemical screening. *Indian Journal of Pharmaceutical Sciences*, *71*(5), 499.

Sharma, D., Misba, L., & Khan, A. U. (2019). Antibiotics versus biofilm: an emerging battleground in microbial communities. *Antimicrobial Resistance & Infection Control*, *8*(1), 1–10.

26 Microplastic Degradation by Actinobacteria

Uttra Chauhan, Gaurav Kumar, and Gaurav Pant

26.1 INTRODUCTION

Microplastic pollution is a significant environmental issue as its degradation is challenging due to high durability and resistance to natural degradation processes. Microplastics are fragments of plastics size smaller than 5 mm and made up of polypropylene (PP), polystyrene (PS), polyethylene (PE), polyesters and polyethylene terephthalate (PET). They are found in air, soil, water, food, human stool and human placenta (Issac et al. 2021). Both ecosystem and human health are adversely affected as microplastic pollution is increasing day by day. Therefore, researchers are establishing cost-effective remediation techniques for the degradation of microplastics. There are various techniques to detect microplastic. Microbes are capable of degrading microplastics; therefore, recent studies have suggested that Actinobacteria may be a potential solution for microplastic degradation. *Actinobacteria* are filamentous, gram-positive, free-living bacteria and are ubiquitously distributed in both aquatic and terrestrial ecosystems. It produces secondary metabolites which are used in manufacturing of antibiotics and antifungals. It has a wide range of applications; they can act as biocontrol agents, biosurfactants, insecticides, enzyme producers, and synthesis nanoparticles. *Actinobacteria* is capable of degrading polylactic acid (PLA) which undergoes fragmentation to form microplastics. So, to degrade microplastics, we have to target PLA. Microbes from the Pseudonocardiaceae family that breakdown PLA include *Nonomuraea*, *Amycolatopsis*, *Thermopolyspora*, *Saccharothrix*, *Lentzea*, *Kibdelosporangium*, *Actinomadura*, *Thermomonospora* and *Streptoalloteichus*. Low-density polyethylene (LDPE), polystyrene (PS) and polylactic acid (PLA) are degraded by *Streptomyces gougerotti*, *Micromonospora matsumotoense* and *Nocardiopsis prasine* (Oliveira et al., 2022). *Amycolatopsis* HT-32 is the first reported PLA degrader, which was isolated from soil by plate count and clear-zone methods (Pranamuda et al., 1997).

26.2 POLY LACTIC ACID DEGRADATION

Actinobacteria has the ability to degrade PLA. Microorganism-producing enzymes such as lipase (esterase), cutinase and proteases (serine proteases) are essential for PLA biodegradation. *Amycolatopsis* is known to degrade PLA using serine protease which is important for PLA degradation. Extracellular PLA depolymerase is secreted by microorganisms and its synthesis is typically accelerated by a variety of inducers, including gelatin, elastin, silk fibroin and other peptides and amino acids. The depolymerase then attacks the intramolecular ester linkages in PLA, producing oligomers, dimers and monomers as a result. When entering microbial membranes, the low molecular weight chemicals use intercellular enzymes to break down into carbon dioxide, water, or methane (Figure 26.1) (Qi et al. 2017).

26.3 METHODOLOGY

26.3.1 Collection of Sample

Actinomycetes are ubiquitous in nature; therefore, samples can be collected from water, soil and sediments (Figure 26.2).

DOI: 10.1201/9781003398400-26

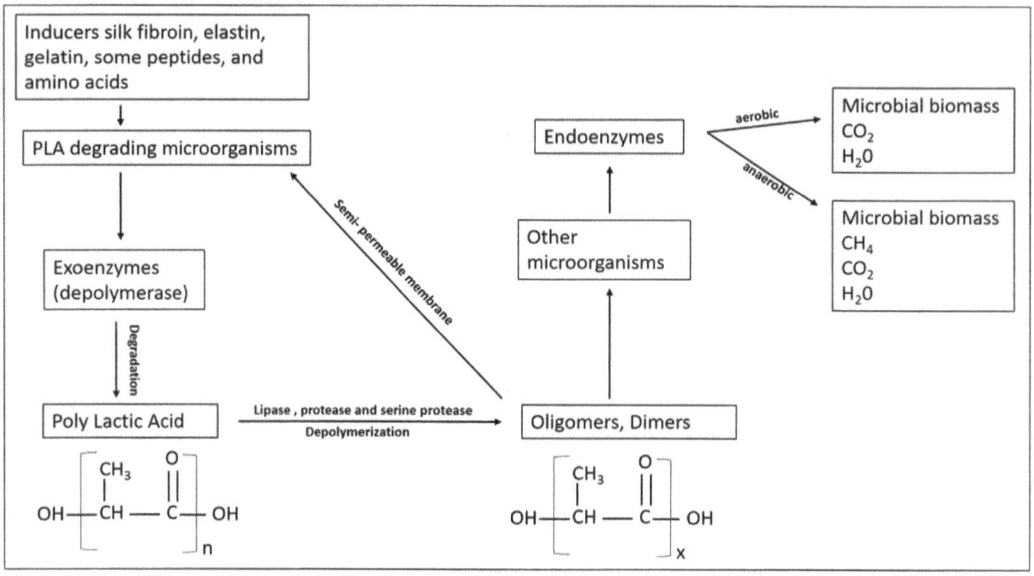

FIGURE 26.1 PLA degradation by microorganism.

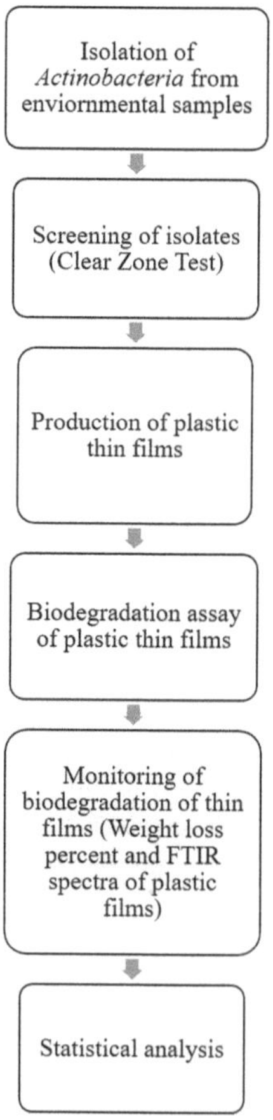

FIGURE 26.2 Procedure of degradation of PLA (microplastic) by *Actinobacteria*.

26.3.2 Isolation and Identification of Actinobacteria

Actinobacteria which is capable of degrading PLA. There are various methods for isolating desirable actinomycetes species from natural habitats (Hayakawa et al., 2018). Actinomycetes can be identified by performing macroscopic and microscopic examination and physiological tests. These techniques are commonly used to differentiate between different species of actinomycetes and to determine their characteristics such as growth patterns, colony morphology and metabolic capabilities with reference to *Bergey's Manual of Systematic Bacteriology*, 2nd Edition, Vol 5, The Actinobacteria, Part A. Culture medium used for the isolation of Actinobacteria was starch casein agar (Masand et al., 2018). The selective isolation methods are (Kumar et al., 2016):

- Nutrient selection in which media is prepared with nutrients specifically utilized by actinomycetes.
- Selective inhibition in which antifungal and antibacterial agents are used to inhibit the growth of other microorganisms.
- Pretreatment of sample, in which soil, marine sample or plant parts were treated with a physical or chemical method in order to decrease the number of non-actinomycetes bacteria or fungi.
- Enrichment method, in which nutrient media can be enriched with certain additional supplements, which favors the growth of actinomycetes or inhibit the growth of other microbes.

26.3.3 Preparation of Emulsified Media

Prepare emulsified media to isolate PLA-degrading Actinobacteria and pour the prepared emulsion media and the minimum medium into Petri dishes after autoclaving. Prepare a serial dilution of soil and water samples containing Actinobacteria and spread on emulsified PLA agar. Then, incubate it for 2 weeks at 25°C

26.3.4 Clear-Zone Test

Actinobacteria having the potential to degrade PLA forms a clear zone (halo) around colonies as a medium containing polymer is used as a carbon source. These strains of Actinobacteria are used for biodegradation assays.

26.3.5 Preparation of Thin Plastic Films

For converting PLA pellets into films by using hydraulic press, 1.5 g of PLA pellets was placed on the bottom of an aluminum plate for melting till it becomes transparent. Then, another aluminum plate is placed on the first one and the polymer was pressed for 5 minutes at 10–20 bar pressure. Before removal of the thin film, it is allowed to cool at room temperature. Now, the film is removed and is ready for use in biodegradation assays. After preparation of thin plastic films, a few films should be exposed to UV at 254 nm for 30 minutes and stored at room temperature for later use. Few films are not treated with UV. Cut films into small rectangles and submerge for 30 minutes in 2% SDS, wash with distilled water and dry at 50°C for overnight. Place the films in 70% ethanol for 30 minutes after drying and dry under laminar hood. Then, films were aseptically transferred to a sterilized flask.

26.3.6 Biodegradation Assays

Prepare a liquid medium with a pre-inoculum (20 mL) and allow it to grow for 2 weeks before using in the biodegradation assays. With a sterile pipette, 10% pre-inoculum biomass for each of the selected actinomycete strains was transferred into a fresh medium. Sterile thin films should be added to culture media and incubated at 30°Cand 100 rpm in an orbital shaker. The control was also maintained in non-inoculated medium. Perform these assays for 60–180 days in triplicates.

26.3.7 Monitoring and Analysis of Thin Films Biodegradation

Remove thin plastic films regularly from the culture with sterile forceps to monitor the plastics biodegradation process. The bacterial biofilm formed is washed off by agitation in 2% (w/v) SDS for 5 hours and then washed with distilled water. Place the films on filter paper and dry it overnight at 500°C. Analysis of plastic films biodegradation is performed by comparing surface chemical structural changes and weight loss by FTIR-ATR and mechanical assays. This assays will be performed in triplicates. Weigh the polymeric films both before and after the biodegradation assay using an analytical weighing balance. Determine the weight loss percentage by comparing the dry weight of the films with the original weight.

Weight loss (%) = $m_{ini} - m_{dry} / m_{ini} \times 100\%$

where, m_{ini} is the original weight of the polymeric film before the degradation assay

m_{dry} is the dry weight of the polymeric film after the biodegradation assay

26.3.8 Fourier Transform Infrared Spectra of Plastic Films

This method is used to analyze the chemical variations in the structure of the plastic films resulting from the biodegradation assays in comparison with the original polymers and the corresponding controls. FTIR spectra of the film samples were recorded in the 400–4000 cm^{-1} range. The y-axis is transmittance (T%), and the x-axis is wave number (cm^{-1}) in the spectra. These assays were performed in triplicate. Statistical analysis is performed to examine the UV radiation pretreatment to the culture medium.

26.4 OBSERVATION AND CALCULATIONS

26.4.1 Clear Zone Test

On emulsified polymer media, a clear zone will be formed by species having the potential to degrade PLA (Figure 26.3).

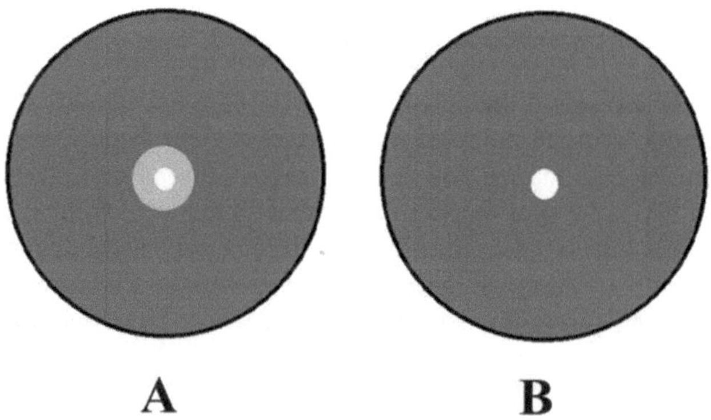

FIGURE 26.3 (a) Species capable of degrading PLA will form halo (clear) zone around colonies; (b) Species unable to degrade PLA will not form halo zone around colonies.

26.4.2 Thin Plastic Films Evaluation

Weight loss of plastic films is seen and calculated with the formula; if the species are able to degrade PLA and the chemical changes in the plastic film is evaluated with the help of FTIR, decrease in absorption bands is monitored. FTIR-ATR is used to observe changes in chemical functional groups present on the surface of the thin plastic films and compared to equivalent controls.

Weight loss (%) $= m_{ini} - m_{dry} / m_{ini} \times 100\%$

where, m_{ini} is the initial weight of the polymeric film before the degradation assay.

m_{dry} is the dry weight of the polymeric film after the biodegradation assay (Figure 26.4).

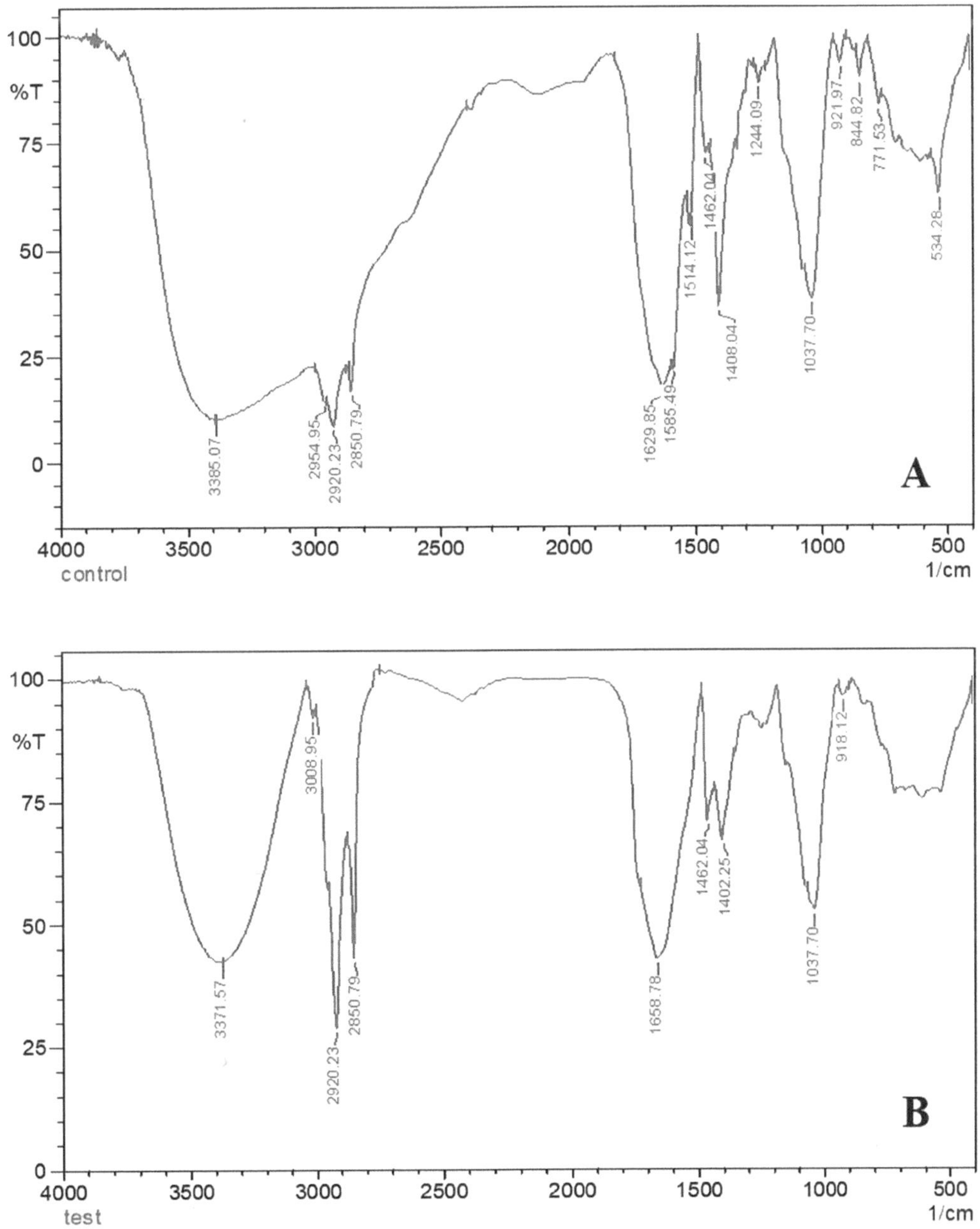

FIGURE 26.4 FTIR spectra compression of control PLA (a) and degraded product (b).

26.5 CONCLUSION

Microplastic pollution has adverse effects on the ecosystem as well as to humans. *Actinobacteria* produces some hydrolytic enzymes having potential to degrade microplastic and are capable of degrading PLA. However, to promote sustainable management of plastic waste, there is a need to focus on the development of microbial-degradation processes and different techniques. Therefore, researchers are planning to establish cost-effective remediation techniques.

REFERENCES

Butbunchu, N. & Pathom-Aree, W. 2019. Actinobacteria as promising candidate for polylactic acid type bioplastic degradation. *Frontiers in Microbiology*. https://doi.org/10.3389/fmicb.2019.02834.

Issac, M. N. & Kandasubramanian, B. 2021. Effect of microplastics in water and aquatic systems. *Environmental Science and Pollution Research International*, 28: 19544–19562. https://doi.org/10.1007/s11356-021-13184-2.

Oliveira, J., Almeida, P.L., Sobral, R.G., Lourenço, N.D. & Gaudêncio, S.P. 2022. Marine-derived Actinomycetes: Biodegradation of plastics and formation of PHA bioplastics-A circular bioeconomy approach. *Marine Drugs* 20: 760. https://doi.org/10.3390/ md20120760.

Pranamuda, H., Tokiwa, Y. & Tanaka, H.1997. Polylactide degradation by an Amycolatopsis sp. *Applied Environmental Microbiology* 63(4):1637–40. doi: 10.1128/aem.63.4.1637-1640.1997.

Qi, Xang., Ren, Yiwei. & Wang, Xingzu. 2017. New advances in the biodegradation of poly(lactic) acid. *International Biodeterioration & Biodegradation*. 117: 215–223. https://doi.org/10.1016/j.ibiod.2017.01.010.

Shilpa, Basak, N. & Meena, S.S. 2022. Microbial biodegradation of plastics: Challenges, opportunities, and a critical perspective. *Frontiers of Environmental Science & Engineering*. 16: 161. https://doi.org/10.1007/s11783-022-1596-6

Zaaba, N.F. & Jaafar, M. 2020. A review on degradation mechanisms of polylactic acid: Hydrolytic, photodegradative, microbial, and enzymatic degradation. *Polymer Engineering & Science*. 60: 2061–2075. https://doi.org/10.1002/pen.25511.

27 Antibacterial and Antibiofilm Activity of Biosynthesized Nanoparticles from Actinomycetes

An Alternative, Environmental-Free, Sustainable Approach

Anand Ravichandran, Abhitha Venkataramakrishnan, Ayyadurai Niraikulm, and Saravanan Periasamy

27.1 INTRODUCTION

Actinomycetes are common and form a robust population in diverse environments, including both terrestrial and aquatic habitats (Bhatti et al., 2017). Actinomycetes are recognized to synthesize bioactive compounds, including pharmacologically potential compounds, antibiotics, enzyme inhibitors, enzymes, and antibiotics (Chun et al., 1997; Labeda et al., 1997; Xu et al., 2004); among the various products, L-asparaginase enzyme has been conventionally utilized as a therapeutic agent in the treatment of various human diseases, primarily acute lymphoblastic leukaemia (DeJong, 1972; Mostafa and Salama, 1979; Narayana et al., 2007). In nature, most of the bacteria can exist as planktonic cells, free-living organisms, or bacterial biofilms with an encased extracellular matrix consisting of exopolysaccharides, proteins, lipids, nucleic acids, extracellular DNA, etc. (Kiedrowski and Horswill, 2011). Biofilm's mode of growth seems to increase survival and propagation rates (Jefferson, 2004). Multispecies biofilms are the most important microorganism lifestyles for protecting against various stresses. Biofilm attachment had a deleterious impact on the global economy, causing serious challenges in maritime transport, health care services, and medicine because of the increasing resistance against antimicrobial molecules, chemical moieties, enhanced gene transfer, altered sustainable biodegradables, and increased production of subordinate metabolites (Balagurunathan et al., 2015). Biofilms are more resistant to multidrug treatment for a variety of infections and diseases associated with both device-related and non-device tissue-associated infections, which are extremely difficult to eradicate or remove. Hundreds of naturally occurring compounds that are nontoxic to bacteria and act as antibiofilms and antibiotics have been discovered. The effectiveness of antimicrobial compounds is currently becoming more significant due to the increase in illnesses that are multidrug resistant. Several studies have also shown that marine Actinomycetes produce bioactive compounds (Braña et al., 2017; Chao et al., 2015; Bae et al., 2015; García et al., 2018).

DOI: 10.1201/9781003398400-27

27.1.1 Nanotechnology Approach in Actinomycetes

In the last 10 years, there has been a lot of interest in novel synthesis methods for nanomaterials such as quantum dots (QDs), metal nanoparticles, graphene, carbon nanotubes (CNTs), and their composites for environmental remediation. Nanomaterials and nanoparticles are created using a wide variety of synthesis processes, including lithographic methods, ball milling, etching, and sputtering (Cao, 2004). Chemical vapour deposition, sol-gel processes, spray pyrolysis, laser pyrolysis, and atomic/molecular condensation are additionally employed in the bottom-up approach, in which nanoparticles are created from simpler molecules (Singh et al., 2018; Rajivgandhi et al., 2022). Although considerable efforts have been made to minimize the toxicity generated by chemical production, green synthesis, a biological production method, is being exploited to minimize the toxicity levels to a significant extent. Green synthesis is critical for avoiding the production of undesirable or hazardous byproducts by developing dependable, long-lasting, and environmentally responsible synthesis procedures (Vimala et al., 2022). Metal nanoparticles have been produced using a green method to accommodate a variety of biological components (e.g., bacteria, fungi, Actinomycetes algae, and plant extracts) (Singh et al., 2018). Consequently, in this chapter, we emphasize the environmentally friendly production of nanoparticles from *Actinomycetes* metabolites and their characterization using various analytical methods and assays.

27.2 MATERIALS AND REAGENTS

1. 1 mM Silver nitrate stock solution
2. 0.22 µm Cellulose acetate syringe filter
3. Sterile glass slides (2 cm * 2 cm)
4. 0.1% Crystal violet solution
5. Flat bottom microtitre plates
6. 2% Sucrose solution
7. Propidium iodide
8. Syto-9 Green
9. 100% Absolute ethanol
10. Milli Q water
11. 1X PBS buffer
12. Sterile Petri dishes
13. Tryptone soy broth
14. Nutrient broth
15. Digital incubator
16. Confocal laser scanning microscope

27.3 CULTURE MEDIA AND ORGANISMS

The following biofilm strains, *Streptococcus salivarius MTCC 13009*, *Staphylococcus pasteuri MTCC 13076*, and *Pseudomonas aeruginosa* MTCC 13077, were isolated from Rajalakshmi Engineering College, Thandalam, Chennai, and the strains were deposited in the Microbial Type Culture Collection, IMTECH, Chandigarh, India. For the synthesis of silver nanoparticles, the ATTC 25421 *Streptomyces* sp. strain was used. *Streptomyces* sp. were grown on R2YE ActinoBASE media, and tryptone soy broth (TSB) media were used for biofilm formation.

27.4 METHODOLOGY

27.4.1 Biosynthesis of Silver Nanoparticles

Actinomycete bioactive metabolites were extracted using simple centrifugation and filtration steps. The potential grown cultures of Actinomycetes were centrifuged at 8,000 g for 15 minutes to remove the bacterial cells.

1. The supernatant was further filtered using a 0.22-μm cellulose acetate syringe filter. The filtered crude with metabolites was stored at −4°C for future use.
2. 1 mM Silver nitrate ($AgNO_3$) was added to the filtered supernatant in the ratio of 1:9. The reaction was conducted at room temperature.
3. The reduction of $AgNO_3$ was observed by a colour change from white to light brown.

27.4.2 Characterization of Biosynthesized Silver Nanoparticles

1. The surface plasmon resonance of biosynthesized silver nanoparticles (AgNPs) was analysed in a UV-visible spectrophotometer by scanning a small aliquot of nanoparticles with 1 mM $AgNO_3$ as a control in the range of 200 and 900 nm.
2. To identify putative biomolecules associated with the bio-reduction of silver and the stability of silver nanoparticles, Fourier transform infrared spectroscopy (FTIR) studies were carried out.
3. A pellet was created by grinding the synthesized silver nanoparticle powder and potassium bromide. The functional groups were observed with a resolution of 4 cm^{-1} ranging from 4,000 to 400 cm^{-1} in FTIR.
4. The size and dispersion of nanoparticles were determined using atomic force microscopy (AFM). To prepare samples for AFM, glass slides were coated with thin films of biosynthesized silver nanoparticles, which were then left there for 24 hours at room temperature before being heated for 2 hours at 80°C.
5. Using the NanoSurf Software, the measurements of the images comprised levelling on the plane, precise modifications to the background, and intermittent stripe elimination with minute cantilever adjustment in AFM.
6. The morphology of silver nanoparticles can be studied using a scanning electron microscope with energy dispersive X-ray analysis (EDX). The samples were coated on the glass slides (2 cm * 2 cm) as air drying and loaded onto a carbon tape-fixed stub.
7. The images were obtained at a magnification of 50,000× and its elemental composition (EDX) was analysed using the QUANTAX EDS software.
8. Transmission electron microscope (TEM) is widely recognized as the most popular technique for measuring and characterizing nanoparticles. TEM provides structure information and images of a nanomaterial with a spatial resolution equal to the level of atomic dimensions.
9. The sample for TEM was created by loading nanoparticles onto a carbon-coated copper grid and the surface morphology of synthesized silver nanoparticles was analysed using Image J 1.4.5.
10. The crystalline structure of nanoparticles was determined using X-ray diffraction analysis. A thin film formed by using the fine powder of AgNPs was mounted on the glass slides and scanned with two angles ranging from 20°C to 80°C at 40 kV and 30 mA to determine the crystalline structure and particle size. SHELXTL software was used for the structure solution and refinement.
11. Zeta potential measurement is a technique for examining the charge on the surface of a substance and targeting the long-term stability of nanoparticles in solution (colloids).

12. The biosynthesized silver nanoparticles were mixed with an equal volume of Milli Q water and centrifuged at 8,000 g for 20 minutes to collect the supernatant, which is then diluted with Milli Q water and sonicated for 15 minutes. The zeta potential can be calculated using optical measurements of particle motion.
13. The boundary of the electric potential of the double layer is called the particles' "zeta potential," and it typically ranges from +100 to –100 mV. The colloidal stability is predicted by the magnitude of the zeta potential. Nanoparticles with zeta potential values greater than +25 mV or less than –25 mV are typically very stable.

27.4.3 Antibacterial and Antibiofilm Assays of the Biosynthesized Silver Nanoparticles

27.4.3.1 Tube Assay

1. The tube biofilm assay provides a qualitative evaluation of biofilm formation.
2. 200 µL of overnight biofilm strains were inoculated in 2 mL of TS broth supplemented with 2% sucrose, which was then incubated for 24 hours at 37°C.
3. After incubation, the spent media were discarded. Biofilms attached to the glass tube were stained using 0.1% crystal violet (Christensen et al., 1982).
4. The biofilm was more adhered to the stain, indicating the strong formation of biofilms.

27.4.3.2 Microtitre Plate Method

1. In this assay, the sterile wells were inoculated with two different sets of biofilms, one as control biofilm strains and another with nanoparticle-treated biofilms.
2. The overnight biofilm strains were subcultured in the freshly prepared media in the microtitre plate.
3. For biofilm formation, a 2% sucrose solution was added to the wells and incubated at 37°C for 24 hours.
4. The biofilm adhered to the microtitre plates, which were stained with 0.1% crystal violet stain for 10 minutes, and the excess dye was discarded and allowed to dry.
5. The biofilm was quantified by adding 100% absolute ethanol to the wells and their OD values were measured at a wavelength of 550 nm using ELISA Plate reader (O'Toole, 2011).

27.4.3.3 Growth Kinetics of Bacterial Strains

1. Biofilm strains were cultured in 100 mL of nutritional broth to evaluate the kinetics of microbial growth in the presence of biosynthesized nanoparticles.
2. As a control, biofilm strains without nanoparticles were used.
3. The growth media were incubated at 37°C. By taking regular readings at 1-hour intervals at 600 nm for 24 hours, the growth rate was estimated using a UV-visible spectrophotometer.

27.4.3.4 *In Vitro* Petridish Biofilm Assay

1. The biofilm strains were inoculated into TSB-containing Petri plates.
2. The biofilm growth medium was supplemented with 2% sucrose.
3. The sterile glass slides were used as a biofilm-forming substrate in the Petri plate.
4. After incubation, the attached biofilm on glass slides was washed with 1X PBS.
5. Propidium iodide and Syto-9 Green fluorescent dyes were used to stain biofilm slides.
6. The glass slides were visualized in confocal laser scanning microscopy (CLSM). The images were obtained as z-stacks using Zeiss Image Viewer.
7. The biofilm architecture of control and treated biofilm strains was analysed from the obtained images using Image J and COMSTAT.
8. The biofilm parameters, such as biovolume, maximum thickness, and roughness coefficient, were analysed from the z-stacks.

27.5 CONCLUSION AND FUTURE PERSPECTIVES

Green synthesis and biogenic nanoparticle synthesis assume a significant role in the current push to create clean, green technologies for the synthesis of nanomaterials. The use of non-microorganisms and reductive bio-excretory processes in the production of new metal nanoparticles is an appealing prospect. Nano-pathogenic microorganisms and their secondary metabolites have been found to be advantageous, being non-carcinogenic and biocompatible during both synthesis and application. The efficacy of these secondary metabolites, when used in the synthesis, produces functionalized nanoparticles with a biological shell and metal core that can synergistically act for the prevention of different diseases. For the required shape, size, and dispersion, the controlled synthesis of nanoparticles can be done using biosynthetic methods in a much easier way and can be utilized for many pharmaceutical and medical applications. The use of secondary metabolites will be advantageous for industrial scale production of nanoparticles. Keeping this in view, we have discussed the synthesis of AgNPs for various possible applications using secondary metabolites as reducing agents for metallic salts.

27.6 TROUBLE SHOOTING GUIDE

Problem	Possible Cause	Troubleshooting
Synthesized nanoparticles turn black	Because of nanoparticle antiaggregation or a problem with the silver nitrate stock solution.	Ensure all the reagents are prepared with milliQ water and are sterile. Cell debris in culture-free supernatants may also have an effect on nanoparticle stability.
Nanoparticles characterization	In characterization, the coating materials should have been interacted.	To avoid a cross-metal reaction, avoid coating nanoparticles with other metals or sputtering.
3-D stack imaging problems	This is because of the error of the laser in the confocal microscope.	Use a confocal microscope and pre-processed laser light. (Keep the laser on before acquiring images.)
Image threshold problems	The acquisition of light from the pinhole	Adjust the eyepiece for focusing rather than the auto-focus.

ACKNOWLEDGEMENTS

We would like to thank DST-SERB File number ECR/2017/001335 and DST-FIST SR/2021/1156 for funding and Rajalakshmi Engineering College for their infrastructure and research facilities. The authors would also like to thank Dr. Clauberg from Denmark Technical University for providing Image J and COMSTAT.

DATA AVAILABILITY

The data used to support the findings of this protocol are available upon reasonable request. The nucleotide sequences of the microbial strains used in this protocol have been deposited in the GenBank database with accession numbers (*S. salivarius* B4TS2 MN474012; *S. pasteuri* B1SA1 MT039534; *P. aeruginosa* B3DPE3 MW332548).

REFERENCES

Bae M, Chung B, Oh KB, Shin J, Oh DC (2015) Hormaomycins B and C: new antibiotic cyclic depsipeptides from a marine mudfat-derived Streptomyces sp. *Mar Drugs* 13(8):5187–5200. https://doi.org/10.3390/md13085187.

Balagurunathan R, Saneetha M, Shanmugasundaram T (2015) Antibiofilm compounds from Actinobacteria. *ENVIS Newslett* 13.

Bhatti AA, Haq S, Bhat RA (2017) Actinomycetes benefaction role in soil and plant health. *Microb Pathog* 111:458–467. https://doi.org/10.1016/j.micpath.2017.09.036.

Braña AF, Sarmiento-Vizcaíno A, Osset M, Pérez-Victoria I, Martín J, De Pedro NN et al (2017) Lobophorin K, a new natural product with cytotoxic activity produced by Streptomyces sp. M-207 associated with the deep-sea coral Lophelia pertusa. *Mar Drugs* 15(5):144. https://doi.org/10.3390/md15050144.

Cao G. *Nanastructures and Nanomaterials: Synthesis, Properties and Applications*. Singapore: World Scientific; 2004.

Chao R, Yuan Y, Zhao H (2015) Recent advances in DNA assembly technologies. *FEMS Yeast Res* 15:1–9.

Christensen GD, Simpson WA, Bisno AL, Beachey EH (1982). Adherence of slime-producing strains of Staphylococcus epidermidis to smooth surfaces. *Infect Immun* 37(1):318–26. https://doi.org/10.1128/iai.37.1.318-326.1982. PMID: 6179880; PMCID: PMC347529.

Chun J, Youn HD, Yim YI, Lee H, Kim MY, Hah YC, et al. (1997) Streptomyces seoulensis sp. nov. *Int J Syst Bacteriol* 47:492–498.

DeJong PJ (1972) L-Asparaginase production by Streptomyces griseus. *Appl Microbiol* 23:1163–1164.

García CM, Bargiela R, Martínez M, Ferrer M (2018) Metagenomic protocols and strategies. *Metagenomics*. https://doi.org/10.1016/B978-0-08-102268-9.00002-1.

Rajivgandhi G,Gnanamangai BM, Prabha TH, Poornima S, Maruthupandy M, Alharbi NS, et al. (2022) Biosynthesized zinc oxide nanoparticles (ZnO NPs) using actinomycetes enhance the anti-bacterial efficacy against K. pneumoniae. *J King Saud Univ Sci* 34(1):101731. https://doi.org/10.1016/j.jksus.2021.101731. https://www.sciencedirect.com/science/article/pii/S1018364721003931

Jefferson K (2004) What drives produce biofilm? *FEMS Microbiol Lett* 236(2):163–73. https://doi.org/10.1016/j.femsle.2004.06.005.

Kiedrowski MR, Horswill AR (2011) New approaches for treating staphylococcal biofilm infections. *Ann N Y Acad Sci* 1241:104–121.

Labeda DP, Lechevalier MP, Testa RT (1997) Streptomyces stramineus sp. nov., a new species of the verticillate streptomycetes. *Int. J. Syst. Bacteriol* 47:747–753.

Mostafa SA, Salama MS (1979) L-asparaginase producing Streptomyces from soil of Kuwait, Zentralbl Bakteriol. *Naturwiss* 134:325–334.

Narayana KJP, Kumar KG, Vijayalakshmi M (2007) L-asparaginase production by Streptomyces albidoflavus. *Indian J Microbiol* 48:331e336.

O'Toole GA (2011) Microtiter dish biofilm formation assay. *J Vis Exp*. (47):2437. https://doi.org/10.3791/2437.

Singh J, Dutta T, Kim KH (2018) 'Green' synthesis of metals and their oxide nanoparticles: applications for environmental remediation. *J Nanobiotechnol* 16:84. https://doi.org/10.1186/s12951-018-0408-4

Vimala R, Sridharan S, Jayapriya M, Govindan R, Kanisha C, Manoharan N, et al. (2022). Antibiofilm activity of extracellular green-synthesized nanoparticles by Actinobacteria. https://doi.org/10.1007/978-1-0716-1728-1_106.

Xu P, Li WJ, Wu WL, Wang D, Xu LH, Jiang CL (2004) Streptomyces hebeiensis sp. nov. *Int J Syst Evol Microbiol* 54:727–731.

28 Extraction and Purification of Rapamycin from *Streptomyces hygroscopicus*

Sanjeev K. Ganesh, Mohanasrinivasan V., and Subathra Devi C.

28.1 INTRODUCTION

Streptomyces hygroscopicus is a gram-positive filamentous actinomycete species known for its ability to produce a wide range of bioactive pharmaceutical compounds. Rapamycin was discovered as a macrocyclic antifungal compound from *S. hygroscopicus*. Rapamycin is a 31-membered macro lactone ring synthesized by polyketide synthase 1. It is a nitrogen-containing macrolide compound with a molecular formula of $C_{51}H_{79}NO_{13}$. Now, rapamycin is used as a drug against cancer, auto-immune diseases, immune-modulator and neurodegenerative diseases [1,2].

In the recent years, demand for rapamycin has increased exponentially due to its pharmacological importance. Calcineurin inhibitors like ascomycin and FK506 are structurally similar to rapamycin. In comparison with calcineurin inhibitors, rapamycin is an mTOR inhibitor which shows very low toxicity and more oral absorption [3]. Most predominantly, rapamycin is used as an immunosuppressant in organ transplantation procedures. It is used to prevent the body's rejection responses against the transplanted graft or organ. In addition to these, rapamycin is currently being studied for its multifaceted application in cancer therapy, prevention of ageing, etc. Rapamycin exhibits very low side effect when compared to other immunosuppressants [4,5].

The maximum amount of rapamycin produced by *S. hygroscopicus* is 42 mg/L. Rapamycin is a hydrophobic compound so that toluene, ethyl acetate, hexane and chloroform are the solvents used to extract rapamycin from the broth culture. Absorption chromatography and flash chromatography are found to be the most suitable methods for the purification of crude rapamycin extract [6–8]. The present report also focused on the extraction and purification of rapamycin from fermentation broth.

28.2 MATERIALS

28.2.1 Spore Preparation

Culture: *S. hygroscopicus*

28.2.2 Inoculum Preparation

S. hygroscopicus spore suspension and YMA (Yeast Mannitol Agar) media

28.2.3 Shake-Flask Fermentation and Preliminary Analysis of Rapamycin

Test organism: 24-hour old *Candida albicans* broth culture

DOI: 10.1201/9781003398400-28

28.2.4 Extraction and Purification of Rapamycin

Toluene, sodium bicarbonate and distilled water

28.2.5 Purification of Rapamycin Using Column Chromatography

Column, silica gel, hexane and acetone.

28.2.6 Determining the Concentration of Extracted Rapamycin Using Ultra-Performance Liquid Chromatography (UPLC)

Methanol, acetonitrile, HPLC grade water, Column: UNISON UK C18, 3 μm, 4.6 mm diameter, 250 mm length and rapamycin standard.

28.3 METHOD

28.3.1 *S. hygroscopicus* Spore Preparation

1. Spore preparation: Inoculate the cells in the sporulation media (4 g/L dextrose, 20 g/L agar, 4 g/L yeast extract and 10 g/L malt extract) [9].
2. Growth conditions: Incubate the plate at 28°C for 20 to 25 days.
3. Spore collection: Add 1 mL of distilled water over the plate to harvest the black-coloured spore suspension. Preserve the spores in 25% glycerine solution under –20°C.

28.3.2 Inoculum Preparation

1. Seed culture (Inoculum) preparation: Inoculate the stored spore suspension in 100 mL grown media (YMA) containing yeast extract 4.0 g/L, malt extract 10 g/L, glucose 4 g/L and agar 15 g/L with pH 6.5–7.0 [10].
2. Growth conditions: Incubate the inoculated media at 28°C and 250 rpm for 7 days.

28.3.3 Shake-Flask Fermentation and Preliminary Analysis of Rapamycin

1. **Production media**: Inoculate 2% (v/v) of the seed culture in the production media containing (g/L): fructose: 22 g, mannose: 5 g, malt extract: 10 g, casein: 0.3 g, $(NH_4)_2SO_4$: 5.3 g, NaCl: 5 g, K_2HPO_4: 4 g, $ZnSO_4{\cdot}7H_2O$: 0.06 g, $MgSO_4{\cdot}7H_2O$: 0.0025 g, $MnSO_4{\cdot}H_2O$: 0.012 g, $FeSO_4{\cdot}7H_2O$: 0.1 g, $CoCl_2{\cdot}6H_2O$: 0.010 g, Na_2SO_4: 0.3 g, $CaCO_3$: 3 g [11].
2. **Growth condition**: Adjust the pH to 7 and incubate the seed culture inoculated media at 25°C and 120 rpm for 7 days.
3. After 7 days of incubation, centrifuge 5 mL of broth culture at 5,000 rpm for 7 minutes and collect the supernatant.
4. Check the antifungal activity of the supernatant against *Candida albicans* using the well diffusion method.

28.3.4 Extraction and Purification of Rapamycin

1. Add equal amount of toluene to 7-day-old rapamycin containing broth culture of *S. hygroscopicus*. Stir for 4 hours at 50°C and 300 rpm.
2. Separate the amber-coloured oily residue formed above the media using a separating funnel. Repeat the extraction process with fresh toluene.

3. Wash the oily residue with 5% same volume sodium bicarbonate followed by washing with double the volume of distilled water. Collect the oily residue using a separating funnel.
4. Mix the oily residue with equal volume of acetone and concentrate the product to 1/4th of the total volume using a rotary evaporator at 40°C.
5. Mix the concentrated residue with charcoal and stir it for 15 minutes at 50°C. Filter and collect the concentrated amber-coloured oily residue.

28.3.5 Purification of Rapamycin Using Column Chromatography

1. Load the collected rapamycin residue on column.
2. Column length: 1,500 mm × diameter 250 mm.
   Mobile phase: 60% hexene and 40% acetone.
   Mesh size: 60–200.
3. Collect the fraction and mix activated charcoal. Stir for 15 minutes at 40°C. Filter and collect the amber-coloured solution.
4. Concentrate and dry the residue at 40°C under vacuum and collect the purified form of rapamycin.

28.3.6 Determining the Concentration of Extracted Rapamycin Using Reversed Phase UPLC

1. Protocol for the analysis of rapamycin through UPLC:
   Buffer A: 80% methanol and 20% Acetonitrile
   Buffer B: HPLC grade water
   Mobile phase: 80% buffer A and 20% buffer B
   Column: UNISON UK C18, 3 µm, 4.6 mm diameter, 250 mm length.
   Flow rate: 1 mL/min
   Injection volume: 20 µL
   Diluent: Methanol
   Temperature: 55°C.
   Wave length: 70 nm
   Approximate retention time: 6.3 minute (Figure 28.1).
2. Concentration of extracted rapamycin can be estimated by plotting a calibration curve using the standard rapamycin solution.
3. Note the peak areas for different concentrations (minimum five) of rapamycin standard.
4. Correlation coefficient of the standard curve should not be less than 0.996.
5. Plot the peak area versus concentration using an Excel sheet.
6. Create the calibration curve using the formula $Y=mX+C$ (where, Y is peak area and X is concentration).
7. Get the value for m and C from the graph.
8. Plot the value of the peak area of extracted rapamycin on the standard calibration curve graph.
9. Calculate the concentration of extracted rapamycin (X) using the formula ($Y=Mx+C$) from the graph.

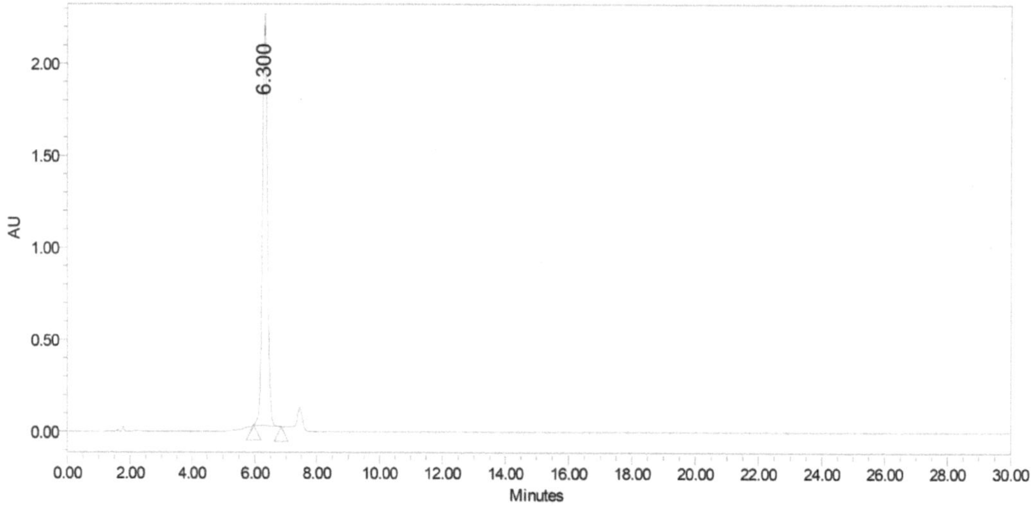

Peak Results

Name	RT	Area	Height	Amount	Units	% Area
1	6.300	24481229	2179539			100.00

FIGURE 28.1 UPLC chromatogram of Rapamycin.

REFERENCES

1. Park SR, Yoo YJ, Ban YH, Yoon YJ. 2010. Biosynthesis of rapamycin and its regulation: past achievements and recent progress. *J. Antibiot.* 63(8):434–41. DOI: 10.1038/ja.2010.71
2. Zou X, Li J. 2013. Precursor engineering and cell physiological regulation for high level rapamycin production by Streptomyces hygroscopicus. *Ann. Microbiol.* 63(4):1371–8. DOI: 10.4014/jmb.1403.03024
3. Weischer M, Röcken M, Berneburg M. 2007. Calcineurin inhibitors and rapamycin: cancer protection or promotion? *Exp. Dermatol.* 16(5):385–93.
4. Ganesh SK, Subathra DC. 2023. Molecular and therapeutic insights of rapamycin: a multi-faceted drug from Streptomyces hygroscopicus. *Mol. Biol. Rep.* 1–19. DOI: 10.1007/s11033-023-08283-x
5. Chen Y, Krol J, Sterkin V, Fan W, Yan X, Huang W, Cino J, Julien C. 1999. New process control strategy used in a rapamycin fermentation. *Process Biochem.* 34(4):383–9. DOI: 10.1016/s0032-9592(98)00108-3
6. Abdel-Fattah YR. Non-conventional method for evaluation and optimization of medium components for Rapamycin production by Streptomyces hygroscopicus. 2008. *J. Microbiol.* 3:405–13.
7. More AS, Gadalkar S, Rathod VK. 2017. Extraction of rapamycin (sirolimus) from Streptomyces rapamycinicus using ultrasound. *Prep. Biochem. Biotechnol.* 47(6):627–32. DOI: 10.1080/10826068.2017.1303609
8. Yen HW, Chiang MH. 2015. Using the mycelium-covered cereals as an efficient inoculation method for rapamycin fermentation in a 15-L fermenter using Streptomyces hygroscopicus. *BIOB.* 2(1):1–7. DOI: 10.1186/s40643-015-0071-7
9. Luthra U, Kumar H, Tripathi A, Talele S, Gupte V. 2019. Study on growth and ascomycin production by Streptomyces hygroscopicus subsp. ascomyceticus. *ISBAB.* 19:101159. DOI: 10.1016/j.bcab.2019.101159
10. Dutta S, Basak B, Bhunia B, Chakraborty S, Dey A. 2014. Kinetics of rapamycin production by Streptomyces hygroscopicus MTCC 4003. *3 Biotech.* 4(5):523–31. DOI: 10.1007/s13205-013-0189-2
11. Rani PB, Kumar BS, Rao AK, Sreenivasrao S, Narasu ML. 2013. Recovery and purification of rapamycin from the culture broth of Mtcc 5681. *Pak J Biol Sci: PJBS.* 16(5):219–25. DOI: 10.3923/pjbs.2013.219.225

29 Isolation and Identification of Squalene from the Ethyl Acetate Extract of *Streptomyces* sp. VITGV156 (MCC 4965)

Veilumuthu Pattaulavar, Sasikumar Sundarasan, and John Godwin Christopher

29.1 INTRODUCTION

The endophytic *Streptomyces* produces secondary metabolites, such as antibiotics, drugs, vitamins, and siderophores. The genome sequencing of *Streptomyces* sp. reveals the presence of huge unexplored biosynthetic gene clusters (BGCs), which can potently produce novel bioactive secondary metabolites [1]. Interestingly, some unique compounds were found in the secondary metabolites. Genome sequencing brings a better understanding of BGCs, which leads to the development of novel pharmaceutical compounds.

Squalene has a number of biological functions, including those related to the prevention of diabetes, the reduction of cholesterol, and the maintenance of skin hydration, and it's a naturally occurring, highly unsaturated triterpene hydrocarbon molecule ($C_{30}H_{50}$) that arises only through mevalonic acid (MVA), a step in the synthesis of cholesterol. Squalene is an intermediate of cholesterol biosynthesis, which forms solely via mevalonic acid (MVA) [2]. Squalene was initially found in shark liver oil. Squalene serves as a precursor to a large number of bioactive compounds, such as steroids and hopanoids. As a result, squalene is largely used in the chemical, food, cosmetic, and pharmaceutical industries. Global squalene demand has increased over the last decade, attracting significant public and scientific attention. The global squalene market demand in 2014 was approximately 2,670 tonnes [3]. Now, the squalene market size is 5,300 tonnes in 2022, with a value of 241.9 million USD, and this is expected to grow up to 356.14 million USD by 2032.

Microbial production of squalene can be a suitable alternative in the near future as the requirement is very high.

29.2 MATERIALS AND METHODS

29.2.1 Requirements

1. Tomato plant samples (*Lycopersicon esculentum* – Arka Rakshak).
2. Surface sterilising agents –70% alcohol, 90% ethanol, 0.9% sodium hypochlorite, 70% ethanol, 10% $NaHCO_3$ and sterile DH_2O.
3. ISP2 (International Streptomyces Project-2) agar medium (yeast extract 4.0 g, malt extract 10.0 g, dextrose 4.0 g, agar 20.0 g, distilled water 1,000.0 mL – pH 7.2).
4. c-TAB and Phenol: chloroform.
5. Ethyl acetate, Na_2SO_4 and MeOH.
6. Rotary vacuum evaporator.
7. Deuterated solvents (DMSO, water, methanol, chloroform).
8. Analysis of GCMS, LC–MS, HR-MS, 2D and 3D NMR.

DOI: 10.1201/9781003398400-29

29.2.2 Bioinformatic Analysis

1. Sequenced whole genome.
2. Collection of complete genome sequence from NCBI (if not sequenced for specific strain).
3. Silent biosynthetic gene cluster analysis using software tools such as the fungal and bacterial versions of antiSMASH (up to six analyses can be freely accessed) at https://fungismash.secondarymetabolites.org or https://antismash.secondarymetabolites.org., respectively.
4. BLAST program for the analysis of NRPS and PKS proteins prediction.

29.2.3 Protocol

29.2.3.1 Extraction of Secondary Metabolites

- Tomato plant samples were surface sterilised using a series of solutions, including 70% alcohol for 1 minute, 90% ethanol for 1 minute, sodium hypochlorite (0.9%) for 4 minutes, 70 ethanol for 30 seconds, 10% $NaHCO_3$ for 5 minutes, and finally rinsed in sterile distilled water for 3 minutes [4].
- The plant parts were cut into small pieces. These cut segments were streaked in ISP2 agar plates for actinomycetes growth [5].
- These inoculated microbial plates were kept at 30°C for 15 days incubation. Then, they were purified by streaking particular isolates over ISP2 agar plates. The separated individual colonies were thus used for further analysis.
- Small-scale cultures of *Streptomyces* sp. were grown side-by-side in the same media and worked up identically [6].
- Seed cultures were prepared by adding fresh spores to a 5-mL ISP2 broth in a 20-mL sterile tube.
- The culture was incubated at 30°C, 150 rpm for 3 days, and then, this log phase culture is diluted with 500 mL of ISP2 broth (in a 1,000 mL Erlenmeyer flask).
- After 7 days, the cells were centrifuged, and the supernatant was extracted twice with equal volumes of ethyl acetate, and the organic phase was dried over Na_2SO_4 and evaporated to dryness in vacuum. Then, it's stored in the refrigerator for further purification.
- The organic phases (extract) were dissolved in 100 μL MeOH and analysed in GC-MS.
- GC-MS chromatogram was compared with NIST database. This gives an outline on the compounds, based on their molecular weight.
- Based on the retention time from GC-MS chromatogram, different fractions can be collected from LC-MS and will be subjected to GCMS for purity checking. Then, it's sent for the molecular mass and structural analysis using 2D and 3D NMR.

29.2.4 Genome Analysis

Genomic DNA was isolated from the actinomycetes culture using c-TAB and phenol:chloroform extraction method followed by RNase treatment. The quality and quantity of the isolated gDNA were checked on NanoDrop. This genomic DNA was sequenced using the Illumina NextGen sequencing method.

- DNA sequence analysis of actinomycetes genome was done using computer-aided bioinformatic tool such as antiSMASH [7] and BLAST program [8].
- Analysis of PKS and NRPS predicted proteins were carried out using programs like ASMPKS [9] and NRPS predictor [10].

TABLE 29.1
Potential BGCs of Secondary Metabolites in Actinomycetes Predicted Using AntiSMASH 6.0 (Descending Order)

S. No.	Scaffolds	Nature of Cluster	Corresponding Known Cluster	Resemblance (%)
1	Core hits −17.2	Terpene	*Geosmin*	100%
2	Core hits −7.1	Ectoine	Ectoine	100%
3	Core hits −12.1	Terpene	Albaflavenone	100%
4	Core hits −18.2	Terpene	Hopene	100%
5	Core hits −12.2	T2PKS	Spore Pigment	66%
6	Core hits -2.2	Terpene	Isorenieratene	63%
7	Core hits -7.2	Melanin	Melanin	60%

TABLE 29.2
Compounds of Extracts Identified in *Streptomyces* sp.

S. No	Chemical Compound	RT	Molecular Weight	Molecular Formula	Area %	Activity	Ref
1	Benzyl alcohol	7.838	108.13	C_7H_8O	0.26	Antimicrobial	[1]
2	1H-Pyrazole, 4,5-dihydro-5,5-dimethyl-4-isopropylidene	9.122	98.14	$C_5H_{10}N_2$	0.96	Antibacterial	[2]
3	Ethanol, 2-(2-butoxyethoxy)-	9.659	162.22	$C_8H_{18}O_3$	1.24	Antimicrobial	[3]
4	Benzoic acid	10.019	122.12	$C_7H_6O_2$	0.43	Antimicrobial	[4]
5	2-Dodecene, (Z)-	10.942	168.31	$C_{12}H_{24}$	0.71	Antibacterial	[5]

29.3 OBSERVATIONS AND RESULTS

Whole genome sequencing of *Streptomyces* VITGV156 revealed all the available biosynthetic gene clusters with the help of the antiSMASH tool. Table 29.1 is filled with antiSMASH detail. The above observation reveals that squalene gene is expressed in standard laboratory conditions. Table 29.2 shows all other compounds from the GC-MS chromatogram, by comparing with NIST database.

REFERENCES

1. Liang, M., L. Liu, F. Xu, X. Zeng, R. Wang, J. Yang, W. Wang, et al. Activating cryptic biosynthetic gene cluster through a CRISPR-Cas12a-mediated direct cloning approach. *Nucleic Acids Research* 50(6) (2022):3581–3592. https://doi.org/10.1093/NAR/GKAC181.
2. Wołosik, K., M. Knaś, A. Zalewska, M. Niczyporuk, and Adrian Wojciech Przystupa. The importance and perspective of plant-based squalene in cosmetology. *Journal of Cosmetic Science* 64(1) (2013): 59–66.
3. Rudzińska, M., P. Górnaś, M. Raczyk, and A. Soliven. Sterols and squalene in apricot (Prunus Armeniaca L.) kernel oils: the variety as a key factor. *Natural Product Research* 31(1) (2016): 84–88. https://doi.org/10.1080/14786419.2015.1135146.
4. Fialho de Oliveira, M., M. G. da Silva, and S. T. Van Der Sand. Anti-phytopathogen potential of endophytic actinobacteria isolated from tomato plants (Lycopersicon esculentum) in southern Brazil, and characterization of Streptomyces Sp. R18(6), a potential biocontrol agent. *Research in Microbiology* 161(7) (September 2010): 565–572. https://doi.org/10.1016/j.resmic.2010.05.008.
5. Veilumuthu, P., and J. G. Christopher. Diversity of Actinomycetes in tomato plants. *Indian Journal of Agricultural Research* 57 (April 5, 2022): 95–102. https://doi.org/10.18805/IJARE.A-5913.
6. Ibnouf, A., and A. Waggiallah. Isolation and extraction of some compounds that act as antimicrobials from Actinomycetes. *Saudi Journal of Biological Sciences* 29(8): 103352.

7. Blin, K., S. Shaw, A. M. Kloosterman, Z. Charlop-Powers, G. P. van Wezel, M. H. Medema, and T. Weber. AntiSMASH 6.0: Improving cluster detection and comparison capabilities. *Nucleic Acids Research* 49(W1) (July 2, 2021): W29–W35. https://doi.org/10.1093/NAR/GKAB335.
8. Yoon SH, SM Ha, J Lim, S Kwon, and J Chun. A large-scale evaluation of algorithms to calculate average nucleotide identity. *Antonie van Leeuwenhoek* 110(10) (October 1, 2017): 1281–1286. https://doi.org/10.1007/S10482-017-0844-4.
9. Tae H, EB Kong, and K Park. ASMPKS: An analysis system for modular polyketide synthases. *BMC Bioinformatics* 8 (September 3, 2007): 327. https://doi.org/10.1186/1471-2105-8-327.
10. Rausch C, T Weber, O Kohlbacher, W Wohlleben, and DH Huson. specificity prediction of adenylation domains in nonribosomal peptide synthetases (NRPS) using transductive support vector machines (TSVMs)." *Nucleic Acids Research* 33(18) (2005): 5799–5808. https://doi.org/10.1093/NAR/GKI885.

30 Docking and ADME Studies Protocol

K. Prem Anand, M. A. Jayasri, and K. Suthindhiran

30.1 INTRODUCTION

The pandemic viruses – corona, influenza, and Ebola – have provided retrospective insights into disease emergence and severity and urged the need to develop new vaccines (Excler et al. 2021). Nosocomial infections are increasing at an alarming rate, which leads to the emergence of multidrug resistance (MDR), pan drug resistance (PDR), and extensive drug resistance (XDR) in clinical isolates and are a serious threat (Magiorakos et al. 2012; Mohammed et al. 2021). The treatment scheme for nosocomial infection-causing bacteria is perplexing, and we are at the edge of botching up our last line of antibiotics. Drug development is highly challenging for antibiotic-resistant microbes due to their defense mechanism – drug inactivation enzymes (ESBL – extended-spectrum beta-lactamase and carbapenemase), efflux pumps, and changes in membrane pores are the vital mechanisms that aid in antibiotic resistance (Aminov 2010). Such resistance is widely produced by nosocomial infection-causing, *Acinetobacter baumannii*, Enterobacteriaceae, *Klebsiella pneumoniae*, *Pseudomonas aeruginosa*, and *Staphylococcus aureus* (Noyal et al. 2009; Chaudhary and Payasi 2014; Chaudhary and Pandya 2015). In addition, *Mycobacterium tuberculosis* (MTB) and human influenza virus (HIV) are of serious concern (Deeks et al. 2015; Koch and Mizrahi 2018).

The drug fabricators have a massive gap in identifying the right target and developing new therapeutic products. Usually, the drug discovery process requires ~1.3 billion dollars and 12–15 years to identify a single drug candidate, even though the drug could fail at any stage (Drews 2000; Kitchen et al. 2004). Though we have thousands of active molecules whose druggable nature is undetermined, screening all molecules through the conventional pipeline is painstaking and ineffective. Marine actinobacteria are no exception in isolating and identifying novel therapeutic compounds. Isolation of secondary metabolites from the microbial origin is an alternative technique for identifying new therapeutic compounds with more significant potential. The computer-aided drug discovery (CADD) process supports faster target site and lead compound identification. Moreover, the CADD process can support the high-throughput screening (HTS) of millions of biomolecules in less time to identify potential lead compounds, which reduces 50% of drug failure in clinical trials (Ebhohimen et al. 2020). Furthermore, the docking analysis can provide crucial information, such as conformational changes, binding affinity, the interaction between active site residues, and protein structural changes due to drug molecules.

The pharmacokinetic profile evaluating the adsorption, distribution, metabolism, and excretion (ADME) analysis determines the fate of the drug inside the body. The drug molecule should be highly active with less toxicity and hold up until the target site; even though it inhibits the disease in various animal models, it could fail drastically. The ADME studies provide insights into the drug's nature, pH, solubility, adsorption location, and degradation due to various cox enzymes. Identifying these effects of the drug inside the body through simulations can allow drug manufacturers to formulate the drug molecules based on their mode of action. In this protocol, we have illustrated the methodology for molecular docking using the AutoDock software package and ADME analysis using the SwissADME server (Daina et al. 2017).

DOI: 10.1201/9781003398400-30

30.2 REQUIREMENTS

1. Windows 10/11 PC
2. MGL Tools [https://ccsb.scripps.edu/mgltools/]
3. Autodock [https://autodock.scripps.edu/]
4. Discovery Studio Visualizer [https://discover.3ds.com/discovery-studio-visualizer-download]
5. Protein-Ligand Interaction Profiler [https://plip-tool.biotec.tu-dresden.de/plip-web/plip/]
6. Protein Data Bank [https://www.rcsb.org/]
7. Drug Bank [https://go.drugbank.com]
8. SwissADME Webserver [http://www.swissadme.ch/]

30.3 PROTOCOLS FOR DOCKING ANALYSIS

30.3.1 Software Setup

1. The MGL Tools v1.5.7 was downloaded from [https://ccsb.scripps.edu/mgltools/] and installed in the system.
2. Similarly, the AutoDock v4 software was downloaded from [https://autodock.scripps.edu/] and installed in the system.
3. The Discovery Studio Visualizer software was downloaded [https://discover.3ds.com] and installed in the system.
4. Once the AutoDock software package has been installed, the **autodock4** and **autogrid4** programs will be generated in the "**C:\Program Files (x86)\The Scripps Research Institute\Autodock\4.2.6,**" which will be mapped during the grid generation and docking process.
5. Once everything is installed, a new folder is created and renamed as "**autodock**" in the "**D:**" drive [D:\autodock]; this folder will be our working directory where the protein, ligand, and other programs will be saved and processed during the docking process.
6. Once created, the **autodock4** and **autogrid4** tools in "**C:\Program Files (x86)\The Scripps Research Institute\Autodock\4.2.6**" are copied and placed inside the "**autodock**" folder (Note 1).
7. Also, the "**AD4.1_bound**" and "**AD4_parameters**" files from [**C:\Program Files (x86)\MGLTools-1.5.7\Lib\site-packages\AutoDockTools**] have been placed inside the "D:\autodock" folder.
8. Once all the essential tools and files (autodock4, autogrid4, AD4.1_bound, and AD4_parameters) are placed inside the working folder ["D:\autodock"], let us set it as the default working direction by opening the AutoDock Tools software and selecting **File → Preferences → Set → Startup Directory → D:\autodock → Set**. Now, the **D:\autodock** will be our default working directory (Figure 30.1a).

30.3.2 Protocol to Perform Docking Using AutoDock

30.3.2.1 Retrieving Protein from RCSB-PDB

1. In this protocol, we have selected the HIV-1 reverse transcriptase enzyme and studied its inhibition using the lamivudine drug, and the drug binding, orientation, and amino acid interaction were identified.
2. The protein molecule "**1TKX**" was searched and retrieved from the Research Collaboratory for Structural Bioinformatics – Protein Data Bank (RCSB-PDB) [https://www.rcsb.org/structure/1TKX].

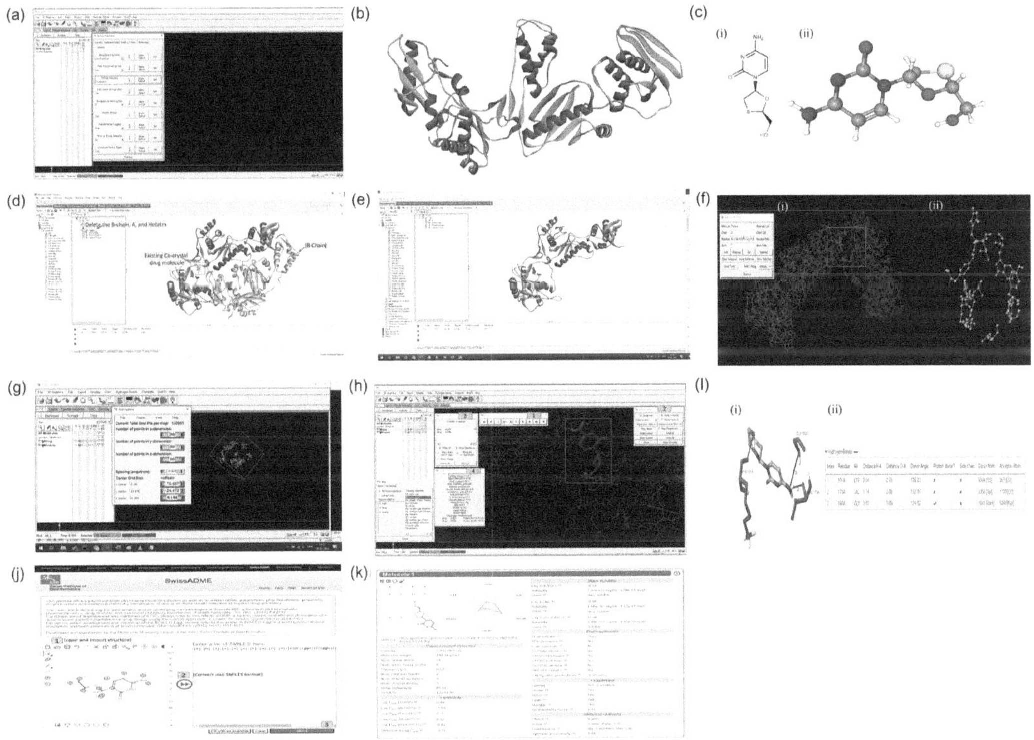

FIGURE 30.1 (a) Setup for creating default working directory; (b) crystal structure of the HIV-1 reverse transcriptase protein – A-chain (1TKX); (c) the lamivudine drug molecule, (i) 2D and (ii) 3D structure; (d) (i) protein structure of HIV-1 reverse transcriptase with the co-crystal drug and B-chain and (ii) protein structure without the drug and B-chain; (e) (i) mapping of active site region using the amino acid string, (ii) highlighted amino acids residues depicting the active site cavity; (f) positioning the grid box around the active site region to specify the drug interaction site; (g) molecular docking results showing the interaction of the drug in the active site; (h) interaction of protein-drug molecules through LYS, GLY, and VAL amino acids through H-bond; (i) importing the ligand molecule for pharmacokinetic analysis using SwissADME webserver; (j) complete pharmacokinetic profile of the lamivudine drug molecule.

3. The protein structure was downloaded in the ".pdb" format as "**1TKX.pdb**" and saved in the "D:\autodock" folder (Figure 30.1b).

30.3.2.2 Retrieving Drug Molecules from PubChem

1. The drug molecule (lamivudine) was searched in the Drug Bank database [https://go.drugbank.com], and the lamivudine drug molecule was retrieved.
2. The ligand structure was downloaded as a 3D conformer in the "**.pdb**" format and saved in the "D:\autodock" folder (Figure 30.1c – (i) and (ii)).
3. While downloading the structure, if it opens as a text file on the webpage, copy the text and paste it into notepad → go to **File** → **Save** → set the "**Save as type**" to "**All Files**" and rename the file as "**ligand.pdb**" → **Save** → by this we can convert the drug molecule from the "**text file**" to "**.pdb**" formats.
4. Once the drug molecule has been downloaded, transfer the "**ligand.pdb**" to the "**D:\autodock**" folder.

30.3.2.3 Pre-Processing Protein Molecule

1. Open the Discovery Studio Visualizer software package and select **File** → **New** → **Molecule Window.**

2. Once the molecule window is opened, insert the protein molecule by selecting the **File → Insert From → File**, and select the protein molecule (1TKX.pdb) downloaded from the protein databank.
3. The protein molecule, "**1TKX**", has been crystalized along with the co-crystal drug molecule (quinolone). The co-crystal drug molecule is removed by selecting the "**Hetatm**" and deleting it.
4. The protein molecule also contains two chains (A and B), where this protocol mainly focuses on the active site present in the A-chain, so the B-chain is selected and deleted.
5. Now, the protein molecule with only the A-chain and co-crystal molecule removed is saved by **File → Save as →** rename as **Protein.pdb →** Save the file to the "D:\autodock" folder (Figure 30.1d – (i) and (ii)).
6. Open the AutoDock Tools software and Open the protein molecule by selecting **File → Read Molecule → Protein.pdb** (select the protein molecule) (Note 2).
7. Select the **Edit → Hydrogen → Add → Polar only** and select OK.
8. To add Kollman charges, select **Edit → Charges → Add Kollman charges**.
9. To add Gasteiger charges, select **Edit → Charges → Compute Gasteiger** (Note 3).
10. Now, let us assign the force field to the protein molecule by selecting **Edit → Atoms → Assign AD4 type**; this step reduces the noises created during the protein interaction with the ligand.
11. Finally, let us export the processed protein file into "**.pdbqt**" format; for that, select, **Grid → Macromolecules → Choose → Select the protein → Select Molecule →** Rename the file as **"Protein.pdbqt"** (Note 4).

30.3.2.4 Processing Ligand Molecule

1. In the AutoDock Tools software, select the **Ligand → Input → Open → Change "files of type" into "PDB files (*.pdb)" →** select the **"ligand.pdb"** file.
2. Select the **Ligand → Torison tree → Choose root**.
3. Again select the **Ligand → Torison tree → Detect root**.
4. Finally, select the **Ligand → Torison tree → Choose Torsions**. This illustrates that the selected ligand molecule has 4 rotatable bonds out of 32, which vary depending on the ligand molecule's structure.
5. Let us save the file into "***.pdbqt**" format; select **Ligand → Output → Save as PDBQT** (Note 5).
6. Now that we have successfully processed the protein and ligand molecules, we proceed with the active site prediction and grid generation.

30.3.2.5 Active Site Prediction and Grid Generation

1. The protein molecule might have multiple active sites, and finding the right active site to activate/inhibit the protein is essential. There are several approaches to finding the active site of the protein: (a) Identify through literature survey, (b) software-based active site detection – meta-pocket, cast-p, and other webservers.
2. Identifyingactivesitesthroughtheliteraturesurveyisthemostcommonandaccuratetechnique though the studies are dedicatedly performed on ligand interaction on active sites. Here, we have the active site residues of "**1TKX**" protein molecule identified through literature survey (**LEU100,LYS101,LYS103,VAL179,TYR181,TYR188,TRP229,LEU234,HIS235, TYR318**) (Note 6).
3. Now, let us prepare the workspace by deleting all the molecules from the AutoDock Tools by selecting **Edit → Delete → Delete all molecules**.
4. Let us import the processed protein molecule by selecting, **Grid → Macromolecules → Open → select the "protein.pdbqt" file** (Note 7).
5. Now, let us open the ligand molecule by selecting **Grid → Set map types → Open ligand**.

6. Once the "**protein.pdbqt**" and "**ligand.pdbqt**" have been imported, let us map the active site using the identified amino acid residues by selecting the **Select → Select From String → Click Molecule List and select the protein → click chain list → Select "Protein A" i.e., A-chain) → paste the amino acid sequence in the Residue → Finally for Atom type the "*" [asterisk symbol] → Add** [Now, the selected amino acid residues will be highlighted stating the active site region] → **Dismiss** (Figure 30.1e – (i)).
7. Let us highlight only the selected amino acid residues surrounding the active site by selecting the **Display → Lines → select "display only" → Ok** (Figure 30.1e – (ii)).
8. Now, let us map the active site using the grid box by selecting, **Grid → Grid box → adjust the x-y-z-dimension to increase the grid box size and Center Grid Box to move the grid box and make sure the active site is entirely inside the grid box → Select File → Close saving current** (Figure 30.1f).
9. Once done, let us export the grid parameter file by selecting, **Grid → Output → Save GPF → Give file name as "grid.gpf".**
10. Finally, let us run the AutoGrid tool by selecting **Run → Run AutoGrid → Program Pathname → browse → autogrid4 and Parameter Filename → Browse → grid.gpf → Launch** (Note 8).
11. Once done, open the "**grid.glg**" log-file, scroll till the end, and verify that the grid generation has been completed. Furthermore, we can proceed with the docking process.

30.3.2.6 Docking

1. To perform docking select, **Docking →Macromolecule → Set Rigid Filename → Select the protein.**
2. Then, import the drug by selecting, **Docking → Ligand → Open → Select the ligand file → Accept**.
3. Select **Docking → Search Parameters → Genetic Algorithm → Accept**.
4. Select **Docking → Docking Parameters → Accept**.
5. Select **Docking → Other Options → AutoDock4.2 Parameters**.
6. Select **Docking → Output → Lamarckan GA →** rename as **"dock.dpf"**.
7. Finally, run the docking by selecting **Run → Run AutoDock → Program Pathname → browse → autodock4 and Parameter Filename → Browse → dock.dpf → Launch** (Note 9).

30.3.2.7 Analysis

1. The docking results were analyzed by selecting, **Analyze → Docking → Open → dock. dlg** (This will open the ligand molecules).
2. Once selected, let us open the protein molecule by selecting, **Analyze → Macromolecules → Choose/open Protein.**
3. Now, let us select, **Analyze → Confirmations → Play, ranked by energy** (This shows all the ten different confirmations of the drug molecule interacted with the protein molecule).
4. Let the drug molecule be in its first conformation and select the "**&**" symbol; this will open the **Set Play Options Tab →** Select **Build H-bonds** and Select **Show Info** (Note 10, Figure 30.1g).
5. The binding energy is crucial information for studying the protein-ligand interaction. Our study results obtained –5.81 as binding energy; if the binding energy is lower, the better the binding energy. Once the binding energy values have been noted, let us save the file as protein and drug complex by selecting the **Set Play Options Tab → Write Complex → rename as "complex.pdbqt"**.
6. Once saved, let us clear the workspace to convert the "complex.pdbqt" file into "complex. pdb" and select **Edit → Delete → Delete all molecules**.

7. Now, let us open the protein-ligand complex by selecting **File → Read Molecule → complex.pdbqt**.
8. Once opened, let us save the file into "***.pdb**" format by selecting **File → Save → Write PDB** (This will convert the "**complex.pdbqt**" file into "**complex.pdb**").
9. Furthermore, software packages can study the protein-ligand interaction; the easiest way is to use the Protein-Ligand Interaction Profiles webserver. Let us open the website by accessing [https://plip-tool.biotec.tu-dresden.de/plip-web/plip/index] and uploading the "**complex.pdb**" file.
10. The results state the interaction of the protein and ligand molecules. This study shows **three hydrogen bonds** formed between the protein and ligand by **LYS, VAL, and GLY** at **101A, 179A, and 190A**, respectively. We can also find the bond angle, distance, and donor and acceptor atom information (Figure 30.1h).
11. Based on the docking analysis, we can conclude that the lamivudine drug molecule showed higher binding affinity (–5.81) with three hydrogen bonds formed by LYS101A, VAL179A, and GLY190A when docked against the HIV-1 reverse transcriptase (1TKX) protein molecule. Study results conclude that lamivudine can be a potential drug candidate to inhibit the HIV-1 reverse transcriptase and control HIV-1.

30.3.3 Protocol for Performing ADME Analysis Using SwissADME Server

1. The ligand molecule to be tested was retrieved as mentioned in Section 30.3.2.2.
2. Open the SwissADME server website [http://www.swissadme.ch/].
3. From the left pane window, select **Open** → Choose the drug molecule (**ligand.pdb**) → Convert the drug molecule into SMILES format → Finally, select "**Run**" (Figure 30.1i).
3. Results state the physicochemical properties, pharmacokinetics, water solubility, druglikeness, lipophilicity, and medicinal chemistry (Figure 30.1j).

30.3.3.1 Findings from SwissADME Analysis

1. The bioavailability radar provides insights into the drug likeliness; the pink region represents optimal range where the solid lines indicate the drug's nature. From the bioavailability radar, we clearly state the selected ligand molecule can be considered a drug molecule.
2. The physicochemical properties state the molecular formula ($C_8H_{11}N_3O_3S$), molecular weight (229.26 g/mol), heavy atoms (15), H-bond acceptors (4) and donors (2), and total polar surface area (TPSA) – 115.67 $Å^2$.
3. The lipophilicity is another physicochemical property identified using particition co-efficient known from Log $P_{o/w}$ (iLOGP – 0.97, XLOGP3 – -0.93, WLOGP – -0.91, MLOGP – -0.61, SILICOS-IT – -0.47), with a consensus log of – -0.39. This indicates the compound has a low lipophilic nature, indicating high water solubility, which increases the bioavailability of the drug in aqueous environment such as oral administration. Whereas the higher lipophilic nature of the drug denotes that the drug can cross lipid membranes and potentially increase the absorption and distribution in the tissues (Note 11)
4. Similarly, the water solubility of the drug molecule is estimated using Log S (ESOL – -0.84, Ali – -1.02, SILICOS-IT – -0.43), indicating the compound lamivudine is highly soluble in water and can be used for appropriate mode of administration which increases the bioavailability of the drug.
5. The pharmacokinetics studies state the drug molecule is highly absorbed in the gastrointestinal tract (GI).
6. Finally, the drug likeliness based on Lipinski, Veber, Egan, and Muegge states the ligand molecule can be considered a drug molecule.

30.4 CONCLUSION

This study explains the protocol for performing molecular docking using the HIV-1 reverse transcriptase enzyme as the target protein and inhibited using the lamivudine drug molecule. The study results suggest the drug molecule showed a significant binding affinity (–5.81), forming three H-bonds on three amino acid residues. Results confirm that the selected drug molecule can possibly inhibit the HIV-1 reverse transcriptase protein. In addition, the ADME analysis identifies the drug's pharmacokinetic nature. They showcased that the drug molecule is highly soluble in Gastrointestinal tract (GI) and satisfies all the pharmacokinetic properties. From the docking and ADME analysis, we could conclude the lamivudine drug molecule can be considered as a potential inhibitor for the HIV-1 reverse transcriptase enzyme.

IMPORTANT NOTES

Note 1. The working directory can be created at any location on the system, depending on the space availability, which promotes easy access to molecules during the docking process.

Note 2. Hence, the default directory has been previously defined; by default, the "D:\autodock" folder will be opened. In addition herein, all the methods performed are carried out using the AutoDock Tools software package.

Note 3. The Gasteiger charges value should always be negative; if it is positive, then there is an issue with the protein molecule, which should be further processed for energy minimization and proceeded further.

Note 4. This step may show a warning message; we can ignore it if it shows any error message; then, we need to investigate the protein molecule further and resolve the error message.

Note 5. In a few instances when we deal with polymer-drug molecules, the number of rotatable bonds will be more than 32. In that case, we can manually set the number of torsions by selecting the **Ligand → Torsion tree → Set the number of Torsions** and set the appropriate number of rotatable bonds.

Note 6. While writing the active site string, we should always mention the amino acid's first three letters in caps, followed by the position of the amino acid, as mentioned.

Note 7. If the dialogue box pops up seeking to preserve the input charges → select "**Yes**". Furthermore, a warning message will appear stating that the Gasteiger charges are already added, and the protein contains no non-bonded atoms; select **"ok"** and ignore the warning message.

Note 8. This process will take a few minutes, depending upon the system configuration. The process running status can be checked using the **"grid.glg"** file by opening the file using notepad software.

Note 9. The docking process will take some time, depending on the system configuration. If the "**AD4.1_bound**" and **"AD4_parameters"** files are not in the same directory, the docking process will not occur and may produce an error and get terminated.

Note 10. This will show the H-bond interaction with the drug, protein molecules, and binding energy values. The lower the binding energy, the better affinity of the drug toward the protein molecule.

Note 11. Lipophilicity and water solubility are important factors in drug development in maintaining the balance of drug polarity, which determines the drug mode of administration, drug permeability across lipid membranes, and absorption and distribution of the drug. Also, the water-soluble drug reduces the accumulation of drugs in lipophilic tissues and increases excretion.

ACKNOWLEDGMENTS

The authors wish to thank the Department of Biotechnology, Govt. of India, under the Aquaculture and Marine Biotechnology scheme (Grant No: BT/PR9733/AAQ/3/649/2013) and Vellore Institute of Technology, Vellore, for providing us with all the necessary financial and infrastructure support.

CONFLICT OF INTERESTS

The authors declare that they have no conflict of interest.

REFERENCES

Aminov RI (2010) A brief history of the antibiotic era: Lessons learned and challenges for the future. *Front Microbiol* 1:1–7. https://doi.org/10.3389/fmicb.2010.00134.

Chaudhary A, Pandya N (2015) Emergence of multidrug resistant *Acinetobacter baumannii* as nosocomial pathogen: Clinical significance and antimicrobial sensitivity. *IOSR J Dent Med Sci* 14:40–44. https://doi.org/10.9790/0853-141064044.

Chaudhary M, Payasi A (2014) Molecular characterization and in vitro susceptibilities of β-lactamase producing *Escherichia coli*, *Klebsiella* species, *Acinetobacter baumannii*, *Pseudomonas aeruginosa* and *Staphylococcus aureus* to CSE1034 and other β-lactams. *Asian Pac J Trop Med* 7:S217–S223. https://doi.org/10.1016/S1995-7645(14)60235-2.

Daina A, Michielin O, Zoete V (2017) SwissADME: A free web tool to evaluate pharmacokinetics, drug-likeness and medicinal chemistry friendliness of small molecules. *Sci Rep* 7:42717. https://doi.org/10.1038/srep42717.

Deeks SG, Overbaugh J, Phillips A, Buchbinder S (2015) HIV infection. *Nat Rev Dis Prim.* https://doi.org/10.1038/nrdp.2015.35.

Drews J (2000) Drug discovery: A historical perspective. *Science* 287:1960–1964. https://doi.org/10.1126/science.287.5460.1960.

Ebhohimen IE, Edemhanria L, Awojide S, et al (2020) Advances in computer-aided drug discovery. In: Egbuna C, Kumar A, Ifemeje JC, Ezzat SM, Kaliyaperumal S (eds), *Phytochemicals as Lead Compounds for New Drug Discovery*. Elsevier: Netherlands, pp. 25–37, https://doi.org/10.1016/B978-0-12-817890-4.00003-2.

Excler JL, Saville M, Berkley S, Kim JH (2021) Vaccine development for emerging infectious diseases. *Nat Med* 27:591–600. https://doi.org/10.1038/s41591-021-01301-0.

Kitchen DB, Decornez H, Furr JR, Bajorath J (2004) Docking and scoring in virtual screening for drug discovery: Methods and applications. *Nat Rev Drug Discov* 3:935–949. https://doi.org/10.1038/nrd1549.

Koch A, Mizrahi V (2018) Mycobacterium tuberculosis. *Trends Microbiol* 26:555–556. https://doi.org/10.1016/j.tim.2018.02.012.

Magiorakos AP, Srinivasan A, Carey RB, et al (2012) Multidrug-resistant, extensively drug-resistant and pandrug-resistant bacteria: An international expert proposal for interim standard definitions for acquired resistance. *Clin Microbiol Infect* 18:268–281. https://doi.org/10.1111/j.1469-0691.2011.03570.x.

Mohammed MA, Salim MTA, Anwer BE, et al (2021) Impact of target site mutations and plasmid associated resistance genes acquisition on resistance of *Acinetobacter baumannii* to fluoroquinolones. *Sci Rep* 11:1–16. https://doi.org/10.1038/s41598-021-99230-y.

Noyal MJC, Menezes GA, Harish BN, et al (2009) Simple screening tests for detection of carbapenemases in clinical isolates of nonfermentative Gram-negative bacteria. *Indian J Med Res* 129:707–712.

31 Methods and Protocols for the Extraction of the Genomic DNA, Gene Cloning, and Heterologous Expression from the Actinomycetes of Extreme Habitats

Hasti Ramavat and Satya P. Singh

31.1 INTRODUCTION

Extremophiles are organisms that thrive in extreme environments. The various kinds of extremities are associated with salinity, pH, temperature, pressure, organic solvent, nutritional availability, and geological conditions. Actinomycetes are reported to live in various extreme habitats, such as marine sediments and highly alkaline and saline environments. The presence of complex stress management mechanisms accounts for survival in these adverse and hostile habitats. The investigations on the diversity of actinomycetes in extreme niches have gained importance as they are known to produce a wide range of different bioactive compounds of commercial importance [1,21,23, 26]. Extremozymes remain active and stable in various extreme conditions of pH, salinity, temperature, hydrostatic pressure, and other extremes. Therefore, the enzymes possessing higher activities in diverse extreme environments are commercially very important [2,10,22,23]. New developments in the field of molecular biology, metabolic engineering, and DNA sequencing have led to an interest in novel extremozymes from actinomycetes. In this context, the straightforward approaches of cloning and expression of recombinant enzymes are highly significant and employed to elevate enzyme production to meet the industrial demand on a large scale at a commercially viable cost. Extremozymes have the ability to meet the requirements of high reproducibility at different extremities [2,3]. The extremozymes possess high potential for various applications, but their commercialization and large-scale applications are still limited [4,5]. Some of the associated problems are related to the long generation time, the lower yield of desired biomolecules, and the lack of knowledge of growth requirements. The main challenge in the industrial application of these extremozymes is the large-scale cultivation of cell factories. In order to maintain the extreme environmental conditions in the fermentation process, the equipment needs to be designed specifically [4,11,16,27]. The recovery and downstream processing of the extremozymes at a large scale are other limitations [4,6]. Therefore, in view of these bottlenecks, the cloning of genes coding for extremozymes and expression into a mesophilic host can provide an alternate way [7].

DOI: 10.1201/9781003398400-31

31.2 MATERIALS

31.2.1 Genomic DNA Extraction

The materials required for genomic DNA extraction are YEME (yeast extract malt extract) broth, NaCl buffer pH9, centrifuge tubes, cooling centrifuge, STE buffer pH8, GET buffer, 20% SDS, lysozyme solution (10 mg/mL in Tris-Cl buffer pH8), P:C:I (25:24:1), C:I (24:1), 3 M potassium acetate, chilled ethanol, TE buffer (pH 8), MiliQ water, –20°C refrigerator, UV-visible spectrophotometer, agarose powder, 1X TAE buffer, EtBr dye, 6X gel loading dye, DNA marker, agarose gel apparatus, gloves, tips, and autopipettes.

31.2.2 PCR Gene Amplification

The materials required for PCR gene amplification are a thermocycler, PCR mastermix, forward primer, reverse primer, template DNA, molecular biology grade water, PCR tubes, gloves, tips, and autopipettes.

31.2.3 Extraction of Plasmids and Restriction Digestion

Escherichia coli Top_{10} cells with the plasmid, LB broth, antibiotics, incubator shaker, microfuge tubes, 10X restriction enzyme (RE) buffer, nuclease-free water, restriction enzyme/s, agarose, and electrophoresis set up are the necessary materials.

31.2.4 Preparation of Competent Cells and Transformation

To prepare competent cells, the materials used are $CaCl_2$, D/w, Whatman filter paper, glycerol, LB broth, ice cubes, and centrifuge tubes

31.2.5 Detection of the Transformants and Gene Expression

To detect the transformants and gene expression, the materials needed are thermocycler, bacterial colony, PCR tube, molecular biology grade water, agarose, UV transilluminator, LB broth, selective antibiotic, orbital shaker, restriction enzyme, X-gal, Petri plates, autopipettes, and gloves.

31.3 METHODS

31.3.1 Extraction of the Genomic DNA

The extraction of the genomic DNA involves the following process.

Growth of the organism in YEME broth with 5% salt and pH9 at 28°C–30°C for 3–4 days [9–11]. Media is supplemented with salt due to the halophilic nature of the actinomycetes [25].

⇩

Cell harvest by centrifugation at 6,708 × *g* (10,000 rpm) for 15 minutes.

⇩

Supernatant is discarded and the cell pellet is suspended in STE buffer (pH8).

⇩

Centrifugation of the cell suspension at 6,708 × *g* (10,000 rpm) for 5–15 minutes and the pellet is retained.

⇩

The pellet is re-suspended in GET buffer (4 M guanidine HCl^{+} 0.01 M EDTA + 0.05 M Tris-Cl pH 8) with SDS (20% w/v) stock solution and freshly prepared lysozyme solution (10 mg/mL in Tris-Cl buffer pH 8), followed by the incubation at 60°C for 2 hours.

⇩

Addition of an equal volume of phenol: chloroform: isoamyl alcohol (25:24:1), followed by centrifugation at (6,708 × *g*) 10,000 rpm for 15 minutes.

⇩

Transfer of the aqueous phase into a fresh tube and addition of an equal volume of C: I chloroform: isoamyl (24:1), followed by centrifugation at 6,708 × *g* for 15 minutes.

⇩

Transfer of the aqueous phase into a fresh tube. The DNA is pooled by adding an equal volume of 3 M potassium acetate and double the volume of chilled ethanol.

⇩

The DNA is then suspended in an appropriate volume of the TE buffer pH-8 or miliQ water and then preserved at −20°C till further use.

⇩

The extracted DNA is analyzed spectrophotometrically by calculating the ratio of 260/280 nm and 260/230 nm to assess the protein impurities and humic acid contamination, respectively. A 260/230 ratio of greater than 2 and a 260/230 ratio near 1.8 indicate the purity of DNA, while lower ratios are indicators of contamination by protein and humic acid. The size and concentration of the extracted DNA are also analyzed on 0.8%–2% (depending on the molecular weight of the DNA) agarose gel electrophoresis [28].

31.3.2 PCR Gene Amplification

In a gradient thermocycler, the PCR technique is generally used to optimize the annealing temperatures for the designed primers. The PCR reaction mixture of 25 μL includes PCR mastermix; forward and reverse primers at the final amount of 25 pmol; and 100 ng of template DNA.

TABLE 31.1
PCR Amplification Conditions and Base Pair Amplification of Different Primer Sets

Primer	Primer Sequences	Initial Denaturation	Denaturation Annealing Extension	Final Extension	No. of Cycles	Amplification	References
U1-F U1-R	5′-AGAGTTTGAT CCTGGCTCAG-3′ 5′-AAGGAGGTGA TCCAGCCGCA-3′	94°C to 10 minutes	94°C to 30 seconds 56°C to 30 seconds 72°C to 1 minute	72°C to 10 minutes	30	1,500 bp	[12]
APrBL NdeI F0 APrBL XhoI R0	5′-CCGCATATGGCGCA GGTTGGAACATTTTG-3′ 5′-CCGCTCGAGCGAGTA GGTCTCTTTTGCAG-3′	95°C to 5 minutes	95°C to 50 seconds 56°C to 45 seconds 72°C to 1 minute 30 seconds	72°C to 7 minutes	35	1,014 bp	[16]
U2-F U2-R	5′-CCAGCAGCCG CGGTAATACG-3′ 5′-ATCGGCTACCT TGTTACGACTTC-3′	94°C to 5 minutes	94°C to 1 minute 55°C to 1 minute 72°C to 1 minute	72°C to 10 minutes	30	1,000 bp	[13]
SPS-5F SPS-5R	5′-GGATCC GCCG CCGAGGACGAC-3′ 5′-GTCGACATGGGA TATTATGAC-3′	95°C to 5 minutes	94°C to 30 seconds 56.1°C to 45 seconds 72°C to 1.5 minutes	72°C to 5 minutes	30	1,200 bp	[11]
Strep-B Strep-E	5′-ACAAGCCCTGG AAACGGGtGT-3′ 5′-CACCAGGAATT CCGATCT-3′	94°C to 8 minutes	95°C to 1 minute 54°C to 40 seconds 72°C to 2 minutes	72°C to 10 minutes	30	520 bp	[14]
Strep-B Strep-F	5′-ACAAGCCCTG GAAACGGGtGT-3′ 5′-ACGTGTGCAGC CCAAGACA-3′	95°C to 8 minutes	95°C to 1 minute 58°C to 40 seconds 72°C to 2 minutes	72°C to 10 minutes	30	1,170 bp	[14]
N-F N-R	5′-CGCATAGGGTGCT GGTGGAAAG-3′ 5′-GAGGTCGGGTTG CAGACTTCG-3′	94°C to 4 minutes	94°C to 30 seconds 56°C to 30 seconds 72°C to 1 minute	72°C to 10 minutes	30	1,120 bp	[15]

The final volume of the reaction mixture is made up to 25 μL with molecular biology grade water. Due to the high G + C content in actinomycetes, the denaturation step is carried out at relatively higher temperatures (Table 31.1).

31.3.3 Extraction of Plasmids and Restriction Digestion [9,11]

Extraction of the plasmid DNA by SDS mini preparation method involves the following process.

E. coli cells having the desired plasmids are inoculated in the LB broth containing respective antibiotics and incubated at 37°C for 12–16 hours.

⇩

Aliquots of the overnight-grown culture are used to extract plasmid [9].

Plasmid DNA is isolated by SDS mini preparation method [9,11].

⇩

E. coli Top_{10} harboring pET 21a+ plasmid is grown at 37°C in LB broth containing ampicillin (50 μg/mL).

⇩

The overnight-grown culture of 3 mL is centrifuged at 10,000 rpm for 1 minute.

⇩

The supernatant is discarded and the pellet is suspended in 1 mL of phosphate buffer saline.

⇩

Centrifuged at 10,000 rpm for 1 minute and the pellet is re-suspend in 150 μL of solution 1.

⇩

To the above mixture, 300 μL of solution 2 is added, vigorously mixed by a vortex, and then incubated on ice for 5 minutes.

⇩

Solution 3 of 200 μL is then added to the above suspension vigorously mixed by a vortex and incubated on ice for 5 minutes.

(Solutions 1, 2, and 3 are prepared as described in reference [9]).

⇩

Centrifuged at 10,000 rpm for 15 minutes to collect the supernatant in a fresh tube.

⇩

An equal volume of chloroform: isoamyl alcohol (24:1) is then added to the supernatant followed by centrifugation at 10,000 rpm for 15 minutes.

⇩

The upper aqueous phase is collected in another tube and half a volume of pre-cooled isopropyl alcohol is added to incubate at room temperature (25°C) for 20 minutes.

⇩

The solution is then centrifuged at 10,000 rpm for 15 minutes, the supernatant is discarded, and the pellets are air dried and dissolved in an appropriate volume of TE buffer.

Restriction digestion [9,11]:

The restriction digestion is carried out in the reaction mixture: 50 ng of pET 21a+ vector/1 μg insert DNA, 2.5 μL 10× RE buffer, and nuclease-free water to make up a final volume of 25 μL.

⇩

The reaction mixture is incubated at 37°C overnight for RE digestion. The selection of the restriction endonuclease is based on the bioinformatics analysis of the cutter and non-cutter sites present in the vector and insert. Besides, the same RE non-cutter site in the gene of interest should also be present in the multiple cloning site (MCS) of the selected vector.

⇩

The status of the digestion is analyzed by loading a 5-μL aliquot of the RE-digested sample along with a marker on agarose gel.

⇩

Finally, the RE-digested vector and the PCR-amplified amplicon are precipitated and purified by ethanol.

31.3.4 Preparation of the Competent Cells and Transformation

The competent cells of the selected bacterial strain are prepared by the procedure mentioned below [19,20]:

Day-wise preparatory steps:

Day 1	⇨	1. Preparation of (10X)1M $CaCl_2$ solution 2. Cells are grown overnight at 37°C
Day 2	⇨	1. The overnight grown cells are sub-cultured in LB broth for 3-4 hrs 2. $CaCl_2$ washing step
Day 3	⇨	1. Transformation by heat shock 2. Spreading of the medium with the recombinant plasmid on LB+ antibiotic plates and incubation at 37°C overnight
Day 4	⇨	The growth is observed for the **isolated colonies**

Prepare the buffer and overnight-grown culture a day prior to the experiment.

⇩

For the preparation of 1 M $CaCl_2$ stock solution (10×), 11.1 *g* $CaCl_2$ was initially dissolved in 80 mL D/W, and finally, the volume was adjusted to 100 mL followed by filtration on the Whatman filter paper.

⇩

Working solution of 0.1 M $CaCl_2$ is prepared to dilute the stock solution, 1:10 in D/W followed by its filtration.

⇩

$CaCl_2$–glycerol mixture is prepared by adding 6 mL of 0.1 M $CaCl_2$ to 9 mL sterile glycerol and 45 mL D/W.

⇩

The overnight-grown culture of *E. coli* is inoculated into 1 mL LB broth and incubated at 37°C in shaking condition for overnight.

⇩

The overnight-grown culture of 1 mL *E. coli* is inoculated to 99 mL of fresh LB broth and incubated at 37°C for 3–4 hours until the absorbance reaches 0.4.

⇩

All the tubes and $CaCl_2$ solution should be kept under chilled condition at 4°C. The culture is filled into the tubes and placed on ice for 20 minutes.

⇩

The cultures are centrifuged at 4,000 rpm for 10 minutes at 4°C. The supernatant is discarded and the pellets are re-suspended in 20 mL of chilled 0.1 M $CaCl_2$ solution followed by its incubation on ice for 30 minutes.

⇩

After centrifugation at 4,000 rpm for 10 minutes at 4°C, the supernatant is discarded and the pellet is re-suspended in 5 mL ice-cold 0.1 M $CaCl_2$ with 15% glycerol.

⇩

Plasmid DNA of 1–5 μL (10 pg–100 ng) is added to the cell suspension and insulated on ice for 30 minutes.

⇩

The cell suspension is then transferred to a water bath at 42°C for 30 seconds and incubated on ice for 2 minutes.

⇩

The pre-warmed LB broth of 1 mL is then added to it and incubated in shaking condition at 37°C for 1–2 hours.

⇩

The cultures are spread in 1:10, 1:100, and 1:100 dilutions on the LB plates containing respective antibiotics and incubated overnight at 37°C.

⇩

The plates are then observed for the isolated colonies on the LB plates with antibiotics.

The competent cells are also prepared by the physical and chemical methods with certain specific modifications [17,18].

31.3.5 Detection of the Transformants and Gene Expression

Confirmation of transformation and detection of gene expression is done by the following strategies [9,16,29].

31.3.5.1 Colony PCR [16,30,31]

The colony PCR is used to screen the intact bacteria for the presence of the desired DNA sequence.. This method is used to screen the colonies after transformation with a recombinant plasmid.

A portion of the colony of the positive clone is taken by a sterile toothpick and dissolved in the PCR tube containing the PCR reaction mixture. The portion of the colony is directly used as a source of the template for the PCR.

⇩

The initial step of the denaturation is carried out at 96°C for 10 minutes.

⇩

The reaction mixture is prepared and optimized for the PCR cycle to specifically amplify the gene of interest.

⇩

The resultant PCR product is analyzed on 1.2% agarose gel as described in the above section.

31.3.5.2 Screening for the Restriction Digestion [16,29]

The bacterial colonies of the positive clone are selected and used to inoculate 5 mL of LB broth-containing medium supplemented with a suitable antibiotic.

⇩

The inoculated culture is incubated overnight in shaking conditions at 37°C. The plasmid extract is carried out as per the method described above.

⇩

The extracted plasmid is digested by the selected REs and analyzed by the agarose gel electrophoresis to confirm the cloning.

31.3.5.3 Blue–White Screening

A reporter gene, such as the LacZ coding for the β-galactosidase, is used. The detection is based on α complementation where a fragment of the gene LacZ α complements with the mutant LacZ gene of the cell.

⇩

The individual cell produces non-functional peptides. But when a plasmid containing LacZ α gene is transformed into the mutant LacZ-containing cells, functional β-galactosidase is produced.

⇩

The cells grown on a plate containing X-gal form blue colonies due to the production of active β-galactosidase.

⇩

Multiple cloning sites are present within the LacZ α gene. Successful ligation disrupts the LacZ α gene, resulting in white colonies. The gene expression is also detected by the expression of the green fluorescent protein. The expression of this protein confirms the expression of our gene of interest due to the fluorescence. Besides, sequence-specific probes with fluorescent dyes are used in quantitative PCR for the analysis of gene expression.

This chapter describes the general protocols for DNA extraction, molecular cloning of the desired gene, and detection of the gene expression with particular reference to halophilic bacteria. The illustrated protocols would be useful to those contemplating the cloning of the genes of extremophilic microorganisms and their expression in the mesophilic host.

ACKNOWLEDGMENTS

HR acknowledges the award of the ShoDH Fellowship to her from the Government of Gujarat. SPS is grateful for the award of the DBT-Multi-Intuitional Project, MoES Net Working Project, UGC-BSR Faculty Fellowship, and International Travel Fellowships from DST-SERB, CSIR, DBT, and UGC. Besides the support from Saurashtra University, the facilities created under the UGC-CAS and DST-FIST are duly acknowledged. The contributions of Dr. Sangeeta Gohel as Co-PI and Dr. Dalip Singh Rathore and Dr. Mahejbin Sheikh as research fellows in the MoES-Project on marine actinomycetes are duly acknowledged.

REFERENCES

1. Gohel, S.D., Sharma, A.K., Dangar, K.G., Thakrar, F.J. and Singh, S.P., 2018. Biology and applications of halophilic and haloalkaliphilic actinobacteria. In *Extremophiles* (pp. 103–136), Rao, S. (eds). CRC Press, Florida.
2. Fenice, M., Khare, S.K. and Gorrasi, S., 2021. Mining, designing, mechanisms and applications of extremophilic enzymes. *Frontiers in Microbiology*, *12*, p. 709377.
3. Yahaya, R.S.R., Normi, Y.M., Phang, L.Y., Ahmad, S.A., Abdullah, J.O. and Sabri, S., 2021. Molecular strategies to increase keratinase production in heterologous expression systems for industrial applications. *Applied Microbiology and Biotechnology*, *105*(10), pp. 3955–3969.
4. Chettri, D., Verma, A.K., Sarkar, L. and Verma, A.K., 2021. Role of extremophiles and their extremozymes in biorefinery process of lignocellulose degradation. *Extremophiles*, *25*(3), pp. 203–219.
5. Sarmiento, F., Peralta, R. and Blamey, J.M., 2015. Cold and hot extremozymes: industrial relevance and current trends. *Frontiers in Bioengineering and Biotechnology*, *3*, p. 148.
6. Gabani, P. and Singh, O.V., 2013. Radiation-resistant extremophiles and their potential in biotechnology and therapeutics. *Applied Microbiology and Biotechnology*, *97*(3), pp. 993–1004.
7. Gomes, J., Gomes, I. and Steiner, W., 2000. Thermolabile xylanase of the Antarctic yeast *Cryptococcus adeliae*: production and properties. *Extremophiles*, *4*(4), pp. 227–235.
8. Albuquerque, P., Ribeiro, I., Correia, S., Mucha, A.P., Tamagnini, P., Braga-Henriques, A., Carvalho, M.D.F. and Mendes, M.V., 2021. Complete genome sequence of two deep-sea streptomyces isolates from Madeira archipelago and evaluation of their biosynthetic potential. *Marine Drugs*, *19*(11), p. 621.
9. Sambrook, J., Fritsch, E.F. and Maniatis, T., 1989. *Molecular Cloning: A Laboratory Manual*. Cold Spring Harbor Lab. Press, Cold Spring Harbor, NY.
10. Gohel, S.D. and Singh, S.P., 2018. Molecular phylogeny and diversity of the salt-tolerant alkaliphilic actinobacteria inhabiting Coastal Gujarat, India. *Geomicrobiology Journal*, *35*(9), p. 775789. https://doi.org/10.1080/01490451.2018.147117.
11. Gohel, S.D. and Singh, S.P., 2012. Cloning and expression of alkaline protease genes from two salt-tolerant alkaliphilic actinomycetes in *E. coli*. *International Journal of Biological Macromolecules*, *50*(3), pp. 664–671.
12. Edwards, U., Rogall, T., Bocker, H., Emde, M. and Bottger, E.C., 1989. Isolation and direct complete nucleotide determination of entire genes. Characterization of a gene coding for 16S ribosomal DNA. *Nucleic Acids Res*earch, *17*, pp. 7843–7853. https://doi.org/10.1093/nar/17.19.7843.
13. Lu, J.J., Perng, C.L., Lee, S.V. and Wan, C.C., 2000. Use of PCR with universal primersand restriction endonuclease digestions for detection and identification f common bacterial pathogens in cerebrospinal fluid. *Journal of Clinical Microbiology*, *38*, pp. 2076–2080.
14. Rintala, H., Nevalainen, A., Ronka, E. and Suutari, M., 2001. PCR primers targeting the 16S rRNA gene for the specific detection of *Streptomycetes*. *Molecular and Cellular Probes*, *15*, pp. 337–347. https://doi.org/10.1006/mcpr.2001.0379.
15. Zhi, X., Yang, L., Wang, Y., Tang, S., Lou, K. and Li, W., 2008. Genus-specific PCR for molecular identification of novel isolates of the genus *Nesterenkonia*. *Acta Microbiologica Sinic*, *48*(5), pp. 644–650.
16. Bhatt, H.B. and Singh, S.P., 2020. Cloning, expression, and structural elucidation of a biotechnologically potential alkaline serine protease from a newly isolated haloalkaliphilic Bacillus lehensis JO-26. *Frontiers in Microbiology*, *11*, p. 941.
17. Nakata, Y., Tang, X. and Yokoyama, K.K., 1997. Preparation of competent cells for high-efficiency plasmid transformation of Escherichia coli. In *cDNA Library Protocols* (pp. 129–137), Cowell, I.G., Austin, C.A. (eds). Humana Press, New Jersey.

18. Green, M.R. and Sambrook, J., 2020. The Inoue method for preparation and transformation of competent *Escherichia coli*: "Ultracompetent" cells. Cold Spring Harbor Protocols, 2020(6), p. 101196.
19. Nishimura, A., Morita, M., Nishimura, Y. and Sugino, Y., 1990. A rapid and highly efficient method for preparation of competent Escherichia coli cells. *Nucleic Acids Research, 18*(20), p. 6169.
20. Chang, A.Y., Chau, V., Landas, J.A. and Pang, Y., 2017. Preparation of calcium competent Escherichia coli and heat-shock transformation. *JEMI Methods, 1*, pp. 22–25.
21. Rathore, D.S., Sheikh, M.A., Gohel, S.D. and Singh, S.P., 2021. Genetic and phenotypic heterogeneity of the Nocardiopsis alba strains of seawater. *Current Microbiology, 78*(4), pp. 1377–1387.
22. Sharma, A.K., Kikani, B.A. and Singh, S.P., 2020. Biochemical, thermodynamic and structural characteristics of a biotechnologically compatible alkaline protease from a haloalkaliphilic, *Nocardiopsis dassonvillei* OK-18. *International Journal of Biological Macromolecules, 153*, pp. 680–696.
23. Thumar, J. and Singh, S.P., 2022. Antimicrobial potential and metabolite profiling ofmarine actinobacteria. In *Actinobacteria* (pp. 241–264), Karthik, L. (eds). Springer, Singapore.
24. Chen, X., Zheng, Z., Li, F., Ma, X., Chen, F., Chen, M. and Tuo, L., 2023. Description and genomic characterization of Nocardioides bruguierae sp. nov., isolated from *Bruguiera gymnorhiza. Systematic and Applied Microbiology, 46*, p. 126391.
25. Kuzmanovic, N., Bunk, B., Sproeer, C., Fruehling, A., Neumann-Schaal, M., Overmann, J. and Smalla, K., 2023. Genomics of the "tumorigenes" clade of the family Rhizobiaceae and description of *Rhizobium rhododendri* sp. nov. *bioRxiv. MicrobiologyOpen, 12*(2), p.e1352.
26. Ibrahimi, M., Loqman, S., Jemo, M., Hafidi, M., Lemee, L. and Ouhdouch, Y., 2023. A facultative predatory potential *Actinobacteria* spp. and prospects in the plant disease biocontrol's advances. *Frontiers in Microbiology, 13*, p. 5402.
27. Raval, V.H., Rawal, C.M., Pandey, S., Bhatt, H.B., Dahima, B.R. and Singh, S.P., 2015. Cloning, heterologous expression and structural characterization of an alkaline serine protease from sea water haloalkaliphilic bacterium. *Annals of Microbiology, 65*(1), pp. 371–381.
28. Leite, D.C., Balieiro, F.C., Pires, C.A., Madari, B.E., Rosado, A.S., Coutinho, H.L. and Peixoto, R.S., 2014. Comparison of DNA extraction protocols for microbial communities from soil treated with biochar. *Brazilian Journal of Microbiology, 45*(1), 175–183. https://doi.org/10.1590/s1517-83822014000100023.
29. Bertero, A., Brown, S. and Vallier, L., 2017. Methods of cloning. In *Basic Science Methods for Clinical Researchers* (pp. 19–39), Jalali, M., Saldanha, F. and Jalali, M. (eds). Academic Press, Cambridge, MA.
30. Azevedo, F., Pereira, H. and Johansson, B., 2017. Colony PCR. In *PCR* (pp. 129–139), Domingues, L. (eds). Springer, New York.
31. Woodman, M.E., 2008. Direct PCR of intact bacteria (colony PCR). *Current Protocols in Microbiology, 9*(1), pp. A–3D.

32 Activation of Silent Genes in Actinomycetes by Different Methods

Veilumuthu Pattaulavar, Ramamoorthy Siva, and John Godwin Christopher

32.1 INTRODUCTION

The Actinomycetes are a phylum of unicellular filamentous gram-positive bacteria. They are of great economic importance to humans. They produce a variety of bioactive secondary metabolites, including enzymes, immunosuppressive drugs, anticancer drugs and antibiotics [1]. Actinomycetes produce nearly 10,000 of the 23,000 recorded bioactive secondary metabolites produced by microbes, accounting for 45% of all bioactive microbial metabolites [2]. Moreover, about 7,600 bioactive compounds are produced by the *Streptomyces* genus alone. The new bioinformatic technologies and methods, such as genome mining, reveal that Actinomycetes can synthesise some 150,000 new antimicrobial compounds than those currently known [3].

The discovery of new bioactive compounds is significantly crucial in applied microbiology, clinical studies and agriculture. Especially, new bioactive compounds from natural (microbial or plant) products are precious sources that help us to create new drugs to cure diseases or to overcome antibiotic resistance. Therefore, strategies to activate or up-regulate the "silent" gene clusters have recently gained attention to enhance natural product production. Several techniques to activate silent BGCs have been established. This methodology focuses on developing effective strategies to boost or trigger the activation of cryptic genes [4]. Elicitors can be any compounds that stimulate the activation or synthesis of a specific metabolite. In particular, amino acids, minerals, carbon and nitrogen sources have served as elicitors in expressing or activating certain silent gene clusters in Actinomycetes [5]. However, it can be difficult to find the right elicitors that can cause a silent gene to express. Recently, a high-throughput platform was created to find Actinomycete elicitors of silent gene clusters, enabling the evaluation of the expression of known cryptic gene clusters of interest [6].

32.2 REQUIREMENTS

1. ISP2 agar medium (malt extract –10.0 g, dextrose –4.0 g, yeast extract –4.0 g, agar –20.0 g and distilled water 1,000.0 mL – pH 7.2).
2. c-TAB and phenol: chloroform
3. Elicitors (glycerol, ethanol, tryptophan and glycine (1 mg/mL))
4. Ethyl acetate, Na_2SO_4 and MeOH
5. Rotary vacuum evaporator
6. Deuterated solvents (DMSO (Dimethyl sulfoxide), water, methanol and chloroform)
7. Instruments – GC – MS, LC-MS, HR-MS, 2D and 3D NMR

DOI: 10.1201/9781003398400-32

32.3 BIOINFORMATICS

1. Complete genome sequence was collected from NCBI.
2. Silent biosynthetic gene cluster analysis can be done using a software tool such as the fungal and bacterial versions of antiSMASH at https://fungismash.secondarymetabolites.org and https://antismash.secondarymetabolites.org, respectively.
3. BLAST program helps to analyse NRPS and PKS predicted proteins which were carried out using programs NRPS and ASMPKS predictor.

32.4 PROTOCOL

1. Actinomycete strains with some unique characters were selected (such as antimicrobials, pigments and enzymes).
2. C-TAB and the phenol:chloroform DNA extraction method was followed. Then, RNase treatment for DNA purification was done. NanoDrop was used to quantify the DNA. This genomic DNA was sequenced using the Illumina NextGen sequencing method.
3. The bioinformatics software "antiSMASH" was used to analyse the Actinomycete chromosomal DNA utilising computer-aided database searching and sequence analysis [7]. The software provides the details of PKS, NRPS, Ripp, ASMPKS and lanthipeptide cluster [8].
4. This result will highlight all the available genes, including silent genes.
5. Seed cultures are prepared by adding fresh spores to a 5-mL ISP2 broth in a 20-mL sterile tube.
6. The culture was incubated at 30°C and 150 rpm for 3 days and then diluted into 500 mL of ISP2 broth (in a 1,000-mL Erlenmeyer flask) to a final concentration of 0.05% (w/v).
7. Elicitors (glycerol/ethanol/tryptophan/glycine) were added to the ISP2 broth, which are well-known with Actinomycetes at the desired concentrations of 5 mL/L [5].
8. The cultures were once again grown at 30°C and 150 rpm.
9. After seven days, cells were removed by centrifuging at 5,000 rpm for 5 minutes, and the resultant supernatant was twice extracted with 1,000 mL of ethyl acetate.
10. The extracted ethyl acetate was dried over Na_2SO_4 and subsequently evaporated to dryness in a vacuum.
11. The extracted concentrated metabolite is stored in the refrigerator until further analysis.
12. This concentrated metabolite is dried in a vacuum evaporator and dissolved in 100 µL MeOH.
13. GC-MS analysis is done, as it uses mass spectrum databases for preliminary chemical identification and peak interpretation.
14. Purification of crude extract through collection of fraction and separations can be done by liquid chromatography mass spectrometry (LC-MS).
15. Fractions collected from the LC-MS were purified; then, the compound was taken to detect its molecular mass and structure [9].
16. To confirm their structures, comprehensive NMR data sets were analysed. All 2D/3D NMR spectra were collected in MeOH-d4 in the triple resonance cryoprobe of an A8 Avance III HD 800 MHz NMR spectrometer (Bruker). NMR data were analysed in MestReNova [10].

32.5 OBSERVATIONS AND RESULTS

Beyond this, the actual potential of the organism is revealed in the whole genome sequence. Different bioinformatic tools can be used, such as antiSMASH (Table 32.1), ARTS, CLUSEAN and SMURF. Based on this result, different elicitors like glycerol, ethanol and amino acids can be used to enhance or activate those silent or hidden genes.

The chromatography of GC-MS shows all the compounds in the media at the given time. Their results are tabulated in Table 32.2.

32.6 RESULTS

TABLE 32.1
Potential BGCs of Secondary Metabolites in Actinomycetes Predicted Using AntiSMASH 6.0 (Descending Order) with Few Examples

S. No.	Scaffolds	Nature of Cluster	Corresponding Known Cluster	Resemblance (%)
1	Core hits –17.2	Terpene	Geosmin	100
2	Core hits –7.1	Ectoine	Ectoine	100
3	Core hits –21.1	T1PKS	Streptovaricin	29
4	Core hits –20.1	T1PKS	Nystatin A1	27
5	Core hits –3.2	T3PKS	Herboxidiene	8
6	Core hits –14.1	Butyrolactone	Prejadomycin/Rabelomycin/Gaudimycin C/Gaudimycin D/Uwm6/Gaudimycin A	6
7	Core hits –21.2	Terpene	Versipelostatin	5

TABLE 32.2
Compounds from GC-MS Identified in *Streptomyces* sp.

S. No	Chemical Compound	RT	Mol. Weight	Molecular Formula	Area %	Nature of the Compound	Activity	Spectrum of Action	Ref
1	Phenol, 2,4-*bis*(1,1-dimethylethyl)	13.803	138.21	$C_9H_{14}O$	1.68	Cyclohexane	Antimicrobia	Broad	[1]
2	*N*-Dimethylhexanamide	15.648	128.17	$C_{10}H_8$	0.41	Methylene	Antimicrobia	Broad	[2]
3	1-Octadecene	15.791	190.12	$C_8H_5F_3O_2$	0.50	Ester	Antimicrobia	Broad	[3]
4	Thiophene, 2-isobutyl-5-propyl-	16.353	164.20	$C_{10}H_{12}O_2$	2.59	Ester	Antimicrobia	Broad	[4]
5	4-Aminothiophenol	17.024	278.5	$C_{17}H_{30}OSi$	1.13	Phenolic	Antimicrobia	Narrow	[5]

REFERENCES

1. Selim, Manal Selim Mohamed, Sayeda Abdelrazek Abdelhamid, and Sahar Saleh Mohamed. "Secondary Metabolites and Biodiversity of Actinomycetes." *Journal of Genetic Engineering and Biotechnology* 19, no. 1 (2021): 1–13. https://doi.org/10.1186/S43141-021-00156-9/.
2. Pham, Janette V., Mariamawit A. Yilma, Adriana Feliz, Murtadha T. Majid, Nicholas Maffetone, Jorge R. Walker, Eunji Kim, et al. "A Review of the Microbial Production of Bioactive Natural Products and Biologics." *Frontiers in Microbiology* 10 (2019): 1404. https://doi.org/10.3389/FMICB.2019.01404.
3. Lacey, Heather J., and Peter J. Rutledge. "Recently Discovered Secondary Metabolites from Streptomyces Species." *Molecules* 27, no. 3 (2022): 887. https://doi.org/10.3390/MOLECULES27030887.
4. Ochi, Kozo. "Insights into Microbial Cryptic Gene Activation and Strain Improvement: Principle, Application and Technical Aspects." *The Journal of Antibiotics* 70, no. 1 (2016): 25–40. https://doi.org/10.1038/ja.2016.82.
5. Zong, Gongli, Jiafang Fu, Peipei Zhang, Wenchi Zhang, Yan Xu, Guangxiang Cao, and Rongzhen Zhang. "Use of Elicitors to Enhance or Activate the Antibiotic Production in Streptomyces." *Critical Reviews in Biotechnology* 42, no. 8 (2022): 1260–1283. https://doi.org/10.1080/07388551.2021.1987856.
6. Okada, Bethany K., and Mohammad R. Seyedsayamdost. "Antibiotic Dialogues: Induction of Silent Biosynthetic Gene Clusters by Exogenous Small Molecules." *FEMS Microbiology Reviews* 41, no. 1 (2017): 19–33. https://doi.org/10.1093/FEMSRE/FUW035.
7. Blin, K., S. Shaw, A. M. Kloosterman, Z. Charlop-Powers, G. P. van Wezel, M. H. Medema, and T. Weber. "AntiSMASH 6.0: Improving Cluster Detection and Comparison Capabilities." *Nucleic Acids Research* 49, no. W1 (2021): W29–W35. https://doi.org/10.1093/NAR/GKAB335.
8. Rausch, Christian, Tilmann Weber, Oliver Kohlbacher, Wolfgang Wohlleben, and Daniel H. Huson. "Specificity Prediction of Adenylation Domains in Nonribosomal Peptide Synthetases (NRPS) Using Transductive Support Vector Machines (TSVMs)." *Nucleic Acids Research* 33, no. 18 (2005): 5799–5808. https://doi.org/10.1093/NAR/GKI885.
9. Park, Jin Soo, Da Eun Kim, Sung Chul Hong, Seung Young Kim, Hak Cheol Kwon, Chang Gu Hyun, and Jaeyoung Choi. "Genome Analysis of Streptomyces Nojiriensis Jcm 3382 and Distribution of Gene Clusters for Three Antibiotics and an Azasugar across the Genus Streptomyces." *Microorganisms* 9, no. 9 (2021): 1802. https://doi.org/10.3390/MICROORGANISMS9091802.
10. Mothana, Azal A., Hassan A. Al-Shamahy, Ramzi A. Mothana, Jamal M. Khaled, Adnan J. Al-Rehaily, Abdullah Y. Al-Mahdi, and Ulrike Lindequist. "*Streptomyces* Sp. 1S1 Isolated from Southern Coast of the Red Sea as a Renewable Natural Resource of Several Bioactive Compounds." *Saudi Pharmaceutical Journal* 30, no. 2 (2022): 162–171. https://doi.org/10.1016/J.JSPS.2021.12.012.

33 Multimodal Spectroscopy-Based Proteomics and Metabolomics Approach to Explore Actinomycetes

Akhiya A. R., Vartika Lohani, Soumen Kundu, Md Quasid Akhter, Balendu Upmanyu, and Swarnendu Bag

33.1 INTRODUCTION

The order of bacteria known as Actinomycetes is well known for its prolific production of bioactive, specialized metabolites. Actinomycetes (Phylum: Actinobacteria), a subclass of bacteria, are one of the most abundant sources of bioactive natural compounds with a variety of biological functions [1]. Actinomycetes continue to be the most abundant source of antibiotics since their discovery in the 1940s and 1950s. Actinomycetes (actinobacteria) have received much more attention nowadays, as these bacteria can produce a variety of natural drugs, natural products, and other bioactive metabolites, including antibiotics, enzyme inhibitors, and enzymes [2]. Actinomycetes produce a variety of metabolic products that exhibit biological processes important to human health, agriculture, and the economy. Approximately 160 antibiotics from various bacteria have been used for human therapy and agriculture, and about 100–120 of these compounds, including streptomycin, erythromycin, gentamicin, vancomycin, and avermectin, are produced mainly by Actinomycetes [3]. The genomes of the Actinomycetes have high guanine and cytosine (G+C) content, complex life cycles, and a filamentous morphology [4].

Bioactive substances are primarily and naturally produced by bacteria. For the creation of new medications, numerous laboratories and researchers have concentrated on studying actinomycete sources from novel ecosystems, including oceans, harsh environments, plants, and lichens. High-throughput (omics) techniques have improved our comprehension of the complex mechanisms controlling the expression of biosynthetic gene clusters (BGCs) producing secondary metabolites [5].

More than 2,000 proteins are produced from the genomic data in a microbial cell, and a significant portion of these can be investigated utilizing proteomics [6]. More than half of the projected proteome for genomes with less than 1,000 genes is thought to be able to be identified from the genome. Similar to how 30% and 10% of the expected proteome from genomes with about 2,500 and 4,000 genes, respectively, can be found. The application of proteomics may be able to reveal information of a significant portion of the microbe's proteome in a microbial genome with 600–7,000 predicted genes, which constitutes a medium-sized, complicated system [7].

With regard to the phenotypic-genotypic dichotomy, the analysis of the cellular proteome occupies a middle ground because the proteins under study are both metabolic by-products and products of genes [8]. The usage of DNA fingerprinting techniques has grown significantly over the past 20 years, and it is now clear that these techniques can be used to identify and categorize microbes. However, in recent years, automation in proteomics technology has increased its throughput and

DOI: 10.1201/9781003398400-33

potential use for a variety of microbiological purposes, including strain typing and epidemiological studies, identification of microbes inhabiting a particular ecosystem, detection of biological warfare agents, detection of water- and food-borne pathogens, detection of blood and urinary tract pathogens, and detection of antibiotic resistance [9]. This chapter provides a quick overview of different isolation techniques and methodologies based on mass spectrometry to identify actinobacteria.

Actinomycetes revealed their potential metabolic landscape and, consequently, a far greater potential to produce specialized metabolites that had been discovered by traditional screening-based methods in the previous 50 years. With the advent of next-generation sequencing (NGS) technologies and the ability to routinely obtain good quality whole genome sequences in fact, the genome sequences showed that a typical Actinomycete has the genetic potential to produce roughly ten times as many secondary metabolites as have been found through screening and chemical analytics. This is because actinomycete species typically contain about 30 secondary metabolite gene clusters on an average [10].

Bacterial natural products become a good source for drug discovery-related research. Actinomycetes, Gram-positive filamentous bacteria, are well known for its high abundance of natural products having varieties of secondary metabolites with high biological importance. More than 70% of the secondary metabolites are produced by genus Streptomyces (under order Actinomycetes). Polyketide and non-ribosomal peptide synthetases are the two main enzymes that produce these large varieties of secondary metabolites [11].

Metabolites from Actinomycetes can be used against bacteria, viruses, fungi, nematodes, and insects, even in cancer therapy due to their anticancer properties. Actinomycetes produce some enzymes that are useful for various fields. For industrial purposes, chitinases, amylases, and cellulases are used; phosphatase and nitrogenase are used for plant growth. They also produce different VOCs (volatile organic compounds) having antimicrobial properties. Even their secreted enzymes digest insect's exoskeleton (chitin) and wood (cellulose, lignin etc.). Actinomycetes colonize different plant's roots and convert different bioorganic compounds to simplified forms (such as amino acids, carbohydrates, vitamins, lipids, and organic acids) that make the soil more nutritive and favourable for better plant growth, which helps plant growth and also acts as pesticides [12].

Naturally derived metabolites found in the Actinomycetes are erythromycin, chloramphenicol, gentamicin, kanamycin, neomycin, streptomycin, tetracyclines, valinomycin, daptomycin, platensimycin, and rifamycin as antimicrobial compound; bleomycin, actinomycin D, anthracyclines, mitomycin, and enediyne as anticancer compound; and avermectins, milbemycins, and spinosyns as antiparasitic or insecticidal compound [13].

Similarly, different bioactive compounds were also found like abyssomicin, actinofuranones A–B, and benzanthraaquinone having both antibacterial and antitumoral activities, and analogs-metacycloprodigiosin, butenolides, mechercharmycin, and diphosphatidylglycerol having only antitumoral activity [14].

Untargeted metabolomics can be a useful tool to identify different metabolites from these bacteria to understand their level of antimicrobial activity. LC-MS- and NMR-based approaches fall under metabolomics study. Recently, liquid chromatography-coupled high-resolution mass spectrometry is frequently used for this kind of study due to high-throughput data and better resolutions. In this chapter, we explained the sample preparation and LC-MS data analysis for untargeted metabolomics [15].

These findings significantly influenced the recent resurgence of interest in the study of natural products and these abundant microorganisms. Furthermore, the introduction of high-throughput (omics) methods, genomics, transcriptomics, proteomics, metabolomics, and computational tools has significantly improved our comprehension of complicated cellular physiology and essential pathways controlling the generation of secondary metabolites [10].

33.2 MATERIALS AND METHODS

33.2.1 Principle for the Isolation of Actinobacteria

- The investigation of the actinomycete population in a unique habitat comes first. All Actinomycetes in this situation should be separated and recognized as pure cultures. The isolation media that are employed must be favourable to the growth of additional Actinomycetes while inhibiting the growth of other microorganisms in order to achieve this goal.
- The second step is to isolate specific Actinomycetes, such as those belonging to a known species or genus or having unique physiological traits including tolerance to chemicals, alkalis, acids, salts, high and low temperatures, and antibiotics.
- The isolation of unidentified Actinomycetes comes in third. Numerous Actinomycetes have so far been isolated and identified from a variety of locations around the globe [16,17].

33.2.2 Collection and Pre-treatment of Samples (Actinobacteria)

Actinomycetes are saprophytes that can be found in a variety of natural environments, such as soil, lakes, oceans, plants, and animals. Novel actinobacteria continue to flourish in soil. Ecological parameters like nutrition, pH, temperature, salinity, aeration, moisture, and organic matter composition have a significant impact on the types and numbers of actinobacteria found in soil and various different substrates. In fact, the selection of environmental samples can have a significant impact on the success of isolating significant quantities of a certain actinobacteria [18].

It is better to get soil samples from pristine areas, such as desert, saline, and alkaline soils, and ancient forests. A total of 5–20 cm deep soil samples are taken and placed in sterile paper or plastic bags. Oceanic Actinomycetes are extensively dispersed, and several natural compounds have been derived from these organisms. With the help of a sampler, sediment from the deep ocean is gathered; these samples are then stored in sterile glass bottles at a temperature of 4°C.

Actinomycetes must first undergo pre-treatment since they grow more slowly than other bacteria and fungi. Pre-treatment regimens often target Actinomycetes by preventing or getting rid of undesirable microbes. Isolation of Actinomycetes from these samples is done via a number of chemical and physical pre-treatments. Since most bacteria are less resistant to desiccation than actinomycete spores, most undesired Gram-negative bacteria, sediments, and lichens can be removed by air-drying the soil sample at room temperature. This prevents the isolation plates from becoming overrun with these bacteria.

Ultrasonic waves can be used to treat sample suspensions for 40' at 180 W. It can increase the proportion of Actinomycetes, decrease the number of unwanted bacteria in the sample, and release the saprophytes fixed by soil granules into the solution. Different chemicals, such as benzethonium chloride, chlorhexidine gluconate, phenol, and SDS, should be used to separate unique actinobacterial taxa based on the variable resistance of actinobacterial spores to treatment. These substances can kill Gram-negative bacteria, endospore-forming bacilli, and pseudomonads for 30 minutes at 30°C while also increasing the frequency of actinomycete and lowering bacterial counts [17,19].

33.2.3 Isolation of Actinobacteria

For this investigation, a significant number of Actinomycetes should be prospectively studied. A sample of actinobacteria should be isolated using dried sandy soils with a pH of 7.0 that have been collected from diverse sites. It is possible to isolate Actinomycetes using a variety of mediums and to test them for antibacterial activity against a wide range of Gram-positive and Gram-negative bacteria as well as yeast. For efficient antimicrobial production, a number of fermentation variables and processes should be improved.

33.2.3.1 Sample Collection

Soil samples from various areas should be collected using sterile polythene bags and other tools. These samples need to be air-dried for a minimum of 1 week before being used for the isolation of Actinomycetes [20,21].

33.2.3.2 Soil Treatment and Isolation

One gram of soil should be dissolved in 9 mL of sterile distilled water. After settling in 10–15 dilutions, 0.1 mL of 10–12, 10–13, 10–14, and 10–15 dilutions should be added to starch casein agar. For easier comprehension, the actinomycete colonies were given names. These colonies should be isolated from the packed plate of bacteria for future research and incubated at 28°C.

Each of these separate actinomycete colonies should be subcultured in the starch casein plate after being divided into several sub-colonies. These plates should be incubated for 7 days at 28°C. Colonies will then form, which should be subcultured in a test tube containing SCA slants, incubated at 28°C for 7 days, and stored as the stock in the refrigerator. The isolates should be subcultured once every 2–3 weeks to keep the cultures from drying off [20,22].

33.2.3.3 SubCulturing of the Samples

The duplicate perpendicular streak method should be utilized for the initial assessment of an isolated strain's ability to inhibit other bacterial species. A Mueller-Hinton agar plate should have the isolated bacteria, streaked along the centre of the plate. The bacteria should then be cultured at 28°C until Actinomycetes are evident as an adhering vertical colony line. Following the incubation, each of the other standard organisms (Staphylococcus aureus, Pseudomonas aeruginosa, Escherichia coli, Proteus vulgaris, Candida albicans, Salmonella typhimurium, Aspergillus niger, Klebsiella pneumoniae, and Bacillus subtilis) should be streaked 1 cm apart and 2 mm on the side of an actinomycete colony on the nutrient agar, vertical to the colony, and then incubated for 24 hours at 37°C. After incubation, plates should be checked to evaluate if the test organisms' growth was hampered. To detect the growth inhibition, the number of test organisms grown on the streak line was used [20,22–24].

33.2.3.4 Extraction of Antibiotics

The broth was taken after incubation and centrifuged for 30 minutes at 4,000 rpm. Ethyl acetate should be added in an equal amount to the obtained supernatant. The two phases should be vortexed for an hour. The aqueous and organic phases of the solutions should be separated, by keeping the solution in a separate funnel for 5 minutes. The upper organic phase was produced and evaporated in a water bath at 40°C, and the lower aqueous phase was removed. The residue was weighed and combined with a little amount of pH 7 phosphate buffer to solubilize crude antibiotic extract [20,25,26].

33.2.3.5 Screening and UV Exposure

Using the agar well diffusion method, a second screening against tested microorganisms was conducted. The Mueller-Hinton broth should contain the test strains, which should then be incubated at 37°C for 4 hours. Following incubation, the broth's turbidity was assessed and contrasted with the McFarland standard of 0.5. Test organisms should be cultured on nutrient agar plates from McFarland-adjusted broth cultures on the lawn or carpet. With a sterile agar borer, 5 mm wells should be made in the nutrient agar plates, and 40 μL of antibiotic fractions extracted with ethyl acetate should be added.

Plates should be left at room temperature for 20–30 minutes to allow antibiotic portions to diffuse before being incubated at 37°C for 24 hours. After 24 hours, plates should be checked, along with the size of the inhibition zone surrounding each well.

On the selected soil isolate, the UV mutagenesis method should be used to obtain hyper-antibiotic producing stable UV mutants (D3). Actinomycete (D-3) isolate should be spread on the surface of actinomycete agar plates. The plates were subsequently exposed to UV at 254 and 324 nm having a distance of 25 cm for 5, 15, 30, 60, 90, 120, 150, 180, 240, 300's, and more, before incubation at a temperature of 27°C for 24 hours [5,20,27].

33.2.4 Proteome Study of Actinobacteria

The term proteome can be simply defined as the complete set of proteins expressed by an organism, and the branch of science that deals with the study of proteome is called proteomics. There are various techniques by which we can perform proteomics such as gel electrophoresis, NMR, and mass spectrometry (LC-MS/MS).

There are many proteins present in the actinobacteria that are involved in various different functions like carbohydrate metabolism, energy metabolism, lipid metabolism, amino acid metabolism, genetic information processing, and, most importantly, xenobiotic biodegradation, for example, dehydrochlorinase, haloalkane dehalogenase, short-chain alcohol dehydrogenase, etc [28]. The term xenobiotic is used for any chemical substance that is foreign to us like drugs, pesticides, cosmetics, and flavourings [29,30]. There has been extensive research on xenobiotic biodegradation these days because these substances, when introduced in the environment, may cause problems like environmental pollution because sometimes, they are not degraded properly [31]. Hence, identification of these proteins which are present in actinobacteria and are helpful in xenobiotic biodegradation can be very important.

An example will explain this in a better way:

By photochemically chlorinating benzene in the presence of UV light, HCH (hexachlorocyclohexane) is produced for commercial use. The five differently stable isomers that make up the technical-grade HCH that is created in this way are given as follows: β (5%–12%), α (60%–70%), δ (6%–10%), γ (10%–12%), and ε (3%–4%). The γ-HCH isomers can be extracted from the technical HCH. This extracted and purified form of γ-isomer is also known as lindane, an insecticide, which is used very commonly. There is a huge amount of HCH musk generated in this process. This musk includes all the three remaining isomers of HCH, which contaminates our environment [32] (Table 33.1).

In this chapter, we are mainly going to focus on the mass spectrometry-based approach for proteomics study of actinobacteria.

33.2.5 Protocol for Protein Extraction for LC-MS/MS Analysis

Cultured cells are briefly rinsed in a washing buffer (25 mM Tris buffer-pH 7 and 2 mM EDTA). Then, the same buffer is used for the resuspension, and then, the cells are smashed with the help of liquid nitrogen. SDS-PAGE is used to determine the protein's integrity, and Bradford is used to determine its concentration. Reducing solution (200 mM DTT, 100 mM Tris, pH 7.8) and alkylation solution (200 mM iodoacetamide, 100 mM Tris, pH 7.8) are used to treat a sufficient amount of proteins. Proteins are precipitated with TCA (10%) after each solution is treated with samples for 1 hour at room temperature. Cleaning of protein pellet is required which can be done by rinsing it with acetone (pre-chilled, –20°C) following centrifugation (16,100 g for 30 minutes at 4°C). The proteins are then trypsinized after being resuspended in 50 mM ammonium bicarbonate at a concentration of 10 µg/L. Finally, digested proteins are purified with the help of a Zip-Tip C18 column and then chilled to –80°C before being analysed by MS [28,38].

TABLE 33.1
Some Actinobacterial Proteins and Their Mechanism of Action in Xenobiotic Biodegradation

S No.	Enzymes Responsible for Xenobiotic Biodegradation	Mechanism of Action
1	Dehydrochlorinase	Upstream HCH degradationa
2	Haloalkane dehalogenase	Upstream HCH degradation
3	Mycothiol S-transferase	Down-stream HCH degradationb
4	Ring-cleaving dioxygenases	Down-stream HCH degradation
5	Non-haem chloroperoxidase	Chloride ion detoxification system
6	Short-chain alcohol dehydrogenase	Upstream HCH degradation

[a] **Upstream HCH Degradation**: 1,3,4,6-tetrachloro-1,4-cyclohexadiene (1,3,4,6-TCDN) is produced by two dehydrochlorination reactions via the intermediate γ-pentachlorocyclohexene (γ-PCCH) [33]. A second metabolite, 2,4,5-trichloro-2,5-cyclohexadiene-1-ol (2,4,5-DNOL), is then used to generate 2,5-dichloro-2,5-cyclohexadiene-1,4-diol (2,5-DDOL) through two rounds of hydrolytic dechlorinations [34]. Then, 2,5-DDOL is changed to 2,5-dichlorohydroquinone (2,5-DCHQ) through a dehydrogenation process [35]. The so-called upstream degradation pathway is finished by the production of 2,5-DCHQ.

[b] **Down-stream HCH Degradation:** 2,5-DCHQ is reductively dechlorinated to produce chlorohydroquinone (CHQ) [36]. The splitting of pathway leads to a minor route, by which hydroquinone (HQ) is formed. It is further ring-cleaved to form γ-hydroxymuconic semialdehyde (γ-HMSA). CHQ is directly ring-cleaved to an acylchloride, which is then converted to maleylacetate (MA) in the major route [37]. Succinyl-coenzyme A and acetyl-coenzyme A (CoA) are formed from MA which are further metabolized in the citric acid cycle.

33.2.6 Protocol for LC-MS/MS Analysis of Protein Extract

Liquid chromatography coupled with high-resolution mass spectrometry is employed to conduct tests using a minimum of three physiologically distinct duplicates. Trifluoroacetic acid is used to dissolve the samples, and a C18 column is used for LC of the peptide samples. The gradient consists of two solutions: Solution B, acetonitrile with 0.1% HCOOH, and Solution A, 0.1% HCOOH (v/v) peptides are eluted at a flow rate of 0.300 μL/min with a gradient of 7% B to 95% B. The mass spectrum is obtained. The setup of the instrument makes it possible to acquire full MS and MS/MS while also identifying the peptides as they are separated chromatographically. For MS/MS fragmentation, the 20 sharpest peaks in each MS spectrum should be selected [28,38].

33.2.7 Data Analysis

Identification and quantification of proteins are performed with the help of Actinobacterial Uniprot Database and MaxQuant (ver. 1.5.5.1) software. Label-free quantification (LFQ) method is preferred for quantifying and normalizing the spectral count. Data processing is done with the help of software Perseus (ver. 1.5.5.3). Protein filtration can be done by eliminating all the contaminants. These proteins are said to be valid only if there are at least two replicates for a single condition. For calculating relative abundance of every protein, we need to (a) transform the protein intensities to a logarithmic scale (base 2) and (b) normalize the protein intensities to protein size. Two sample T-tests [FDR(q) <0.05 and log2 fold change >2] are applied to check differences in protein expression. One-way ANOVA ($q<0.05$ and log2 fold change >2) can be applied to check the variations in protein expression. Later on, Tukey's HSD post hoc test can be used to identify means of specific groups which are different since all the different pairs of means can be compared by applying this test [39].

33.2.8 Metabolite Extraction

To perform untargeted metabolomics of bacteria, extraction of metabolites is needed. Sufficient volume of bacteria culture was taken and placed them on ice. Counting should be done for normalization purposes. Quenching should be done rapidly with dry ice or ethanol and placed it on ice. Then, centrifugation should be done at 3,220×g for 10 minutes at 4°C. After the centrifugation is done, the cells will form the pellet. Washing that cell pellet with sterile saline is mandatory at 3,220×g for 3 minutes each. The metabolite extraction should be done with 3X volume of extraction buffer {i.e., chloroform: methanol: water (previously chilled at −80°C)}. Internal standards can be used if required. The samples were snap frozen immediately in liquid nitrogen, and then, thaw process should be done on ice. The freeze-thaw process helps to rupture the cells that cause the cellular metabolites to come into the solution. Then, the samples should be centrifuged again at 14,000×g for 10 minutes. The supernatant needs to be transferred to a fresh tube and lyophilized in a vacuum concentrator. The lyophilized metabolites can be stored at –80°C until LC-MS runs [40].

33.2.9 LC-MS Analysis of Metabolites

For LC-MS-based untargeted metabolomics analysis, the lyophilized samples should be reconstituted in sufficient volume of 20% methanol/water (v/v) and stored at –80°C until run. Samples must be analysed on any high-performance liquid chromatography-coupled high-resolution mass spectrometry like Orbitrap or TOF (Time of Flight) instrument with an ESI (electrospray ionization) source using both HILIC and C18 column to get both polar and non-polar metabolites, respectively. Polarity switching should be used which means both positive and negative polarity should be applied while running the samples in each column. The liquid chromatography solvent should contain 20 mM ammonium bicarbonate in water as solvent A and acetonitrile as solvent B. The LC gradient can be optimized based on the type of instrument used and the flow rate can be determined by the pore diameter of the column used. Injection volume can be adjusted based on the signal intensity, and the method runtime can be 15–30 minutes long. A pooled sample as a quality control (QC) can be used to check the instrument response and stability. Pure standards can also be used along with the samples to aid in metabolite identification [41].

33.2.10 Data Processing and Statistical Analysis

Untargeted metabolomics data analysis can be performed in mzMatch and Ideom. It annotates raw MS data files into identified metabolite lists based on exact mass and chromatographic retention time. Quantification of metabolites can be done using peak intensity or height. Variance should be analysed in MetaboAnalyst to check if the data is univariate or multivariate. Data normalization can be done by log transform. Multivariate analysis can be done by principle component analysis. Significantly altered metabolites can be identified using ANOVA test for one-way analysis ($p<0.05$; FDR ≤ 0.1). Metabolic pathway analysis can be done based on the identified metabolites that are statistically significant [11].

33.2.11 Sample Processing for NMR

For NMR-based untargeted metabolomics study, the lyophilized samples should be reconstituted in sufficient volume of heavy water (D2O) under an inert environment due to its reactivity with air. The resuspended sample can be introduced to the instrument for NMR spectroscopy.

33.2.12 NMR Data Analysis for Metabolites

The approach followed for metabolite data analysis in NMR is chemometric analysis as explained.

33.2.12.1 Chemometric Analysis

It is a conventional approach in global metabolomics that is untargeted. It involves retrieving information from measurements done on chemical systems using mathematical procedures. Before analysing the data, various preprocessing steps are undertaken such as solvent peak removal, peak alignment, and baseline correction followed by multivariant analysis. However, there are challenges. Data needs to be scaled and filtered as samples are classified based on minor spectral features. Moreover, imperfect peak analysis for samples where peak position is sensitive to variables such as pH, temperature, and concentration of macromolecules poses another formidable challenge (Figure 33.1).

33.2.12.2 Outlier Removal

The capability to identify outliers is a crucial aspect of chemometric data analysis. It is a very important and delicate procedure. An outlier is a sample that deviates from the norm and may have a negative impact on the model. The Grubbs test is used to detect outliers. There could be many reasons for the appearance of outliers, like improper sample preparation or handling, or error due to faulty instruments. There are two types of outliers: positive and negative. Only the positive outliers that are the true outliers are removed. If a false outlier occurs, then, it is due to faulty calibration model that is required to be recalculated. Sometimes, an outlier is introduced due to the presence of insufficient samples of a particular kind. In that event, similar samples must be introduced into our experiment to give a clear and better picture of the result.

33.2.12.3 PreProcessing of NMR Data

Preprocessing involves removal of variation within data not concerned with any analytical information.

Common preprocessing methods are as follows:

- **Phase Correction:**

  The existence of phase errors is one of the issues with NMR data. The pulse sequence delays and off-resonance effects are the two main causes of the phase problems. To safeguard the receiver component from the damaging effects of the pulse, brief time gaps of few tens of microseconds are added between the closing of transmitters following execution of every pulse and detection by the receiver to capture FID. The obtained spectra will have both absorptive and dispersive signals as a result of these delays, and the peaks will have few portions that are beneath the baseline and do not have their ideal Lorentzian shapes. The signals are out of phase for a second cause, which is off-resonance effects brought on by defective pulses. For the nuclei that vibrate beyond the frequency of excitation range of the pulse, if the pulse power is insufficiently low to equally excite every nucleus in the sample, the magnetization will not perfectly flip on the XY-plane after 90 pulses. This causes a phase discrepancy between these off-resonance nuclei and the nuclei that are correctly stimulated by the pulse and are in resonance. Phase errors of this type are typically minimal in 1 H NMR and straightforward to fix because proton nuclei have low-frequency dispersion. It is common practice to phase the spectrum, which involves both 0-order and first-order phase correction. The first-order phase correction is required to correct more complex phase problems because it is linearly correlated with frequency. The 0-order phase correction, unlike first-order phase correction, is not dependent upon frequency and may fix phase defects that equally influence each and every peak [42].
- **Baseline Correction:**

  Handling potential fluctuations in the spectra's baseline is crucial in NMR spectroscopy. Baseline differences between the samples can impede exploratory research as a

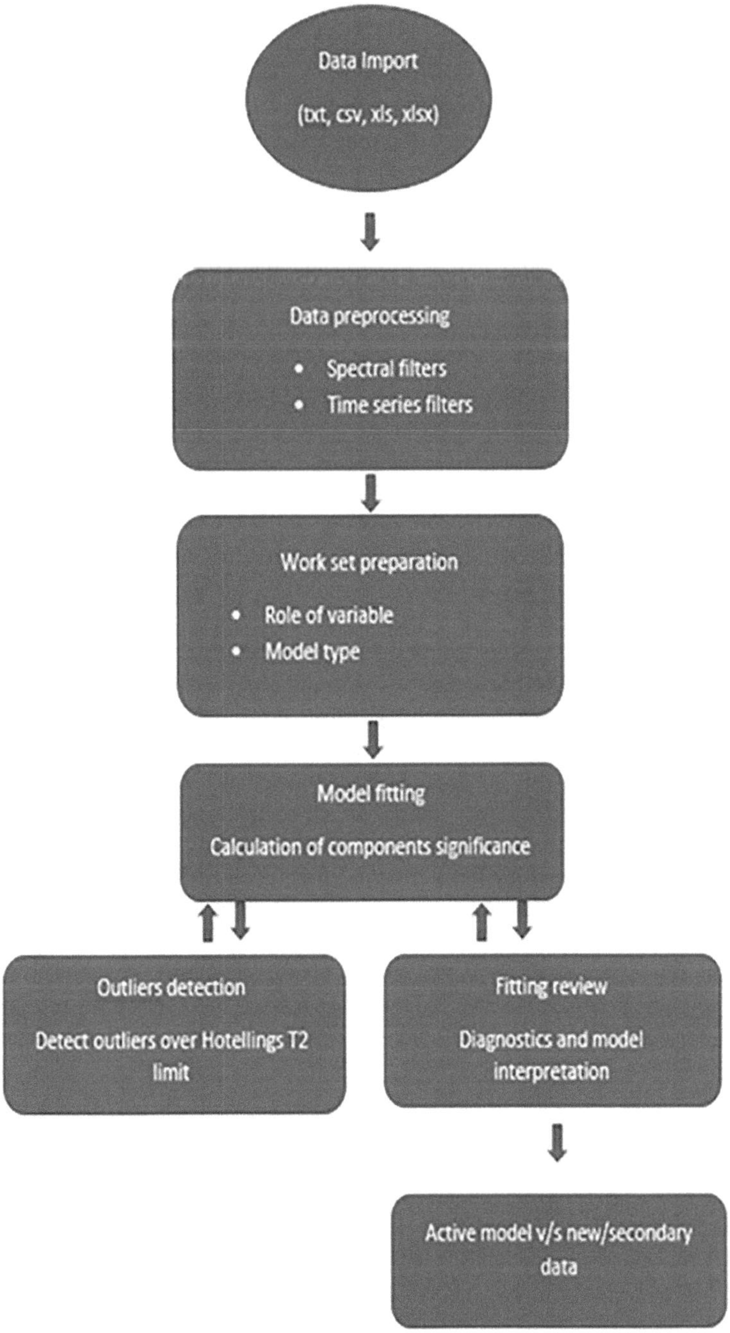

FIGURE 33.1 Workflow of the data analysis.

typical preprocessing approach (centring) will be unable to eliminate the impacts, which can result in erroneous quantification. Baseline distortions can impede the best information retrieval, particularly in metabolomics research where tiny peaks can be crucial. By adjusting an exponential or polynomial or sine function to the baseline, removal techniques attempt to model the baseline. Following baseline correction, the faulty baseline is removed from the sample.

- **Reference Deconvolution:**

  Errors due to faulty instruments (such as field instability, amplitude errors, pulse phase, and magnetic field inhomogeneity) are systematic and independent of the frequency of the signals they impact. Reference deconvolution fixes these kinds of errors.

  Reference deconvolution converts the entire dataset into a form, and it was expected to be without errors due to instrumental imperfection by using a highly resolved signal from the experimental dataset as the reference signal, comparing it to its theoretically expected "ideal" form. Reference deconvolution is quick and simple to use, and it can significantly enhance the outcomes of multivariate analysis of NMR data. In vivo/in vitro NMR research, where sample heterogeneity causes signal widening, can also benefit greatly from reference deconvolution.
- **Alignment:**

  A change in the location of a signal of a chemical compound between chains of similar samples can be caused by small changes in the sample's pH and temperature. Multivariate models become unnecessarily complicated and difficult to interpret as peaks of a particular compound change between samples. Multivariate models can only be used effectively with NMR datasets if the data are well-reproducible and peak position variations of the same chemical compound in different samples are nominal or corrected. It is therefore important to correct the peaks of spectra before multivariate analysis. The icoshift method is one popular method for aligning NMR data. Icoshift adjusts signals in characterized phantom spans between a progression of spectra, by utilizing the most extreme relationship between the ghastly stretches as the rules. Icoshift is quick and can align multiple spectra at once because it uses a fast Fourier transform algorithm. Giskedegrd and co. compared how well five distinct warping algorithms, one of which was icoshift, aligned spectral peaks, specifically for NMR data with varying degree of misalignment. Giskedegrd and co. icoshift was recommended for the NMR dataset due to its speed and reproducibility [43].
- **Normalization:**

  Normalization is necessary because reproducible shim and sample filling conditions are frequently difficult to achieve between samples. Normalization can preserve fascinating chemical and biological data while removing systematic bias. P. Ebrahimi et al. allowed the maximum shift to be 550 points done in relation to an internal standard that can be a spectra-present chemical or artificial signal. The compounds tetramethyl silane (TMS), 3-(trimethylsilyl) propionic acid (TSP) and 4,4-dimethyl-4-silapentane-1-sulfonic acid (DSS), which are frequently added to NMR samples. They serve as an internal standard for the calibration and quantification of the chemical shift axis. There are various approaches to normalization, each of which makes distinct assumptions regarding the nature of the data. Probably, the most well-known standardization techniques are (a) unit test power total, (b) unit test vector length (Euclidean standard), and (c) probabilistic remainder/middle overlay change. 1 H NMR studies using probabilistic quotient normalization are extremely reliable. Normalization to unit length was commonly used in the past. It was because of two reasons. First, it was assumed that every sample in the sample set had nearly identical proton intensity. Second, normalization in accordance with the internal standard was susceptible to sampling errors. 1 H NMR is also used to prognosticate lipoprotein distributions using the ERETIC (electric reference to access in vivo concentrations) method. PULCON relates the outright forces of two spectra that are estimated in various arrangement conditions in light of the natural properties of the examples. PULCON is compatible with all NMR spectrometers and does not require any particular instruments or software. Normalization strategies might be different for different applications. For instance, proper normalization procedures are required for the NMR measurement of urinary samples due to the inherent normalization issue caused by the diverse sample dry matter.

- **Means Focusing:**
  This preprocessing technique focuses on the mean of the signal intensities. Following mean focusing, the signal intensities vary around zero rather than their earlier mean value. It works by removing the common spectral offset from every spectrum that highlights relevant variations. The elements in each column are subtracted from the mean value of the particular column to obtain a mean-centred dataset.
- **Scaling:**
  Scaling involves dividing each variable by a scaling factor to adjust fold differences between the variables. It can make it easier to interpret the data by highlighting important chemical and biological information in the data. Especially for metabolomics data, Pareto scaling and autoscaling are the two most widely used techniques.
  The scaling factor in autoscaling is the variable's standard deviation; on the other hand, in Pareto scaling, it is the square root of the standard deviation. Autoscaling causes variance of all variables to be the same, making them all equally important. However, for variables devoid of any chemical information, this can also lead to a relatively large inflation of measurement errors.
  Autoscaling is commonly employed for metabolic profiling in metabolomics. On the other hand, Pareto scaling is mostly used for metabolic fingerprinting. Pareto scaling lets you saddle baseline noise and small noisy signals.
  Multivariant approaches are broadly categorized into supervised and unsupervised.
- **Unsupervised Analysis:**
  In unsupervised analysis, any information about our sample of interest such as disease control is not known. It is mostly used for generation of hypotheses. Principal component analysis (PCA) is the most widely used method for unsupervised analysis. However, other methods such as factor analysis, hierarchical clustering analysis (HCA), and k-nearest neighbour analysis are also present. Using PCA, the metabolites can be visualized as a set of principal components (ranked). Score plots in PCA are used to visualize variance, and using loading plots, we can see the possible causes of the variance. If the variables identities are not known, further analysis is done to confirm their identities. Although PCA is most widely used, the most useful method for comparing small variables is HCA.
  In PCA, a multivariate dataset (data with multiple variables) is converted into three new variables:
  -Principal components (PC's)
  -Scores (show variation in concentration)
  -Loadings (indicates whether preprocessing is required or not)
  Using these three new variables, the hidden variation is revealed for further examination. It has excellent data presentation capabilities and is used to extract and visualize systematic variation in a data matrix.
  The principal component that is calculated captures variance within the data that is removed to determine a new principal component. In this way, two PCs are determined. First PC captures maximum variation in the dataset. Second PC reproduces the leftover variance. Both of them are later compared.
  There are various software programs like MetaboAnalysts for doing PCA. However, programming languages like R are also used.
- **Supervised Analysis:**
  In supervised analysis, sample information such as disease or control is known. It is used for developing predictive models that use sample information as dependent variables and metabolites as independent variables. The most widely used method in supervised analysis is partial least squares analysis (PLS). However, other methods such as soft independent modelling of class analogies (SIMCA), logistic regression, neural networks, and random forests are also used for supervised analysis in metabolomics.

Objective of PLS is regression analysis. PLS regression is analogous to PCA. However, there are differences; PLS regression is a kind of data analysis where we know which data is important and which is not, whereas PCA is performed without the use of a priori knowledge of what is important. Alternatively, PCA is unlabelled whereas PLS is labelled. PLS is an extension of PCA. PLS is a predictive regression technique (unlike PCA that is exploratory). It uses PC's to correlate them with the information we want to predict such as physical property and calculates latent variables. However, the spectral data's latent variables are determined in such a manner that only the information required to predict any chemical property and physical property is extracted.

SIMPA-P+ software is used for NMR metabolite data analysis by doing PCA and PLS. SIMCA-P+ is a package mainly utilized for PCA and PLS regression (PLS-R) methods. The SIMCA-P+ data analysis workflow is as follows.

SIMCA-P+ accepts raw datasets as text or excel or csv files that contain a lot of zeros or non-numeric characters.

In order to ensure that the result is reliable, the input data must have less than 50% missing values in the observations.

The software will automatically tell you to include or exclude the variable with zero-valued observation, and the tolerance for missing values can be adjusted.

A scaling weight of 1 is given to a variable with no value.

Before beginning the analysis, a number of parameters must be identified, including the data type, the roles of the variables, and the indication of data inclusion or exclusion.

Many sorts of diagnostics and translation can be carried out using SIMCA-P+, for example, loading plots, score plots, distance to the model (DModX), and Hotelling's T2 and for a significant corroborative information investigation [44] (Figure 33.2).

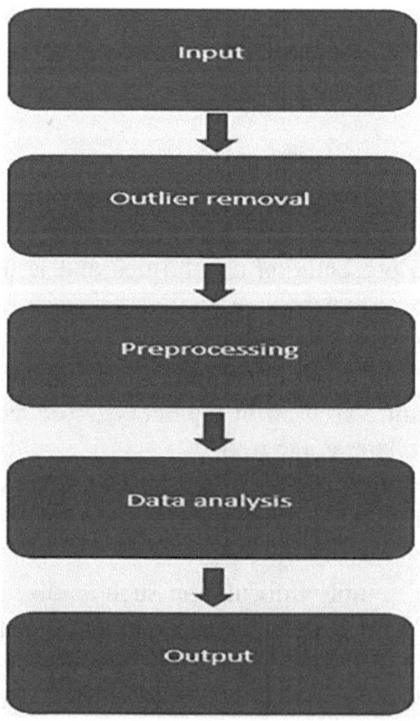

FIGURE 33.2 Workflow of SIMCA-P+ software.

33.3 CONCLUSION

We conclude that LC-MS- and NMR-based techniques can be used for proteomics and metabolomics research in Actinomycetes. Introduction of high-throughput (omics) methods, genomics, transcriptomics, proteomics, metabolomics, and computational tools has significantly improved our comprehension of complicated cellular physiology and essential pathways controlling the generation of secondary metabolites. In a microbial cell, the genetic information generates more than 2,000 proteins, and a sizable subset of them can be studied using proteomics. Due to their anticancer properties, metabolites from Actinomycetes can be employed in cancer therapy as well as against bacteria, viruses, fungi, nematodes, and insects. Untargeted metabolomics can be a helpful method to distinguish between these bacteria's metabolites and gauge the degree of their antimicrobial activity.

REFERENCES

1. De Simeis, D., & Serra, S. (2021). Actinomycetes: A Never-Ending Source of Bioactive Compounds – An Overview on Antibiotics Production. *Antibiotics (Basel, Switzerland)*, 10(5), 483. Robertsen, H. L., & Musiol-Kroll, E. M. (2019). Actinomycete-Derived Polyketides as a Source of Antibiotics and Lead Structures for the Development of New Antimicrobial Drugs. *Antibiotics (Basel, Switzerland)*, 8(4), 157.
2. Jagannathan, S. V., Manemann, E. M., Rowe, S. E., Callender, M. C., & Soto, W. (2021). Marine Actinomycetes, New Sources of Biotechnological Products. *Marine Drugs*, 19(7), 365.
3. Barka, E. A., Vatsa, P., Sanchez, L., Gaveau-Vaillant, N., Jacquard, C., Meier-Kolthoff, J. P., Klenk, H. P., Clément, C., Ouhdouch, Y., & van Wezel, G. P. (2015). Taxonomy, Physiology, and Natural Products of Actinobacteria. *Microbiology and Molecular Biology Reviews*, 80(1), 1–43.
4. Graves, P. R., & Haystead, T. A. (2002). Molecular Biologist's Guide to Proteomics. *Microbiology and Molecular Biology Reviews*, 66(1), 39–63.
5. Manzoni, C., Kia, D. A., Vandrovcova, J., Hardy, J., Wood, N. W., Lewis, P. A., & Ferrari, R. (2018). Genome, Transcriptome and Proteome: The Rise of Omics Data and Their Integration in Biomedical Sciences. *Briefings in Bioinformatics*, 19(2), 286–302.
6. Bush, W. S., & Moore, J. H. (2012). Chapter 11: Genome-Wide Association Studies. *PLoS Computational Biology*, 8(12), e1002822.
7. Chandramouli, K., & Qian, P. Y. (2009). Proteomics: Challenges, Techniques and Possibilities to Overcome Biological Sample Complexity. *Human Genomics and Proteomics*, 2009, 239204.
8. Singhal, N., Kumar, M., Kanaujia, P. K., & Virdi, J. S. (2015). MALDI-TOF Mass Spectrometry: An Emerging Technology for Microbial Identification and diagnosis. *Frontiers in Microbiology*, 6, 791.
9. Gomez-Escribano, J. P., Alt, S., & Bibb, M. J. (2016). Next Generation Sequencing of Actinobacteria for the Discovery of Novel Natural Products. *Marine Drugs*, 14(4), 78.
10. Sebak, M., Saafan, A. E., AbdelGhani, S., Bakeer, W., El-Gendy, A. O., Espriu, L. C., Duncan, K., & Edrada-Ebel, R. (2019). Bioassay- and Metabolomics-Guided Screening of Bioactive Soil Actinomycetes from the Ancient City of Ihnasia, Egypt. *PLoS One*, 14(12), e0226959.
11. Silva, G. C., Kitano, I. T., Ribeiro, I. A. F., & Lacava, P. T. (2022). The Potential Use of Actinomycetes as Microbial Inoculants and Biopesticides in Agriculture. *Frontiers in Soil Science*, 2, 833181.
12. Jose, P. A., Maharshi, A., & Jha, B. (2021). Actinobacteria in Natural Products Research: Progress and Prospects. *Microbiological Rresearch*, 246, 126708.
13. Selim, M. S. M., Abdelhamid, S. A., & Mohamed, S. S. (2021). Secondary Metabolites and Biodiversity of Actinomycetes. *Journal, Genetic Engineering & Biotechnology*, 19(1), 72.
14. Bose, U., Hewavitharana, A. K., Ng, Y. K., Shaw, P. N., Fuerst, J. A., & Hodson, M. P. (2015). LC-MS-Based Metabolomics Study of Marine Bacterial Secondary Metabolite and Antibiotic Production in *Salinispora arenicola*. *Marine Drugs*, 13(1), 249–266.
15. Subramani, R., & Aalbersberg, W. (2013). Culturable Rare *Actinomycetes*: Diversity, Isolation and Marine Natural Product Discovery. *Applied Microbiology and Biotechnology*, 97(21), 9291–9321.
16. Jiang, Y., Li, Q., Chen, X., & Jiang, C. (2016). Isolation and Cultivation Methods of *Actinobacteria*. In *Actinobacteria-Basics and Biotechnological Applications*, 39–57. In-Tech.
17. Devanshi, S., Shah, K. R., Arora, S., & Saxena, S. (2021). Actinomycetes as an Environmental Scrubber. In *Crude Oil-New Technologies and Recent Approaches*. IntechOpen.

18. Hug, J. J., Bader, C. D., Remškar, M., Cirnski, K., & Müller, R. (2018). Concepts and Methods to Access Novel Antibiotics from Actinomycetes. *Antibiotics (Basel, Switzerland)*, 7(2), 44.
19. Ibnouf, E. O., Aldawsari, M. F., & Ali Waggiallah, H. (2022). Isolation and Extraction of Some Compounds That Act as Antimicrobials from Actinomycetes. *Saudi Journal of Biological Sciences*, 29(8), 103352.
20. Sharma, S. (2016). Isolation and Screening of Actinomycetes from Soil Sample of Dal Lake (Kashmir) Against Selected Pathogens. *International Journal of Engineering Science & Advanced Research*, 2, 51–54.
21. Budhathoki, S., & Shrestha, A. (2020). Screening of Actinomycetes from Soil for Antibacterial Activity. *Nepal Journal of Biotechnology*, 8(39, 102–110.
22. Ibnouf, E. O. (2021). Screening of O-7 Isolate Actinomycete Producing Antimicrobials in Different Growth Conditions against Selected Pathogens. *International Journal of Pharmaceutical and Phytopharmacological Research*, 11(2), 13–23.
23. Ganesan, P., Host, R., Antony, D., Appadurai, D. R., Gandhi, M. R., Paulraj, M. G., Ignacimuthu, S., & Al-Dhabi, N. A. (2017). Isolation and Molecular Characterization of Actinomycetes with Antimicrobial and Mosquito Larvicidal Properties. *Beni-Suef University Journal of Basic and Applied Sciences*, 6(2), 209–217.
24. Gopinath, B. V., Vootla, P. K., Jyothi, R., Reddy, K. S. (2013). Antimicrobial Activity of Actinomycetes Isolated from Coal Mine Soils of Godavari Belt Region, A.P., India. *Asian Journal of Biological Sciences*, 4(4), 518–523.
25. Ghorbani, N. R., Greiner, R., Alikhani, H.A., Hamedi, J., & Yakhchali, B. (2013). Distribution of Actinomycetes in Different Soil Ecosystems and Effect of Media Composition on Extracellular Phosphatase Activity. *Journal of Soil Science and Plant Nutrition*, 13(1), 223–236.
26. Butler, M. J., Takano, E., Bruheim, P., Jovetie, S., Marinelli, F., & Bibb, M. J. (2003). Deletion of Scb A Enhancer Antibiotic Production in *Streptomyces lividans. Applied Microbiology and Biotechnology*, 61, 512–516.
27. Sineli, P. E., Herrera, H. M., Cuozzo, S. A., & Costa, J. S. D. (2018). Quantitative Proteomic and Transcriptional Analyses Reveal Degradation Pathway of γ-Hexachlorocyclohexane and the Metabolic Context in the Actinobacterium *Streptomyces* sp. M7. *Chemosphere*, 211, 1025–1034.
28. Patterson, A. D., Gonzalez, F. J., & Idle, J. R. (2010). Xenobiotic Metabolism: A View through the Metabolometer. *Chemical Research in Toxicology*, 23(5), 851–860.
29. Idle, J. R., & Gonzalez, F. J. (2007). Metabolomics. *Cell Metabolism*, 6(5), 348–351.
30. Nagata, Y. (2020). Microbial Degradation of Xenobiotics. *Microorganisms*, 8(4), 487.
31. Lal, R., Pandey, G., Sharma, P., Kumari, K., Malhotra, S., Pandey, R., …... & Oakeshott, J. G. (2010). Biochemistry of Microbial Degradation of Hexachlorocyclohexane and Prospects for Bioremediation. *Microbiology and Molecular Biology Reviews*, 74(1), 58–80.
32. Imai, R., Nagata, Y., Fukuda, M., Takagi, M., & Yano, K. (1991). Molecular Cloning of a *Pseudomonas paucimobilis* Gene Encoding a 17-Kilodalton Polypeptide That Eliminates HCl Molecules from -Hexachlorocyclohexane. *Journal of Bacteriology*, 173, 6811–6819.
33. Nagata, Y., Nariya, T., Ohtomo, R., Fukuda, M., Yano, K., &Takagi, M. (1993). Cloning and Sequencing of a Dehalogenase Gene Encoding an Enzyme with Hydrolase Activity Involved in the Degradation of γ-Hexachlorocyclohexane in *Pseudomonas paucimobilis. Journal of Bacteriology*, 175, 6403–6410.
34. Nagata, Y., Ohtomo, R., Miyauchi, K., Fukuda, M., Yano, K., & Takagi, M. (1994). Cloning and Sequencing of a 2,5-Dichloro-2,5-Cyclohexadiene-1,4-Diol Dehydrogenase Gene Involved in the Degradation of -Hexachlorocyclohexane in *Pseudomonas paucimobilis. Journal of Bacteriology*, 176, 3117–3125.
35. Miyauchi, K., Suh, S. K., Nagata, Y., & Takagi, M. (1998). Cloning and Sequencing of a 2,5-Dichlorohydroquinone Reductive Dehalogenase Gene Whose Product Is Involved in Degradation of -Hexachlorocyclohexane by *Sphingomonas paucimobilis. Journal of Bacteriology*, 180, 1354–1359.
36. Miyauchi, K., Adachi, Y., Nagata, Y., & Takagi, M. (1999). Cloning and Sequencing of a Novel Meta-Cleavage Dioxygenase Gene Whose Product Is Involved in Degradation of -Hexachlorocyclohexane in *Sphingomonas paucimobilis. Journal of Bacteriology*, 181, 6712–6719.
37. Costa, J. S. D., Herrero, O. M., Alvarez, H. M., & Leichert, L. (2015). Label-Free and Redox Proteomic Analyses of the Triacylglycerol-Accumulating *Rhodococcus jostii* RHA1. *Microbiology*, 161(3), 593–610.
38. Chhun, A., Sousoni, D., Aguiló-Ferretjans, M. D. M., Song, L., Corre, C., & Christie-Oleza, J. A. (2021). Phytoplankton Trigger the Production of Cryptic Metabolites in the Marine Actinobacterium *Salinispora tropica. Microbial Biotechnology*, 14(1), 291–306.

39. Maifiah, M. H., Creek, D. J., Nation, R. L., Forrest, A., Tsuji, B. T., Velkov, T., & Li, J. (2017). Untargeted Metabolomics Analysis Reveals Key Pathways Responsible for the Synergistic Killing of Colistin and Doripenem Combination against *Acinetobacter baumannii. Scientific Reports*, 7, 45527.
40. Ceraso, J. M., Smith, P. B., Landers, J. S., & Dye, J. L. (1977). A Sodium-23 Nuclear Magnetic Resonance Study of the Exchange Kinetics of Sodium(1+) Ion with 2,2,2-Cryptate Complexes in Water, Ethylenediamine, Tetrahydrofuran, and Pyridine. *The Journal of Physical Chemistry*, 81(8), 760–766. DOI: 10.1021/j100523a016
41. Ishak, N. A., Tahir, N. I., Mohd Sa'id, S. N., Gopal, K., Othman, A., & Ramli, U. S. (2021). Comparative Analysis of Statistical Tools for Oil Palm Phytochemical Research. *Heliyon*, 7(2), e06048. DOI: 10.1016/j.heliyon.2021.e06048.
42. Pathmasiri, W., Kay, K., McRitchie, S., & Sumner, S. (2020). Analysis of NMR Metabolomics Data. In: Li, S. (eds), *Computational Methods and Data Analysis for Metabolomics. Methods in Molecular Biology*, vol. 2104. Humana, New York, NY.

34 Isolation of DNA and RNA from Actinobacteria

G. R. Shree Kumari, Ranjitha Dhevi V. Sundar, K. Ajith Kumar, and Lokesh Ravi

34.1 INTRODUCTION

A microworld of organisms exists, playing a key role in recycling biomaterials that are refractorily produced, by decomposing/degrading complex mixture of polymers found in dead animals and plants. These microworld of organisms are distributed widely both in aquatic and terrestrial systems. These group of organisms are identified to exhibit or share characteristics of both fungi and bacteria. These microworld of organisms are recognized as actinobacteria, which are predominantly found in soil as they play a dominant role in polymer degradation or plastic degradation. As these group of microorganisms produce secondary metabolites, which are recognized as bioactive metabolites, they are exploited to a greater extent as the secondary metabolites are valuable biotechnologically [1]. These biologically active metabolites produced by actinobacteria with enzymes have significant economic value as they are considered as therapeutic agents. They serve as excellent antibiotics and have anti-cholesterol, anti-aging, immunosuppressant properties, etc. [2]. They smell earthy or produce geo-smell, a volatile compound which is a key characteristic of actinobacteria [3].

Belonging to the phylum "Actinomycetes," they are chiefly recognized as Actinomycetes. Morphologically, they are dominantly aerobic, filamentous, and Gram-positive bacteria. They are generally found spreading as colonies in soil and water plants. In nature, actinobacteria are ubiquitously present with high consistency, and when they are cultured in the laboratory, they adhere stiffly to the surface of the agar, by producing hyphae along with conidia/sporangia as characteristic features such as fungi in the nutrient medium or culture medium [2,4,5].

34.2 ISOLATION OF DNA FROM ACTINOBACTERIA [6,7]

34.2.1 Materials

- Pure actinobacterial culture
- Weighing balance
- Centrifuge
- Eppendorf tubes
- Pipettes
- Beaker
- Conical flask
- Micro pestles
- Liquid nitrogen
- TE buffer (10 mM Tris and 1 mM EDTA)
- Lysozyme
- Proteinase K
- Phenol: chloroform (v/v)
- RNase solution (20 µg/mL)

DOI: 10.1201/9781003398400-34

- Sodium dodecyl sulfate (10%)
- Ethanol
- Cuvette
- UV spectrophotometer
- Agarose
- Electrophoresis

34.2.2 Procedure

1. The pure actinobacterial culture was aseptically inoculated from the mature slant culture into the nutrient broth (100 mL) and incubated in a rotary shaker at 180 rotations per minute (rpm) at 37°C for 2 days.
2. At 10,000 rpm, the cultures were centrifuged for 10 minutes.
3. The supernatant was discarded, and thetube with pellet was kept in the liquid nitrogen for 4–5 minutes for the culture to freeze.
4. Using micro pestles, the frozen sample was ground and transferred into a fresh Eppendorf containing 500 µL of TE buffer (10 mM Tris and 1 mM EDTA (*Ethylenediaminetetraacetic* acid)) with lysozyme supplemented (20 mg/mL).
5. The tube was incubated at 37°C for 30 minutes.
6. A total of 20 µL of 10% sodium dodecyl sulfate (SDS) (w/v) and proteinase K (20 µL) were added into the tube and incubated for 30 minutes at 55°C.
7. Lysate was extracted with an equal measurement of phenol: chloroform solution (1:1) at 10,000 rpm for 5 minutes.
8. On a fresh tube, the aqueous phase was carefully transferred and 20 µL of RNase solution (20 µg/mL) was added to prevent RNA contamination and then incubated for 1 hour at 37°C.
9. Once again, extraction was carried out with phenol: chloroform s (1:1) at 10,000 rpm for 10 minutes at 16°C.
10. To a fresh tube, aqueous phase was transferred and DNA was precipitated by adding ethanol (70%–90%) and kept at –20°C for 30 minutes.
11. Centrifugation was done at 10,000 rpm for 10 minutes to get pellet. The pellet was washed twice with 90% ethanol and dissolved in 80 µL of TE buffer. The gel electrophoresis of the DNA is shown in Figure 34.1.

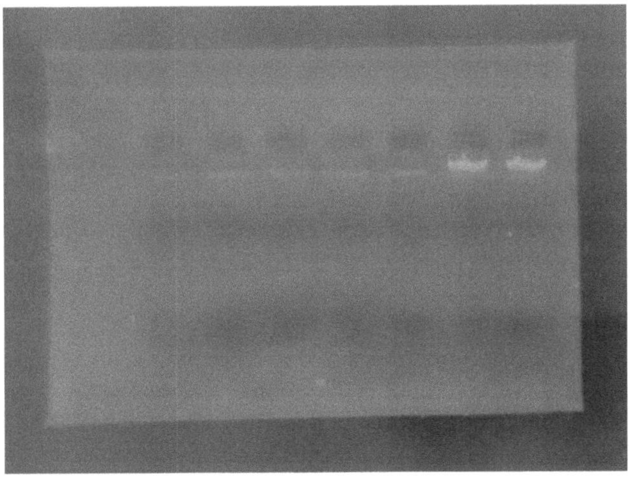

FIGURE 34.1 Gel electrophoresis of genomic DNA under UV transilluminator.

12. The DNA concentration was determined by calculating the absorbance at 260/280 nm using UV spectrophotometer.
13. Using agarose gel electrophoresis (0.8%), the purity of the DNA was analyzed.

34.3 ISOLATION OF RNA FROM ACTINOBACTERIA [8]

34.3.1 Materials

- Pure actinobacterial culture
- Weighing balance
- Centrifuge
- Eppendorf tubes
- Pipettes
- Beaker
- Conical flask
- Micro pestles
- Vortex mixer
- Powder-free gloves
- DEPC-treated water
- 70% ethanol
- Isopropyl alcohol
- TRIzol reagent
- Chloroform
- UV spectrophotometer
- Cuvette

34.3.2 Procedure

34.3.2.1 Homogenization

1. In 1 mL of TRIzol reagent, about 50–100 mg of sample was homogenized with motor and pestle.

34.3.2.2 Phase Separation

2. The homogenized sample was incubated at 15°C–30°C for 5 minutes to allow the thorough separation of complexes of nucleoprotein.
3. Chloroform (200 μL)/TRIzol (1 mL) was added, vortexed thoroughly, and incubated for 2–3 minutes at 15°C–30°C.
4. At 12,000 rpm, the samples were centrifuged for 15 minutes at 2°C–8°C.
5. After centrifugation, the mixture was divided into phenol-chloroform phase, aqueous upper phase (colorless), a lower phase (red), and interphase.
6. RNA was remained in the upper (aqueous) phase. The volume of upper phase is around 60% of the reagent (TRIzol) volume used for homogenization.

34.3.2.3 Precipitation of RNA

7. The aqueous phase was carefully transferred into a new Eppendorf tube. By mixing with 0.5 mL of isopropyl alcohol per 1 mL of TRIzol reagent, the RNA was precipitated.
8. At 15°C–30°C, the samples were incubated for 10 minutes, followed by centrifugation at 12,000 rotations per minute at 2°C–8°C for 10 minutes.
9. Before centrifugation, the precipitated RNA was generally invisible forming a gel-like pellet on the tube (side and bottom).

34.3.2.4 Washing of RNA

10. Supernatant was removed, and with 70% ethanol, the RNA pellet was washed by adding 70% ethanol (1 mL)/TRIzol reagent (1 mL) which was initially used for homogenization.
11. The sample was vortexed, and it was centrifuged at 7,500 rotations per minute at 2°C–8°C (5 minutes).
12. Above washing procedure was repeated to remove the leftover ethanol.

34.3.2.5 Redissolving RNA

13. For 5–10 minutes, the pellet RNA was air-dried.
14. RNA dissolved in DEPC-treated water by solution passing few times over pipette tip.

34.3.2.6 Spectrophotometric Examination

15. RNA (1 µL) was diluted with DEPC-treated water (39 µL) in the dilution (1:40).
16. Sample concentration and purity were determined by measuring OD at 260 and 280 nm.

34.4 APPLICATIONS

1. Genome data mining analysis of actinobacteria ideally delineates the exploitation of genomic information to discover the biosynthetic pathways of natural products and their possible interactions. This technique is more or less dependent on bioinformatic and computational tools. Some of the studies have incorporated this technique into understanding the tolerance to several abiotic and biotic stresses like tolerance toward heavy metal stress and drought on some species of actinobacteria [9].
2. Transcriptomic analysis provides insight into the transcriptome of a cell of actinobacteria, sum of all transcriptomes by using high-end sophisticated sequencing methods and tools [10].
3. Phylogenetic analysis of actinobacteria provides the evolutionary developmental study and gives a clear and accurate analysis of the origin and evolution of actinobacteria by molecular characterization [11].
4. Since actinobacteria produce secondary metabolites, biosynthetic gene clusters (BGCs) to see if there are any specific contribution of these secondary metabolites produced by actinobacteria are involved in competition with different microbes [9].

34.5 CONCLUSION

Further studies can be conducted postisolation of DNA and RNA from actinobacteria such as genome data mining analysis, transcriptomic analysis, and phylogenetic analysis, using biotechnologically available methods and other molecular techniques.

REFERENCES

1. Anandan R, Dharumadurai D, Manogaran GP (2016) *An Introduction to Actinobacteria*. Intech Open.
2. Devi CS, Keziah SM, Naine SJ, Mohanasrinivasan V (2022) *Actinobacteria: Microbiology to Synthetic Biology*. Singapore: Springer. https://doi.org/10.1007/978-981-16-5835-8
3. Gerber NN, Lechevalier HA (1977) Production of geosmin in fermentors and extraction with an ion-exchange resin. *Appl Environ Microbiol* 34:857–858.
4. Hayakawa M, Yoshida Y, Iimura Y (2004) Selective isolation of bioactive soil actinomycetes belonging to the *Streptomyces violaceusniger* phenotypic cluster. *J Appl Microbiol* 96:973–981.
5. Ilić SB, Konstantinović SS, Todorović ZB (2005) UV/Vis analysis and antimicrobial activity of *Streptomyces* isolates. *Med Biol* 12:33.
6. Ravi VA, Jayanna NB, Sushma K, Naga SBV, Bodaiah B, Sudhakar P (2018) Rapid isolation of genomic DNA from mangrove soil actinobacteria of coastal Andhra Pradesh. *Int J Recent Sci* 6:6645–6648.

7. Kumar V, Singh G, Sardar B, Singh B (2010) An improved method for isolation of genomic DNA from filamentous actinomycetes Antibacterial and anticancer metabolites from actinomycetes view project. *J Sci Eng Tech Mgt* 1–2.
8. Sreevidya M, Gopalakrishnan S, Kudapa H, Varshney RK (2016) Exploring plant growth- promotion actinomycetes from vermicompost and rhizosphere soil for yield enhancement in chickpea. *Brazilian J Microbiol* 47:85–95.
9. Undabarrena A, Ugalde JA, Seeger M, Cámara B (2017) Genomic data mining of the marine actinobacteria Streptomyces sp. H-KF8 unveils insights into multi-stress related genes and metabolic pathways involved in antimicrobial synthesis. *PeerJ* 5:e2912. https://doi.org/10.7717/peerj.2912.
10. Huang K, Zhang B, Chen Y, Liu ZQ, Zheng YG (2021) Comparative transcriptome analysis of *Streptomyces nodosus* mutant with a high-yield amphotericin B. *Front Bioeng Biotechnol* 8:1–14.
11. Verma M, Lal D, Kaur J, Saxena A, Kaur J, Anand S, Lal R (2013) Phylogenetic analyses of phylum Actinobacteria based on whole genome sequences. *Res Microbiol* 164:718–728.

35 Extraction of DNA from Actinomycetota and Amplification of the 16S rRNA Gene

Manik Prabhu Narsing Rao and Arinthip Thamchaipenet

35.1 INTRODUCTION

Deoxyribonucleic acid (DNA) is often referred to as the "blueprint of life" because it contains all the information that an organism requires to function (Bukyya et al. 2021). DNA extraction is a primary and crucial step in molecular biology that is used in many areas of the biological sciences (Tan and Yiap 2009). It was Friedrich Miescher who accidentally obtained the crude extract of DNA while investigating proteins from leukocytes and coined the term nuclein which is now known as DNA (Miescher-Rüsch 1871). There are several methods for extracting DNA, including chemical, thermal, enzymatic, mechanical, or a combination of the aforementioned techniques (Elkins 2013; Dong et al. 2019). The common steps in DNA extraction methods include cell lysis, purification, and DNA recovery (Barbosa et al. 2016).

In the 1980s, in the laboratories of Woese and others, a new standard for identifying bacteria began to develop using the part of the DNA (16S rRNA gene sequences) (Woese and Fox 1977). 16S rRNA gene sequences have been used in bacterial taxonomy due to its presence in almost all bacteria (conserved), its function over time has not changed (less mutation), and it is large enough for informatics purposes (about 1,500 bp) (Janda and Abbott 2007). The present section discusses some of the most common methods used for extracting DNA from *Actinomycetota*, as well as the steps involved in the amplification of the 16S rRNA gene.

35.2 DNA EXTRACTION BY PHENOL–CHLOROFORM METHOD

The phenol–chloroform DNA extraction method is a type of liquid–liquid extraction technique developed by Barker (1998). This technique separates molecules based on their solubility. Proteins and non-nucleic acid cellular components were denatured and digested using sodium dodecyl sulfate (SDS) and proteinase K. Phenol:chloroform:isoamyl alcohol (25:24:1) promote the separation of lipids and cellular debris into the organic phase, while DNA remains in the aqueous phase after centrifugation. Using ethanol precipitation, DNA can be recovered and concentrated from the aqueous phase. A detailed flow chart for DNA extraction from *Actinomycetota* members (Chen et al. 2016) is mentioned in Figure 35.1.

35.3 DNA EXTRACTION USING CHELEX 100

The protocol for DNA extraction using Chelex 100 was developed by Walsh et al. (2013). Chelex 100 is a chelating resin with a high affinity for polyvalent metal ions and is composed of styrene divinylbenzene copolymers containing paired iminodiacetate ions (Singer-Sam et al. 1989). This method involves heat denaturation of cells, and Chelex resin prevents DNA degradation caused by

DOI: 10.1201/9781003398400-35

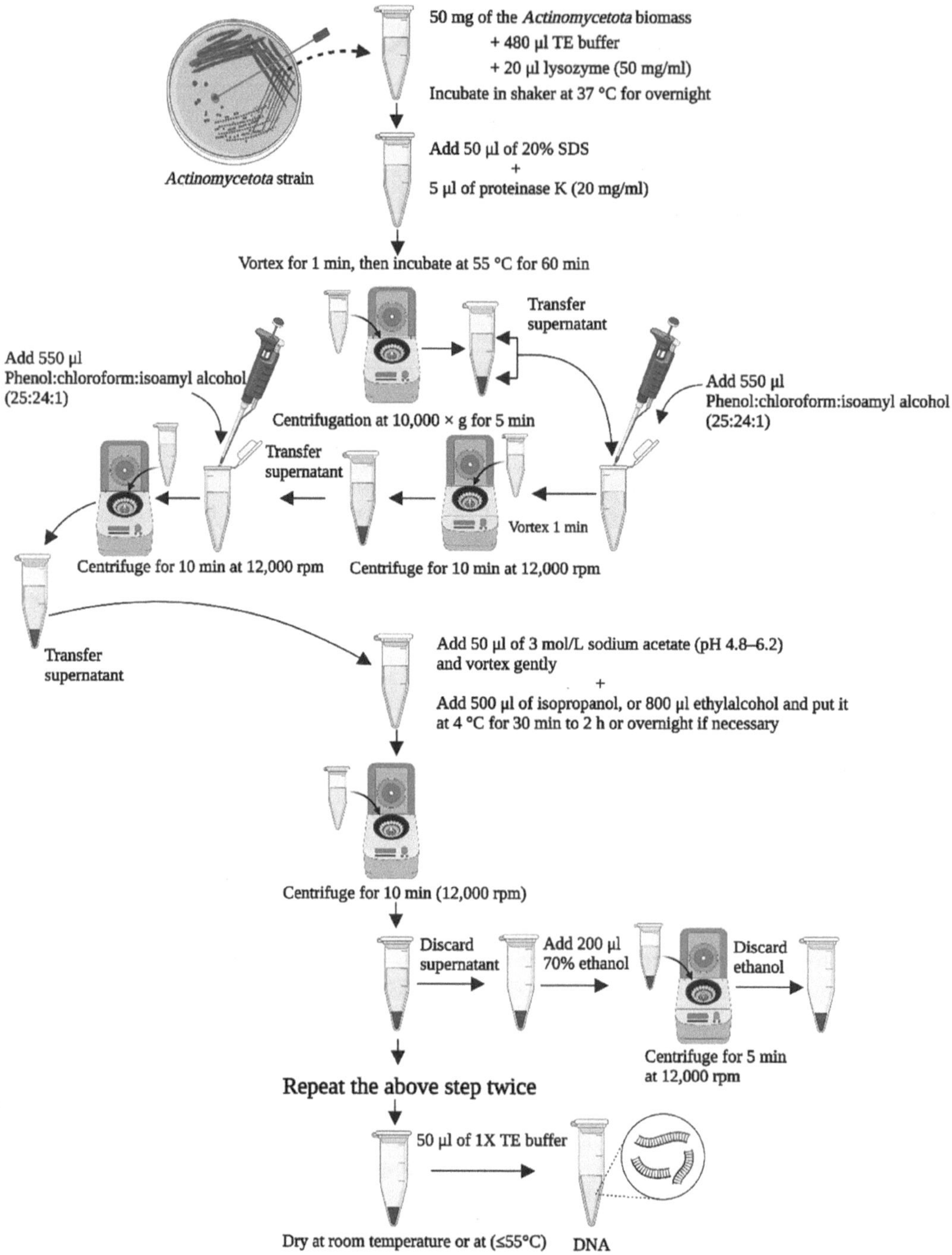

FIGURE 35.1 Protocol for DNA extraction from *Actinomycetota* members by the phenol–chloroform method.

degradative enzymes (DNases) by arresting cations like magnesium ions (an important co-factor for DNase action). High temperatures cause the release of DNA into the solution and facilitate the binding of Chelex resin to magnesium ions. Magnesium ions act as cofactors for the activation of deoxyribonucleases. DNA degradation will be prevented because magnesium ions are no longer able to bind to deoxyribonucleases (Singh et al. 2018). The detailed flow chart of DNA extraction using Chelex 100 is mentioned in Figure 35.2.

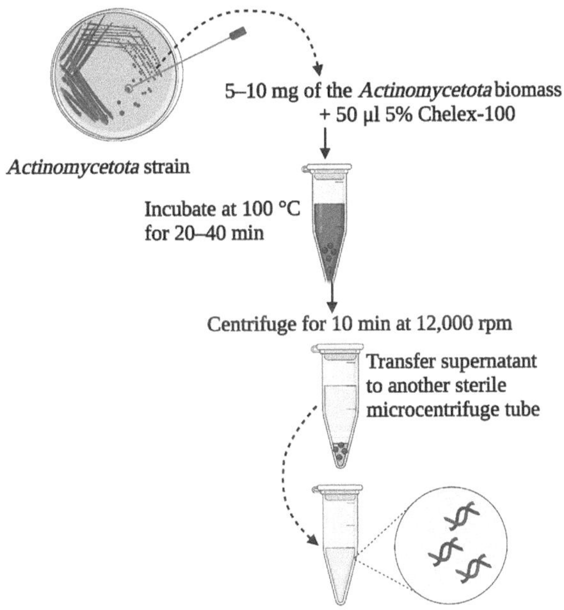

FIGURE 35.2 Protocol for DNA extraction using Chelex 100.

35.4 DNA EXTRACTION USING MICROWAVE

Microwave irradiation was best known as a sterilization tool, but it has recently been used to accelerate the release of genomic DNA (Melendez et al. 2016). In this method, cells were washed with 1 mL PBS (pH 8.0) and 1 mL washing buffer [50 mM Tris/HCl, pH 7.7; 25 mM EDTA; 0.1% SDS; 0.1% polyvinylpyrrolidone (PVP)]. The cells were re-suspended in 50 µL lysis buffer (50 mM Tris/HCl, pH 8.0; 25 mM EDTA; 3% SDS; 1.2% PVP), which were then heated at 700 W in a microwave oven for 45 s. About 400 µL warm extraction buffer [10 mM Tris/HCl, pH 8.0, 1 mM EDTA, 0.5 M sodium acetate, 1.2% PVP (pre-warmed at 65°C)] needs to be added immediately and mixed with the cell suspension. An equal volume of phenol/chloroform was used for extraction. The DNA was precipitated with isopropanol and washed with 70% ethanol and dissolved using 20 µL TE solution (Dong et al. 2019).

35.5 AMPLIFICATION OF THE 16S rRNA GENE

Polymerase chain reaction (PCR) was developed to amplify a specific target DNA sequence and consists of three main steps. In the first step, the DNA sample is heated (at 94°C–98°C for 3–8 minutes) which pre-denatures the double-stranded DNA and splits it into two single strands. In the second step, the sample is heated (at 94°C–98°C for 30–60 seconds) to denature the double-stranded DNA continually, and then the temperature is decreased to approximately 52°C–65°C, depending on the annealing temperature of primers, to allow the primers to bind or anneal with a specific site in the single template strands. In the last step, the temperature is then raised to 72°C, allowing the DNA polymerase to react with the addition of dNTPs to create a new strand of DNA (Chen et al. 2016). Then, the reaction is repeated in several cycles (chain reaction) to obtain a good amount of PCR product. In general, primers used for the amplification of 16S rRNA genes of *Actinomycetota* members include 8–27f (5′-CCGTCGACGAGCTCAGAGTTTGATCCTGGCTCAG-3′) and 1523–1504r (5′-CCCGGGTACCAAGCTTAAGGAGGTGATCCAGCCGCA-3′) (Dong et al. 2019). The PCR mix (50 µL) consists of 5.0 µL 10×PCR buffer, 4.0 µL 2.5 mM dNTPs, 1.0 µL 8–27f primer (25 pmol/µL), 1.0 µL 1523–1504r primer (25 pmol/µL), 1.0 µL DNA template (1 µg/mL), 0.3 µL *Taq*

DNA polymerase (5 U/μL), and 37.7 μL dd H_2O (Chen et al. 2016). The PCR conditions include denaturation at 95°C for 5 minutes, followed by 35 cycles of denaturation at 95°C for 1 minute, annealing at 56°C for 1 minute, and extension at 72°C for 3 minutes. At the end of the cycles, the reaction mixture was kept at 72°C for 5 minutes and then cooled to 4°C (Dong et al. 2019).

35.6 EXPECTED RESULTS

The amplified 16S rRNA gene will be around 1,500 bp size.

ACKNOWLEDGMENT

All figures are created with BioRender.com.

REFERENCES

Barbosa, C., Nogueira, S., Gadanho, M., and Chaves, S. 2016. DNA extraction: finding the most suitable method. In: *Molecular Microbial Diagnostic Methods: Pathways to Implementation for the Food and Water Industries*. Elsevier Inc. pp. 135–154.

Barker, K. 1998. Phenol-Chloroform Isoamyl Alcohol (PCI) DNA Extraction.

Bukyya, J.L., Tejasvi, M.L.A., Avinash, A., Chanchala, H.P., Talwade, P., Afroz, M.M., Pokala, A., Neela, P.K., Shyamilee, T.K., and Srisha, V. 2021. DNA profiling in forensic science: a review. *Global Medical Genetics*. 8(4):135–143.

Chen, X., Jiang, Y., Li, Q., Han, L., and Jiang, C. 2016. Molecular phylogenetic identification of *Actinobacteria*. In: D. Dharumadurai, & Y. Jiang (eds.), *Actinobacteria-Basics and Biotechnological Applications*. pp. 141–174. Intechopen.

Dong, Z.Y., Narsing Rao, M.P., Wang, H.F., Fang, B.Z., Liu, Y.H., Li, L., Xiao, M., and Li, W.J. 2019. Transcriptomic analysis of two endophytes involved in enhancing salt stress ability of *Arabidopsis thaliana*. *The Science of the Total Environment*. 686:107–117.

Elkins, K.M. 2013. *Forensic DNA Biology*. Academic Press

Janda, J.M., and Abbott, S.L. 2007. 16S rRNA gene sequencing for bacterial identification in the diagnostic laboratory: pluses, perils, and pitfalls. *Journal of Clinical Microbiology*. 45 (9):2761–2764.

Melendez, J.H., Santaus, T.M., Brinsley, G., Kiang, D., Mali, B., Hardick, J., Gaydos, C.A., and Geddes, C.D. 2016. Microwave-accelerated method for ultra-rapid extraction of *Neisseria gonorrhoeae* DNA for downstream detection. *Analytical Biochemistry*. 510:33–40.

Miescher-Rüsch, F. 1871. *Ueber die chemische Zusammensetzung der Eiterzellen*. 4: 441–60. DOI: 10.1017/S000708742000062X.

Singer-Sam, J., Tanguay, R.L., and Riggs, A.D. 1989. Use of Chelex to improve the PCR signal from a small number of cells. *Amplifications*. 3:11

Singh, U.A., Kumari, M., and Iyengar, S. 2018. Method for improving the quality of genomic DNA obtained from minute quantities of tissue and blood samples using Chelex 100 resin. *Biological Procedures Online*. 20 (1):12.

Tan, S.C., and Yiap, B.C. 2009. DNA, RNA, and protein extraction: the past and the present. *Journal of Biomedicine & Biotechnology*. 2009:574398.

Walsh, P.S., Metzger, D.A., and Higuchi, R. 2013. Chelex 100 as a medium for simple extraction of DNA for PCR-based typing from forensic material. *BioTechniques*. 54 (3):134–139.

Woese, C.R., and Fox, G.E. 1977. Phylogenetic structure of the prokaryotic domain: the primary kingdoms. *Proceedings of the National Academy of Sciences of the United States of America*. 74 (11):5088–5090.

36 Optimization of Colony Polymerase Chain Reaction for the Identification of Actinobacteria by Sequencing of the 16S rDNA

Mustafa Oskay

36.1 INTRODUCTION

Actinobacteria produce industrially useful metabolites such as antibiotics and enzymes and on the other hand they have ecological importance such as biodegradation. Among actinobacteria, members of the genus *Streptomyces* are the most isolated in soil and the most studied because they produce a large number of important metabolites. Because of these intensive studies, new strains and accordingly new secondary metabolites, especially antibiotics and natural products, were discovered. Actinobacteria produce more than 75% of antibiotics used today. However, the resistance of pathogenic bacteria to most of the antibiotics used in the treatment necessitates the screening of new antibiotic-producing actinobacteria [1,2]. At this stage, sequence analysis of the products obtained by PCR using specific primers for preliminary information purposes is necessary for the identification of new actinobacteria. The cell wall composition of actinobacteria varies considerably between genera but is different and more complex compared to other Gram-positive bacteria. Due to the lack of an efficient and fast DNA extraction procedure for actinobacteria, it may be difficult to isolate appropriate amounts and purity of template DNA from actinobacteria for PCR applications, and extra pre-lysis solution or applications may be required [3].

The utilization of direct colony PCR in molecular genetic studies has produced better and quicker results as compared to the conventional methods. In this protocol, it was aimed to optimize colony PCR and PCR parameters without genomic DNA extraction as a template for the detection of specific DNA sequences in actinobacteria.

36.2 MATERIAL, REAGENTS, AND PRIMERS

PCR buffer (10×, with or without magnesium chloride ($MgCl_2$), supplied with DNA polymerase)

Deoxynucleotide 5′-triphosphates (dNTP) mix (10 mM, supplied with DNA polymerase)

Oligonucleotide primer 1 stock (20 μM, forward primer; 27F (5′-AGAGTTTGATCMTGGCT CAG-3′))

Oligonucleotide primer 2 stock (20 μM, reverse primer; 1525R (5′-AAGGAGGTGWTCCARCC-3′))

Taq DNA polymerase (5 U/μL, ThermoScientific, EP0702, if possible heat stable)

PCR grade water (H_2O) (or double-distilled H_2O, sterile)

$MgCl_2$ (25 mM, if required)

Dimethyl sulfoxide (DMSO, 100%, sterile)

Triton-x100 (100%, sterile)

DOI: 10.1201/9781003398400-36

DNA Ladder (100 bp, Geneaid DL007)
DNA loading dye (6×, Biotechrabbit, BRO800301, for sample loading)
DNA dye (Safe DNA gel stain, SYBR, Invitrogen)
Ethanol (70%, room temperature)
Disposable gloves
Ethanol-resistant marker
0.2 mL thin-walled PCR tubes
PCR tube rack
A set of micropipettes at different concentrations
PCR tube holder (if possible)
2-mL microcentrifuge tube (for the preparation of master mixture)
Toothpicks, sterile
Thermal cycler (Applied Biosystems Veriti Thermal Cycler, USA)
UV transilluminator (gLite gel scanner)
Additional reagents and equipment for agarose gel electrophoresis

36.3 STRAINS AND GROWTH MEDIA

The following actinobacteria kept in our laboratory were used in this protocol: *Streptomyces* sp. (60 strains), *Micromonospora* sp. (10 strains), *Nocardia* sp. (5 strains), *Nocardiopsis* sp. (3 strains), *Actinomadura* sp. (2 strains), *Actinopolyspora* sp. (2 strains), *Saccharopolyspora* sp. (1 strain), *Rhodococcus* sp. (1 strain), and type strains (*Streptomyces griseus* ATCC 21897 and *Streptomyces antibioticus* ATCC 10382). These strains were obtained from the microorganism culture collection of Manisa Celal Bayar University, Biology Department. Actinobacteria were stored at –80°C as glycerol cultures were developed in Yeast Extract-Malt Extract Agar (YEMA) medium at 28°C and made ready for colony PCR applications (Figure 36.1).

36.4 PROCEDURE

1. For all applications, always wear fresh disposable gloves and clean workbenches with 70% ethanol to avoid contaminating the reaction mixture or reagents.

   If possible, to lower the risk of contamination, set up PCR mixtures and work in a laminar flow cabinet equipped with a UV lamp and set up the PCR mixture, perform thermal cycling, and analyze PCR products in separate areas. Always have spares of extremely sterile pipettes, tips, PCR tubes, or other materials available.

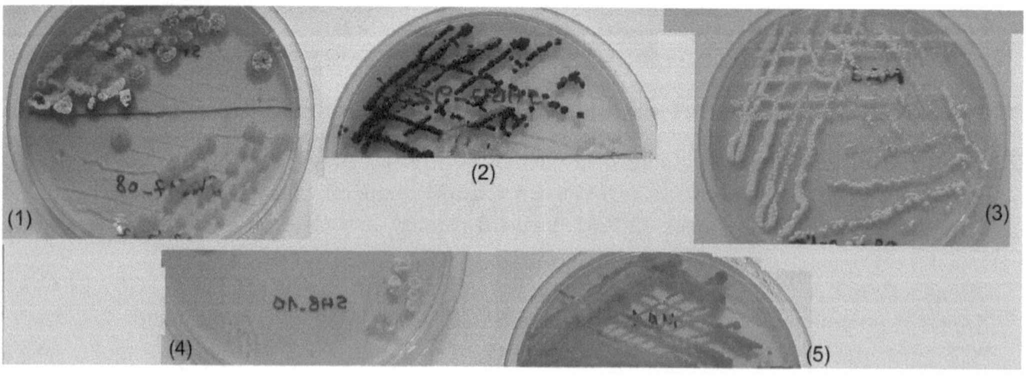

FIGURE 36.1 Representative colony appearances of some actinobacteria used in the protocol: (1) *Streptomyces* sp., (2) *Micromonospora* sp., (3) *Nocardia* sp., (4) *Actinomadura* sp., (5) *Rhodococcus* sp. Ten-day-old colonies of actinobacteria growth in the YEMA.

2. Prepare mycelial growth of actinobacteria on YEMA or appropriate agar media (such as glycerol asparagine agar, starch casein agar, or other International Streptomyces Project (ISP) medium [4].

   The growth of actinobacteria preserved at 4°C over 3 months may be used without any problem. However, some pigment or antibiotic-producing actinobacteria may have a problem. In case of an encounter, using them when they form visible colonies (3–5 days) provides more successful results.
3. Gently vortex and spin down all solutions for PCR amplification after thawing if stored at −20°C.
4. Pipette the following PCR reagents in the following order into a 2-mL microcentrifuge tube for preparing a master mixture: PCR grade H_2O, PCR buffer, dNTPs, $MgCl_2$, primer 1, primer 2, DMSO, Triton-x100, and *Taq* DNA polymerase (see Table 36.1 for colony PCR components and concentrations).

   Gently mix the individual solutions of the PCR reagents by pipetting up and down at least 15–20 times or gently vortex, and care should be taken to avoid introducing bubbles. The volume of the master mix prepared will depend upon the number of reactions and the volume of each reaction. To set up parallel reactions and to minimize the possibility of pipetting errors, prepare a sufficient master mix for the number of reactions plus 10% extra (e.g. 550 µL for ten samples of 50 µL each). Consider also negative and positive control reactions. Although the PCR buffer (DreamTaq Buffer) includes $MgCl_2$ at a concentration of 20 mM, used in this protocol supplied with the DNA polymerase, if not, the final appropriate concentration of $MgCl_2$ is 1.5 mM, which is generally ideal for colony PCR.
5. Mark the individual 0.2 mL thin-walled PCR tubes with an ethanol-resistant marker and dispense appropriate volumes of master mix into them.

   Always keep working on ice. A 96-well plate as a holder for the 0.2 mL thin-walled PCR tubes should be useful.
6. Using a sterile toothpick, pick up a small amount of mycelia of actinobacteria colony directly from the agar plate and dip it into the individual PCR tubes containing the master mix.

   Microscopic amounts taken with a toothpick are sufficient; do not wash the tips by stirring the reaction mixtures. Too many mycelia may cause amplification inhibition from other actinobacterial metabolites. It will be helpful to observe under a 4× or 10× objective lens of the microscope or with the help of a magnifying glass to ensure that there

TABLE 36.1
Optimized Colony PCR Reaction Components and Concentrations

Reaction Reagents	Concentration of Stock Solutions	Final Concentration	In Volume (µL)
DMSO	100%	5%	2.50
Triton-x100	100%	0.5%	0.25
PCR Buffer	10×	1.0×	5.00
$MgCl_2$	25 mM	1.5 mM	3.00
dNTP's	10 mM	0.2 mM	1.00
Forward primer	20 µM	0.5 µM	1.25
Reverse primer	20 µM	0.5 µM	1.25
Taq DNA Polymerase	5 U/µL	2.0 U	0.40
PCR grade H_2O	-	-	35.35
Total volume	-	---	50

TABLE 36.2
Optimized Colony PCR Cycling Conditions

Cycle Step	Temperature (°C)	Time	Number of Cycles
Initial denaturation	95	5 minutes	1
Denaturation	95	60 seconds	34
Annealing	49	40 seconds	
Extension	72	90 seconds	
Final extension	72	8 minutes	1
Hold	4	∞	1

are no other contaminating bacteria while touching the colony. In addition, it is recommended to decide when the colony should be selected by examining the similarity of the colonies belonging to the actinobacteria strain.

7. Put caps on the PCR tubes, place them in a thermal cycler, firmly close the lid, and start the cycling program immediately according to the optimized thermal cycling conditions outlined in Table 36.2.

   When using a thermal cycler that does not contain a heated lid, overlay the reaction mixture with 20–25 µL of mineral oil. The annealing temperature should be 5°C lower than the melting temperature (T_m) of the primers. Differences in T_m between the two primers should not exceed 5°C. Annealing for 40 seconds is normally sufficient. The annealing temperature should be optimized stepwise in 1°C–2°C increments. When additives other than those used in this protocol which change the melting temperature of the primer-template complex are used (e.g. glycerol or betaine), the annealing temperature must also be adjusted. The number of cycles may vary depending on the amount of mycelia in the PCR mixture. About 25–40 cycles are sufficient. T_m of the primers can be calculated using the following equation:

   $T_m = 4\ (G + C) + 2\ (A + T)$

   where G, C, A, and T represent the number of respective nucleotides in the primer.
8. When the program has finished, the 0.2 mL thin-walled PCR tubes may be removed and stored at 4°C for the short term. PCR amplicons can be detected by an agarose gel electrophoresis with DNA dye and bands to be visualized with a UV illuminator.

36.5 REPRODUCIBILITY AND CONTROL TESTING

Sensitivity and reproducibility in colony PCR were tested in parallel reactions with a total of 21 strains of the target (*Nocardia asteroides*, *Rhodococcus ruber*, *Actinopolyspora erythraea*, *Actinomadura cremea*, *Streptomyces thermocarboxydus*, *Streptomyces hygroscopicus*, *Streptomyces cellulosae*), including ATCC-type strains (*Streptomyces griseus* ATCC 21897 and *Streptomyces antibioticus* ATCC 10382) and non-target species (*Lysinibacillus fusiformis*, *Bacillus amyloliquefaciens*, *Bacillus velezensis*, *Bacillus mycoides*, *Alkalihalobacillus alcalophilus*, *Bacillus licheniformis*, *Halomonas* sp., *Halobacillus* sp., *Acinetobacter* sp., *Aeromonas* sp., and *Escherichia coli*). In addition, actinobacterial DNA was isolated using the commercial product: WizardR Genomic DNA Purification Kit (Promega, A1120) by following the conditions specified by the Promega instructions.

36.6 DETECTION OF PCR PRODUCTS BY AGAROSE GEL ELECTROPHORESIS

Depending on the size of the amplicons, a 0.8%–1.5% agarose gel can be used for colony PCR product. The resulting PCR products were checked on 1% agarose gels containing 10 µL of SYBR (Invitrogen) DNA dye added, using Tris Borate EDTA (TBE) buffer at 100 V for 30 minutes.

Afterwards, the gels were examined in a UV transilluminator (gLite gel scanner) and pure and clear bands-forming samples were stored at –20°°C for sequence analysis.

36.7 FINDINGS AND RECOMMENDATIONS

This protocol describes the comprehensive information on the colony PCR amplification of high G–C content DNA from a range of actinobacteria strains. A large number of preliminary trials were carried out for optimization of the colony PCR conditions, and these trials are not given here because no results or partial products were obtained. For amplification of the DNA region encoding the 16S rRNA gene of actinobacteria, two universal primers 27F (5′-AGAGTTTGATCMTGGCTCAG-3′) and 1525R (5′-AAGGAGGTGWTCCARCC-3′) [5] were used. Optimized reaction mixes and PCR conditions for colony PCR are given in Tables 36.1 and 36.2, respectively.

The colony PCR technique described here resulted in the successful amplification of actinobacteria strains and worked at 98% efficiency. As predicted, when optimized parameters and primers for the actinobacteria were used, approximately a 1,500 bp PCR product could be detected for all of the tested strains (Figure 36.2).

However, paying attention to some important points while applying the protocol is necessary. These important points are included as much as possible in the procedure and troubleshooting (Table 36.3) sections. When using actinobacteria cultures that have been kept for more than 3 months, 80% efficiency can be obtained from the colony PCR technique (Figure 36.3). Therefore, young actinobacteria cultures are more suitable for the colony PCR protocol. In addition, for colony PCR studies of actinobacteria, 5% DMSO and 0.5% Triton-x100 as additives have to be included in the normal reaction components [6–9]. If they are not added, the cells taken from the actinobacteria mycelia do not lyse during the reaction and no PCR product is formed. At high rates of these two chemicals, the PCR is inhibited, and at low concentrations, a partial product is formed. Very rarely, pigment or antibiotic production of some of the actinobacteria does not give results in colony PCR applications. Presumably, the pigment or antibiotic produced inhibits the PCR. The PCR conditions given in the method are the optimum conditions determined after many experiments.

It turned out that the initial denaturation temperature and time are very important in the amplification of specific regions of actinobacteria by colony PCR. At first denaturation at 94°C for a maximum of 3 minutes for the colony PCR of actinobacteria, cell lysis does not occur and no product is formed. Therefore, adding 2 U of the polymerase enzyme instead of 1.25 U to the reaction mixture

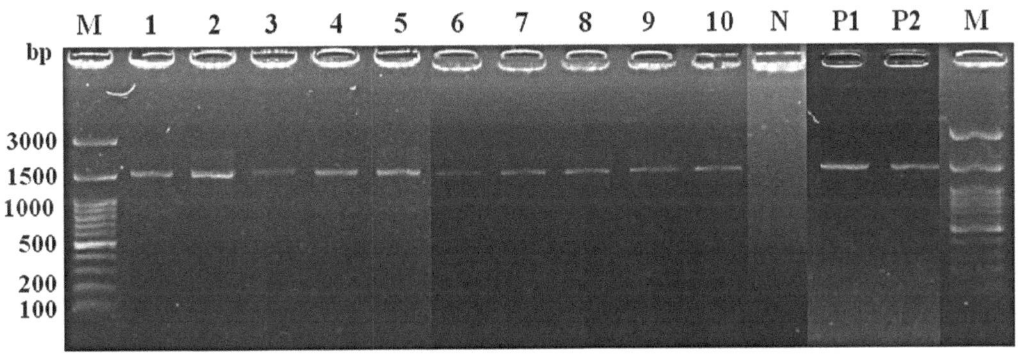

FIGURE 36.2 Representative results of colony PCR. Lane M: 100 bp ladder, Lane N: negative control, Lane P1: positive control-*Streptomyces griseus* ATCC 21897, Lane P2: positive control-commercial kit (Promega, A1120), Lane 1: *Streptomyces hygroscopicus*, Lane 2: *Streptomyces thermocarboxydus*, Lane 3: *Streptomyces cellulosae*, Lane 4: *Nocardia asteroids*, Lane 5: *Actinopolyspora erythraea*, Lane 6: *Actinomadura cremea*, Lane 7: *Rhodococcus ruber*, Lane 8: *Saccharopolyspora* sp., Lane 9: *Micromonospora* sp., Lane 10: *Nocardiopsis* sp.

TABLE 36.3
Troubleshooting Guide for Direct Colony PCR of Actinobacteria

Problem	Possible Cause	Comments and Suggestions
No product detected in colony PCR	Pipetting error, missing or contamination of reagents	Check the concentrations and storage conditions of the reagents. Discard all the reagents of the assay; repeat the protocol using new reagents
	The thermal cycler was not calibrated	Recalibrate the cycler according to the manufacturer's instructions.
	Insufficient amount of starting mycelia or amount of mycelia too high	Adjust the amount of mycelia, repeat the colony PCR
	Colony of actinobacteria too old or contaminated or metabolites of actinobacteria inhibited PCRs	Prepare new culture of actinobacteria on suitable media or culture of actinobacteria till they form visible colonies (3–5 days), repeat the colony PCR
	Initial denaturation temperature or time low	Actinobacterial DNA has G–C-rich regions that pose some challenges to PCR. Adjust the Initial denaturation temperature or time according to the protocol and repeat the colony PCR
	Poor quality of *Taq* DNA polymerase	If possible, use PCR-certified *Taq* DNA Polymerase or use specific DNA polymerases designed for G–C-rich template or add more DNA polymerase up to 2.5 U (because 5% DMSO may reduce the activity of *Taq* DNA polymerase) to the individual PCR
	Annealing temperature not optimal	Ensure that the annealing temperature specified in the protocol or recalculate the T_m of the primers used for optimal annealing or manipulate the annealing temperature of the cycling program
	Mg^{2+} concentration is not optimal, especially too much Mg^{2+}	For optimal results, the final reaction concentration of Mg^{2+} should be between 0.5 and 5.0 mM. Always start with the Mg^{2+} concentration provided in the protocol or adjust the Mg^{2+} concentration by titrating
Non-specific products or undesired amplicons detected (ladder effect of PCR bands on an agarose gel)	Primers have highly repetitive sequences when amplifying genomic DNA or primers may not amplify efficiently	Choose the advisable PCR conditions in the protocol. Redesign the more specific primers to the outcome of the problem
	Non-specific binding of primers to genomic DNA	Ensure that the annealing temperature is 5°C below primers T_m. If so, decrease the annealing temperature by just 1°C–2°C and repeat the colony PCR
	Too high or too low dNTP or instability of dNTPs	Adjust the concentrations of dNTP or use new stocks of dNTPs
	Colony of actinobacteria contaminated	(see above) repeat the colony PCR
Products did not predict lengths	Primers do not amplify the desired DNA sequences	Design new specific primers or see also the problem above (non-specific products detected) and repeat the colony PCR

(Continued)

TABLE 36.3
Troubleshooting Guide for Direct Colony PCR of Actinobacteria

Problem	Possible Cause	Comments and Suggestions
Formation of secondary structures	Self-annealing of primers to each other, rather than the genomic DNA	Choose the advisable PCR conditions in the protocol. Ensure that the annealing temperature is correct. Design new specific primers
	T_m for each primer is drastically different	T_m for both primers should not differ more than 5°C–6°C. Modify the T_m of primer or change the primers and repeat the colony PCR
	Excessive dNTP or primers concentrations	Check and adjust the concentrations of dNTP and primers
Occur of primer dimers, a small product of less than 100 bp appeared on the agarose gel	Excessive primer concentration over the template concentration	Check and adjust the concentrations of primers and also DMSO concentrations

and first denaturation at 95°C for 5 minutes, the protocol gives the best results. Temperatures above this value considerably reduce the activity of the polymerase enzyme [6–11]. Although six different annealing temperatures (54°C, 52°C–48°C) were used, 51 and 50 were good for amplification of the target and 49°C was best in stability and specificity (data not shown). In addition, the type of commercial DNA polymerase used in the colony PCR affects the quality of the PCR product. In the activities of four different *Taq* DNA polymerases used in the protocol (here, their trade names were not given due to ethical issues), a *Taq* polymerase was the most stable and gave 100% results; the efficiency of other three polymerase was 70%, 40%, and 30%, respectively. It is estimated that one of the factors that increase the efficiency of the *Taq* polymerase in colony PCR is the 0.5% (v/v) Nonidet P40 that it contains [3,9,11].

Sensitivity and reproducibility of the protocol were controlled in parallel reactions using a total of 21 strains of the target (actinobacteria strains used in Section 36.5), including type strains *Streptomyces griseus* ATCC 21897 and *Streptomyces antibioticus* ATCC 10382 and non-target species (other different bacteria strains such as *Bacillus amyloliquefaciens*, *Bacillus velezensis*, *Halomonas* sp., *Acinetobacter* sp., *Aeromonas* sp. and *Escherichia coli*). In addition, actinobacterial DNAs were isolated using the commercial kit and compared for suitability of the protocol (Figure 36.3). To rule out false-positive results (undesired products on an agarose gel) that may occur because of contamination from PCR components or other bacteria that are near to the actinobacteria colony, it is prudent to use proper protocol, including a negative and positive control when possible [6,7,11]. Results showed that the 16S rRNA PCR products obtained by colony PCR were consistent with reproducibility testing data and suitable for use in sequence determination processes. Furthermore, the protocol provided herein has proven to be suitable and usable in other different microorganisms for colony PCR. However, not many bacteria have been studied on this subject and more data is needed.

36.8 CONCLUSION

The colony PCR protocol for rapid detection and identification of actinobacteria PCR components, concentrations, and conditions was optimized. Findings confirm that a major advantage of the colony PCR procedure does not require DNA extraction and purification for identifying and determining the phylogenetic diversity of actinobacteria that have high G–C content. In addition, this protocol may be useful for characterization by colony PCR amplification of a range of other bacterial species. PCR technique proposed in this protocol is easy to use and interpret, significantly reduces the

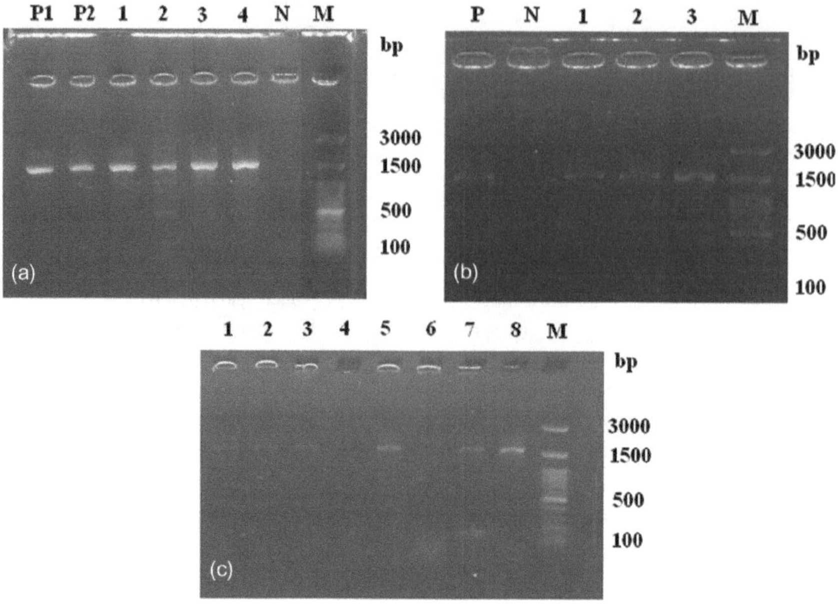

FIGURE 36.3 Representative results of colony PCR. (a) Reproducibility testing; Lane M: 100 bp ladder, Lane N: negative control, Lane P1: positive control-*Streptomyces antibioticus* ATCC 10382, Lane P2: positive control-commercial kit (Promega, A1120), Lane 1: *Streptomyces hygroscopicus*, Lane 2: *Nocardia asteroids*, Lane 3: *Actinomadura cremea*, Lane 4: *Rhodococcus ruber.* (b) Representative results of colony PCR for other bacteria; Lane M: 100 bp ladder, Lane N: negative control, Lane P: positive control-commercial kit (Promega, A1120), Lane 1: *Bacillus licheniformis*, Lane 2: *Halomonas* sp., Lane 3: *Escherichia* coli. (c) Representative results of colony PCR for about 4 months culture of actinobacteria; Lane M: 100 bp ladder, Lane 1: *Streptomyces* sp., Lane 2: *Micromonospora* sp., Lane 3: *Nocardia* sp., Lane 4: *Nocardiopsis* sp., Lane 5: *Actinomadura* sp., Lane 6: *Actinopolyspora* sp., Lane 7: *Saccharopolyspora* sp., Lane 8: *Rhodococcus* sp.

time and cost of the screening process from a lot of environmental actinobacteria isolates, and is a reproducible and efficient method for easily identifying the actinobacteria.

36.9 TROUBLESHOOTING

Before changing anything given in this protocol, be sure that an erroneous result was not due to human error. Begin by confirming all reagents were not contaminated and added to a PCR. For questions not addressed here, please contact the author by email.

ACKNOWLEDGMENTS

Partial financial support was received from Manisa Celal Bayar University under project number 2015-075 within the scope of scientific research projects.

CONFLICT OF INTEREST

The author has no conflict of interest to declare that is relevant to the content of this protocol chapter.

DATA AVAILABILITY

The data used to support the findings of this protocol is available upon reasonable request. The nucleotide sequences of some microbial strains obtained in this protocol have been deposited in the GenBank database with accession numbers which are given in parenthesis: two strains of *Streptomyces cellulosae* (OP893632 and OP893633), *Streptomyces hygroscopicus* (OP893634), *Streptomyces thermocarboxydus* (OP893635), *Actinomadura cremea* (OP893636), *Streptomyces tendae* (OP893637), *Actinopolyspora erythraea* (OP893638), *Streptomyces coelicolor* (OP893639), *Nocardia asteroids* (OP893640), *Rhodococcus ruber* (OP893641), two strains of *Alkalihalobacillus alcalophilus* (OP891017 and OP891018), *Bacillus mycoides* (OP891019), *Bacillus cereus* (OP891020), *Bacillus velezensis* (OP891021), *Bacillus amyloliquefaciens* (OP891022), *Lysinibacillus fusiformis* (OP891021), and *Bacillus licheniformis* (MG722770).

REFERENCES

1. Oskay, M. 2009. Comparison of *Streptomyces* diversity between agricultural and non-agricultural soils by using various culture media. *Scientific Research and Essays* 4: 997–1005.
2. Neethu, K., Karthik, L. and Li, Z. 2022. Nonribosomally and ribosomally synthesized bioactive peptides (NRPS and RiPPs) from actinobacteria. In *Actinobacteria*, ed. L. Karthik, 87–99. Springer Nature Singapore.
3. Gathogo, E.W., Waugh, A.C., Peric, N., Redpath, M.B. and Long, P.F. 2003. Colony PCR amplification of actinomycete DNA. *Journal of Antibiotics* 56: 423–424.
4. Shirling, E.T. and Gottlieb, D. 1966. Methods for characterization of *Streptomyces* species. *International Journal of Systematic Bacteriology* 16: 313–340.
5. Lane, D.J. 1991. 16S/23S rRNA sequencing. In *Nucleic Acid Techniques in Bacterial Systematics*, eds. E. Stackebrandt and M. Goodfellow, 115–148. John Wiley & Sons.
6. Woodman, M.E., Savage, C.R., Arnold, W.K. and Stevenson, B. 2016. Direct PCR of intact bacteria (colony PCR). *Current Protocol of Microbiology* 42: A–3D.
7. Lorenz, T.C. 2012. Polymerase chain reaction: basic protocol plus troubleshooting and optimization strategies. *Journal of Visualized Experiments* 22: e3998.
8. Roux, K.H. 2009. Optimization and troubleshooting in PCR. In *Cold Spring Harbor Protocols*, eds. C. W. Dieffenbach and G. S. Dveksler, 1–6. CSHL Press. doi: 10.1101/pdb.ip66.
9. Śpibida, M., Krawczyk, B., Olszewski, M. and Kur, J. 2017. Modified DNA polymerases for PCR troubleshooting. *Journal of Applied Genetics* 58: 133–142.
10. Knebelsberger, T. and Stöger, I. 2012. DNA extraction, preservation, and amplification. *Methods of Molecular Biology* 858: 311–38.
11. Ishikawa, J., Tsuchizaki, N., Yoshida, M., Ishiyama, D. and Hotta, K. 2000. Colony PCR for detection of specific DNA sequences in actinomycetes. *Actinomycetologica* 14: 1–5.

37 Identification of Actinomycetes by Polyphasic Taxonomy

Balaji Muralikrishnan

37.1 INTRODUCTION

Species identification is essential to keep a record of the phenotypic and genotypic characterization of an organism which helps in understanding them and their relationship with the environment. The growing number of novel genus and species of actinomycetes over time corroborates with the observation of Pace (1997) and Baltz (2007) that only 1% of the total microbiome of the planet is discovered. Apart from their ecological significance, the documentation provides valuable information on the potential of these isolates to produce valuable bioactive compounds. Actinomycetes have been an important resource for identifying useful bioactive compounds that have ranged from life-saving antibiotics to useful dyes. Over ten thousand bioactive compounds have been identified so far and they are in use for human welfare. It constitutes over 45% of all known biologically active compounds and 90% of usable antibiotics. Therefore, the characterization of isolates avoids the problems and financial loss related to rediscovery of the same known compounds.

37.2 MATERIALS AND METHODS

The materials required are *Streptomyces* sp., OA161, International Streptomyces Project (ISP)-2 medium, ISP-5 medium, ISP-4 medium, oat meal agar, ISP-7 medium, tris-saturated phenol, chloroform, isopropanol, ethanol, agarose, Silica 60 F254, silica TLC (Thin-Layer Chromatography) plates, 6N HCl, methanol, inoculation loop, primers given in Table 37.1, Sigma Taq DNA polymerase, dNTPs, 10X Taq DNA polymerase buffer, DMSO (Dimethyl sulfoxide), phosphate-buffered saline, proteinase K, lysozyme, RNase, 15% sodium hydroxide (w/v) in 50% methanol (v/v), coverslips, cellulose aluminium TLC (Thin-Layer Chromatography) sheets and pyridine, C-18 reverse

TABLE 37.1
List of Primers

Gene	Primer	Sequence (5′ to 3′)	Amplified Fragment Length (bp)	Annealing Temperature (°C)	Extension Time (seconds)
16S rRNA	FP	AGAGTTTGATCCTGGCTCAG	1,432	55	90
	RP	ACGGCTACCTTGTTACGACT			
gyrB	FP	GAAGTCATCATGACCGTTCTGCAYGCN GGNGGNAARTTYGA	1,200	58	70
	RP	AGCAGGGTACGGATGTGCGAGCCRTCNA CRTCNGCRTCNGTCAT			
recA	FP	CAAGGCAGACAAGCAAAAGGCG	963	50	60
	RP	TTGTCCTCGATCTCATAGGC			

Abbreviations: Y=C or T, R=A or G, N=A or T or C or G.

DOI: 10.1201/9781003398400-37

phase Nano LC column, Dionex HPLC (High Performance Liquid Chromatography) system and other chemicals are mentioned in the experiments.

37.3 INITIAL MORPHOLOGICAL IDENTIFICATION OF ACTINOMYCETES

37.3.1 Preparation of Pure Culture

On a solid media, actinomycetes colonies have a leathery appearance and resist sticking onto the inoculation loop, although some of the Micromonospora do have sticky property. Such colonies develop a white or grey powder coating on maturation indicating the spore formation.

1. Using a sterile inoculation loop, the powdery spores should be harvested and inoculated onto ISP-2 solid medium. Quadrant streak has to be performed to get individual colonies.
2. Incubate the inoculated petri dish at 28°C for 10 days and observe for tiny leathery colonies (*An example of a purified culture is shown in* Figure 37.1).

37.3.2 Gram Staining (to Confirm Filamentous Morphology)

1. *Use any Gram staining kit.* On top of a clean glass slide, place a drop of sterile water. Patch a loop full of culture from a single colony onto the slide. *Ensure that there are no bigger clumps while forming loop spread.*
2. Heat fix for 15 seconds in and out of the flame, and after air drying for 20 seconds, place a drop of crystal violet and allow it to get adsorbed for 1 minute. Wash the excess stain with running tap water.
3. Place a drop of iodine over the stained surface and allow it to perform the mordant action for 30 seconds. Again, wash the slide with running water and now add a drop or two of absolute ethanol to remove the dye from the cells. Wash again with running tap water.
4. After that, counter stain it with safranin (a drop over the stained region) and wash the excess stain just like the way crystal violet was removed in previous steps.
5. Mount the slide and observe it under the microscope for filamentous bacteria as shown in Figure 37.2.

Once it is confirmed as actinomycetes, genomic DNA can be isolated to perform genotypic analysis.

37.4 GENOMIC DNA ISOLATION

Streptomyces and other actinomycetes have relatively thicker cell walls, and it would be different to break them using conventional techniques. Therefore, it is advised to either freeze fracture the cell walls or break the cell walls using zirconium beads.

FIGURE 37.1 Photographs of isolated colonies of actinomycetes.

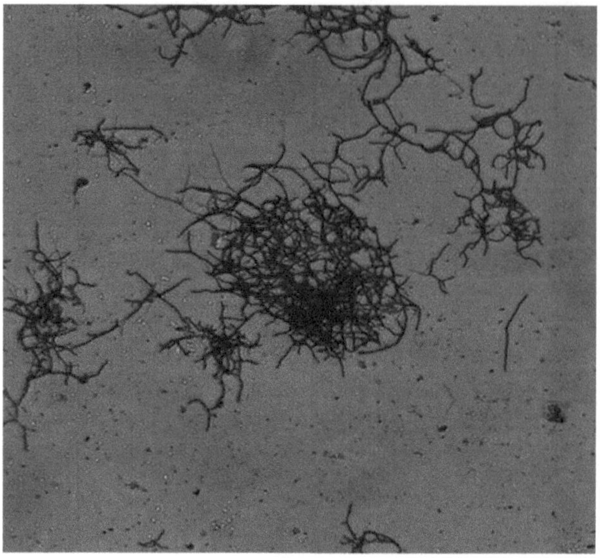

FIGURE 37.2 Gram staining shows the filamentous nature of the actinomycetes.

1. A 48-hour-old liquid culture should serve as a starting material for isolation of DNA. Therefore, for this, a single colony from the pure culture has to be inoculated into 20 mL of ISP-2 medium and incubated at 200 rpm for 48 hours prior to the DNA isolation.
2. On the day of isolation, the culture should be transferred and centrifuged at 5,000 rpm for 10 minutes. Discard the supernatant and using a sterile loop transfer the culture to a sterile pestle and motor.
3. Now, add 5–10 mL liquid nitrogen to freeze the cell pellet and using a pestle grind it several times until it loses temperature and liquefies. Add 500 μL of 1X PBS (phosphate-buffered saline) and 50 μL of lysozyme (20 mg/mL) to the grind and mix them well using a micropipette.
4. After incubating them at 37°C for 1 hour, add 20 μL of proteinase and 50 μL of 10% SDS and incubate at 55°C for 30 minutes.
5. Later, 600 μL of phenol is added and vortexed. The mixture was centrifuged and only the aqueous layer was transferred and re-extracted with an equal volume of chloroform. *These steps can be repeated to achieve pure DNA.*
5. The aqueous layer from the previous step is added with 2.5 volumes of cold absolute ethanol and 10% sodium acetate solution (3M) to precipitate the genomic DNA.
6. Incubation at –20°C improves the DNA yield; even otherwise, handling at room temperature also gives a good concentration of DNA (1–2 mg/mL) which is amicable for downstream PCR (Polymerase Chain Reaction) and sequencing application.
7. After precipitation, add 200 μL of 70% ethanol to wash off impurities by centrifuging for 10,000 rpm for 2 minutes.
8. Discard the ethanol and air dry the pellet and then keep inside a Laminar hood for 15 minutes.
9. Now, add 50 μL of TE (Tris-EDTA) buffer (pH-8) to the dried pellet and dissolve the precipitated DNA (*If it remains insoluble, placing it at 4°C is recommended).*
10. The DNA has to be estimated using a nano-drop, and a 260/280 value of 1.8 would ensure its purity.
11. Agarose gel electrophoresis is performed to ensure the isolated product is DNA by running the sample at 80 V for 45 minutes. A thick band just below the loading well is an ideal representation of good genomic DNA.

37.4.1 16S rRNA Gene Sequencing and Multi-locus Sequence Typing

The primers used for the amplification of 16S rRNA gene are given in Table 37.1.

37.4.1.1 PCR

1. The reaction mix for a 50 µL amplification reaction has been given below.

10× PCR buffer with $MgCl_2$ (Sigma)	10 µL
Template DNA (50–200 ng)	1 µL
Forward primer (20 mM)	1 µL
Reverse primer (20 mM)	1 µL
dNTP mix (5 mM)	2 µL
Taq DNA polymerase (Sigma)	1 U/µL
Sterile MilliQ water	34 µL
Total	50 µL

The reaction conditions are as follows: initial denaturation at 95°C for 5 minutes, followed by 30 cycles of denaturation at 95°C for 30 seconds, annealing at 54°C for 30 seconds, and extension at 72°C for 45 seconds. A final extension was performed at 72°C for 7 minutes.

For multi-locus sequence typing, the amplification reaction conditions remained the same except that the annealing temperature and primers changed according to the details given in Table 37.1. The products were electrophoresed on a 1% agarose gel containing ethidium bromide and visualized under UV light. The PCR product was purified using GE Healthcare gel band purification kit and sequenced directly using a Taq BigDye Terminator Cycle Sequencing Kit in ABI-automated DNA sequencer (Applied Biosystems). All the genes were sequenced with different internal sequencing primers to cross-check the sequence fidelity. The sequences were compared with the help of Ez-Taxon and BLAST databases, respectively.

(*An example of Ez-Taxon results is shown in* Figure 37.3).

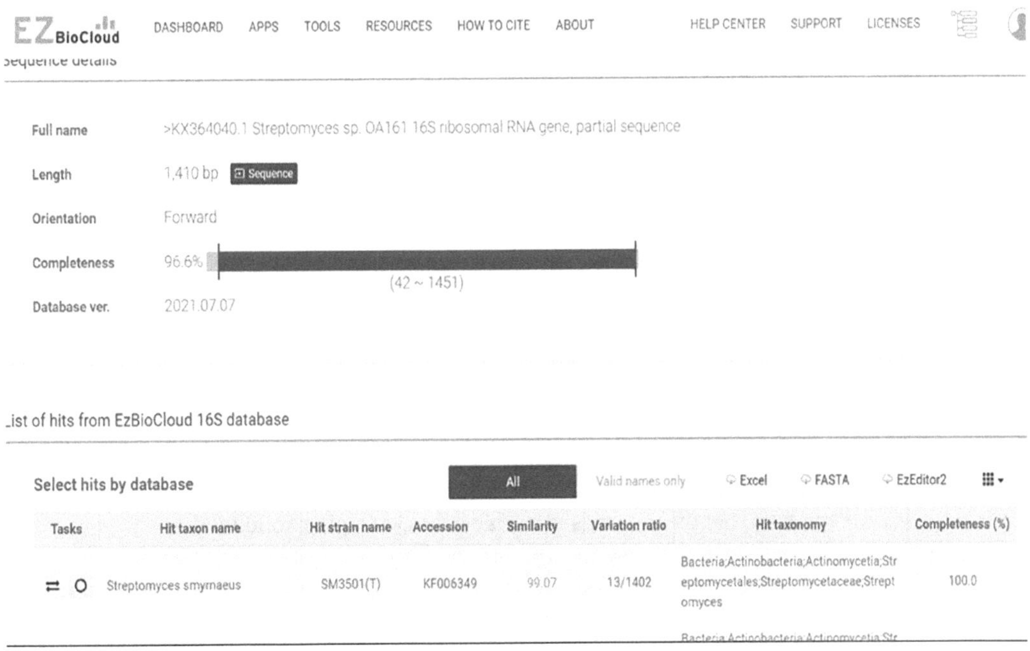

FIGURE 37.3 The results of sequence similarity of 16S rRNA gene of *Streptomyces* sp., OA161.

37.4.2 Phylogenetic Analysis

After identifying similar DNA sequences in the NCBI (National Center for Biotechnology Information) database using BLAST analysis, the similar sequences are either downloaded and assigned their identities (name of the species) or the accession numbers are listed. While selecting the sequences with which the phylogeny would be performed, it is essential to avoid different strains of the same species. An outgroup organism has to be selected and assigned to show the difference in clade formation in a phylogenetic tree. It is simply used to root the phylogeny.

37.4.2.1 Phylogenetic Construction Using MEGA 5 Software

1. Open the MEGA 5 software, and from the alignment menu, choose 'Do BLAST search' and in that choose 'blastn'.
2. In the search box, either the sequences can be pasted in the FASTA (First database similarity search tool) format with sequence IDs or accession numbers can be provided.
3. Once the sequences are imported, sequence alignment has to be performed by clicking the 'align' tab. While doing so, it is advised to select 'MUSCLE' algorithm instead of 'ClustalW' as the former is superior to the latter.
4. After selecting 'MUSCLE', click 'ok' and an alignment file '.mas' which has to be exported in '.meg extension' and stored in the hard disk.
5. This '.meg' file is opened in MEGA 5 software again as this is the only readable form (.mas cannot be read by MEGA 5).
6. After opening the file, find the model's menu and click 'find best DNA models' and click 'compute'.
7. Next, from the phylogeny menu, choose 'construct/test Maximum Likelihood Tree' and a preference dialogue box will open. In that, three parameters have to be changed,
   i. Model – WAG model
   ii. Rates among sites – G+1
   iii. Gap/missing data treatment – Partial deletion. Click compute to generate a tree, and this will be displayed on the same page. This tree has to be exported as a '.mts' file extension.
8. Later, a bootstrap test is run using the 'Bootstrap method' tool found under the phylogeny test menu. Any number between 100 and 2,000 can be for running the test; however, larger the number, better the reliability of the tree. Therefore, 1,000 should be preferred, and this would take some extra time to compute.
9. The tree appears after the computation with number at its every node. These numbers signify the evolutionary relationship between the members of the tree and those with <50% are considered less reliable.

(An example of a phylogenetic tree for Streptomyces sp., OA161 is given in Figure 37.4)

37.5 CHEMOTAXONOMIC CHARACTERISTICS

37.5.1 Growth Assessment and Utilization of Carbon Sources

1. Prepare different growth media (ISP-2, ISP-4, ISP-7, actinomycetes isolation agar, ISP-5, tryptone soy medium, oat meal agar, and humic acid-containing solid media.
2. Perform a quadrant streak from the mother culture with a loop of spores in all the media.
3. Incubate at 28°C for 2 weeks. Monitor for the tiny colonies maturing into spore-forming spread-out ones. The relative time period for this development is recorded for a particular strain and rated from '++++ for exuberant growth' to '+ for poor growth'.

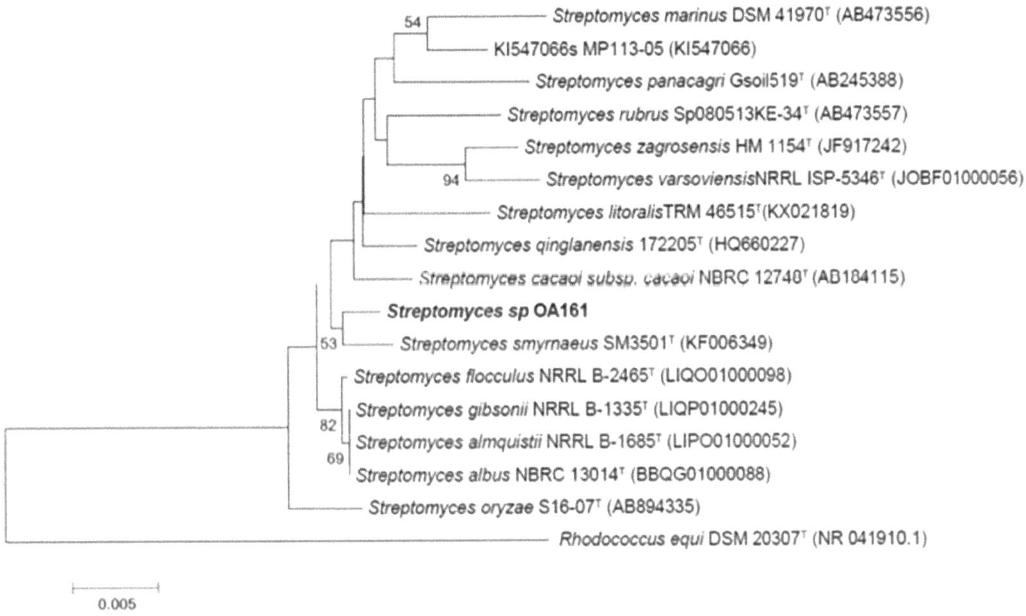

FIGURE 37.4 Evolutionary relationship of taxa.

4. The morphology and colour of the substrate and aerial mycelium are noted at the end of 2 weeks. The best medium for growth is identified and subsequently used in other experiments below.
5. Similarly, using the HiCarbo kit (Himedia, India), several carbon sources can be tested that support the actinomycetes growth with simple inoculation on the solid media-containing wells and observing the change of colour from red to yellow. Red indicates non-utilization while yellow shows utilized carbon source.
6. Likewise, the API ZYM Kit (Biomerieux) identifies the number of enzymes produced by the isolate that are secreted out into the growth medium.

37.5.1.1 Assessment of Salt and Temperature Tolerance

1. Salt tolerance is identified by growing the isolate in 0%, 2%, 4%, 10%, and 15% sodium chloride-containing growth medium, and the ability to form colonies represents salt tolerance.
2. Growth below 4% represents slight halophile; below 10% is considered moderate halophile while 15% and above represents extreme halophile.
3. As salt tolerance test was performed, the isolates were grown on solid media plates but incubated at different temperatures (4°C, 25°C, >50°C (grown in liquid media)). Growth at 4°C represents psychrotolerant while growth only at 25°C and between are mesophiles. Growth above 50°C is classified as thermophiles.

37.5.2 Fatty Acid Methyl Ester (FAME) Analysis

1. The isolate has to be exuberantly grown in tryptic soy agar for 7 days at 28°C.
2. The grown colonies are to be harvested (two–three loop full cultures) into a sterile Eppendorf.
3. 1 mL of 15% sodium hydroxide (w/v) in 50% methanol (v/v) should be added to it and thoroughly mixed using vortex and the mixture should be boiled in water bath for 5 minutes.

4. After cooling down to room temperature, 2 mL of 55% 6 N HCl in methanol has to be added to methylate the fatty acids.
5. This total mixture has to be extracted with 1.5 mL of hexane: ethyl butyl ether (1:1). The separated organic phase should be used as injection liquid to GC-MS. A polar stationary phase is used and a non-polar mobile phase is used.
6. A 20-minute run was performed to analyse the fatty acid profiles of the isolate. Nitrogen and hydrogen were used as carrier gases at 1.5 and 2 mL/min, respectively.
7. The fatty acid standards are run to match profiles of that to the query isolate.
8. The m/z values are matched to their respective fatty acids using conversion to equivalent chain length (ECL) by the MIDI Sherlock software that compares data from its own database.

(*A typical FAME analysis result is presented in Figure 37.5*)

37.5.3 Analysis of Menaquinones and Total Carbohydrates of the Isolate

37.5.3.1 Menaquinone Profiling

1. The isolate has to be exuberantly grown at ISP-2 medium for 4 days at 28°C.
2. Pellet the culture using centrifugation and lyophilize the culture to powder form.
3. Extract 100 mg of the lyophilized culture using 1 mL chloroform: methanol (2:1, v/v) by vortex.
4. Keep it in a rocker for 1 hour and pellet down the cell debris by centrifugation and transfer the organic layer to a clean tube and evaporate the same to concentrate the organic layer.
5. Separate the concentrated organic extract using thin-layer chromatography. Silica F254 plates are used as stationary phase and hexane: diethyl ether is used as mobile phase.

ECL Deviation: 0.004 Reference ECL Shift: 0.004 Number Reference Peaks: 15
Total Response: 625438 Total Named: 556595
Percent Named: 88.99% Total Amount: 535238

Matches:

Library	Sim Index	Entry Name
ACTIN1 3.80	0.306	*Streptomyces lavendulae*

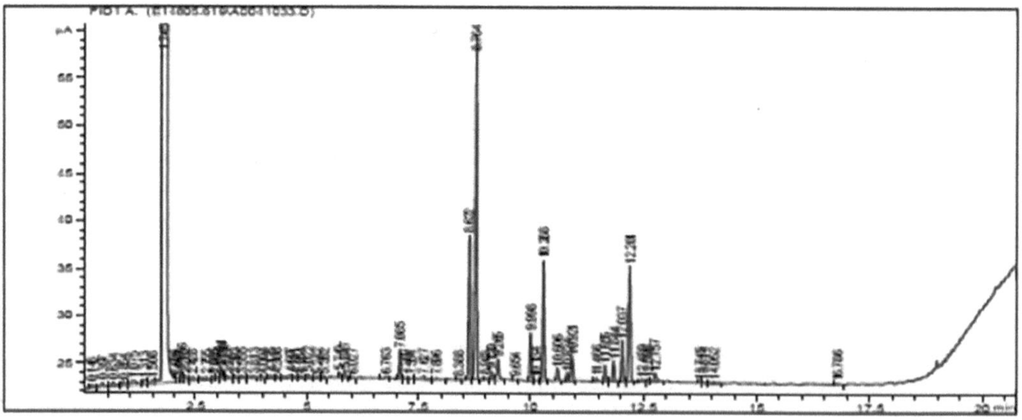

FIGURE 37.5 Gas chromatogram shows the spectrum of fatty acids from the cell wall of the isolate *Streptomyces* sp., OA161 and its MIDI SHERLOCK prediction.

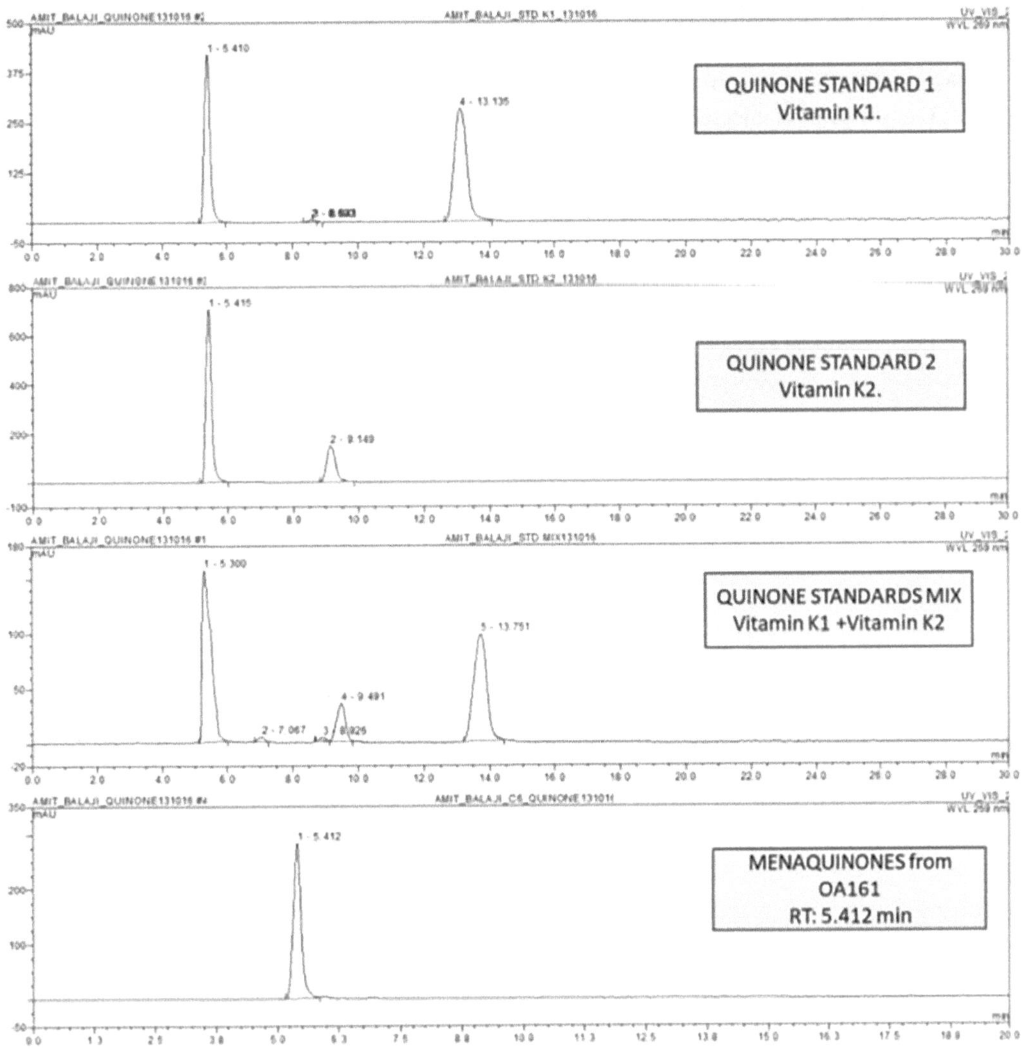

FIGURE 37.6 HPLC chromatogram showing quinone standards and menaquinones of isolate OA161.

6. Once the run is over, shine longer UV light over the plate and identify brown spots. Mark and scrap the brown spots and re-extract them with chloroform: methanol (2:1, v/v).
7. After passing through a 0.2 μm filter, use 10 μL of this extract for HPLC analysis.
8. C8 reverse phase column should be used as the stationary phase and water–methanol gradient should be used as the mobile phase.
9. HPLC method (mobile phase – 50% methanol in 0.1% ammonium fluoride at 0 minute linearly increased to 100% methanol in 0.1% ammonium fluoride at 30 minutes).
10. Vitamin K1 and vitamin K2 are standards. The HPLC profile of the samples is compared with that of the standards to interpret the results.

(An example of the menaquinone profiling is given in Figure 37.6)

37.5.3.2 Total Carbohydrate Profiling

1. Similar to menaquinone profiling, boil 25 mg of the wet weight of isolate OA161in 0.25 N HCl for 15 minutes and filter the extract through a 0.2-µm syringe filter.
2. 5 µL of the same is injected to Dionex CarboPac column and an isocractic run is performed with 30 mM potassium hydroxide solution for 20 minutes at a flow rate of 0.5 mL/min.
3. 5 mM sugar standards are prepared and injected subsequently to compare the profile using the retention time as shown in Figure 37.7.

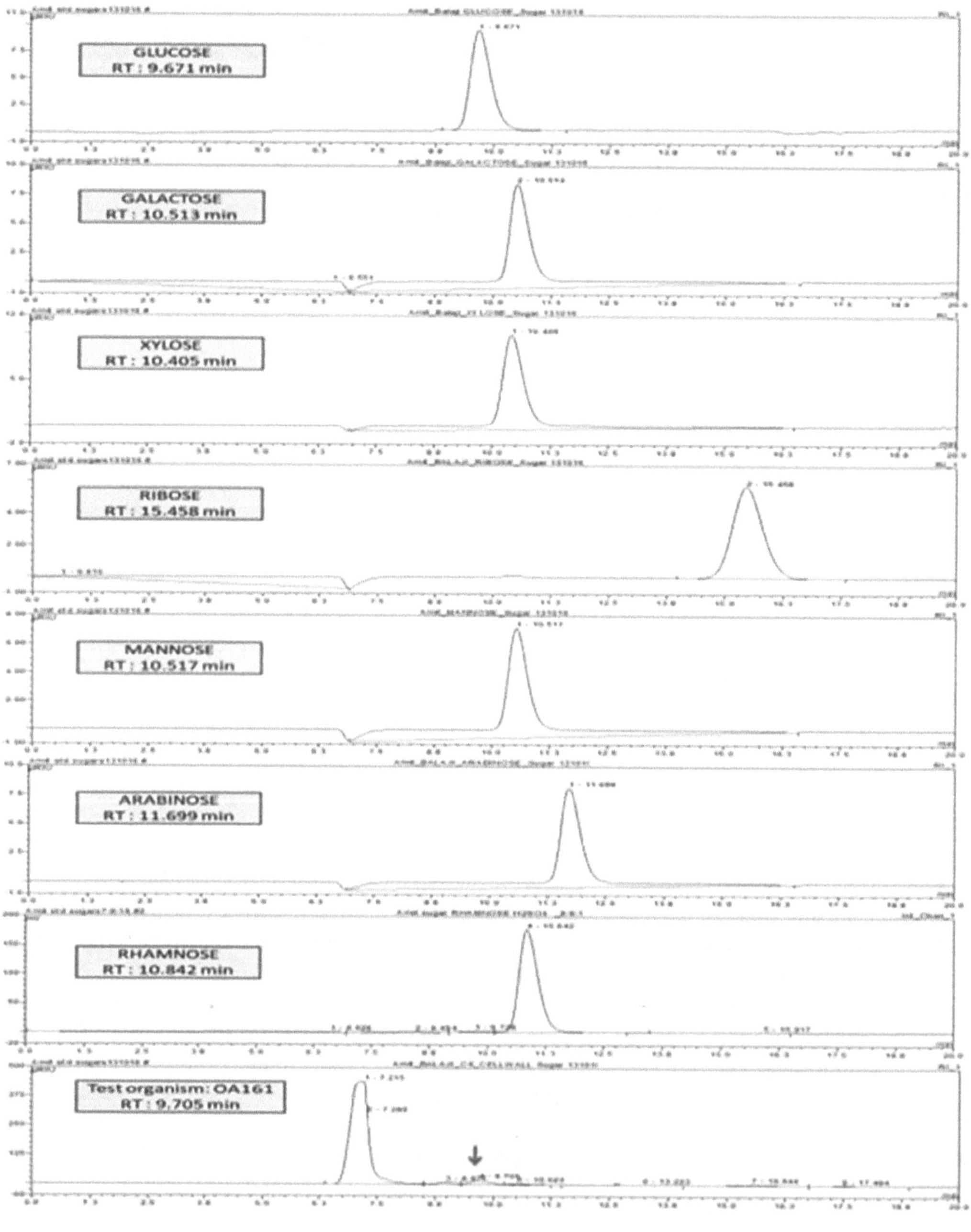

FIGURE 37.7 HPLC chromatogram showing the standard sugars and the sugars of the isolate OA161.

37.5.4 LL-Diaminopellic Acid Production

(This test is very specific to identify Streptomyces when initial characteristics are confirmed as actinomycetes)

1. As described under menaquinone profiling, prepare a lyophilized culture and use 200 mg of the lyophilized powder for extraction.
2. Add 5 mL of 6 N HCl to the 200 mg lyophilized culture and heat the mixture using a boiling water bath for 30 minutes.
3. Pellet the cell debris using centrifugation and use 5 µL of the supernatant for thin-layer chromatography: stationary phase – cellulose aluminium sheet; mobile phase – methanol: pyridine:6N HCl: water (80:10:3:17, v/v).
4. After running the TLC, dry the plate and spray with ninhydrin in acetone (0.2% w/v). Dry the plate at 100°C for 2 minutes and look for chartreuse colour to identify the diaminopellic acid. (A Rf value of 0.2 represents meso form and a Rf value of 0.3 represents LL form of diaminopellic acid.)

37.6 MORPHOLOGICAL ANALYSIS

37.6.1 Scanning Electron Microscopy

(Evaluation of spore morphology is critical data to identify actinomycetes. The spore morphology can be dumbbell, bead, cylindrical, elliptical, or star, and many more shapes and sizes are present. This morphology can be visualized by scanning electron microscopy).

1. Using a sterile inoculation loop, transfer the actinomycetes culture grown on an ISP-2 solid medium (7 days) to a clean tube and add 200 µL of the ISP-2 medium. Mix them well and transfer the 200 µL to a gelatin-coated cover slip.
2. Allow it to settle down for 30 minutes and gently decant the medium.
   (Now, your coverslip has spores and mycelia attached to the coverslip and those unbound are washed away. It is essential to have a thin and less populated coverslip to acquire high-resolution images)
3. Add 100–200 µL of 2.5% glutaraldehyde (in 1X PBS) to the coverslip to cover the region.
4. Incubate the coverslip overnight inside a sterile Petri dish at 4°C.
5. Next day, remove the excess glutaraldehyde by decanting, add 150 µL of 50% ethanol, and incubate it for 10 minutes and decant the ethanol. Subsequently, repeat this procedure with 70% ethanol; 90% ethanol; 100% ethanol.
   (This is a dehydration step)
6. The coverslip was air-dried for 5 minutes, and the samples were gold-sputtered and imaged using a scanning electron microscope.

(An example of SEM image of a Streptomyces strain OA161 is given in Figure 37.8.)

37.7 CONCLUSION

In chronology, actinomycetes have been misplaced under Fungi because of their filamentous nature and formation of mycelia and spores. Later, the identification of 16S rRNA gene helped systematics experts reclassify them to bacteria. There have been at least a million strains isolated around the globe and many are not annotated. Redundancy has been the major issue in disinterest in defining strains to species level. However, in order to reduce this redundancy, a polyphasic approach is necessary to ensure the authenticity of the strain. This includes molecular, biochemical and morphological taxonomical analysis. International systematics have laid stringent criteria to approve an isolate

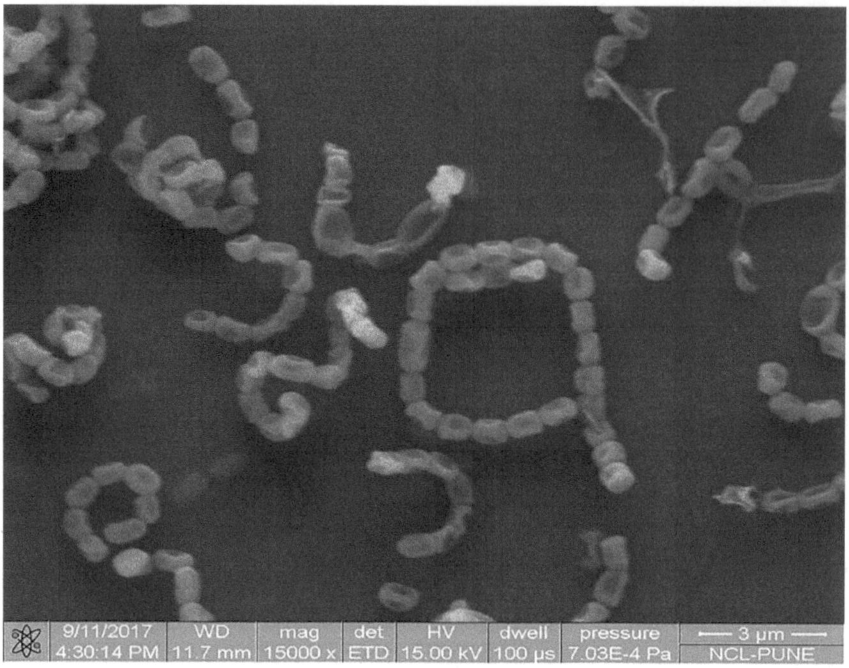

FIGURE 37.8 Scanning electron photomicrograph showing the spore morphology of the isolate OA161.

as a novel species, and to meet the same, most of the required experiments and their procedures are given in this chapter; nevertheless, it is not exhaustive.

ACKNOWLEDGMENTS

This work was supported by CSIR-OSDD, India, and intramural funds of Rajiv Gandhi Centre Biotechnology, Department of Biotechnology, India.

CONFLICT OF INTEREST

The author has no conflict of interest to declare that is relevant to the content of this protocol chapter.

DATA AVAILABILITY

The data used to support the findings of this protocol is available upon reasonable request. The nucleotide sequences of some microbial strains obtained in this protocol have been deposited in the GenBank database with accession number (KX KX364040.1).

REFERENCES

Baltz, R.H. "Antimicrobials from actinomycetes: back to the future." *Microbe* 2 (2007): 6–7.
Pace, N.R. "A molecular view of microbial diversity and the biosphere." *Science* 276.5313 (1997): 734–740.

38 Phylogenetic Analysis of Actinomycetes

D. M. Yashaswini, Andrea Michelle, U. Monisha, and Lokesh Ravi

38.1 INTRODUCTION

Phylogeny constitutes the evolutionary history of a group of organisms. Phylogenetics is the biology that deals with the study of this shared history and the accompanying changes in and among these organisms. Phylogenetic analysis is done to understand the changes that occur among these organisms, which leads to a better understanding of the relationship, shared traits, the adaptation of these organisms to the surroundings over time, etc. It is necessary and is employed as the taxa being studied are not statistically independent and are connected through historical relationships [1]. Logical and valid interpretations of the biological patterns among these taxa require an estimate of these historical relationships. It can be called the comparison of information about different aspects of the groups of organisms being studied.

These phylogenetic relationships are diagrammatically represented for ease of studying and understanding; this representation is called a phylogenetic tree. The construction of this tree requires the genetic molecular information of the taxa being studied. It can be constructed using sequence alignment tools. There are different types of phylogenetic trees; they can be rooted (in the case of a common ancestral organism) or unrooted. They can be cladograms (the length of all the branches is the same but the time taken is not indicated) or phylograms (branch lengths are different based on the evolutionary change that takes place) [2,3]. These trees can be constructed in different shapes; they can be circular, straight, rectangular, or curved (shown in Figure 38.1a).

38.1.1 Components of a Phylogenetic Tree: (Shown in Figure 38.1b)

- Branches
- Nodes (points of divergence of lineage)
- Root
- Taxa (Taxon refers to an organism at the end of each branch)
- Clade (an ancestral group with all its related taxa) [3]

38.1.2 Applications of Phylogenetic Analysis Include

- It is employed in the studies of epidemiology to trace out the roots and origin of a particular disease.
- Extant genes can be derived from the reconstruction of ancestral genes [1].
- Earlier phylogenetic analysis was based on the comparison of morphological characters, but now, it has extended and depends on information gained from genetic materials such as DNA, RNA, and proteins [4].
- Helps in learning the bio-ecography of these taxa throughout time.
- Phylogenetic analysis of some extremely conserved genes helps in establishing and studying the relationships between all living species [2].
- Differences and similarities between the related organisms can be studied [4].

DOI: 10.1201/9781003398400-38

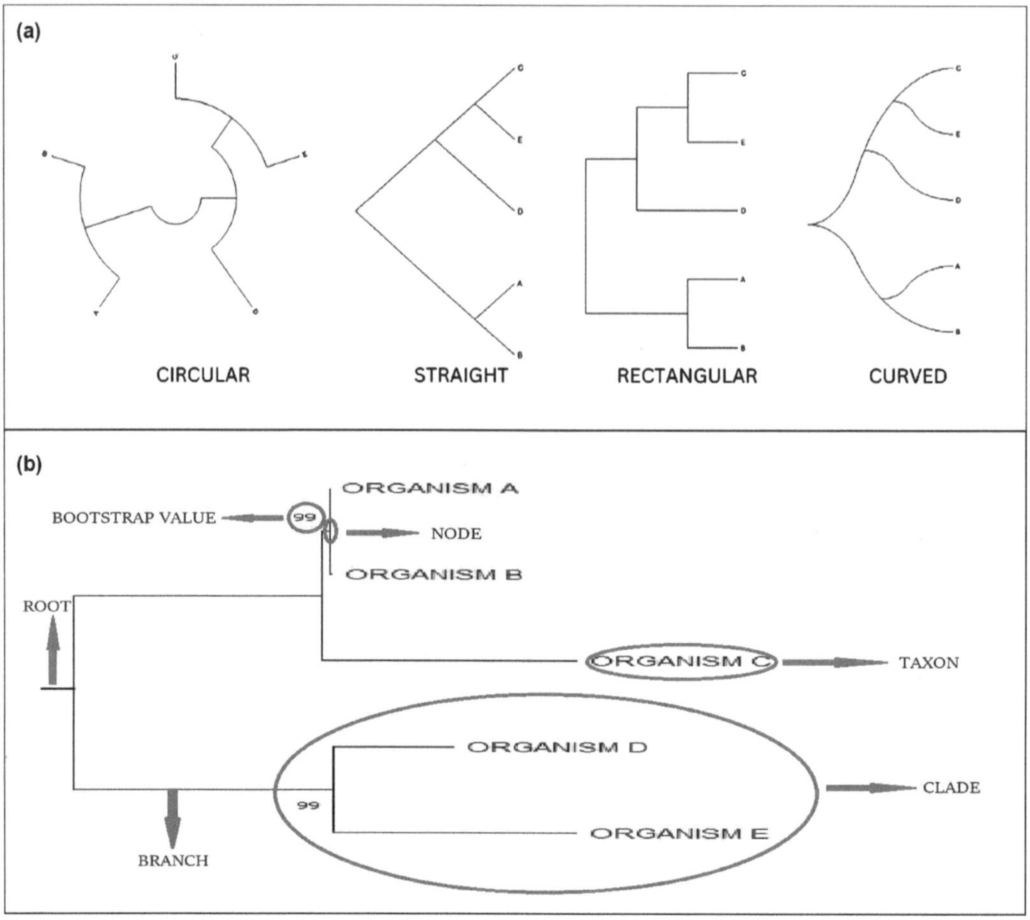

FIGURE 38.1 (a) Types of the phylogenetic tree based on shapes; (b) parts of a phylogenetic tree.

38.1.3 Advantages of Phylogenetic Trees

- A better understanding of the evolution of species is gained by studying phylogenetic trees [4].
- A lot of information is represented in a single diagram, making it available in the same place.
- The closeness of the related species can be understood easily.
- Tree construction methods such as maximum likelihood trees can be used for dissimilar sequences and still understand the relationship between such taxa [4].
- Mutations are taken into account, giving a more accurate analysis.
- Mutations in genomic sequences are responsible for the development of cancer and its progression; these phenomena across populations can be described using different evolutionary pathways, and the application of a phylogenetic tree to this can be used in finding similarities in cancer subtypes. Identification and discovery of genes associated with each cancer subtype help in classifying the cancer subtypes based on their mutations [3].
- Time-resolved phylogenetic trees have been constructed to observe and study the trends in the transmission and evolution of pathogens and infections.
- The tree's topological features can give information about the evolutionary process of the taxa (pathogens in this case) and some basic dynamics involved in an epidemic.
- Topology can also be used to learn about the population size, the surroundings, etc. [3].

The study of actinomycetes and their phylogeny can be simplified and done accurately by employing phylogenetic trees. The protein and DNA sequences required for the construction of these trees can be obtained from numerous verified databases such as GenBank and UniProt. Blast can be used to find the closely related species which can be subjected to the studies. Analytical tools like MegaX can be used in the sequence alignment and the building of the phylogenetic trees.

38.1.4 MEGA

MEGA (Molecular Evolutionary Genetic Analysis) is a statistical analysis software developed in the year 1993 by the Pennsylvania State University under the guidance of evolutionary biologist, Masatoshi Nei. It is a multiple sequence alignment tool used to build phylogenetic trees and works by importing data from other software (CLUSTAL W) and databases (BLAST). Multiple sequence alignment (MSA) is a method of inferring evolutionary relationships between different organisms by aligning similar-sized nucleotide or protein sequences of the organisms [5].

38.2 PROCEDURE

Phylogenetic and molecular evolutionary analyses were conducted using MEGA version 11.

Step 1: Download and install the MEGA software.
Step 2: Open the MEGA tool and select the Align option (Figure 38.2a).
Step 3: Select the 'Edit/build alignment' option (Figure 38.2b).
Step 4: In the alignment editor window that opens, select the 'create a new alignment' option and press OK (Figure 38.2a).
Step 5: Select the type of sequence alignment (in this case, we are selecting DNA) (Figure 38.2a).
Step 6: Copy your unknown or the initial FASTA sequence.
Step 7: Open the web explorer in the alignment explorer window and select the 'do BLAST' search option. The BLAST page opens.
Step 8: Paste the copied FASTA sequence of the unknown organism on the BLAST page.
Step 9: Scroll down, select the 'highly similar sequence' (megablast) option, press Blast, and run the blast (Figure 38.2b).
Step 10: Once the results are obtained, select a few sequences among them with varying similarity percentages. preferably select organisms from different species.
Step 11: To add the sequences of the selected organisms to the alignment explorer, click on the accession number and select 'GenBank' in the new tab.

In the drop-down that appears, select 'FASTA (text)' and select the 'Add to alignment' option in the open tab.

This way, the sequence gets added to the alignment explorer.

Repeat the same steps for all the selected organisms (Figure 38.3a and b).

Step 12: After the addition of the sequences from all the selected organisms, close the web browser and keep the alignment explorer open.
Step 13: Copy the test file/unknown sequence, paste it in the alignment explorer and name it test file/file-1, etc./unknown as per convenience (other sequences can also be renamed as per convenience).
Step 14: To align all the sequences, click on alignment in the alignment explorer window; in the drop-down, select 'align by clustalW', and in the pop-up, select the 'select all' option.
Step 15: Select 'OK' for the confirmation message (Figure 38.4a).
Step 16: In the 'clustalW option', click on 'OK'. Once the alignment process completes, save the file. To do this, click on 'data' in alignment explorer and select 'save session'.

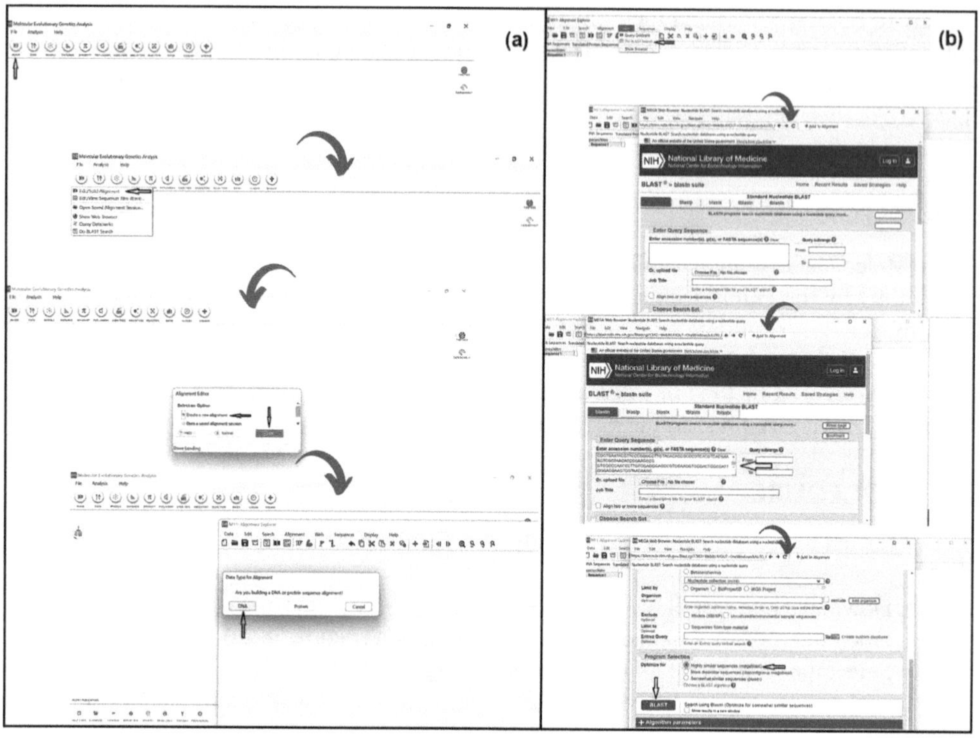

FIGURE 38.2 (a) Schematic representation of MEGA operation; (b) schematic representation of MEGA operation.

Assign a name to the file and click on save (Figure 38.4b).
Now, close the alignment explorer.

Step 17: In the 'Molecular genetic analysis window', click on the option called 'Phylogeny'.
Step 18: Select 'construct/Test maximum likelihood tree' in the drop-down.
Step 19: Select the file from the previously saved session and click on open.
Step 20: Click yes in the confirmation pop-up (Figure 38.5a).
Step 21: In analysis preferences, under 'PHYLOGENY TEST', select the test of phylogeny to be the 'Bootstrap method'. Set the number of Bootstrap replications to 1000, click 'OK', and run the process. (It might take some time) (Figure 38.5b)
Step 22: A fully constructed phylogenetic tree appears, which can be saved in any desirable format (Figure 38.6).

38.3 CONCLUSION

The protocol focuses on the use of MEGA software to build a phylogenetic tree to analyze the evolutionary relationship in *actinobacteria* (*Streptomyces* sp.). Phylogenetic trees can be used to study evolution in depth by means of a simple diagrammatic tree. MEGA software makes the building of a phylogenetic tree easier by use of its in-built web browser option which allows the user to BLAST the sequences and save the session. *Streptomyces* is a species of bacteria that is known for its high amounts of secondary metabolite production. Researchers suggest that they can cause diseases in humans and plants. *Actinomycetes* belong to a monophyletic clade. Phylogenetic analysis using the given protocol shows that the unknown sequence (*Streptomyces globosus*) is closely related to *Streptomyces katrae* and least related to *Streptomyces yangpuensis*.

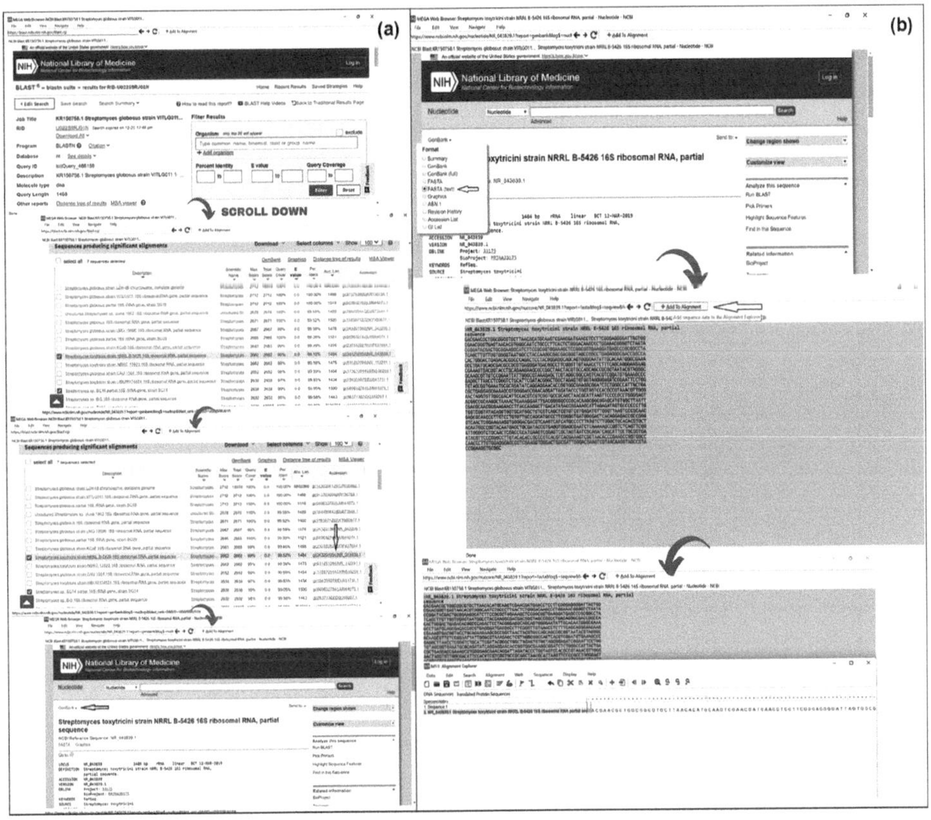

FIGURE 38.3 (a) Schematic representation of MEGA operation; (b) schematic representation of MEGA operation.

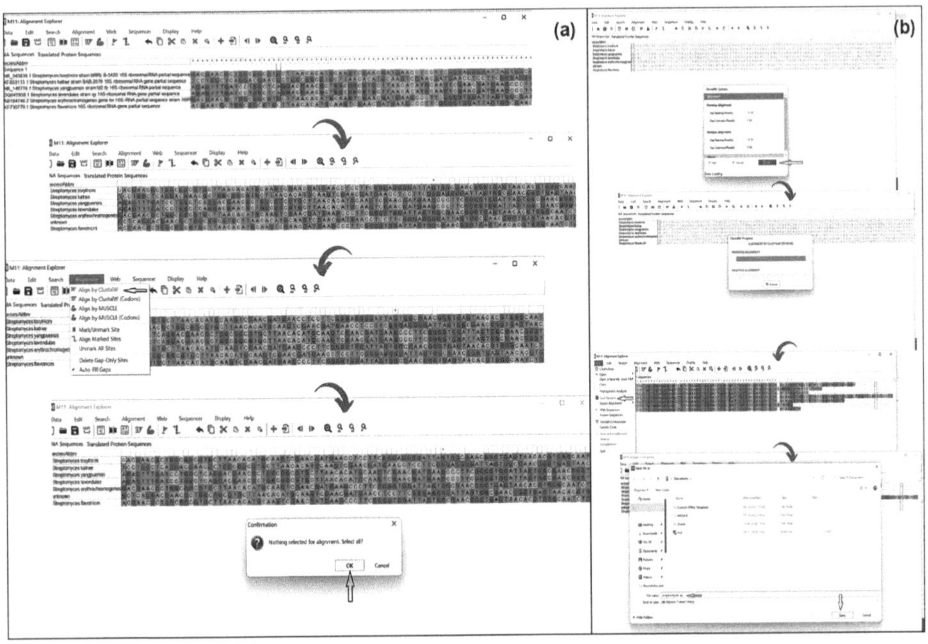

FIGURE 38.4 (a) Schematic representation of MEGA operation; (b) schematic representation of MEGA operation.

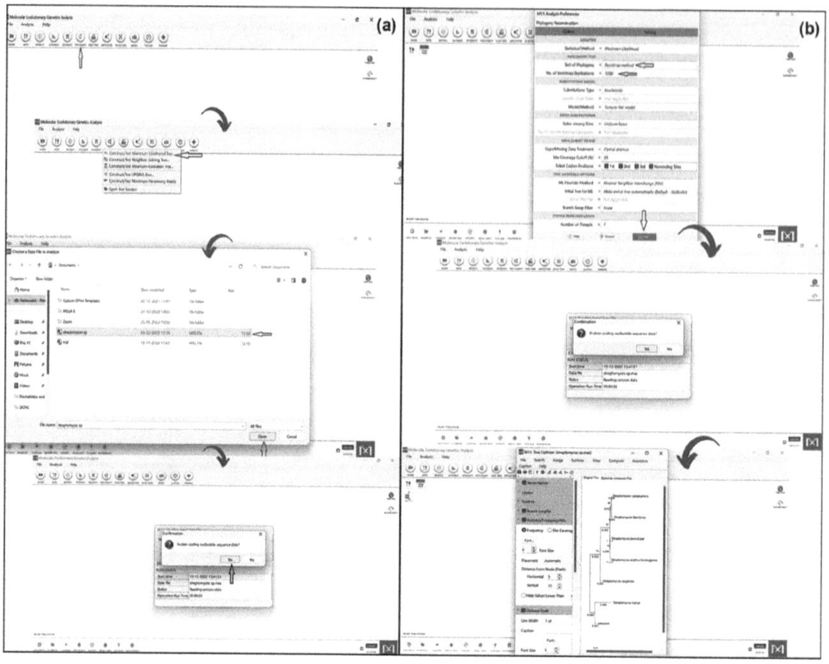

FIGURE 38.5 (a) Schematic representation of MEGA operation; (b) schematic representation of MEGA operation.

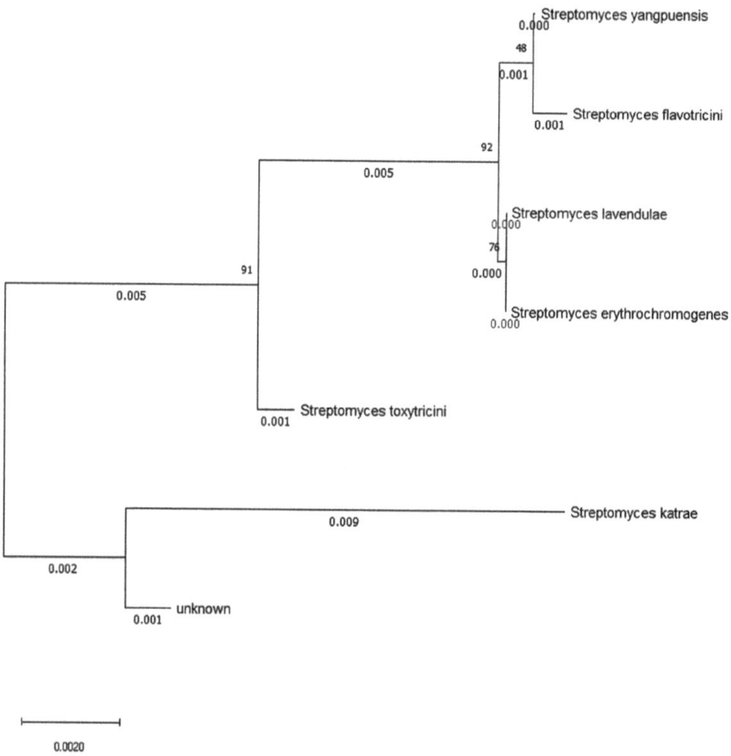

FIGURE 38.6 The obtained phylogenetic tree. This is the original tree obtained by following the given procedure indicating that our test file sequence organism is of *Streptomyces* species in *actinomycetes* and *Streptomyces katrae* is its closest relative, whereas *Streptomyces yangpuensis* is its farthest relative in the selected organisms.

REFERENCES

1. Hillis DM. Primer Phylogenetic Analysis. *Current Biology* 7.3 (1997): R129–R131.
2. Kayondo HW, Ssekagiri A, Nabakooza G, et al. (2021) Employing phylogenetic tree shape statistics to resolve the underlying host population structure. *BMC Bioinformatics*. https://doi.org/10.1186/s12859-021-04465-1
3. Morrison DA (1996) Phylogenetic tree-building. *International Journal for Parasitology* 26(6):589–617.
4. Munjal G, Hanmandlu M, Srivastava S (2019) Phylogenetics algorithms and applications. In: *Ambient Communications and Computer Systems: RACCCS-2018*, edited by Daniel H. de la Iglesia, Juan F. de Paz Santana. Springer Singapore, pp 187–194.
5. Nute M, Saleh E, Warnow T (2019) Evaluating statistical multiple sequence alignment in comparison to other alignment methods on protein data sets. *Systematic Biology* 68:396–411.

39 Genetic Profiling of Actinobacterial Genomes through Polymerase Chain Reaction-based Techniques

Bhatele Pranita and Khanna Asha

39.1 INTRODUCTION

Conventional methods of discrimination between very closely related strains are not reliable. Molecular profiling of commercially important strains is necessary for the conservation of genetic pool as well as for exploiting them for human welfare. Here, we discuss two polymerase chain reaction-based techniques of genetic profiling or generating a molecular fingerprint of actinobacterial genomes.

Random amplification of polymorphic DNA (RAPD) is a PCR-based technique which uses arbitrary primers that bind to the nonspecific sites on the DNA and amplify the DNA. It helps to identify genetic differences without prior knowledge of the sequence. A 10-mer primers are used for the amplification of DNA segments. The segment will not amplify in case of point mutations where the primer sequence does not match with the particular DNA, thus generating polymorphic fingerprints.

Amplified fragment length polymorphism (AFLP) is a DNA fingerprinting technique, combining the reliability of RFLP and the power of PCR. In this technique, a subset of restriction fragments from a digest of genomic DNA are selectively amplified which generates a unique fingerprint for a particular genome [1]. It is highly reproducible and has high discriminatory power [2]. The steps involved are restriction of the DNA and ligation of oligonucleotide adaptors, selective amplification of restriction fragments and analysis of amplified fragments. Total genomic DNA is digested with two restriction enzymes, a frequent cutter with 4 bp recognition site viz. MseI or TaqI and a rare cutter with 6 bp recognition site such as HindIII or EcoRI.

Adaptors specific to a part of the restriction enzyme recognition site are then ligated to both the cut ends of restriction fragments.

Specifically designed primers are used that bind to adaptor sequence and selectively amplify only those restriction fragments that contain a selective base at their 3′ end.

The PCR conditions are so stringently set that allow only the perfectly matched sequences to amplify and thus the amplified bands exhibit polymorphism [2].

These techniques have been used extensively to detect genetic variability in diversity studies of various organisms [3–5]. For both the above-mentioned techniques, RAPD and AFLP, prior sequence knowledge is not required.

39.2 MATERIALS

DNA Isolation – Nutrient broth medium (beef extract 2 g/L, yeast extract 2 g/L, peptone 5 g/L, and sodium chloride 8 g/L, pH 7.1 after sterilization), liquid nitrogen, Tris-EDTA (TE) buffer, lysozyme (20 mg/mL) SDS (w/v), proteinase K, phenol:chloroform solution (v/v, 1:1), ethanol, and RNase solution (20 μg/mL).

DOI: 10.1201/9781003398400-39

TABLE 39.1
PCR Primers and Sequences Used for RAPD Amplification [6]

Primer	Sequence 5′-3′
1247	AAGAGCCCGT
1253	GTTTCCGCCC
1281	AACGCGCAAC
1283	GCGATCCCCA
1290	GTGGATGCGA
D8635	GAGCGGCCAAAGGGAGCAGAC
D9355	CCGGATCCGTGATGCGGTGCG
NP-2	CGGGGGACTGTTGGGCGCCATCT
A-1	GAGGCCCTTC
70-31	GCCCCTCTTG
70-32	GGACCGACTG
70-33	CTGTCGGCTC
70-34	GGACCGCTAG
70-35	CATGTCCGCC
70-36	GCACGTGAGG
70-37	CTATCGCCGC
70-38	GAGAGGGAGG
70-39	CCGGGGTTAC
70-40	CGCAGACCTC
10-40	CGCAGACCTC
CLF-4	AGACACGGCCCAGACTCCTA
Streptomyces specific	GCGTCGAATTAAGCCACA

Random amplification of polymorphic DNA – 100 ng of genomic DNA, MilliQ water, PCR buffer, 1.5 mM $MgCl_2$, dNTPs (0.1 mM each), Taq polymerase (1.5 unit), 50 mM KCl, 10 mM Tris-HCl (pH 9.0), 0.1% (wt/vol) Triton X-100, and 200 nM primer [6].

Primers – A list of PCR primers and sequences is given in Table 39.1.

Amplified fragment length polymorphism – 100–400 ng of genomic DNA, restriction enzymes (TaqI and EcoRI), EcoRI and TaqI adaptors (Table 39.1), T4 DNA ligase, MilliQ water, PCR buffer, $MgCL_2$ (25 mM), dNTPs (0.2 mM each), Taq polymerase (1 unit) and Eco RI and TaqI primers (sequence below) carrying one selective nucleotide, and Phylogeny Inference Program package (PHYLIP), version 3.6a [7].

39.3 PRIMER SEQUENCES FOR PRE-AMPLIFICATION

EcoRI primer sequence 5′-GAC TGC GTA CCA ATT CA-3′
TaqI primer sequence 5′-GAT GAG TCC TGA CCG AC-3′

39.4 METHOD

DNA Isolation – Inoculate cells from a mature slant culture of actinobacteria aseptically into a 250 mL flask containing 30–40 mL of nutrient broth medium and incubate in a rotary shaker at 30°C for 2 days at 180 rpm.

Centrifuge the culture at 10,000 rpm for 10 minutes.

Transfer 0.1 g of mycelium to sterile porcelain dish and crush with liquid nitrogen.

Transfer the crushed mycelium into a fresh tube containing 500 μL of TE buffer supplemented with lysozyme (20 mg/mL).

Incubate the tube at 37°C for 30 minutes.

Add 20 μL of 10% SDS (w/v) and 20 μL of proteinase K into the tube and incubate at 55°C for 30 minutes.

Cool the lysate and extract with equal volume of phenol:chloroform solution (v/v, 1:1) at 10,000 rpm for 5 minutes.

Transfer the aqueous phase carefully to a fresh tube and precipitate the DNA by adding 70%–90% ethanol and keep at –20°C for 30 minutes.

Centrifuge at 10,000 rpm for 10 minutes and discard the supernatant. Wash the pellet twice with 90% ethanol and dissolve the pellet in TE buffer.

Add 20 μL of RNase solution (20 μg/mL) to obtain RNA-free DNA and incubate at 37°C for 1 hour.

Extract the sample once again with equal volume of phenol:chloroform and collect the precipitate as above [8].

The purity and concentration of genomic DNA were checked by spectrophotometric methods, nanodrop or a Biophotometer (Eppendorf AG 22331, Hamburg).

Random amplification of polymorphic DNA – Prepare reaction mixtures consisting of 1 mL of template DNA (100 ng), 50 mM KCl, 10 mM Tris-HCl (pH 9.0), 0.1% (wt/vol) Triton X-100, 0.1 mM dNTPs, 1.5 mM $MgCl_2$, 200 nM primer and 1.5 U of Taq polymerase (Promega).

Set the amplification cycle with an initial denaturation of 4 minutes at 95°C and then 40 cycles of 40-second denaturation at 94°C, 45 seconds at 38°C for primer annealing, and 1.5 minutes at 72°C for primer extension. Complete the reaction sequence with a 5-minute extension and cooling to 4°C [6].

Electrophoresis and photography – Check the PCR products for DNA profile by standard electrophoresis procedures with 1.5% w/v agarose gel (Promega, USA) in 0.5X TBE buffer. Take a picture of the gel using a gel documentation unit [9].

Data analysis of RAPD profiles – Use an engineer's disk with a magnifying lens to analyze the RAPD photographs. Generate a binary matrix based on the fingerprints for each primer as 1 and 0 binary form depending on the presence or absence of the DNA band [10].

Use SPSS software (www.spss.com) [11] for further analysis.

Use the Unweighted Pair Group Mathematical Analysis (UPGMA) to generate the dendrograms.

39.5 AMPLIFIED FRAGMENT LENGTH POLYMORPHISM

Restriction digestion of genomic DNA and adaptor ligation – Digest the genomic DNA with TaqI (5 units in a 10-μL volume, 2 hours at 65°C) and EcoRI (5 units in a 20-μL volume, 2 hours at 37°C).

Ligate the EcoRI and TaqI adaptors (Table 39.2) to the digested DNA with the help of T4 DNA ligase (1 unit in a 10-μL reaction volume, 3 hours at 37°C). Use MilliQ water to dilute the ligated mixture 10 times to generate the template for PCR amplification.

Amplification of the ligated DNA – It will be accomplished in two steps: Pre-amplification and selective amplification [12].

Pre-amplification – Prepare a 25-μL PCR reaction mix consisting of 5 μL ligated DNA, buffer, $MgCL_2$ (25 mM), dNTPs (0.2 mM each), Taq polymerase (1 unit), and Eco RI and TaqI primers carrying one selective nucleotide (50 ng each).

Amplification cycle – 30 cycles of denaturation at 94°C for 30 seconds, followed by annealing at 50°C for 60 seconds and extension at 72°C for 60 seconds.

Selective amplification – Prepare a 12.5-μL PCR reaction mix containing 10× buffer, $MgCl_2$ (1.5 mM), dNTPs (0.2 mM of each) Taq polymerase (1 unit), labeled Eco RI selective primer (5 ng), and TaqI selective primer (30 ng) (Table 39.2).

Data analysis – Record the data in a digital manner where a "+" indicates the presence of a band and a "-" indicates the absence of a band.

Use Phylogeny Inference Program package (PHYLIP), version 3.6a [7] for further analysis of the data.

TABLE 39.2
Sequence of Adaptors and Oligonucleotide [12]

Adaptors	Name	Sequence
Adaptors	Eco R1 top strand	5′-CTCGTAGACTGCGATCC-3′
EcoRI	Eco R1 bottom strand	5′-AATTGGTACGCAGTCTAC-3′
Adaptors	TaqI top strand	5′-GACGATGAGTCCTGAC-3′
TaqI	TaqI bottom strand	5′-CGGTCAGGACTCAT-3′
Primers	E01	5′-GAC TGC TA ATT CA -3′
Eco RI	E32	5′-GAC TGC TA ATT CAA C-3′
	E33	5′-GAC TGC TA ATT CAA G-3′
	E35	5′-GAC TGC TA ATT CAC A-3′
	E38	5′-GAC TGC TA ATT CAC T-3′
	E39	5′-GAC TGC TA ATT CAG A -3′
	E42	5′-GAC TGC TA ATT CAG T -3′
	E44	5′-GAC TGC TA ATT CAT C-3′
	E45	5′-GAC TGC TA ATT CAT G -3′
Primers	T01	5′-GAT GAG TCC TGA CCG AA-3′
TaqI	T02	5′-GAT GAG TCC TGA CCG AC-3′
	T32	5′-GAT GAG TCC TGA CCG AAA C-3′ 5′-GAT GAG TCC TGA CCG AAA G-3′
	T33	5′-GAT GAG TCC TGA CCG AAC A-3′ 5′-GAT GAG TCC TGA CCG AAC T-3′
	T35	5′-GAT GAG TCC TGA CCG ACA C-3′ 5′-GAT GAG TCC TGA CCG ACA G-3′
	T38	5′-GAT GAG TCC TGA CCG ACC T-3′
	T48	5′-GAT GAG TCC TGA CCG ACC AC-3′
	T49	
	T50	
	T51	

REFERENCES

1. Vos P, Hogers R, Bleeker M, Reijans M, van de Lee T, Hornes M, Frijters A, Pot J, Peleman J, Kuiper M, et al. 1995. AFLP: a new technique for DNA fingerprinting. *Nucleic Acids Res.* DOI: 10.1093/nar/23.21.4407.
2. Janssen P, Coopman R, Huys G, Swings J, Bleeker M, Vos P, Zabeau M and Kersters K. 1996. Evaluation of the DNA fingerprinting method AFLP as a new tool in bacterial taxonomy. *Microbiology.* DOI: 10.1099/13500872-142-7-1881.
3. Bhatele P, Dhawan S S, Nagesh A, Tiwari S and Randhawa G S. 2022. Amplified fragment length polymorphism (AFLP) analysis in Cyamopsis spp. for genetic diversity study. *Res J Biotech.* DOI: 10.25303/1701rjbt143148.
4. Kuruvadi N A, Bhatele P, Choudhary M, Tripathi S K, Dhugga K, Gill K S and Randhawa G S. 2013. Genetic diversity study of cluster bean (*Cyamopsis tetragonoloba* (L.) Taub) landraces using RAPD and ISSR markers. *Int J Adv Biotechnol Res.* 4:460–471.
5. Kuruvadi N A, Bhatele P, Tanwar U K, Tripathi S K, Dhugga K S, Gill K S and Randhawa G S. 2014. Identification and characterization of EST-SSR markers in cluster bean (Cyamopsis spp.). Crop Sci. DOI: 10.2135/cropsci2013.08.0522.
6. Roberts M A and Crawford D L. 2000. Use of randomly amplified polymorphic DNA as a means of developing genus- and strain-specific *Streptomyces* DNA probes. *Appl Environ Microbiol.* DOI: 10.1128/AEM.66.6.2555-2564.2000.
7. Felsenstein J. 2001. PHYLIP Phylogeny Inference Package. Version 3.6 (Alpha2).
8. Kumar V, Bharti A, Gusain O and Bisht G S. 2010. An improved method for isolation of genomic DNA from filamentous actinomycetes. *J Sci Engg Tech Mgt.* 2:10–13.
9. Mahasneh A A, Odat J D, Al-Joubori B M and Saadoun I. 2021. Phenotypic and molecular analysis of dominant occurring antibiotic active-producing *Streptomyces* soil flora in Northern Jordan. *Saudi J Biol Sci.* DOI: 10.1016/j.sjbs.2021.04.048.

10. Caetano-Anollés G and Gresshoff P M. 1994. DNA amplification fingerprinting using arbitrary mini-hairpin oligonucleotide primers. *Nat Biotechnol.* DOI: 10.1038/nbt0694-619.
11. Sneath P H A and Sokal R R. 1973. *Numerical Taxonomy: The Principles and Practice of Numerical Classification.* W.H. Freeman.
12. Shete A, Ghole V, Raina M and Aich B. 2014. A polymerase chain reaction based DNA fingerprinting technique: amplified fragment length polymorphism for molecular typing of perchlorate reducing bacteria. *J Environ Prot.* DOI: 10.4236/jep.2014.512109.

40 *In Silico* Analysis of Phylogenetic Relationship and Non-synonymous Polymorphisms in *Actinobacteria*

K. Swetha, Shasmeen Shora, P. Chandra Sekar, Ipsita Debnath, M. Joyce Nirmala, and R. Nagarajan

40.1 INTRODUCTION

Nature is a major pool for most of the biologically active molecules. The phylum *Actinobacteria* belongs to the earliest prokaryotic lineages, comprised of bacteria with higher GC content in their genomic DNA (Gao and Gupta 2012). Being the first prokaryotes, they are widely distributed throughout the biosphere, including the extremobiosphere. *Actinobacteria* create around 10,000 bioactive compounds, most of which belong to the genera *Isoptericola*, *Microbispora*, *Nonomuraea*, *Rhodococcus*, *Microbacterium* and *Nocardiopsis* (Girão et al., 2019). The majority of bioactive chemicals in this phylum are novel and have therapeutic potential, acting as antibacterial, antiviral, anticancer and growth-stimulating agents with more than 70% already in clinical use (Al-Shaibani et al., 2021). Antimicrobial resistance (AMR), on the other hand, is one of the alarmingly increasing global public health threats which requires prompt action as it is an impediment to effective treatment and prevention of diseases brought on by infection-causing bacteria and is anticipated to result in over 10 million deaths by 2050 (Goel et al., 2021). *Actinomycetes* act as the indispensable factor of confronting this concern as they are well known as an eternal source of antibiotics with most antimicrobials in use today being obtained from it. The next-generation sequencing (NGS) techniques and microbial genome mining have changed the field of drug development and optimization, heralding in a new era of antibiotic discovery through identifying crucial genes.

In particular, the sequences of OxyA, OxyB and OxyC constitute members of monooxygenases known as cytochrome P450 enzymes (P450s), which utilize cysteine thiolate-ligated heme moiety to influence complex oxidative processes in microbes (Greule et al., 2022). Constructing phylogenetic trees is an evolutionary relationship analysis tool that enabled us to better understand how molecular machines, cellular processes and other emergent aspects of microbes have evolved (Bozcal et al., 2018). Similarly, single nucleotide polymorphism refers to sequence changes caused when a single nucleotide in the genome differs in paired chromosomes (SNPs). SNPs are utilized as high-resolution markers in gene mapping related to diseases *versus* normal features, aiding in the creation of prospective medication therapies owing to the advancements in DNA technology. SNPs present in the coding region can influence amino acid sequence of the protein coded from nucleotide sequences, which can either positively or negatively influence the amino acid sequence and affect the protein function (Wakui, 2013).

DOI: 10.1201/9781003398400-40

40.2 MATERIALS AND METHODS

40.2.1 Software and Servers Required

a. NCBI – National Centre for Biotechnology Information
b. MEGA – Molecular Evolutionary Genetics Analysis
c. iTOL – Interactive Tree of Life
d. Clustal omega
e. JALVIEW
f. PhD-SNP
g. I-Mutant

40.3 WORKFLOW AND RESULTS

40.3.1 Collection of Sequence

Totally 31 FASTA sequences of OxyA, OxyB and OxyC genes from three species of bacteria were retrieved that includes,

1. *Streptomyces* species
2. *Amycolatopsis* species
3. *Nonomuraea* species

40.3.2 National Centre for Biotechnology Information (NCBI) Database

The NCBI database is used for advance science and health research by providing access to biomedical and genomic information which is easily accessed through https://www.ncbi.nlm.nih.gov/.

- In NCBI, select the all-data base option to search for nucleotide with specific organisms, e.g.: 'oxyA AND Actinobacter'.
- Select the files and click send to choose file destination, change the format type (FASTA) and create file and download.

40.3.3 FASTA Format

- Open the downloaded FASTA sequence file in notepad and save all of the sequences in a single FASTA file.
- Replace the organism name in the information about your protein (after the > symbol).
- For example: *Streptomyces* – organism name, 265 – strain number (>Streptomyces.sp_265).
- Further, choose the file and save as (e.g.: allfasta_sequence) in notepad with '.fasta extension'.

40.3.4 Molecular Evolutionary Genetics Analysis (MEGA) Software

MEGA is an exclusive software in constructing, visualizing and saving the phylogenic tree. It can easily convert alignment files to other formats. MEGA is available as a standalone version made for comparative study of homologous gene sequences from gene families or from other species (Gao et al., 2022). The ClustalW algorithm was employed in the MEGA (v.11) software which performs multiple alignments of the obtained sequences with a concentration on inferring evolutionary relationships and patterns of DNA evolution.

- The current available version of MEGA 11 (64bit) for Windows and Linux can be obtained from https://www.megasoftware.net/

- Download the mega software and install in your working directory.
- Open the mega software click 'aligns – edit/ built new alignment -create new alignment-select DNA (or) protein' based on your sequence.
- 'Choose edit – insert' sequence from file (open all FASTA sequence) and the alignment explorer file is open.
- Delete the empty sequence line and click data – choose export alignment – mega format save the filename in.mng extension.

40.3.5 Phylogenetic Tree

Phylogenetic tree, also known as phylogeny, is an illustration of the evolutionary relationship among various species, organisms or genes that have evolved from one another. Phylogenies provide valuable source of information for classifying organisms, consolidating information about biological diversity and providing insight into evolutionary processes.

- In the software, 'choose phylogeny', it will show the different phylogeny tree option.
- From which, click any one of the tree method and open the 'mega format sequence file'.
- The analysis preference is displayed from which click 'default ok'. It will process the sequence and show in tree format.
- Change the tree format in traditional, radiation and circle type.
- Select 'file – export current tree '(Newwick)' and finally 'save the sequence'.
- Itol – Interactive tree of life – open the web server, and based on rectangle tree, rooted and unrooted tree with colour arrange of unique organisms can be highlighted.
- Herewith, we mention three illustrations of the above processes.
- A) The phylogenetic tree with UPGMA (Unweighted Pair-Group Method using Arithmetic Averages) method in curved format can be represented as branched tree. Finding a non-member sequence with the minimum average differentiation over the cluster members permits the iterative formation and development of clusters (Letunic and Bork, 2021).
- B) The phylogeny could be represented as a construct likelihood method and tree in circular format, accessibility design colour and style annotations for specific nodes and word description. It is based on the intrinsic node attributes and automatic annotations (Letunic and Bork, 2021).
- C) The phylogenetic tree with construct neighbour joining tree could be represented in the root format. Depending on the protein sequence, the colour could be changed and highlighted (Letunic and Bork, 2021).

40.4 RETRIEVAL OF PROTEIN SEQUENCE FROM NCBI

- Open NCBI and choose all-data base to 'nucleotide'.
- Search – 'oxyA AND Actinobacteria'.
- Choose the folder and select the protein ID, which shows the protein FASTA sequence.
- Copy the FASTA sequence and paste in notepad.
- Rename the first line with protein ID, e.g.: (>AAS77615.1_Streptomyces).
- Save the file in the name of (all_protein sequence).txt.

40.4.1 Multiple Sequence Alignments

Clustal omega is a multiple sequence alignment program. It creates divergent multiple sequence alignments that are biologically significant. Phylogenetic trees must be constructed using the fundamental bioinformatics technique of multiple sequence alignment of DNA/RNA sequences. As a result of advancements in DNA sequencing, the volume of bioinformatics data is constantly expanding, necessitating the generation of models for many tools (Wang et al., 2022). A new version of the popular Clustal series of multiple sequence alignment programmes named Clustal Omega is available. It can process very huge volumes (many tens of thousands) of DNA/RNA or protein sequences as it uses the mBED method to create guide trees. For our study, we selected eight sequences of *Streptomyces* and *Amycolatopsis* species.

Step 1:

- In MEGA tool, select type of data as protein instead of DNA/RNA.
- Paste the protein sequences or upload (all protein sequence.txt).
- The tool aligns up to a maximum of 4,000 sequence (or) a file size of 4 MB.

Step 2: default parameter

Step 3: Set the output format, tick the box – be notified by email option, enter your mail ID and title.

Submit the sequence.

- The Clustal omega results are viewed in your uploaded mail ID – click the link to view the alignment.
- Click the show colour option for better visualization.
- An * (asterisk) denotes places where a single, complete conserved residue is present.
- A: (colon) shows preservation between sets with extremely similar feature.
- A. (period) indicates preservation between groups of weakly similar properties.
- Click result viewer, which would show the jalview tap.
- For example, *Streptomyces* multiple sequence alignments are obtained as a result of Clustal omega. It would show the mutation types depending on the asterisk, colon and period (Procter et al., 2021).
- In another example, *Amycolatopsis* multiple sequence alignments are obtained through Clustal omega result viewer. The most mutated sequence alignment can also be highlighted (Procter et al., 2021).

40.4.2 Jalview

It is a free cross-platform tool for editing, visualizing and analysing multiple sequence alignments. The program was originally developed by the University of Oxford and European Bioinformatics Institute (EBI).

- Select the Clustal omega result in jalview tap. It would show the result in jalview format.
- Select: click the select all option.
- Format: change wrap format.
- Colour: click text colour – change the text colour that is required and select the colour based on your visual – (in this case, we chose "percentage identify").
- Annotations: Show auto-calculated annotation – show consensus logo.
- File: click export image and save the image in PNG format.

Neighbour joining/average distance file:

- Calculate: calculate tree/PCA, click neighbour joining/ average distance and click calculate. This would show the tree.
- File: save the image in PNG format (or) crop the image with the help of snipping tool.
- Save the image in PNG format.

- Jalview shows the sequence alignment and replaced amino acid with highlighted white colour. This shows the quality, consensus and conservation (Tenea et al., 2020).
- The amycolatopsis mostly shows in missense mutation format, and the occupancy of the sequence of amino acid interaction with other amino acid (Tenea et al., 2020).
- The average distance shows the average length of the four bacterial sequences (Sánchez-Sanz et al., 2016).
- The neighbour joining distance reveals the weight and distance of the bacterial species (Sánchez-Sanz et al., 2016).

40.5 SIFT (SORTING INTOLERANT FROM TOLERANT)

SIFT is considered to be a prevalent method for analysing missense variation (Alanazi et al., 2022). The variants are based upon the cutoff of 0.05. On the basis of the physical characteristics of amino acids and sequence homology, the SIFT approach is developed to predict how coding variations would affect protein function. We can access the tool through the following link **https://sift.bii.a-star.edu.sg/www/SIFT_seq_submit2.html**

1. Tolerable (SIFT score > 0.05)
2. Damaging (SIFT score ≤ 0.05)
   - Input protein sequence in FASTA format (or) copy the sequence from notepad and paste the protein sequence.
   - Mention the substitution of residues, e.g.: native amino acid -T, and replacement position in column 4 and replaced amino acid A '(T4A)'.
   - Enter your mail ID and give submit.
   - Result would be delivered through email.
   - Click the prediction result to find the sequence represent position and tolerated score as shown in Table 40.1.

ACKNOWLEDGEMENT

We gratefully acknowledge the support of Pratiksha Trust, Bangalore, towards this research.

TABLE 40.1
SIFT Prediction Score and Median Sequence Conservation of MSA

S. No	Replaced Amino Acid	Median Sequence Conservation	Position	Score
Streptomyces	A20V	3.12	6	1.00
	E30T	3.40	4	0.43
	R39Q	3.12	6	0.43
	A36D	3.12	6	0.03
	Y35R	3.34	5	0.38
Amycolatopsis	L32E	3.11	5	0.35
	S38P	3.11	5	0.01
	P39H	3.11	5	0.26
	L42V	3.11	5	0.45

REFERENCES

Alanazi IO, Alamery SF, Ebrahimie E, Mohammadi-Dehcheshmeh M. 2022. "Splice-disrupt genomic variants in prostate cancer". *Molecular Biology Reports* (6):4237–4246. doi:10.1007/s11033-022-07257-9.

Al-Shaibani MM, Radin Mohamed RMS, Sidik NM, Enshasy HAE, Al-Gheethi A, Noman E, et al. 2021. "Biodiversity of secondary metabolites compounds isolated from phylum actinobacteria and its therapeutic applications". *Molecule (Basel, Switzerland)* 26(15). doi:10.3390/molecules26154504.

Bozcal E, Eldem V, Aydemir S, Skurnik M. 2018. "The relationship between phylogenetic classification, virulence and antibiotic resistance of extraintestinal pathogenic *Escherichia coli* in İzmir province Turkey". *PeerJ* 6. doi:10.7717/peerj.5470.

Gao B, Gupta RS. 2012. "Phylogenetic framework and molecular signatures for the main clades of the phylum Actinobacteria". *Microbiology and Molecular Biology Reviews: MMBR* 76(1):66–112. doi:10.1128/MMBR.05011-11.

Gao XY, Liu GC, Zhang JX, Wang LH, Xu C, Yan ZA, et al. 2022. "Pharmacological properties of ginsenoside". Frontiers in Pharmacology 13. doi:10.3389/fphar.2022.754191.

Girão M, Ribeiro I, Ribeiro T, Azevedo IC, Pereira F, Urbatzka R, et al. 2019. "Actinobacteria isolated from *Laminaria ochroleuca:* a source of new bioactive compounds". *Frontiers in Microbiology* (10). doi:10.3389/fmicb.2019.00683.

Goel N, Fatima SW, Kumar S, Sinha R, Khare SK. 2021. "Antimicrobial resistance in biofilms: exploring marine actinobacteria as a potential source of antibiotics and biofilm inhibitors". Biotechnology Reports (Amsterdam, Netherlands) (30). doi:10.1016/j.btre.2021.e00613.

Greule A, Izoré T, Machell D, Hansen MH, Schoppet M, De Voss JJ, et al. 2022. The cytochrome P450 OxyA from the kistamicin biosynthesis cyclization cascade is highly sensitive to oxidative damage". *Frontiers in Chemistry* 10. doi:10.3389/fchem.2022.868240.

Letunic I, Bork P. 2021. "Interactive Tree Of Life (iTOL) v5: an online tool for phylogenetic tree display and annotation". Nucleic Acids Research 49. doi:10.1093/nar/gkab301.

Procter JB, Carstairs GM, Soares B, Mourão K, Ofoegbu TC, Barton D, et al. 2021. "Alignment of biological sequences with jalview". Methods in Molecular Biology (Clifton, NJ) 2231:203–224. doi:10.1007/978-1-0716-1036-7_13.

Sánchez-Sanz G, Tywoniuk B, Matallanas D, Romano D, Nguyen LK, Kholodenko BN, et al. 2016. "SARAH domain-mediated MST2-RASSF dimeric interactions". *PLoS Computational Biology* 12(10):e1005051. doi:10.1371/journal.pcbi.1005051.

Tenea GN, Hurtado P, Ortega C. 2020. "A novel Weissella cibaria strain UTNGt21O isolated from wild solanum quitoense fruit: genome sequence and characterization of a peptide with highly inhibitory potential toward gram-negative bacteria". *Foods (Basel, Switzerland)* 9(9):1242. doi:10.3390/foods9091242.

Wakui M. 2013. "Analysis of single nucleotide polymorphisms (SNPs)". *Rinsho Byori, The Japanese Journal of Clinical Pathology* 11:1008–1017.

Wang Z, Tan J, Long Y, Liu Y, Lei W, Cai J, et al. 2022. "SaAlign: Multiple DNA/RNA sequence alignment and phylogenetic tree construction tool for ultra-large datasets and ultra-long sequences based on suffix array". *Computational and Structural Biotechnology Journal* 20:1487–1493. doi:10.1016/j.csbj.2022.03.018.

41 Isolation of Metagenomic DNA and Data Analysis

Shabiha Nudrat Hazarika, Kangkon Saikia, and Debajit Thakur

41.1 INTRODUCTION

Metagenomics and next-generation sequencing (NGS) techniques have become a rapidly growing field of research to study the microbial world. The metagenomic approach includes collecting and analyzing the complete genome contained in an environmental sample (soil, leaves, and roots). The extraction of metagenomic DNA from plant sources such as leaves and roots has never been easier. The quality and quantity of DNA required for DNA library preparation and sequencing by NGS are very high. Still, high levels of phenolics and polysaccharides in recalcitrant plant species have caused significant difficulty (Porebski et al., 1997). Phenolics (terpenoids and tannins) oxidize upon release from leaf tissues and bind to the DNA. Polysaccharides co-precipitate with genomic DNA and prevent NGS library preparation (Sharma et al., 2002). Both contaminants make it difficult to use DNA for PCR or sequencing, limiting research into their genomes.

Most metagenomic DNA extraction methods from plant tissues use the hexadecyltrimethylammonium bromide (CTAB)-based method (Doyle and Doyle, 1987), intended to solve the challenges of phenolics and polysaccharides. For NGS, the extracted DNA must be of high quality with a purity of approx. 1.8–2.0 at 260/280 nm absorbance ratio. The DNA must be free from contaminants such as RNA, polysaccharides, or proteins. As such, the method described in this chapter has been modified to include polyvinylpyrrolidone (PVP) in the extraction buffer and high salt solutions to extract metagenomic DNA from recalcitrant plant species. The method described does not require long incubations, multiple clean-up, and precipitation steps, making it a quick, easy, and reliable DNA extraction method for NGS sequencing.

The plant microbiota is a rich source of micro-organisms that promote plant growth and protect against disease and environmental disturbances (Doumbou et al., 2005; Shrivastava and Kumar, 2018). With their importance in agriculture, industry, and medicine, high-throughput techniques, such as NGS, for detecting and analyzing plant microbiota have become increasingly necessary. One such important group of microbes is Actinobacteria, ubiquitous in nature and most abundant in soil. Actinobacteria have the capacity to produce various metabolites with biological functions, such as antimicrobial agents and plant-growth promoters (Hazarika et al., 2022). Both culturable and un-culturable methods study free-living soil Actinobacteria and symbiotic Actinobacteria that live within plants as endophytes. Only 1% of the microbial diversity has been cultivated using the culture-dependent method, and 99% of the microbes that cannot be grown may help in the exploration of new secondary metabolites and antibiotics (Assis et al., 2014). The metagenomics approach can explore the microbial genetic pool that classical microbial cultivation techniques cannot study. The metagenomics employs the PCR-based culture-independent technique, which has been used for a more comprehensive exploration of microbial diversity, including both culturable and un-culturable parts of the microbial communities (Singh and Dubey, 2018).

A plant-microbiome-based study employing 16S rRNA gene sequencing has provided information on Actinobacteria in plant tissues. To obtain accurate 16S rRNA gene amplicons, universal/specific conserved broad-range metagenomic primers, including adaptors with indices, are used

DOI: 10.1201/9781003398400-41

(Kumaishi et al., 2022). For the specific differentiation of Actinobacteria in the sample, hypervariable regions V3-V4 are amplified. Using the Illumina HiSeq platform for NGS has provided a great opportunity to analyze the actinobacterial community compared to the sequencing of the 16S rRNA gene of culturable Actinobacteria.

Since metagenomics contributes to research areas, including ecology, biodiversity, identification of natural products, and medicine, it will become an important technique for studying microbial ecology and uncovering the rare microbes in the environment.

41.2 MATERIALS

41.2.1 Samples: Leaf and Root Sample

41.2.2 Consumables

- Sodium hypochlorite
- Nutrient agar media
- Ethylenediaminetetraacetic acid disodium salt dihydrate (EDTA)
- Polyvinylpyrrolidone (PVP)
- Trizma base
- β-Mercaptoethanol
- Sodium chloride (NaCl)
- Hexadecyltrimethylammonium bromide (CTAB)
- RNase A
- Proteinase K
- Phenol: chloroform: isoamlyalcohol (P: C: I)
- Chloroform: isoamyl alcohol (C: I)
- Isopropanol
- Ethanol (molecular grade)
- Agarose

41.2.3 Equipment

- Weighing balance
- Water bath
- Centrifuge
- Nanodrop spectrophotometer
- Gel electrophoresis unit
- Gel documentation system

41.2.4 Others

- Sterile distilled water
- Petri dish
- Liquid nitrogen
- Mortar and pestle
- Spatula
- Pipettes (100–1,000, 20–200, 2–10 μL)
- Microtips (100–1,000, 20–200, 2–10 μL)
- Centrifuge tubes (50, 15, 5 mL)
- Measuring cylinder
- Disposable Gloves

TABLE 41.1
Reagents Required for Metagenomic DNA Extraction

Name of Buffer	Components	Stock Concentration	Working Concentration or Ratio	Volume or Weight in 100 mL
Extraction buffer	Tris-Cl, pH 8.0	1 M	100 mM	10 mL
	EDTA, pH 8.0	0.5 M	20 mM	4 mL
	NaCl	5 M	1.4 M	28 mL
	CTAB	-	2%	2 g
	β-Mercaptoethanol	-	-	2 mL
	PVP	-	-	2 g
NaCl solution	NaCl	2 M	2 M	11.688 g
TE buffer	Tris-Cl, pH 8.0	1 M	10 mM	1 mL
	EDTA	0.5 M	0.1 mM	20 μL

41.2.5 Reagents

The required reagents are listed in Table 41.1.

41.3 PROTOCOLS

41.3.1 Sample Collection

Collect leaf and root samples in ziplock bags. Store the samples in a box with ice packs and transport them to the laboratory for further sample processing.

41.3.2 Surface Sterilization

i. Weigh 1–2 g of leaf and root samples.
ii. Wash the samples with tap water to remove any debris. Further, rinse the samples with sterile distilled water.
iii. Rinse the samples with 70% ethanol for 30 seconds.
iv. Surface sterilize the samples for 5 minutes using 1.5% sodium hypochlorite.
v. Wash the samples three to five times with sterile distilled water.
vi. Determine the efficacy of surface sterilization by spreading 100 μL aliquots of distilled water on nutrient agar (NA) plates taken from the last wash of the samples.

41.3.3 Plant Metagenomic DNA Isolation

i. Under aseptic conditions, macerate the surface-sterilized frozen samples to a fine powder in a chilled mortar pestle using liquid nitrogen.
   (Before grinding, pre-chill the mortar and pestle to prevent the thawing of frozen tissues).
ii. Take 1 g of the ground powder in a 50-mL centrifuge tube and suspend in 10 mL of extraction buffer (see Table 41.1 for the reagents and buffers required for metagenomic DNA extraction).
   (Before suspending the ground powder to extraction buffer, pre-heat the water bath to 65°C).
iii. Incubate the tubes at 65°C water bath for 3 hours (leaf) and 4 hours (root) with regular inversion at an interval of 20 minutes.

iv. Centrifuge the tubes at 10,000 rpm for 20 minutes at 4°C (pellet and remove un-lysed plant tissue). Decant the supernatant to a fresh 50 mL centrifuge tube.
v. Add 5 μL RNase A (100 μg/mL), mix by pulse vortexing for 15 seconds, and incubate at 60°C water bath for 30 minutes. (*Pre-heat the water bath to 60°C).*
vi. Add 5 μL Proteinase K (100 μg/mL), mix by pulse vortexing for 15 seconds, and incubate at 60°C water bath for 15 minutes.
vii. Add an equal volume of Phenol:Chloroform:Isoamlyalcohol (25:24:1) and keep in ice for 5 minutes. Centrifuge at 12,000 rpm for 5 minutes at 4°C.
viii. Carefully transfer the upper aqueous phase to a fresh 15 mL centrifuge tube. Add an equal volume of chloroform:isoamlyalcohol (24:1) and mix gently by inversion. Keep the tubes in ice for 2–3 minutes. (*Repeat twice for root samples*).
ix. Centrifuge the tubes at 12,000 rpm for 10 minutes at 4°C. Again, carefully transfer the upper aqueous layer to 15 mL centrifuge tubes.
x. Add an equal volume of 2M NaCl to the supernatant, and then add 0.6 volume of chilled isopropanol to the supernatant-NaCl mixture. Place the tubes at –20°C freezer and incubate for 1 hour.
xi. After incubation, centrifuge the tubes at 12,000 rpm for 20 minutes at 4°C to pellet the DNA. Discard the supernatant.
xii. Add 500 μL of 80% chilled ethanol to the pellet and centrifuge the tubes at 10,000 rpm for 10 minutes at 4°C. Carefully decant the supernatant and air-dry the DNA pellet at room temperature.
xiii. Suspend the dried DNA pellet in 100 μL of TE buffer.

41.3.4 Quality and Quantity Assessment

The quality of extracted metagenomic DNA has to be determined by visualization of DNA on 0.8% agarose gel and using a Nanodrop spectrophotometer. Pure DNA is generally accepted to have a 260/280 ratio of ~1.8–2.0, a single peak at 260 nm, and no band shearing or RNA contamination in agarose gel visualization.

41.3.5 Amplification and Sequencing

i. PCR amplify the 16S rRNA gene V3 and V4 regions (amplicon of 1,500 bp) from genomic DNA using region-specific primers with overhang adapters. The full-length primer sequences targeting this region are:
   16S Amplicon PCR Forward Primer = 5′
   TCGTCGGCAGCGTCAGATGTGTATAAGAGACAGCCTACGGGNGGCWGCAG
   16S Amplicon PCR Reverse Primer = 5′
   GTCTCGTGGGCTCGGAGATGTGTATAAGAGACAGGACTACHVGGGTATCTAATCC
   The Illumina overhang adapter sequences added to locus-specific sequences are:
   Forward overhang: 5′ TCGTCGGCAGCGTCAGATGTGTATAAGAGACAG
   Reverse overhang: 5′ GTCTCGTGGGCTCGGAGATGTGTATAAGAGACAG
ii. Use the KAPA HiFi Hot Start Kit for PCR to create an amplicon of size ~460 bp. Perform PCR Clean-Up using AMPure XP beads to purify the 16S V3 and V4 amplicon according to the 16S metagenomics protocol.
iii. Perform a subsequent limited cycle amplification to add multiplexing indices and Illumina sequencing adapters using the Nextera XT Index Kit.
iv. Perform PCR Clean-Up 2 using AMPure XP beads to clean up the final library before quantification.
v. Normalize libraries, pool, and sequence on the MiSeq system using v3 reagents.

(Using paired 300-bp reads and MiSeq v3 reagents, full-length reads of the V3 and V4 regions can be obtained. For metagenomic samples, >100,000 reads per sample are sufficient to survey the bacterial composition fully).

41.3.6 Data Analysis

The NGS data is usually represented in FASTQ format containing numerous reads of definite length (usually 150–300 bases). However, depending on different sequencing platforms and library, the reads were represented either in a paired-end format containing both forward and reverse reads or a single-end format containing either forward or reverse reads. The paired and single-end reads are sometimes multiplexed by containing barcode sequences used in sequencing, which should be removed manually (demultiplexed) to get the actual sequence data. The quality of FASTQ files can be initially visualized using quality control (QC) tools such as MultiQC (Ewels et al., 2016). The crucial step in QC is to investigate the presence and removal of adapters, trimming indexes as necessary. However, trimming should be avoided unless the data comprises the intensity of reads with a quality score of 20 or less. Notably, the degree of trimming of the sequences may alter the species diversity and feature count in amplicon sequence variant (ASV) or molecular technology unit (MTU) classification, resulting in data loss. One of the most frequently used quality correction tools for NGS data is Trimmomatic (Bolger et al., 2014). Metagenomic workflow such as QIIME2 (Caporaso et al., 2010) contains integrated QC tools for analysis, visualization, and correction of FASTQ file quality. Once the quality of multiple sequences is checked and corrected, the next phase is to generate ASVs or MTUs, followed by taxonomic classification and statistical analysis.

There are several tools and online servers for the analysis of metagenomic data, which take input of FASTQ files of multiple samples and allow to compute through numerous tools in a workflow (Figure 41.1). Workflows like QIIME2 and MG-RAST server (Keegan et al., 2016) are commonly used. There are various packages which can also be used for metagenomic analysis using R programming. Moreover, various R packages also enable the FASTQ files to be imported, quality checked, and corrected, and subsequent analysis can be carried out. For statistical analysis of metagenomic data, packages such as Phyloseq (McMurdie et al., 2013) and vegan are two of the

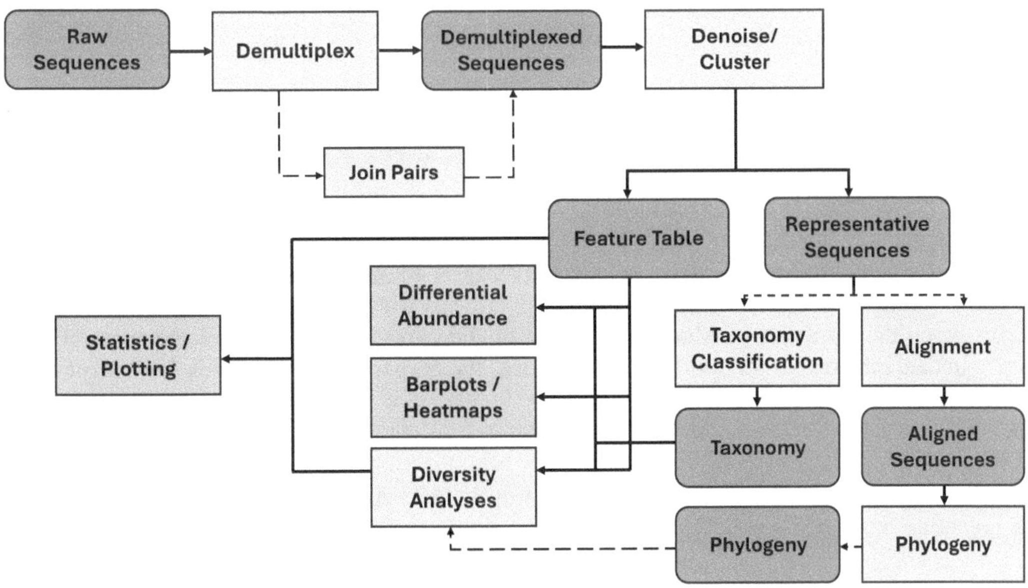

FIGURE 41.1 A basic metagenomic data analysis workflow.

most popularly known. The data from QIIME2 or MG-RAST are often exported in a format for subsequent statistical analysis using these packages.

In the following section, we represent various steps to analyze metagenomic data, including taxonomic classification and diversity analysis using QIIME2 (version 2022.8.3). The input sequences used in this case contain paired-end reads which are already demultiplexed and quality checked using some of the previously mentioned tools.

Navigate to your working directory, copy all the sample sequences (FASTQ files), and create the required sub-directories for storing the outputs and logs generated by multiple tools.

1. Prepare a metadata file as **`metadata.tsv`** (Table 41.2), which will contain sample IDs and absolute file (FASTQ) location of forward and reverse reads. (*The metadata file can also contain various variables, which can be facilitated for subsequent statistical analysis*).

   Use the following command for importing all sample FASTQ files, including the metadata to a QIIME2-specific file format. This takes an input of **`metadata.tsv`** file and produces an output of **`paired-end-sequences.qza`**. in respective directories.

```
qiime tools import \
  --type 'SampleData[PairedEndSequencesWithQuality]' \
  --input-path "data/metadata.tsv" \
  --output-path "sequences/paired-end-sequences.qza" \
  --input-format PairedEndFastqManifestPhred33V2
```

2. In this step, some additional quality assessment using the Deblur tool has to be done, and DADA2 has to be used for ASV generation. After preparation of the input file, these can be visualized and processed for quality analysis using a series of QIIME2 tools like **`demux`**, `deblur` and **`dada2`**. First, the summarized sequence quality of the prepared input sequences of all the samples can be visualized using the following command:

```
qiime demux summarize \
  --i-data "sequences/paired-end-sequences.qza" \
  --o-visualization "sequences/demux-summary.qzv"
```

   Note: The file **`demux-summary.qzv`** can be visualized in QIIME2 view portal at `https://view.qiime2.org/`.

   Now that the summarized sequence quality is analyzed, it can be proceed to ASV generation using `dada2`. In the same step, one can trim the reads at a specific frame or based on quality score. More detailed information about `dada2` is available at QIIME2 documentation. In the following code block, the prepared sequence files were taken as

TABLE 41.2
Format to Prepare Metadata File for Metagenomic Data Analysis

Sample-id	Forward-Absolute-Filepath	Reverse-Absolute-Filepath	Variable_1	Variable_2
S1	$PWD/sequences/1F.fastq.gz	$PWD/sequences/1R.fastq.gz		
S2	$PWD/data/sequences/ 2F.fastq.gz	$PWD/data/sequences/2R.fastq.gz		

inputs, and the ASV feature table was generated as feature-table.qza and ASV sequences as `feature-sequences.qza`.

```
qiime dada2 denoise-paired \
 --i-demultiplexed-seqs "sequences/paired-end-sequences.qza "
\
 --p-n-threads 30 \
 --o-representative-sequences "dada2/feature-sequences.qza" \
 --o-table "dada2/feature-table.qza" \
 --o-denoising-stats "dada2/feature-stats.qza"
```

The descriptive statistics, such as number of features, sequences and ASVs, can be further visualized using the following command:

```
qiime feature-table summarize \
--i-table "all/dada2/feature-table.qza" \
--m-sample-metadata-file "data/metadata.tsv" \
--o-visualization "all/dada2/feature-summary.qzv"
```

Use QIIME2 view to plot the results online. Once the feature table and representative sequence of ASVs are generated, either taxonomic identification or diversity and other statistical analysis can be individually processed.

3. In this step, first perform taxonomic classification of the ASVs using the SILVA database. Use the small ribosomal subunit dataset provided in SILVA SSU version 138.1. QIIME2 uses machine learning to train the available model using this dataset (SILVA SSU 138.1 in this case) and assign taxonomy to the ASVs using representative sequences as the test data. The latest version of the classifiers for SILVA database (`silva-138-99-nb-weighted-classifier.qza`) can be downloaded from QIIME2 official website or SILVA database website. The classification results will be saved in a new taxonomy directory with a file named `taxonomy _ silva.qza`.

```
qiime feature-classifier classify-sklearn \
  --i-classifier "taxonomy_classifiers/silva-138-99-nb-weight-
ed-classifier.qza" \
  --i-reads "dada2/feature-sequences.qza" \
  --p-n-jobs 20 \
  --o-classification "taxonomy/taxonomy_silva.qza"
```

Note: The `-p-n-jobs` parameters can be adjusted as per the system hardware configuration.

Furthermore, the taxonomic abundance can be visualized as bar plots using the following command.

```
qiime taxa barplot \
```

```
--i-table "dada2/feature-table.qza" \
--i-taxonomy "taxonomy/taxonomy_silva.qza" \
--m-metadata-file "data/metadata.tsv" \
--o-visualization "taxonomy/taxa-bar-plot.qzv"
```

Note: Use QIIME2 view to plot the visualization file graphically.

The taxonomic classification often identifies sequences from chloroplast and mitochondrial origin. These must be removed using the following command before subsequent statistical analysis.

```
qiime taxa filter-seqs \
  --i-sequences " dada2/sequences.qza" \
  --i-taxonomy "taxonomy/taxonomy_silva.qza" \
  --p-exclude mitochondria,chloroplast \
  --o-filtered-sequences "dada2/mt_cp_filter/sequences.qza"
```

Similarly, for the feature table, use the following code block:

```
qiime taxa filter-table \
  --i-table "dada2/feature-table.qza" \
  --i-taxonomy "taxonomy/taxonomy_silva.qza" \
  --p-exclude mitochondria,chloroplast \
  --o-filtered-table " dada2/mt_cp_filter/feature-table.qza"
```

4. Once the taxonomy results and ASVs are generated, these can be further analyzed for alpha and beta diversity. The data can also be exported for analysis using R programming using popular metagenomic analysis packages like Phyloseq and vegan. The workflow of diversity analysis includes three steps: first generation of a taxonomic tree, second computation of distance and phylogenetic matrices, and finally calculation of diversity. Alpha and beta diversity can be calculated using either phylogenetic or non-phylogenetic methods. The proper use cases of various tools in this regard can be better understood from QIIME2 documentation.

   Generate a taxonomic tree using sequence alignment by MAFFT (Katoh et al., 2002) and tree construction using RAXML (Stamatakis, 2014). The following code block performs the same using an available QIIME2 module from the toolbox.

```
qiime phylogeny align-to-tree-mafft-raxml \
--i-sequences "dada2/mt_cp_filter/sequences.qza" \
--output-dir "taxonomy/raxml_tree"
```

   Following block of code generates all the core matrices required for diversity and other statistical analysis.

```
qiime diversity core-metrics-phylogenetic \
```

```
  --i-table "dada2/mt_cp_filter/feature-table.qza" \
  --i-phylogeny "taxonomy/raxml_tree/rooted_tree.qza" \
  --p-sampling-depth 1200 \
  --m-metadata-file "data/metadata.tsv" \
  --output-dir "diversity/core-metrics-results"
```

Now, using the taxonomy tree and ASV feature table, we can compute and visualize the taxonomic alpha diversity, initially by analyzing alpha rarefaction plots. The output results `alpha-rarefaction.qzv` produced by the following code can be visualized using QIIME2 view.

```
qiime diversity alpha-rarefaction \
  --i-table "dada2/mt_cp_filter/feature-table.qza" \
  --i-phylogeny "taxonomy/raxml_tree/rooted_tree.qza" \
  --p-max-depth 6000 \
  --m-metadata-file "data/metadata.tsv" \
  --o-visualization "diversity/alpha-rarefaction.qzv"
```

Note: The selection of `--p-max-depth` parameter is crucial in this step and can be deduced by carefully analyzing feature table statistics and taxonomy data.

Use the following code block to interpret alpha diversity (within samples) by the alpha group significance using one of the computed matrices viz. faith_pd vector, evenness vector, shannon vector, etc. In this case, the results were demonstrated by using the shannon vector. Use QIIME2 view tool for visualization output file `shannon-group-significance.qzv`.

```
qiime diversity alpha-group-significance \
  --i-alpha-diversity "diversity/core-metrics-results/shan-
non_vector.qza" \
  --m-metadata-file "data/metadata.tsv" \
  --o-visualization "diversity/alpha-significance/shan-
non-group-significance.qzv"
```

For analysis of beta diversity, one of the core matrices must be used containing computed unweighted unifrac distance between samples within a group (as given in metadata). The results will produce an output of principal coordinate analysis (PCoA) which can be further visualized in the QIIME2 view. Use a definite metadata group instead of `METADATA-CATEGORY`, as described in the metadata file.

```
qiime diversity beta-group-significance \
  --i-distance-matrix "diversity/core-metrics-results/un-
weighted_unifrac_distance_matrix.qza" \
```

```
--m-metadata-file "data/metadata.tsv" \
--m-metadata-column {METADATA-CATEGORY} \
--o-visualization "diversity/beta-group-significance/un-
weighted-unifrac-{METADATA-CATEGORY}-significance.qzv \
--p-pairwise
```

41.4 FINDINGS

This protocol for the isolation of metagenomic DNA successfully extracted the metagenomic DNA. Figure 41.2 shows a representative image of metagenomic DNA extracted from plant tissue. Table 41.3 shows a representative quantity of metagenomic DNA extracted from leaf and root samples of plant tissue.

41.5 CONCLUSION

The described protocol used for the isolation of metagenomic DNA from plant tissues rich in phenols and polysaccharides is an improved method. This protocol minimizes the inclusion of multiple clean-up and precipitation steps. Using this protocol, metagenomic DNA with stringent quality can be extracted for NGS library generation and sequencing using the Illumina MiSeq platform. The method has been developed for all plant species and tissues, such as leaves and roots. The DNA obtained from recalcitrant plant species can be utilized for sequencing and bioinformatic investigations with important contributions in many fields like microbial ecology, biodiversity, medicine, and bioremediation.

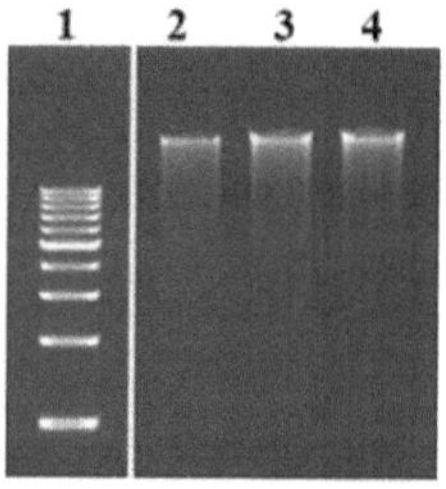

FIGURE 41.2 Representative image of extracted metagenomic DNA in 0.8% agarose gel. Lane 1- Ladder 1 kb, Lane 2- Root sample, Lane 3, and Lane 4- Leaf sample.

TABLE 41.3
Representative Quantity of Extracted Metagenomic DNA Quantified Using Nanodrop Spectrophotometer

Sl. No.	Sample	Nucleic Acid	Unit	A260 (Abs)	A280 (Abs)	260/280	260/230	Sample Type	Factor
1	Leaf-1	234.5	ng/μL	4.691	2.531	1.85	0.87	DNA	50
2	Leaf-2	187.5	ng/μL	3.751	1.899	1.98	1.46	DNA	50
3	Root-1	207.5	ng/μL	4.15	2.351	1.77	1.2	DNA	50
4	Root-2	160.5	ng/μL	3.209	1.698	1.89	1.23	DNA	50

ACKNOWLEDGMENTS

The authors thank the Director, IASST, Guwahati, Assam, India, for providing the infrastructure and resources to carry out this work. The authors also want to acknowledge Macrogen Inc. Seoul, South Korea, for NGS sequencing.

FUNDING SOURCE

This work was supported by DST, a Government of India in-house R&D core budget-funded project for the year 2021–2023 (IASST/1238/21–22/1973–1981 dated 09/06/2021).

CONFLICT OF INTEREST

The authors have no conflict of interest to declare that is relevant to the content of this chapter.

REFERENCES

Assis, D.A., Rezende, R.P., and Dias, J.C. (2014). Use of metagenomics and isolation of Actinobacteria in Brazil's Atlantic Rainforest Soil for antimicrobial prospecting. *ISRN Biotechnol.* 2014, 909601. doi: 10.1155/2014/909601.

Bolger, A.M., Lohse, M., and Usadel, B. (2014). Trimmomatic: a flexible trimmer for Illumina sequence data. *Bioinformatics*. 30(15), 2114–2120. doi: 10.1093/bioinformatics/btu170.

Caporaso, J.G., Kuczynski, J., Stombaugh, J., Bittinger, K., Bushman, F.D., Costello, E.K., et al. (2010). QIIME allows analysis of high-throughput community sequencing data. *Nat Methods*. 7(5), 335–336. doi: 10.1038/nmeth.f.303.

Doumbou, C.L., Hamby Salove, M.K., Crawford, D.L., and Beaulieu, C. (2005). Actinomycetes, promising tools to control plant diseases and to promote plant growth. *Phytoprotection*. 82(3), 85–102. doi: 10.7202/706219ar.

Doyle, J.J. and Doyle, J.L. (1987). A rapid DNA isolation procedure for small quantities of fresh leaf tissue. *Phytochem Bull.* 19, 11–15.

Ewels, P., Måns, M., Sverker, L., and Max, K. (2016). MultiQC: summarize analysis results for multiple tools and samples in a single report. *Bioinformatics*. doi: 10.1093/bioinformatics/btw354.

Hazarika, S.N., Saikia, K., and Thakur, D. (2022). Characterization and selection of endophytic actinobacteria for growth and disease management of Tea (*Camellia sinensis* L.). *Front Plant Sci.* 13, 989794. doi: 10.3389/fpls.2022.989794.

Katoh, K., Misawa, K., Kuma, K., and Miyata, T. (2002). MAFFT: a novel method for rapid multiple sequence alignment based on fast Fourier transform. *Nucleic Acids Res.* 30(14), 3059–3066. doi: 10.1093/nar/gkf436.

Keegan, K.P., Glass, E.M., and Meyer, F. (2016). MG-RAST, a metagenomics service for analysis of microbial community structure and function. *Methods Mol Biol.* 1399, 207–233. doi: 10.1007/978-1-4939-3369-3_13.

Kumaishi, K., Usui, E., Suzuki, K., Kobori, S., Sato, T., Toda, Y., et al. (2022). High throughput method of 16S rRNA gene sequencing library preparation for plant root microbial community profiling. *Sci Rep.* 12(1), 19289. doi: 10.1038/s41598-022-23943-x.

McMurdie, P.J., and Holmes, S. (2013). Phyloseq: an R package for reproducible interactive analysis and graphics of microbiome census data. *PLoS One*. 8(4), e61217. doi: 10.1371/journal.pone.0061217.

Porebski, S., Bailey, L.G., and Baum, B.R. (1997). Modification of a CTAB DNA extraction protocol for plants containing high polysaccharide and polyphenol components. *Plant Mol Biol Report.* 15(1), 8–15. doi: 10.1007/bf02772108.

Sharma, A.D., Gill, P.K., and Singh, P. (2002). DNA isolation from dry and fresh samples of polysaccharide-rich plants. *Plant Mol Biol Report.* (20), 415a–415f. doi: 10.1007/bf02772129.

Shrivastava, P., and Kumar, R. (2018). Actinobacteria: eco-friendly candidates for control of plant diseases in a sustainable manner. In Bhim Pratap Singh, Vijai Kumar Gupta and Ajit Kumar Passari., editors. *New and Future Developments in Microbial Biotechnology and Bioengineering*), pp. 79–91. Elsevier.

Singh, R., and Dubey, A.K. (2018). Diversity and applications of endophytic actinobacteria of plants in special and other ecological niches. *Front Microbiol.* 9, 1767. doi: 10.3389/fmicb.2018.01767.

Stamatakis, A. (2014). RAxML version 8: a tool for phylogenetic analysis and post-analysis of large phylogenies. *Bioinformatics*. 30(9), 1312–1313. doi: 10.1093/bioinformatics/btu033.

42 Practical Methodology of Illumina Sequencing Investigating Actinobacteria Genomes

Alaa F. Elsayed and Dina H. Amin

42.1 INTRODUCTION

All next-generation sequencing (NGS) techniques, such as Illumina and Ion Torrent, rely on the production of a 'library' utilizing native or amplified DNA. Adapters are ligated to the ends of each fragment in a traditional technique after DNA fragmentation and fragment size selection. This is usually followed by a DNA amplification step. The library is then put into a flow cell and sequenced using massive parallel sequencing processes (Head et al., 2014) as shown in Figure 42.1. Each reaction usually entails the integration of fluorescently tagged deoxynucleotides in an elongating DNA chain immobilized on a surface in a stepwise, polymerase-mediated process. The cost-effective and quick resequencing of numerous landmark genomes demonstrated the strength of the new technologies (Ng et al., 2009).

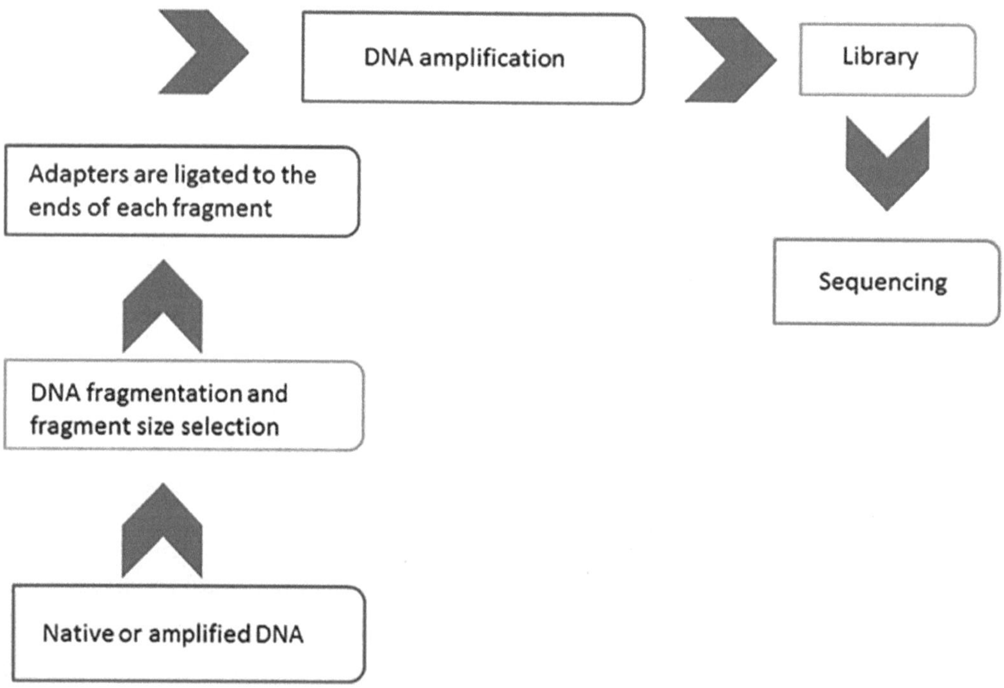

FIGURE 42.1 A flow chart showing the basic principle of next-generation sequencing technology.

DOI: 10.1201/9781003398400-42

Due to the use of bridge PCR for clonal amplification, only Illumina NGS platforms are capable of paired-end sequencing (Mayer et al., 2011). When compared to single-end sequencing technologies, paired-end sequencing allows users to sequence the DNA fragment from both ends, resulting in more coverage, more reads, and more data. Due to the greater likelihood of alignment to a reference, paired-end sequencing produces high-quality sequence data (Mayer et al., 2011).

Illumina has recently been employed in various studies for whole-genome sequencing. Recently, a research used Illumina sequencing to make genome-based classification for *Streptomyces pinistramenti* sp. nov., a novel Actinomycete isolated from a pine forest soil in Poland (Świecimska et al., 2022). Comparative genomics resulted in the identification of *Streptomyces radicis* sp. nov., a novel endophytic actinomycete from roots of plants in Thailand (Kuncharoen et al., 2022). Another research used Illumina whole-genome sequencing to reveal an unexpected diversity among actinomycetoma pathogens which can help in future rapid diagnostic tools and the improvement of treatment outcomes (Watson et al., 2022). It is also used to discover novel alkaloids from marine *Actinobacteria* (De Rop et al., 2022) and genome mining for biosynthetic gene clusters in the novel *Pseudonocardia humida* sp. nov., an Actinomycete isolated from mangrove soil (Zan et al., 2022).

42.2 GENOME SEQUENCING IN THE THIRD GENERATION SEQUENCING ERA

Currently, short read Illumina sequencing technology is responsible for 82% of the bacterial genomes in NCBI Reference Sequence Database (RefSeq). Illumina's dominance was even more striking among surveillance genomes (Segerman, 2020). The Illumina/Solexa platforms are by far the most popular among researchers, owing to their low sequencing costs (Hodzic et al., 2017).

42.3 ILLUMINA WHOLE-GENOME SEQUENCING METHODOLOGY

42.3.1 Reagents

Reagents should be stored according to the manufacturer's instructions:

1. MiSeq Reagent Kit Options:
   - v2 Nano, 300 Cycles – Illumina, Cat# MS-103-1001
   - v2 Nano, 500 Cycles – Illumina, Cat# MS-103-1003
   - v2 Micro, 300 Cycles – Illumina, Cat# MS-103-1002
   - v2 300 Cycles – Illumina, Cat# MS-102-2002
   - v2 500 Cycles – Illumina, Cat# MS-102-2003
   - v3 600 Cycles – Illumina, Cat# MS-102-3003

   1.1. Box 1 of 2. Store at −15°C to −25°C, away from light
   - MiSeq Cartridge
   - Hybridization Buffer (HT1)

   1.2. Box 2 of 2. Store at 2°C–8°C
   - Incorporation Buffer (PR2)
   - Flow Cell
2. Ethanol, molecular-grade, 95%–100% (Fisher Scientific, Cat# BP2818-500 or equivalent)
3. Ethanol, lab-grade, 70% or equivalent for disinfection purposes (Fisher Scientific, Cat# 04–355-309 or equivalent)
4. 100 µL Tween 20
5. Water, molecular-grade (Fisher Scientific, Cat# BP24701 or equivalent)
6. Sodium hypochlorite, 4%–4.99% (Sigma-Aldrich, Cat# 239305-25ML or equivalent)
7. Sodium hydroxide, 1N (Sigma-Alrich, Cat# S2770-100ML or equivalent) pH should be ≥ 12.5.
8. Tween 20 (Sigma-Aldrich, Cat# P9416-50ML or equivalent)

42.3.2 Equipment

1. MiSeq Desktop Sequencer
2. Micropipettes, capable of volumes from 1 to 1,000 μL
3. Heat block
4. Ice buckets
5. Pipet-aid
6. Microcentrifuge
7. Microplate centrifuge or equivalent

42.3.3 Procedure

1. **Getting the MiSeq Reagent Cartridge Ready**
   1.1. Allow enough time for the cartridge to thaw and keep in mind that the HT1 buffer (hybridization buffer) is in the same box as the cartridge and must also be thawed (on ice or at 2°C–8°C).
   1.2. Thaw the reagent cartridge and HT1 and keep them on ice or at 2°C–8°C until the sample is ready to be loaded.
2. **Denaturation of the DNA from the Pooled Library**
   2.1. Preheat a heat block to 96°C for 1 minute.
   2.2. Get the pooled library plate/tubes out of the freezer and place them on ice to defrost. For 1 minute, centrifuge at 800–1,200 rpm (about 280 × g).
   2.3. Thaw a 100-μL aliquot of 1 N NaOH on ice after removing it from the freezer.
   2.4. Dilute the 100-μL aliquot of 1 N NaOH to 0.2 N by adding 400 μL molecular-grade water. To mix, invert the tube many times.
   2.5. In a fresh 1.5-mL microcentrifuge tube, combine 5 μL of the pooled sample library and 5 μL of 0.2 N NaOH.
   2.6. Denature the dsDNA by incubating for 5 minutes at room temperature.
   2.7. Immediately after the incubation, add 990 μL of pre-chilled HT1 to the tube holding 10 μL of denatured pooled library.
   2.8. Chill the denatured DNA until you're ready to move on to the next stage.
3. **Dilute to the final loading concentration desired**
   3.1. The final loading concentration is set empirically for each instrument and user, and it is linked to getting the best cluster density (CD).
   3.2. In the sample tube, thoroughly mix the solution.
   3.3. In a heat block, heat the denatured DNA library to 96 1°C for 2 minutes to guarantee that all dsDNA in the sample is denaturated.
   3.4. Cool for at least 5 minutes in an ice-water bath (prepared by blending about three parts ice and one part water) before loading. The DNA library can be kept on ice or at 2°C–8°C until it's time to load it (around 30 minutes).
4. **Configuring the MiSeq for the run**
   4.1. Before starting the run, make sure the instrument has enough free storage space (100 GB).
   4.2. Check that the Sample Sheet (and, if required, the Sample Plate) configuration in the Illumina Experiment Manager (IEM) on the instrument is complete.
   4.3. Select "Sequence" from the MiSeq Control Software box and instantly begin Preparing and Loading the Flow Cell.
5. **Flow Cell Preparation and Loading**
   5.1. Get the Flow Cell and Incorporation buffer (PR2) from the MiSeq Reagent Box v2 that are kept at 2°C–8°C.
   5.2. Carefully remove the flow cell from the container while wearing clean, powder-free gloves, and without touching the flow cell's glass.

5.3. Lightly rinse the flow cell assembly with molecular-grade water, making sure to fully clean both the glass and plastic casing of any excess salts.
5.4. Using a lint-free wipe, dry the flow cell and cartridge with care around the black port gasket. Ensure that all extra fluid is removed.
5.5. Clean the flow cell glass with ethanol and a clean piece of lens paper, making sure it is clear of streaks, lint, and tissue fibers.
5.6. Using the lens paper, wipe away any excess alcohol and visually inspect the flow cell ports for obstructions, that the gasket is properly placed around the flow cell ports, and that the glass is free of smudges or debris.
5.7. Open the clamp and remove the previously used flow cell by raising the flow cell compartment door and pressing the release button.
5.8. Make that the flow cell stage is lint free. Allow to dry if necessary after wiping with an alcohol wipe or similar.
5.9. Place the flow cell on the stage, close the flow cell clamp, close the compartment door, and choose "Next" in the MiSeq Control Software.

6. **Incorporation Buffer Loading**

6.1. Remove the lid from the capped bottle and gently flip it to mix the Incorporation buffer (PR2).
6.2. Raise the right-hand sipper handle until it locks into place after opening the reagent compartment door.
6.3. Recap the MiSeq Wash Buffer bottle and keep it aside for later use.
6.4. Place the PR2 bottle in the same location as the MiSeq Wash Buffer bottle.
6.5. If necessary, empty the contents of the trash bottle into an approved chemical waste container.
6.6. Lower the sipper handle slowly. Assemble the sippers and place them in the PR2 and trash bottles.
6.7. Verify that the RFID of the PR2 bottle was successfully read in the lower-right corner of the screen.
6.8. On MiSeq Control Software (MCS), select "Next."

7. **Inserting the sample into the cartridge and loading the instrument**

7.1. Remove the thawed reagent cartridge from storage, making sure it's completely dry and the reagents have been mixed by gentle inversion.
7.2. Tap the bottom of the reservoirs with the reagent cartridge on a hard surface to collect the contents.
7.3. Pierce the foil seal above the reservoir labeled "Load Sample" using a clean 1,000 µL pipette tip.
7.4. Pour 600 µL of the denatured DNA library into the "Load Sample" reservoir and gently tap the contents to the bottom using a hard surface.
7.5. Double-check that the MiSeq screen says "load reagent cartridge" to ensure that the chiller's sippers have been raised.
7.6. Remove the wash tray from the reagent chiller and, if necessary, use absorbent wipes to dry the bottom of the chiller compartment.
7.7. Slide the reagent cartridge into the chiller until it stops, holding the end with the Illumina label.
7.8. Close the chiller door and check the screen to ensure the reagent cartridge's RFID has been successfully read.
7.9. Shut the door to the reagent chamber.
7.10. Select "Change Sample Sheet" and then "Browse" to find the suitable sample sheet on the next screen.
7.11. Click "OK" after selecting the desired sample sheet.

7.12. Click "Next" to go to the screen where you can review the run parameters. Review the run parameters to make sure they're proper (Experiment Name, Analysis Workflow, Read Length, etc.) before moving on to the Pre-Run Check.

7.13. Press the "Start Run" button.

7.14. Run statistics (Q30, cluster density, clusters passing filter, expected yield) can be stored in the Workbook when the run is completed.

8. Using the MiSeq to Export Data

8.1. Data can be located in the Illumina MiSeqOutput Run folder (select the most recent run folder) Data Intensities BaseCalls on the D: drive. The raw read data in the BaseCalls folder is in fastq.gz format. For paired readings, each isolate will have two files (R1 and R2). This information can be transferred to a computer workstation by copying it to an external hard disk. Alternatively, to transfer files directly from the MiSeq instrument, it can be connected to a local network share drive.

9. Reagent Disposal and Post-Run Washes

9.1. When the run is over, go to the bottom right-hand side of the screen and click "Start Wash."

9.2. To proceed with a Post-Run bleach wash, make sure the "Perform optional template line wash" item is selected. To minimize nucleic acid carry-over from prior runs, bleach (sodium hypochlorite) is utilized.

9.3. Use molecular-grade water to make a new dilution of molecular-grade sodium hypochlorite (NaOCl).

9.5. Fill the remaining reservoirs with a Tween 20 wash solution containing 0.5% Tween 20.

9.6. Wipe away any excess moisture from the wash cartridge's surface using a lint-free wipe.

9.7. Remove the used cartridge by opening the reagent compartment and chiller doors. Place the cartridge in a safe place to be disposed of.

9.8. Place the wash tray in the reagent chiller and slide it in until it stops and then shut the reagent chiller door.

A. Place the tube in the MiSeq wash cartridge's position 17.

B. Check that the MiSeq tube's neck is flush with the wash cartridge.

9.9 Remove the PR2 bottle and replace it with a MiSeq Wash Buffer bottle by raising the sipper handle (containing 300–500 mL of 0.5% Tween 20 solution).

9.10. Remove the waste container and dispose of the contents in a formamide-safe chemical waste receptacle.

9.11 Replace the reagent compartment door, lower the sipper arm, and close the reagent compartment door after emptying the waste container.

9.12 To begin the wash, select "Next." It will take about 20 minutes to complete this task.

9.13. To return to the MiSeq Control Software home screen, select "Done" on the screen once the Post-Run bleach wash is complete.

9.14. A second Post-Run wash with molecular-grade water must be performed after the bleach wash to guarantee that any trace amount of sodium hypochlorite is eliminated from the template line (Ng et al., 2009) (11).

42.4 CONCLUSION

In this review, we highlighted the method of Illumina sequencing as a common preferred whole-genome sequencing methodology. We clarified the principle of the next-generation technologies, including Illumina, and we introduced the detailed practical procedures of Illumina genome sequencing as a recommended tool for identifying the whole genomes of Actinobacteria.

REFERENCES

De Rop, A. S., Rombaut, J., Willems, T., De Graeve, M., Vanhaecke, L., Hulpiau, P.,... & Soetaert, W. K. 2022. Novel alkaloids from marine *Actinobacteria*: discovery and characterization. *Marine Drugs* 20(1): 6. https://doi.org/10.3390/md20010006

Head, S. R., Komori, H. K., LaMere, S. A., Whisenant, T., Van Nieuwerburgh, F., Salomon, D. R., Ordoukhanian, P. 2014. Library construction for next-generation sequencing: overviews and challenges. *Biotechniques* 56(2): 61–77. https://doi.org/10.2144/000114133

Hodzic, J., Gurbeta, L., Omanovic-Miklicanin, E., Badnjevic, A. 2017. Overview of next-generation Sequencing platforms used in published draft plant genomes in light of genotypization of immortelle plant (*Helichrysium arenarium*). *Medical Archives* 71(4): 288–292. https://doi.org/10.5455/medarh.2017.71.288-292

https://www.cdc.gov/pulsenet/pdf/PNL38-WGS-on-MiSeq-508.pdf

Kuncharoen, N., Yuki, M., Kudo, T., Okuma, M., Booncharoen, A., Mhuantong, W., Tanasupawat, S. 2022. Comparative genomics and proposal of *Streptomyces radicis* sp. nov., an endophytic *actinomycete* from roots of plants in Thailand. *Microbiological Research* 254: 126889. https://doi.org/10.1016/j.micres.2021.126889

Mayer, P., Farinelli, L., Kawashima, E. H. 2011. Method of nucleic acid amplification. *U.S. Patent* 7,985,565.

Ng, S. B., Turner, E. H., Robertson, P. D., Flygare, S. D., Bigham, W., Lee, C., ... & Shendure, J. 2009. Targeted capture and massively parallel sequencing of 12 human exomes. *Nature* 461(7261): 272–276. https://doi.org/10.1038/nature08250

Segerman, B. 2020. The most frequently used sequencing technologies and assembly methods in different time segments of the bacterial surveillance and RefSeq genome databases. *Frontiers in Cellular and Infection Microbiology* 571. https://doi.org/10.3389/fcimb.2020.527102

Świecimska, M., Golinska, P., Goodfellow, M. 2022. Genome-based classification of *Streptomyces pinistramenti* sp. nov., a novel *Actinomycete* isolated from a pine forest soil in Poland with a focus on its biotechnological and ecological properties. *Antonie van Leeuwenhoek* 115(6):783–800. https://orcid.org/0000-0001-9154-8191

Watson, A. K., Kepplinger, B., Bakhiet, S. M., Mhmoud, N. A., Goodfellow, M., Fahal, A. H., Errington, J. 2022. Systematic whole-genome sequencing reveals an unexpected diversity among actinomycetoma pathogens and provides insights into their antibacterial susceptibilities. *bioRxiv* 16(7): e0010128. https://doi.org/10.1101/2022.01.03.474876

Zan, Z. Y., Ge, X. F., Huang, R. R., Liu, W. Z. 2022. *Pseudonocardia humida* sp. nov., an *Actinomycete* isolated from mangrove soil showing distinct distribution pattern of biosynthetic gene clusters. *Current Microbiology* 79(3): 1–8. https://doi.org/10.1007/s00284-022-02784-x

43 Digital DNA-DNA Hybridization for the Identification of *Actinomycetota* Species

Manik Prabhu Narsing Rao and Awalagaway Dhulappa

43.1 INTRODUCTION

The first modern attempt to classify bacteria was based on phenotypic characteristics (Hugenholtz et al., 2021). During the 1960s, there was widespread agreement that bacteria could be best classified by comparing their genomes. Initially, overall base compositions of DNA (mol% G+C values) were used (Lee et al., 1956). Bacteria with significantly different mol% G+C values suggest that they did not belong to the same species (Rosselló-Mora and Amann, 2001). A single value derived from the analysis of DNA base compositions only allowed for very superficial comparisons, necessitating the use of much more precise methods (Brenner et al., 1969). Thus, a method like DNA-DNA hybridization (DDH) was developed (Brenner et al., 1969) and microorganisms could be considered a single species if they shared DDH values above 70% (Wayne et al., 1987). In the 1970s, Woese and Fox's groundbreaking study established ribosomal RNA genes as standard phylogenetic markers in taxonomic studies (Woese and Fox, 1977). Attempts were made to determine relationships between distantly related bacteria by cataloging ribosomal ribonucleic acids (rRNA) (Stackebrandt et al., 1985). The present 16S rRNA gene sequence identity cut-off of 98.7% has been recommended for species delineation, and strains showing 98.7% or higher 16S similarity will be selected for calculating DDH values (Chun et al., 2018). Although DDH has long been considered the "gold standard" for bacterial species identification, its procedures are known to be labor-intensive, prone to experimental error, and inconsistent in terms of reproducibility (Hu et al., 2022). To overcome this drawback, digital DNA-DNA hybridization (dDDH) was developed (Auch et al., 2010) and its results were proven to correlate to DDH's 70% threshold (Meier-Kolthoff et al., 2013). In this section, we will discuss the dDDH analysis procedure along with the cut-off value for species identification.

43.2 GENOME-TO-GENOME DISTANCE CALCULATOR

The dDDH analysis can be performed using the software genome-to-genome distance calculator (GGDC) (Figure 43.1) (https://ggdc.dsmz.de/ggdc.php) (Meier-Kolthoff et al., 2021, 2013). The tool gives users the option to select different parameters to determine the intergenomic distances between genomes and converts those distances into similarity values that can be compared to DDH (Auch et al., 2010). At present, GGDC tool offers six local-alignment programs (NCBI-BLAST (Altschul et al., 1990), WU-BLAST (Altschul et al., 1990), BLAT (Kent, 2002), BLASTZ (Schwartz et al., 2003), BLAST+ (Camacho et al., 2009), and MUMmer (Kurtz et al., 2004)) but BLAST+ is the recommended local-alignment program. Two algorithms (greedy selection and greedy-with-trimming) are used to perform the distance transformation, and three specific distance formulas (formulas 1, 2, and 3) are calculated (with formula 2 recommended). The results are emailed to the user.

 DOI: 10.1201/9781003398400-43

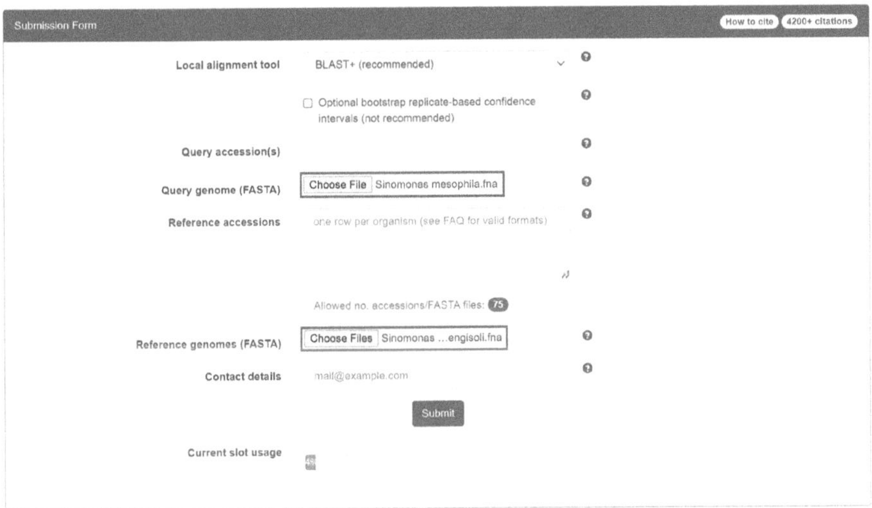

FIGURE 43.1 The website screenshot of the genome-to-genome distance calculator. The black highlighted region to upload genome sequences.

43.3 EXPECTED RESULTS

The dDDH value between *Sinomonas mesophila* and *Sinomonas notoginsengisoli* was less than 70%, which was lower than the recommended cut-off level (70%) for species delineation (Auch et al., 2010) (Figure 43.2).

=-=

Sinomonas_mesophila (query) vs. Sinomonas_notoginsengisoli (reference):

=-=

Formula: 1 (HSP length / total length)
Distance: 0.5957
DDH estimate (GLM-based): 25.10% [21.8 - 28.7%]
Probability that DDH > 70% (i.e., same species): 0.01% (via logistic regression)
Probability that DDH > 79% (i.e., same subspecies): 0% (via logistic regression)

- -

Formula: 2 (identities / HSP length) (RECOMMENDED)
Distance: 0.1994
DDH estimate (GLM-based): 22.00% [19.7 - 24.4%]
Probability that DDH > 70% (i.e., same species): 0% (via logistic regression)
Probability that DDH > 79% (i.e., same subspecies): 0% (via logistic regression)

- -

Formula: 3 (identities / total length)
Distance: 0.6763
DDH estimate (GLM-based): 23.40% [20.6 - 26.5%]
Probability that DDH > 70% (i.e., same species): 0% (via logistic regression)
Probability that DDH > 79% (i.e., same subspecies): 0% (via logistic regression)

- -

FIGURE 43.2 The screenshot results emailed to the user. The formula: 2 (black highlighted region) is the recommended value to estimate the DDH values.

REFERENCES

Altschul, S.F., Gish, W., Miller, W., Myers, E.W., and Lipman, D.J. 1990. Basic local alignment search tool. *Journal of Molecular Biology*. 215(3), 403–410.

Auch, A.F., von Jan, M., Klenk, H.P., and Göker, M. 2010. Digital DNA-DNA hybridization for microbial species delineation by means of genome-to-genome sequence comparison. *Standards in Genomic Sciences*. 2(1), 117–134.

Brenner, D.J., Fanning, G.R., Rake, A.V., and Johnson, K.E. 1969. Batch procedure for thermal elution of DNA from hydroxyapatite. *Analytical Biochemistry*. 28(1), 447–459.

Camacho, C., Coulouris, G., Avagyan, V., Ma, N., Papadopoulos, J., Bealer, K., and Madden, T.L. 2009. BLAST+: architecture and applications. *BMC Bioinformatics*. 10, 421.

Chun, J., Oren, A., Ventosa, A., Christensen, H., Arahal, D.R., da Costa, M.S., Rooney, A.P., Yi, H., Xu, X. W., De Meyer, S., and Trujillo, M.E. 2018. Proposed minimal standards for the use of genome data for the taxonomy of prokaryotes. *International Journal of Systematic and Evolutionary Microbiology*. 68(1), 461–466.

Hu, S., Li, K., Zhang, Y., Wang, Y., Fu, L., Xiao, Y., Tang, X., and Gao, J. 2022. New insights into the threshold values of multi-locus sequence analysis, average nucleotide identity and digital DNA-DNA hybridization in delineating *Streptomyces* species. Frontiers in Microbiology. 13, 910277.

Hugenholtz, P., Chuvochina, M., Oren, A., Parks, D.H., and Soo, R.M. 2021. Prokaryotic taxonomy and nomenclature in the age of big sequence data. *The ISME Journal*. 15(7), 1879–1892.

Kent, W.J. 2002. BLAT–the BLAST-like alignment tool. *Genome Research*. 12 (4):656–664.

Kurtz, S., Phillippy, A., Delcher, A.L., Smoot, M., Shumway, M., Antonescu, C., and Salzberg, S.L. 2004. Versatile and open software for comparing large genomes. *Genome Biology*. 5(2), R12.

Lee, K.Y., Wahl, R., and Barbu, E. 1956. Contenu en bases puriques et pyrimidiques des acides désoxyribonucléiques des bactéries. *annales de l institut pasteur*. 2, 212–224.

Meier-Kolthoff, J.P., Auch, A.F., Klenk, H.P., and Göker, M. 2013. Genome sequence-based species delimitation with confidence intervals and improved distance functions. *BMC Bioinformatics*. 14, 60.

Meier-Kolthoff, J.P., Carbasse, J.S., Peinado-Olarte, R.L., and Göker, M. 2021. TYGS and LPSN: a database tandem for fast and reliable genome-based classification and nomenclature of prokaryotes. *Nucleic Acids Research*. 50(D1), D801–D807.

Rosselló-Mora, R., and Amann, R. 2001. The species concept for prokaryotes. *FEMS Microbiology Reviews*. 25(1), 39–67.

Schwartz, S., Kent, W.J., Smit, A., Zhang, Z., Baertsch, R., Hardison, R.C., Haussler, D., and Miller, W. 2003. Human-mouse alignments with BLASTZ. *Genome Research*. 13(1), 103–107.

Stackebrandt, E., Ludwig, W., and Fox, G.E. 1985. 3 16 S ribosomal RNA oligonucleotide cataloguing. In: Bergan T., (ed.) *Methods in Microbiology*, vol. 18. Academic Press, pp. 75–107.

Wayne, L.G., Brenner, D.J., Colwell, R.R., Grimont, P.A.D., Kandler, O., Krichevsky, M.I., Moore, L.H., Moore, W.E.C., Murray, R.G.E., Stackebrandt, E., Starr, M.P., and Truper, H.G. 1987. Report of the ad hoc committee on reconciliation of approaches to bacterial systematics. *International Journal of Systematic and Evolutionary Microbiology*. 37(4), 463–464.

Woese, C.R., and Fox, G.E. 1977. Phylogenetic structure of the prokaryotic domain: the primary kingdoms. *Proceedings of the National Academy of Sciences of the United States of America*. 74(11), 5088–5090.

44 antiSMASH

Genome Mining and Genome-Wide Analysis of Gene Clusters in Actinopolyspora *Species*

Bhagwan Narayan Rekadwad and Kasbe Prajapati Sudhakar

44.1 INTRODUCTION

Actinobacteria are a valuable resource for the pharmaceutical industry, as their genomes have been mined for numerous useful natural compounds [1,2]. Antibiotics and Secondary Metabolite Analysis Shell (antiSMASH) bacterial version is a bioinformatics tool used for the identification and analysis of secondary metabolite biosynthetic gene clusters in bacteria [3–8]. It is well known that Actinobacteria are capable of producing a wide variety of secondary metabolites, some of which include antimicrobials, anticancer agents, anti-inflammatory chemicals, and other physiologically active substances [9,10]. As a result of this, Actinobacteria have emerged as a useful resource for the research and development of new therapeutic agents as well as the discovery of natural products and important topical and intravenous drugs. In addition, it has been demonstrated that certain species of Actinobacteria, for example, the novel halophilic *Actinopolyspora* sp. YIM90600, produce erythromycins H and I, a clinically important antibacterial drug [11]. The genus Actinopolyspora have 14 species (*Actinopolyspora alba, Actinopolyspora algeriensis, Actinopolyspora biskrensis, Actinopolyspora dayingensis, Actinopolyspora erythraea, Actinopolyspora halophila, Actinopolyspora iraqiensis, Actinopolyspora lacussalsi, Actinopolyspora mortivallis, Actinopolyspora mzabensis, Actinopolyspora righensis, Actinopolyspora saharensis, Actinopolyspora salinaria, and Actinopolyspora xinjiangensis*) and two subspecies (*Actinopolyspora lacussalsi* subsp. *lacussalsi* and *Actinopolyspora lacussalsi* subsp. *righensis*). Of these, *Actinopolyspora dayingensis* (species designation), *Actinopolyspora lacussalsi* subsp. *lacussalsi* and *Actinopolyspora lacussalsi* subsp. *righensis* (subspecies designations) names were not validly published as per standing nomenclature (see order https://lpsn.dsmz.de/order/actinopolysporales). Members of the genus *Actinopolyspora* and *Actinopolyspora*-like strains could be useful as probiotics [12,13], as biocontrol agents [14,15], or to help break down environmental pollutants [16]. In order to find a biosynthetic gene cluster for the discovery of a natural product from the *Actinopolyspora* species, this chapter describes identification and methods. Furthermore, this protocol will assist young microbiologists in genome mining and annotation in locating biosynthetic gene clusters by offering a comprehensive guide using a reference bacteria *Actinopolyspora mortivallis* DSM 44261. The user can use this protocol to perform analyses of multiple bacteria separately.

DOI: 10.1201/9781003398400-44

44.2 MATERIALS

44.2.1 Isolation of Bacterial Pure Culture and DNA Extraction

Axenic culture of *Actinopolyspora* species can be isolated from samples or one can obtain axenic pure culture desired bacteria from International Culture Collections. In this case, axenic culture of bacterial type strain *Actinopolyspora mortivallis* DSM 44261 can be obtained from Leibniz Institute DSMZ, German Collection of Microorganisms and Cell Cultures GmbH (see https://www.dsmz.de/). According to the manufacturer's instructions, a QIAGEN DNA extraction kit (QIAamp DNA Micro Kit (50 preps)) can be used to extract DNA from an axenic pure culture of bacteria. It contains sufficient chemicals for 50 reactions: 50 QIAamp MinElute Columns, Proteinase K, carrier RNA, buffers, and collection tubes for 50 DNA preparations (2 mL).

44.2.2 Whole-Genome Sequencing

Whole-genome sequencing of bacteria can be performed either using nanopore sequencing technology (https://nanoporetech.com/applications). With the help of nanopore sequencing technology, researchers can resolve repetitive and complex structural variations using lengthy nanopore DNA sequencing reads. Nanopore sequencing has indeed revolutionized the field of genomics by significantly simplifying de novo genome assembly. Nanopore sequencing facilitates integrated linking and phasing analyses, enabling precise delineation of closely related species within metagenomic datasets and accurate discrimination between plasmid and chromosomal sequences. Additionally, it offers real-time sequencing of whole microorganisms using only a single read and many more things [17,18]. Moreover, one can use Pacific Biosciences of California, Inc. (PacBio, https://www.pacb.com/) and Illumina Inc. (https://www.illumina.com/) sequencing platforms for genome sequencing.

44.2.3 Biosynthetic Gene Clusters

Biosynthetic gene clusters of secondary metabolites can be detected using the antiSMASH bacterial version (https://antismash.secondarymetabolites.org/). To perform analysis, a complete genome sequence is needed. Genome sequencing can either be obtained through in-house sequencing methods using the protocols mentioned above or it can be downloaded from the NCBI website (https://www.ncbi.nlm.nih.gov/genome/). In this protocol, we have used genome sequences of *Actinopolyspora mortivallis* DSM 44261 (NZ_KB913024.1) downloaded from the NCBI genome repository.

44.3 METHODS

44.3.1 Methodology for Analysis

Genome of *Actinopolyspora mortivallis* DSM 44261 (NZ_KB913024.1) was used, and graphical representations of ectoine gene clusters (region 1 and region 15) taking part in various cellular processes were represented [19]. Steps to perform antiSMASH analysis using bacterial genome to determine the presence of several gene clusters encoding secondary metabolites are outlined on Rekadwad et al. (2024). Rekadwad et al. (2024) outlined overview of antiSMASH analysis [19] data input window has been displayed in the overview of the bacterial version of antiSMASH. This window allows users to either upload the genome sequence in FASTA format or obtain the genome sequence if it has already been deposited in NCBI. As a result of the analysis, the user has the choice to be sent an email notifying them when the genome has been analysed. Users also have the option to bookmark job links or to copy and save a job link in order to view it in further detail later. Under either scenario, the data can be seen for a full month whenever it is most convenient for the user. Extra features for a genome analysis are choice-based and can be activated or deactivated depending on the desired analysis. These extra features include Known Cluster Blast, Cluster Blast, Sub Cluster Blast, MIBiG cluster comparison, Active Site Finder, RRE Finder, Cluster Pfam analysis, Pfam-based GO term annotation, TIGRFam analysis, and TFBS analysis.

44.3.2 Results and Discussion - Analysis of Output

The antiSMASH output shows analysed genomes for various secondary metabolites ranged from region 1 to 25. Each region will have a separate cluster of genes. SAM-dependent methyltransferase, trypsin-like serine protease, diaminobutyrate acetyltransferase, diaminobutyrate-2-oxoglutarate transaminase, ectoine synthase, ADP-ribosylglycohydrolase family protein, and BTAD domain-containing putative transcriptional regulator are all encoded by region 1. Similarly, Region 2 has NRPS-independent-siderophore and NRPS-like, Region 3 has RiPP-like, Region 4 has butyrolactone, Region 5 has T1PKS, Region 6 has terpene, Region 7 has butyrolactone and hglE-KS, Region 8 has NRPS, Region 9 has butyrolactone, Region 10 has nucleoside, Region 11 has NRPS-like, Region 12 has terpene and CDPS, Region 13 has NRPS and betalactone, Region 14 has lanthipeptide class-I, Region 15 has ectoine, Region 16 has terpene, Region 17 has NRPS-like, Region 18 has T1PKS, Region 19 has lanthipeptide class-I and terpene, Region 20 has NRPS, ladderane, and arypolyene, Region 21 has RRE-containing, Region 22 has thiopeptide, Region 23 has NRPS-like and NRPS, Region 24 has siderophore, and Region 25 has linaridin.

44.3.2.1 Analysis of Ectoine Gene Cluster/Region in Genome *Actinopolyspora mortivallis* DSM 44261

a. Isolate and cultivate bacterial pure cultures from environmental samples.
b. Find out the optimum temperature, pH, and nutrient media for growth. *Actinopolyspora mortivallis* DSM 44261 grows well on Luria Bertani Agar (Miller Luria Bertani Agar) at 30°C and a neutral pH.
c. Extraction, purification of DNA, and whole-genome sequencing are as mentioned earlier ("see Sections 2.1 and 2.2").
d. Upload the *Actinopolyspora mortivallis* DSM 44261 FASTA sequence for analysis to the antiSMASH webserver.
e. Once analysis is done by the web server, notification of the output is displayed by the system.

a. Isolate and cultivate bacterial pure culture from environmental samples.
b. Find out optimum temperature, pH and nutrient media for growth. In case of *Actinopolyspora mortivallis* DSM 44261, it grows well in Luria Bertani Agar (Miller Luria Bertani Agar) at 30°C temperature and neutral pH.
c. Extraction, purification of DNA, and whole-genome sequencing as mentioned in former earlier ("see Sections 2.1 and 2.2").
d. Upload FASTA format sequence of *Actinopolyspora mortivallis* DSM 44261 in antiSMASH webserver for analysis.
e. Once analysis is done by the webserver, notification of output is displayed by the system.
f. Output results can be viewed on the same page, one after another. For example, output has colour codes displayed at various positions. Furthermore, the examined genome contains a variety of regions. Region 1 and Region 15 reserved for ectoine comprise core biosynthetic genes, transport-related genes, regulatory genes, resistance genes, TTA codons, binding sites, additional biosynthetic genes, and other genes.

44.4 CONCLUSION

Actinopolyspora is a genus of Actinobacteria, gram-positive bacteria ubiquitous in soil and other natural environments that can produce many secondary metabolites for biomedical applications. They possess several anti-inflammatory, anti-tumour, and antibacterial chemicals. These bacteria may develop novel biochemicals that could treat antimicrobial-resistant bacteria. To find new natural product medications from *Actinopolyspora* species, genomes have been observed for gene clusters using the Antibiotics and Secondary Metabolite Analysis Shell (antiSMASH) bacterial version, a bioinformatics tool. It has been shown that *Actinopolyspora* has various metabolic pathways, secondary metabolite gene clusters, and the presence of functional elements in the form of regions 1–25. A methodology has been discussed to study regions found in the annotated genome of *Actinopolyspora* by giving an example in the form of regions 1 and 2.

ACKNOWLEDGEMENTS

BNR acknowledges part of funds received for research from Yenepoya (Deemed to be University) (Project No.: YU/SeedGrant/104-2021)

CONFLICT OF INTEREST

The author has no conflict of interest to declare that is relevant to the content of this protocol chapter.

DATA AVAILABILITY

The data used to support the findings of this protocol is available within this article.

REFERENCES

1. De Simeis, D., and Serra, S. 2021. *Actinomycetes*: A never-ending source of bioactive compounds-an overview on antibiotics production. *Antibiotics* 10(5): 483. doi:10.3390/antibiotics10050483
2. Barka, E. A., Vatsa, P., Sanchez, L., Gaveau-Vaillant, N., Jacquard, C., Meier-Kolthoff, J. P., Klenk, H. P., Clément, C., Ouhdouch, Y., and van Wezel, G. P. 2015. Taxonomy, physiology, and natural products of Actinobacteria. *Microbiol Mol Biol Rev* 80(1): 1–43. doi:10.1128/MMBR.00019-15
3. Medema, M. H., Blin, K., Cimermancic, P., de Jager, V., Zakrzewski, P., Fischbach, M. A., Weber, T., Takano, E., and Breitling, R. 2011. antiSMASH: rapid identification, annotation and analysis of secondary metabolite biosynthesis gene clusters in bacterial and fungal genome sequences. *Nucleic Acids Res* 39: W339–W346. doi:10.1093/nar/gkr466
4. Blin, K., Medema, M. H., Kazempour, D., Fischbach, M. A., Breitling, R., Takano, E., and Weber, T. 2013. antiSMASH 2.0–a versatile platform for genome mining of secondary metabolite producers. *Nucleic Acids Res* 41: W204–W212. doi:10.1093/nar/gkt449
5. Weber, T., Blin, K., Duddela, S., Krug, D., Kim, H. U., Bruccoleri, R., Lee, S. Y., Fischbach, M. A., Müller, R., Wohlleben, W., Breitling, R., Takano, E., and Medema, M. H. 2015. antiSMASH 3.0–a comprehensive resource for the genome mining of biosynthetic gene clusters. *Nucleic Acids Res* 43(W1): W237–W243. doi:10.1093/nar/gkv437
6. Blin, K., Wolf, T., Chevrette, M. G., Lu, X., Schwalen, C. J., Kautsar, S. A., Suarez Duran, H. G., de Los Santos, E. L. C., Kim, H. U., Nave, M., Dickschat, J. S., Mitchell, D. A., Shelest, E., Breitling, R., Takano, E., Lee, S. Y., Weber, T., and Medema, M. H. 2017. antiSMASH 4.0-improvements in chemistry prediction and gene cluster boundary identification. *Nucleic Acids Res* 45(W1): W36–W41. doi:10.1093/nar/gkx319
7. Blin, K., Shaw, S., Steinke, K., Villebro, R., Ziemert, N., Lee, S. Y., Medema, M. H., and Weber, T. 2019. antiSMASH 5.0: updates to the secondary metabolite genome mining pipeline. *Nucleic Acids Res* 47(W1): W81–W87. doi:10.1093/nar/gkz310
8. Blin, K., Shaw, S., Kloosterman, A. M., Charlop-Powers, Z., van Wezel, G. P., Medema, M. H., and Weber, T. 2021. antiSMASH 6.0: improving cluster detection and comparison capabilities. *Nucleic Acids Res* 49(W1): W29–W35. doi:10.1093/nar/gkab335
9. Selim, M. S. M., Abdelhamid, S. A., and Mohamed, S. S. 2021. Secondary metabolites and biodiversity of actinomycetes. *J Genet Eng Biotechnol* 19(1): 72. doi:10.1186/s43141-021-00156-9
10. Teng, P., Shao, H., Huang, B., Xie, J., Cui, S., Wang, K., and Cai, J. 2023. Small molecular mimetics of antimicrobial peptides as a promising therapy to combat bacterial resistance. *J Med Chem.* doi:10.1021/acs.jmedchem.2c00757
11. Huang, S. X., Zhao, L. X., Tang, S. K., Jiang, C. L., Duan, Y., and Shen, B. 2009. Erythronolides H and I, new erythromycin congeners from a new halophilic actinomycete *Actinopolyspora* sp. YIM90600. *Org Lett* 11(6): 1353–1356. doi:10.1021/ol900143j
12. Xie, F., and Pathom-Aree, W. 2021. Actinobacteria from desert: diversity and biotechnological applications. *Front Microbiol* 12: 765531. doi:10.3389/fmicb.2021.765531

13. Sarkar, G., and Suthindhiran, K. 2022. Diversity and biotechnological potential of marine *Actinomycetes* from India. *Indian J Microbiol* 62: 475–493. doi:10.1007/s12088-022-01024-x
14. Cray, J. A., Connor, M. C., Stevenson, A., Houghton, J. D., Rangel, D. E., Cooke, L. R., and Hallsworth, J. E. 2016. Biocontrol agents promote growth of potato pathogens, depending on environmental conditions. *Microbial Biotech* 9(3): 330–354. doi:10.1111/1751-7915.12349
15. Silva, G. C., Kitano, I. T., Ribeiro, I. A. F., and Lacava, P. T. 2022. The potential use of *Actinomycetes* as microbial inoculants and biopesticides in agriculture. *Front Soil Sci* 2: 833181. doi:10.3389/fsoil.2022.833181
16. Al-Mueini, R., Al-Dalali, M., Al-Amri, I. S., and Patzelt, H. 2007. Hydrocarbon degradation at high salinity by a novel extremely halophilic actinomycete. *Environ Chem* 4(1): 5–7. doi:10.1071/EN06019
17. Khrenova, M. G. Panova, T. V. Rodin, V. A. Kryakvin, M. A. Lukyanov, D. A. Osterman, I. A., and Zvereva, M. I. 2022. Nanopore sequencing for de novo bacterial genome assembly and search for single-nucleotide polymorphism. *Int J Mol Sci* 23: 8569. doi:10.3390/ijms23158569
18. Bouchiat, C., Ginevra, C., Benito, Y., Gaillard, T., Salord, H., Dauwalder, O., Laurent, F., and Vandenesch, F. 2022. Improving the diagnosis of bacterial infections: evaluation of 16s rRNA nanopore metagenomics in culture-negative samples. *Front Microbiol* 13: 943441. doi:10.3389/fmicb.2022.94344
19. Rekadwad, B. 2024. Figure 1. Steps to perform antiSMASH analysis for bacteria. *Zenodo*. doi:10.5281/zenodo.11093490

45 Transcriptome Analysis of Actinomycetes RNA Sequencing Data

A Bioinformatics Protocol for Beginners

Sankaranarayanan Gomathinayagam, Srivarshini Sankar, Nalini Easwaran, Loganathan Karthik, and Kodiveri Muthukaliannan Gothandam

45.1 INTRODUCTION

Transcriptome analysis is a powerful approach that allows researchers to gain insights into gene expression patterns and regulatory mechanisms within an organism. Actinomycetes, a diverse group of Gram-positive bacteria, have attracted significant attention due to their remarkable capability to produce bioactive compounds with diverse biological activities. RNA sequencing (RNA-seq) has emerged as a transformative technology in transcriptome analysis, enabling comprehensive and high-throughput profiling of the entire transcriptome of actinomycetes.

Actinomycetes, including well-known genera such as *Streptomyces* and *Actinomyces*, are renowned for their ability to produce a wide range of bioactive secondary metabolites, including antibiotics, anticancer agents, immunosuppressants, and enzymes. Understanding the genetic and molecular basis of secondary metabolite biosynthesis in actinomycetes is crucial for harnessing their potential in drug discovery, biotechnology, and agriculture.

RNA-seq has revolutionized transcriptome analysis by offering a comprehensive view of gene expression, alternative splicing, post-transcriptional modifications, and non-coding RNA species. By leveraging high-throughput sequencing technology, RNA-seq provides quantitative and qualitative information about the transcriptome of actinomycetes, shedding light on the complex regulatory networks governing secondary metabolite biosynthesis, cellular processes, and environmental responses.

The workflow of transcriptome analysis from actinomycetes RNA sequencing typically involves several key steps. Firstly, total RNA is extracted from the actinomycete culture or tissue of interest, ensuring the preservation of RNA integrity. Subsequently, the extracted RNA undergoes library preparation, which involves reverse transcription, cDNA library construction, and sequencing adaptors ligation. The prepared libraries are then subjected to high-throughput sequencing, generating millions of short reads that represent the transcriptome of the actinomycete organism.

Following sequencing, the obtained RNA-seq reads undergo quality control and pre-processing steps, including read trimming, removal of low-quality reads, and filtering of rRNA and adapter sequences. The processed reads are then mapped to a reference genome or transcriptome assembly using specialized alignment algorithms. This alignment step enables the quantification of gene expression levels and the detection of alternative splicing events or novel transcripts.

DOI 12432545124234-45

Transcript assembly tools are employed to reconstruct complete or partial transcripts, providing insights into the structure and abundance of expressed genes. Differential gene expression analysis is performed to identify genes that exhibit significant expression changes under different experimental conditions or treatments.

The following protocol is aimed at introducing basic know-hows of processing and analyzing the transcriptome data from bacteria at a beginner level.

45.2 MATERIALS

45.2.1 Hardware Requirements

A standard desktop with minimum 4-core CPU, minimum 16 GB RAM, 500 GB hard disk/SSD, with latest version of Ubuntu installed/any Unix-based environment.

45.2.2 Software Requirements

45.2.2.1 Software Dependencies

- fastqc (v0.11.9)
- trimmomatic (v0.39)
- hisat2 (v2.2.1)
- stringtie (v2.1.1)
- Trinotate (v3.2.1)
- R (latest version)

45.2.2.2 R Packages

- tximport
- DESeq2
- To install the R packages, use the following commands in the R environment.

```
install.packages("tximport")
install.packages("DESeq2")
```

45.3 METHODS

45.3.1 Quality Control

1. Install FastQC: FastQC [1] is a widely used tool for quality control of high-throughput sequencing data. Before running FastQC, it needs to be installed on the system. Installation instructions can typically be found on the FastQC website or in the tool's documentation.
2. Run FastQC on the input fastq files: Once FastQC is installed, the next step is to run it on the input fastq files. Fastq files contain the raw sequencing data in a specific format, including the sequence reads and corresponding quality scores. FastQC analyzes these files to generate a comprehensive quality report.

Running FastQC involves providing the path to the input fastq files as a parameter to the tool. FastQC on the input files is run using the following command:

```
fastqc -o output_dir <path/to/fastq/files>/*
```

The * indicates that all the fastq files within that directory should be processed. The -o output_dir option specifies the directory where the FastQC output files will be saved.

45.3.2 Pre-processing

1. Install Trimmomatic: If Trimmomatic [2] is not already installed on the system, it needs to be installed before proceeding with the analysis. Installation instructions can typically be found on the Trimmomatic website or in the tool's documentation.
2. Create a directory to store trimmed fastq files: Before running Trimmomatic, it is recommended to create a separate directory to store the trimmed fastq files. This directory will help organize and manage the processed data separately from the raw data.
3. Trim the input fastq files using Trimmomatic: Trimmomatic is a flexible and widely used tool for trimming sequencing reads. It employs various algorithms and techniques to remove low-quality bases, adaptors, and other artifacts from the reads.

```
mkdir trimmed_fastq_files

trimmomatic PE -phred33 -trimlog trimming.log <path/to/fastq/
files>/*_1.fastq <path/to/fastq/files>/*_2.fastq trimmed_fastq_files/
trimmed_1.fastq
trimmed_fastq_files/unpaired_1.fastq trimmed_fastq_files/trimmed_2.
fastq trimmed_fastq_files/unpaired_2.fastq
SLIDINGWINDOW:4:20 MINLEN:25

Trimmomatic will process the input fastq files based on the
specified options and parameters, and produce trimmed output fastq
files, as well as unpaired files for reads that have been separated
during the trimming process.
```

45.3.2.1 Reference Genome Alignment

1. Install HISAT2: If HISAT2 [3] is not already installed on the system, it needs to be installed before proceeding with the alignment. Installation instructions can typically be found on the HISAT2 website or in the tool's documentation.

```
hisat2 -x reference_genome -1
trimmed_fastq_files/trimmed_1.fastq -2
trimmed_fastq_files/trimmed_2.fastq -S alignments.sam
```

2. Build the HISAT2 index for the reference genome: Before aligning the reads, a HISAT2 index needs to be constructed for the reference genome. The index facilitates efficient mapping of the reads to the reference genome during the alignment process.

It's important to note that the reference genome used for alignment should be appropriate for the actinomycetes species or organism under investigation. The reference genome may be obtained from public databases or generated through sequencing and assembly of the actinomycetes genome.

45.3.2.2 Transcript Assembly

1. Install StringTie: If StringTie [4] is not already installed on the system, it needs to be installed before proceeding with the transcript assembly. Installation instructions can typically be found on the StringTie website or in the tool's documentation.
2. Perform transcript assembly using StringTie: After the alignment step, which generated the SAM file (alignments.sam), StringTie can be used to assemble transcripts from the aligned reads.

The command to perform transcript assembly using StringTie is as follows:

```
stringtie alignments.sam -G reference_annotation.gtf -o
transcripts.gtf -p 4
```

Transcript assembly is a crucial step in transcriptome analysis as it helps to identify and characterize the expressed transcripts, including known and potentially novel isoforms. The assembled transcripts serve as a valuable resource for downstream analyses, such as gene expression quantification, differential expression analysis, and functional annotation.

45.3.2.3 Import Transcript-Level Counts

1. Install R and the required R packages (tximport): R is a programming language widely used for statistical computing and data analysis. Before proceeding, make sure R is installed on your system. Additionally, you need to install the tximport package, which facilitates the import of transcript-level counts into R.
2. Create a text file named "reference_transcripts.tx2gene" and populate it with the required data: In order to map transcript-level counts to gene-level counts, you need to create a text file named "reference_transcripts.tx2gene". This file contains the mapping information between transcript IDs and gene IDs. Each line of the file represents a transcript-gene mapping, with the transcript ID and gene ID separated by a tab or another delimiter.
3. Run Rscript to import transcript-level counts and generate a count matrix: The script provides a command to run Rscript, which invokes an R script to perform the import of transcript-level counts and generate a count matrix. Here's the command

```
Rscript -e 'library(tximport); tx2gene <- read.delim("reference_
transcripts.tx2gene", header=FALSE, stringsAsFactors=FALSE); files
<- c("<path/to/tximport/count_matrix>"); sampleNames <- c("sample1",
"sample2"); txi <- tximport(files, type = "kallisto", tx2gene =
tx2gene, reader = read_tsv); write.table(txi$counts, "count_matrix.
csv", sep=",", quote=FALSE, col.names=NA)'
```

By running the provided Rscript command, the transcript-level counts are imported into R, and a count matrix is generated, which represents the gene-level expression profiles of the samples. This count matrix serves as the input for downstream analyses such as differential expression analysis.

45.3.2.4 Differential Expression Analysis

1. Prepare the metadata.csv file with sample information: The metadata file provides information about the samples used in the analysis, including their condition, treatment, replicates, or any other relevant information. Create a CSV file named "metadata.csv" and populate it with the sample information. The first row should contain column headers, and subsequent rows should contain the sample information. Make sure the column names in the metadata file match the sample names used in the count matrix.

2. Run Rscript to perform differential expression analysis using DESeq2: The script provides a command to run Rscript, which invokes an R script to perform the differential expression analysis using DESeq2 [5,6]. Here's the command:

```
Rscript -e 'library(DESeq2); countMatrix <- as.matrix(read.
csv("count_matrix.csv", row.names=1)); colData <- read.csv("metadata.
csv", row.names=1); dds <- DESeqDataSetFromTximport(txi,
colData=colData, design=~condition); dds <- DESeq(dds); res <-
results(dds); write.csv(res, "differential_expression_results.csv")'
```

By running the provided Rscript command, the differential expression analysis is performed using DESeq2. These results provide information about genes that exhibit significant changes.

45.3.2.5 Annotation

1. Install Trinotate: If Trinotate [7] is not already installed on your system, it needs to be installed before proceeding with the annotation. Installation instructions can typically be found on the Trinotate website or in the tool's documentation.
2. Run Trinotate for transcript annotation: Once Trinotate is installed, you can run it to annotate the assembled transcripts. The specific command to run Trinotate depends on the input files and options you want to use.

```
Trinotate transcript.fasta --gene_trans_map gene_to_trans_map
--transcript_fasta transcript.fasta --sqlite Trinotate.sqlite
-- CPU 4
```

Transcript annotation using Trinotate provides valuable information about the functional characteristics and biological roles of the assembled transcripts. It helps in understanding the potential functions and pathways associated with the expressed transcripts, enabling further downstream analysis and interpretation of the transcriptome data.

REFERENCES

1. Andrew, S. (2010). FastQC: a quality control tool for high throughput sequence data. Retrieved from https://www.bioinformatics.babraham.ac.uk/projects/fastqc/
2. Bolger, A. M., Lohse, M., & Usadel, B. (2014). Trimmomatic: a flexible trimmer for Illumina sequence data. *Bioinformatics*, 30(15), 2114–2120. doi:10.1093/bioinformatics/btu170
3. Kim, D., Paggi, J. M., Park, C., Bennett, C., & Salzberg, S. L. (2019). Graph-based genome alignment and genotyping with HISAT2 and HISAT-genotype. *Nature Biotechnology*, 37(8), 907–915. doi:10.1038/s41587-019-0201-4
4. Pertea, M., Pertea, G. M., Antonescu, C. M., Chang, T. C., Mendell, J. T., & Salzberg, S. L. (2015). StringTie enables improved reconstruction of a transcriptome from RNA-seq reads. *Nature Biotechnology*, 33(3), 290–295. doi:10.1038/nbt.3122.
5. Soneson, C., Love, M. I., & Robinson, M. D. (2015). Differential analyses for RNA-seq: transcript-level estimates improve gene-level inferences. *F1000Research*, 4, 1521. doi:10.12688/f1000research.7563.1
6. Love, M. I., Huber, W., & Anders, S. (2014). Moderated estimation of fold change and dispersion for RNA-seq data with DESeq2. *Genome Biology*, 15(12), 550. doi:10.1186/s13059-014-0550-8
7. Bryant, D. M., Johnson, K., DiTommaso, T., et al. (2017). A tissue-mapped axolotl de novo transcriptome enables identification of limb regeneration factors. *Cell Reports*, 18(3), 762–776. doi:10.1016/j.celrep.2016.12.063

46 Single-Cell RNA Sequencing of the Actinomycetes

Akshatha Prasanna, Aishwarya Das, and T. S. Sai Pawan

46.1 INTRODUCTION

Bacteria belonging to the phylum Actinomycetes are characterized by their high GC content [1]. They are gram-positive, filamentous bacteria that form branching networks of hyphae and are often found in soil and aquatic environments. Actinomycetes display an extensive array of physiological and metabolic traits, including the manufacturing of extracellular enzymes and multiple secondary metabolites [2]. Actinomycetes are important in many environmental and industrial processes, including the decomposition of organic matter and the production of antibiotics, enzymes, and other compounds used in the food, pharmaceutical, and chemical industries. Some species of Actinomycetes are pathogenic and can cause infections in humans and animals, while others are used as probiotics to promote health.

The study of all the transcripts, or RNA molecules, generated by an organism's genes is known as transcriptomics. Transcriptomics research can be used to comprehend the genetic underpinnings of actinomycete synthesis of antibiotics and other biological processes. Researchers often utilize methods like RNA sequencing to recognize and quantify the transcripts present in the bacteria under various conditions. Transcriptomics studies can be used to investigate the response of Actinomycetes to different environmental conditions or treatments, such as exposure to antibiotics or other chemicals. By comparing the transcriptomes of Actinomycetes under different conditions, researchers can gain insights into how these bacteria adapt and respond to their environment and how they might be influenced by factors such as stress or changes in nutrient availability.

A substantial advancement in transcriptome research has been made possible by the recent developments in single-cell RNA sequencing (scRNA-seq) technologies. This technology will enable the analysis of gene expression at the single-cell level. This allows to examine the transcriptomes (the collection of all RNA molecules generated by a cell, including mRNA, rRNA, and tRNA) of particular cells in a population using the single-cell RNA sequencing (scRNA-seq) approach [3]. This is significant because it provides light on expressions of the gene patterns of certain cells within a population, which is vital for comprehending the function of particular cell types and how they interact with one another. The transcriptomes of specific cells within a population of *Actinomyces odontolyticus*, a bacterium linked to periodontal disease, have been examined in actinobacteria using scRNA-seq. The study discovered that each cell in the population had a very diversified transcriptome and expressed a particular set of genes.

46.1.1 History of scRNA Sequencing

Single-cell RNA sequencing was first compared to microarray analysis, a study on a four-cell stage blastomere [4]. Unfortunately, the capacity to expand multiplex numbers was constrained at the time since scRNA sequencing libraries had to be manually created in each tube. A commercially accessible apparatus that reduced single-cell separation and library creation into a two-step procedure was made available in 2014 after the first multiplexed scRNA sequencing approach was created in 2011 [5]. Gierahn et al. released the seq-well single-cell library production technique at the beginning of 2017. This approach to creating single-cell libraries is effective, affordable, and user-friendly, and it has the potential to be extensively used in the future.

DOI: 10.1201/9781003398400-46

46.1.2 Single-Cell RNA Sequencing

Single-cell RNA-seq is a technique used to identify and quantify mRNA molecules in individual cells, allowing for a more detailed understanding of gene expression and population diversity than traditional bulk RNA-seq. The process involves isolating single cells, recording their transcripts, and creating sequencing libraries that assign the transcripts to specific cells.

scRNA-seq on bacterial population provides a wide range of potential research subjects such as microbial evolution and adaptation, microbial population dynamics and to study the mechanisms behind drug resistance in bacteria, which can help inform the development of new antibiotics and other treatments.

46.2 STEPS INVOLVED IN THE PROCESS OF ISOLATING RNA FROM INDIVIDUAL CELLS

Following are the steps involved in scRNA-Seq protocols.

46.2.1 Isolation of Single Cell and RNA

Single-cell isolation and capture is a crucial process that involves the extraction of individual cells from tissues to obtain precise genetic and biochemical data. This technique not only facilitates the investigation of various molecular and genetic mechanisms but also ensures the collection of high-quality cells. By isolating and capturing single cells, researchers can avoid any potential biases that may arise from analyzing a heterogeneous population of cells. The process of cell isolation can be accomplished through various methods, such as isolating complete cells, cells with specific nuclei or organelles, or even cells that express particular flag proteins.

46.2.2 Reverse Transcription (RT)

This stage is critical because it affects the proportion of the cell's total RNA population that will eventually be subjected to sequencer analysis. The efficiency of the synthesis of full-length cDNA and the creation of libraries skewed toward the 3′ or 5′ end of genes may be affected by the processivity of reverse transcriptases and the priming techniques used.

46.2.3 Amplification

Once the cDNA has been synthesized via reverse transcription, it has to be amplified to make sufficient copies for use in subsequent operations. In vitro transcription (IVT) and polymerase chain reaction (PCR) are commonly utilized methods for amplifying cDNA.

46.2.4 Library Preparation

The core steps in the RNA sequencing procedure are library creation, template creation, and sequencing. The approach is completed by using sequencing controls and data analysis techniques, which produce meaningful annotated data. Ends of reverse-transcribed RNA into cDNA are ligated using adaptors, followed by amplification and purification of cDNA using primers that are complementary to the adaptors.

46.3 DATA ANALYSIS TOOLS FOR ANALYZING SCRNA SEQUENCING DATA

10× Genomics created the analytic pipeline known as Cell Ranger for clustering and data visualization of scRNA sequencing data. It also offers a collection of algorithms for quality assurance, alignment, and quantification of gene expression. Due of its ease of use and availability as a commercial product, the 10× Genomics Chromium platform, which utilizes microdroplet-based high-throughput single-cell RNA sequencing, is widely utilized [6].

Smart-seq2 is a well-known protocol for single-cell RNA sequencing that employs a combination of microplates and cell-sorting technologies. This method enables the assessment of gene expression at a single-cell level, enabling the generation of full-length cDNA from each cell and producing vast amounts of data. It is a platform to enhance the sensitivity of each individual cell to the gene level, which is typically employed in conjunction with a microdroplet-based technology [7,8]. The TAS-Seq method for single-cell RNA sequencing employs a terminator-based solid phase cDNA amplification and sequencing strategy that is made possible by the use of nano wells or beads. TAS-Seq has demonstrated great flexibility to these variances, in contrast to conventional approaches that employ terminal transferase, which can be confounded by variations in the reaction. This enables TAS-Seq processing simpler and more dependable which makes it a potential tool in the field of scRNA-seq [9].

A R tool called **Seurat** is used to examine and display single-cell data, including scRNA-seq data. It comprises multiple approaches for grouping and analyzing gene expression patterns together with tools for quality control, standardization, and dimensionality reduction.

With an emphasis on differential expression analysis and the recognition of cell states and transitions, **Monocle** is an R package for analyzing and displaying data from scRNA-seq experiments. It contains techniques for finding differentially expressed genes and building gene regulatory networks, as well as tools for dimensionality reduction, clustering, and visualization.

High-dimensional data analysis methods include uniform manifold approximation and projection (UMAP), principal component analysis (PCA), t-distributed stochastic neighbor embedding (t-SNE), and scvis. These methods enable the reduction of high-dimensional data into lower-dimensional spaces while maintaining its important features. The data may be efficiently seen and explored using the resulting low-dimensional representations, which also highlight significant patterns and relationships that would not have been obvious in the initial high-dimensional environment. These techniques must be used in order to comprehend complicated datasets more thoroughly and to help find clusters and other types of structures in the data [10–12].

SC3 and Seurat are two clustering techniques that use statistical analysis of gene expression data and commonly used markers to group cell subpopulations [13]. In addition to these clustering techniques, a number of specific tools, including as SCDE, BCseq, MAST, and Census, have been created to carry out differential expression gene (DEG) analysis. In order to examine the genetic basis of cellular heterogeneity and obtain knowledge of the biological mechanisms that control cellular differentiation and function, researchers employ these approaches to discover genes that are differently expressed between various cell types [14–17].

Artificial intelligence (AI) techniques, such as regression, classification, dimensionality reduction, and clustering, have become important tools in the analysis of single-cell RNA sequencing data. These methods can help researchers predict clinical outcomes and gather more specific data on cell types, differentially expressed genes, biomarker expression patterns, tracking lineage transition, and characterizing disease subtypes. By identifying patterns and relationships within large datasets, AI algorithms can provide new insights into the complex biological processes that govern cellular differentiation and function. As the field of single-cell RNA sequencing continues to evolve, AI techniques are likely to play an increasingly important role in the scRNA-seq data processing and interpretations [18] (Table 46.1).

46.3.1 ScRNA-Seq of Actinomycetes Species

46.3.1.1 Isolate and Prepare Single Cells

1. Obtain a sample of Actinomycetes cells, either from a culture or from a tissue sample.
2. To obtain single-cell suspensions for downstream applications, various techniques such as FACS, microfluidics, or manual picking are used to isolate individual cells. These methods

TABLE 46.1
Summary of Tools Available ScRNA-Seq Data Analysis

Sl. No	Tool	Features	References
1	Cellenics	Cloud-based and open source for single-cell data storage and analysis. The interface of Cellenics is divided up into four sections: data administration, data processing, data exploration, and plots and tables.	[19]
2	Scatter and SinQC	Programs designed specifically for scRNA-Seq data, provide improved quality assurance	[20,21]
3	HISAT and STAR	Mapping of billions of reads	[22,23]
4	SAVER	UMI-based analytic pipeline, it provides an accurate transcript count estimation for 5′ or 3′ end-transcript sequencing with extremely little technical noise	[24]
5	DESeq2 and TMM	Data normalization and differential gene expression	[25]
6	10× Genomics	Clustering and data visualization	[6]
7	Smart-seq23	Microdroplet-based technology	[7,8]
8	Seurat	Grouping and analyzing gene expression patterns	[13]
9	Monocle	R package for analyzing and displaying data. It contains techniques for finding differentially expressed genes and building gene regulatory networks	[26]
10	SC3	Clustering algorithms for grouping of cell subpopulations	[12]
11	SCDE, BCseq, MAST, and Census	DEG analysis	[14–17]

can enable the identification and isolation of specific cell types or subpopulations from complex biological samples. Once single cells are obtained, they can be processed for single-cell RNA sequencing.

3. Enzymatic lysis: This technique includes breaking down the cell wall with enzymes like lysozyme or pronase in order to liberate the contents of the cell. This technique may be less likely to result in RNA degradation and is gentler than chemical lysis.

46.3.1.2 Reverse Transcription and Amplification of cDNA

1. Use reverse transcriptase and a reverse transcription primer to reverse transcribe the RNA from each single cell into complementary DNA (cDNA).
2. The cDNA is often amplified using PCR for generating sufficient material for sequencing. Depending on the platform and library preparation technique, single-cell RNA sequencing often takes a modest quantity of cDNA, typically in the range of picograms to nanograms. For instance, while other platforms, such as the 10× Genomics Chromium Single Cell Gene Expression platform, normally require 10–100 ng of cDNA per sample, the Smart-seq2 technique only needs 1–10 pg of cDNA per cell [27,28].
3. This step can be performed using a commercial kit or a custom protocol (High-Capacity cDNA Reverse Transcription Kit).

46.3.1.3 Library Preparation and Sequencing

To conduct next-generation sequencing (NGS) on cDNA from single-cell libraries, the first step is to amplify the cDNA using PCR. After amplification, the cDNA must be modified to enable NGS. This modification process usually involves attaching adapters, fragmenting the cDNA, and amplifying it again.

Once the cDNA is prepared for sequencing, a high-throughput sequencing platform such as Illumina or PacBio is used to sequence single-cell libraries. This process involves breaking the cDNA into small fragments, which are then sequenced simultaneously. The resulting sequence data is then analyzed to identify the unique patterns of gene activity displayed by individual cells.

46.3.2 Single-Cell RNA Sequencing (ScRNA-Seq) Data Analysis of Actinomycetes

46.3.2.1 Alignment

1. Use software such as BWA or Bowtie to a reference genome and align the sequenced reads.
2. This step helps to accurately identify the genes represented by the reads and accurately quantify their levels of expression.
3. Analysis of the raw sequencing data is required in order to weed out any low-quality reads and match the reads to a reference genome.

46.3.2.2 Quality Control (QC)

1. Removing poor readings from the dataset.
2. Filter out undesirable cells using QC metrics such as the quantity of distinct genes identified within each cell, the overall count of molecules detected in a cell, and the proportion of sequences that align with the mitochondrial genome.
3. Visualize the QC metrics to help identify and exclude problematic cells.
4. Examining QC metrics and filtering cells according to user-defined criteria are both made simple by the Seurat software. Finding the total amount of molecules in a cell, the number of distinct genes present in each cell, and the percentage of reads aligning with mitochondrial DNA are some of the frequently used QC metrics. The Seurat's PercentageFeatureSet() function may be used to calculate the proportion of transcripts that are related to mitochondrial genes. Gene uniqueness and the total number of molecules can be calculated using the CreateSeuratObject() function.

46.3.2.3 Normalize the Data

1. To account for variations in sequencing depth and cell-specific biases, data is normalized.
2. The data has to be normalized once undesirable cells have been eliminated from the dataset. Global-scaling normalization approach called "**LogNormalize**" multiplies the measurements of each cell's feature expression by the total expression, log-transforms the result, and then normalizes the measurements for each cell's feature expression by the entire expression is used.

46.3.2.4 Select Features (Genes) with Significant Cell-to-Cell Variation

1. Select genes that show high levels of expression in some cells but low levels in others.
2. This step helps to bring out biological signals in the data and focus the analysis on genes that are likely to be biologically relevant.
3. A common method for feature selection is the "FindVariableFeatures" function in Seurat which identifies genes with a significant cell-to-cell variation by modeling links between the data's mean and variance. These are genes that exhibit high expression levels in certain cells while being expressed at low levels in others, are thought to be important for revealing biological signals in the dataset, and are often included in downstream analysis.

46.3.2.5 Scale the Data

1. Scale the data using a linear transformation such as "ScaleData" from Seurat package.
2. This step equalizes the weight of all genes in the analysis, preventing the dominance of highly expressed genes.
3. Scaling also helps to improve the performance of techniques for dimensional reduction like PCA.
4. The **ScaleData()** method shifts each gene's expression such that it has a mean value of 0 across all cells and scales each gene's expression so that the variation between cells is 1.

46.3.2.6 Perform Dimensional Reduction

1. Use a method such as PCA to diminution of dimensionality of the data and visualize the relationships between cells.
2. This step helps to identify patterns and trends in the data.
3. PCA is then performed on the normalized data. By default, only the pre-selected variable features are used as input; however, a different selection can be defined using the "features" parameter. **VizDimReduction(), DimPlot(),** and **DimHeatmap()** are just a few of Seurat's helpful methods for displaying the cells and features that make up the PCA.

46.3.2.7 Evaluate the Dataset's "Dimensionality"

Seurat uses the scores of cells to cluster them together in PCA, where each PC represents a "meta feature" that merges data from a group of related characteristics as a way to reduce the technical noise in any single feature for scRNA-seq data. The JackStraw procedure is used to build a "null distribution" which determines feature scores; one approach involves randomly permuting a segment of the data and subsequently repeating PCA. PCs with a considerable enrichment of low p-value attributes are ones that should be looked for. To compare the distribution of p-values, the JackStrawPlot() method is being used. The heuristic technique creates an "Elbow plot" (ElbowPlot() function), which ranks the principal components according to the proportion of variation each one explains.

46.3.2.8 Cluster the Cells

1. Use a clustering method to group cells into clusters based on their gene expression patterns.
2. This step helps to identify subpopulations of cells that may be biologically or functionally distinct.
3. Using Euclidean distance in a principle component analysis (PCA) space, a K-nearest neighbors (KNN) network can be built, and the edge weights connecting two cells can then be changed depending upon how much their neighborhood overlaps. The dataset's dimensionality is an input to the FindNeighbors() function, which is used in this stage. To cluster the cells and maximize a shared modularity function, modularity optimization techniques like the Louvain algorithm or SLM are used. The FindClusters() function is employed to carry out this process, with the resolution parameter influencing the level of detail in the subsequent clustering outcome.

46.3.2.9 Run Non-linear Dimensional Reduction (UMAP/tSNE)

To view and study the datasets, Seurat provides a number of non-linear dimensionality reduction algorithms, including tSNE and UMAP. These techniques are made to recognize the data manifold so that comparable cells may be grouped together in a small-dimensional space. On this dimension reduction plots, cells inside the aforementioned graph-based clusters should co-localize. Using the same PCs that were used for the clustering analysis as input to the UMAP and tSNE.

46.3.2.10 Finding Differentially Expressed Features (Cluster Biomarkers)

Seurat can be used to discover markers that categorize clusters based on differential expression. By default, it compares all other cells to a single cluster to identify positive and negative indicators. This procedure is automated for all clusters using the **FindAllMarkers()** function in Seurat package, but you may alternatively compare clusters to one another or to all cells.

46.3.2.11 Assigning Cell Type Identity to Clusters

The **FeaturePlot()** function in Seurat R package may be used to investigate cell type identification by plotting the expression of several detected markers by cluster. The processed data may then be examined to find novel genetic components, recognize patterns in gene expression, and distinguish genetic differences across cells. A number of bioinformatics tools, including **DESeq2** and **EdgeR**, can be used for this. Furthermore, it is possible to visualize and analyze the study findings to learn more about the biology under investigation using tools like **Cytoscape** and **ggplot2**

46.3.2.12 Visualize the Data

Utilize software such as Seurat or Monocle to show the information and the relations between the cells. Understanding the data's structure, trends, and differentially expressed genes is aided by this phase.

46.4 CONCLUSION

The transcriptomes of individual cells within populations of actinobacteria can be studied using scRNA-seq, revealing key details about the biology and behaviur of these microorganisms. Bulk RNA-seq techniques are able to identify the general expression of genes in large cell populations but are unable to distinguish the transcriptional profiles between different bacterial populations. Initial attempts to generate a worldwide gene expression profile for one or a few bacterial cells were not successful in accurately predicting the physiological status of individual bacteria within a population. Thus, scRNA technologies have the potential to be used in understanding the biology of bacteria, as well as its applications in medicine, such as exploring the immune system, recognizing and classifying cell types, comprehending cellular function, examining the development, diagnosing diseases, and discovering new drugs. Overall, scRNA-seq is a valuable tool for understanding the transcriptional heterogeneity and regulation of gene expression in Actinomycetes. It has the potential to identify novel genes and pathways involved in the production of bioactive compounds and to guide the development of new therapeutics based on these bacteria.

REFERENCES

1. Hazarika SN, Thakur D. Actinobacteria (2020). Beneficial microbes in agro-ecology. Chapter 21 - Actinobacteria. Academic Press. ISBN 9780128234143, 443–476
2. Schrempf, H. (2001). Recognition and degradation of chitin by streptomycetes. *Antonie van Leeuwenhoek*, 79(3–4), 285–289.
3. Svensson, V., Vento-Tormo, R., & Teichmann, S. A. (2018). Exponential scaling of single-cell RNA-seq in the past decade. *Nature Protocols*, 13(3), 599–604.
4. Tang, F., Barbacioru, C., Wang, Y., Nordman, E., Lee, C., Xu, N., Wang, X., Bodeau, J., Tuch, B. B., Siddiqui, A., Lao, K., & Surani, M. A. (2009). mRNA-seq whole-transcriptome analysis of a single cell. *Nature Methods*, 6(5), 377–382.
5. Gierahn, T. M., Wadsworth, M. H. 2nd, Hughes, T. K., Bryson, B. D., Butler, A., Satija, R., et al. (2017). Seq-well: portable, low-cost RNA sequencing of single cells at high throughput. *Nature Methods*, 14(4), 395–398.
6. Lafzi, A., Moutinho, C., Picelli, S., & Heyn, H. (2018). Tutorial: guidelines for the experimental design of single-cell RNA sequencing studies. *Nature Protocols*, 13, 2742–2757.
7. Picelli, S., Bjorklund, A. K., Faridani, O. R., Sagasser, S., Winberg, G., & Sandberg, R. (2013). Smart-seq2 for sensitive full-length transcriptome profiling in single cells. *Nature Methods*, 10, 1096–1098.
8. Tabula Muris Consortium et al. (2018). Single-cell transcriptomics of 20 mouse organs creates a Tabula Muris. *Nature*, 562, 367–372.
9. Ono, K. (1990). Inhibitory effects of various 2′,3′-dideoxynucleoside 5′-triphosphates on the utilization of 2′-deoxynucleoside 5′-triphosphates by terminal deoxynucleotidyl transferase from calf thymus. *Biochimica et Biophysica Acta (BBA) – Gene Structure and Expression*, 1049(1), 15–20.
10. Chen, G., Schell, J. P., Benitez, J. A., Petropoulos, S., Yilmaz, M., Reinius, B., Alekseenko, Z., Shi, L., Hedlund, E., Lanner, F., Sandberg, R., Deng, Q., & Teichmann, S. A. (2016). Single-cell analyses of X chromosome inactivation dynamics and pluripotency during differentiation. *Genome Research*, 26(9), 1342–1354.
11. Becht, E., McInnes, L., Healy, J., Dutertre, C.-A., Kwok, I. W. H., Ng, L. G., Ginhoux, F., & Newell, E. W. (2019). Dimensionality reduction for visualizing single-cell data using UMAP. *Nature Biotechnology*, 37(1), 38–44.

12. Kiselev, V. Y., Kirschner, K., Schaub, M. T., Andrews, T., Yiu, A., Chandra, T., Natarajan, K. N., Reik, W., Barahona, M., & Green, A. R. (2017). SC3: consensus clustering of single-cell RNA-seq data. *Nature Methods*, 14(5), 483–486.
13. Satija, R., Farrell, J. A., Gennert, D., Schier, A. F., & Regev, A. (2015). Spatial reconstruction of single-cell gene expression data. *Nature Biotechnology*, 33(5), 495–502.
14. Kharchenko, P. V., Silberstein, L., & Scadden, D. T. (2014). Bayesian approach to single-cell differential expression analysis. *Nature Methods*, 11, 740–742.
15. Chen, L., & Zheng, S. (2018). BCseq: accurate single cell RNA-seq quantification with bias correction. *Nucleic Acids Research*, 46, e82.
16. Finak, G., McDavid, A., Yajima, M., Deng, J., Gersuk, V., Shalek, A. K., Slichter, C. K., Miller, H. W., McElrath, M. J., Prlic, M., et al. (2015). MAST: a flexible statistical framework for assessing transcriptional changes and characterizing heterogeneity in single-cell RNA sequencing data. *Genome Biology*, 16, 1–13.
17. Qiu, X., Hill, A., Packer, J., Lin, D., Ma, Y.-A., & Trapnell, C. (2017). Single-cell mRNA quantification and differential analysis with census. *Nature Methods*, 14(3), 309–315.
18. Del Giudice, M., Peirone, S., Perrone, S., Priante, F., Varese, F., Tirtei, E., Fagioli, F., & Cereda, M. (2021). Artificial intelligence in bulk and single-cell RNA-sequencing data to foster precision oncology. *International Journal of Molecular Sciences*, 22(9), 4563.
19. Cellenics. https://github.com/hms-dbmi-cellenics.
20. McCarthy, D. J., Campbell, K. R., Lun, A. T. L., & Wills, Q. F. (2017). Scater: pre-processing, quality control, normalization and visualization of single-cell RNA-seq data in R. *Bioinformatics*, 33(8), 1179–1186.
21. Jiang, P., Thomson, J. A., & Stewart, R. (2016). Quality control of single-cell RNA-seq by SinQC. *Bioinformatics*, 32(15), 2514–2516.
22. Kim, D., Langmead, B., & Salzberg, S. L. (2015). HISAT: a fast spliced aligner with low memory requirements. *Nature Methods*, 12(4), 357–360.
23. Dobin, A., & Gingeras, T. R. (2015). Mapping RNA-seq reads with STAR. *Current Protocols in Bioinformatics*, 51(1), 11.14.1–11.14.19.
24. Huang, M., Wang, J., Torre, E., Dueck, H., Shaffer, S., Bonasio, R., Murray, J. I., Raj, A., Li, M., & Zhang, N. R. (2018). SAVER: gene expression recovery for single-cell RNA sequencing. *Nature Methods*, 15(7), 539–542.
25. Love, M. I., Huber, W., & Anders, S. (2014). Moderated estimation of fold change and dispersion for RNA-seq data with DESeq2. *Genome Biology*, 15, 550.
26. Trapnell, C., Cacchiarelli, D., Grimsby, J., Pokharel, P., Li, S., Morse, M., Lennon, N. J., Livak, K. J., Mikkelsen, T. S., & Rinn, J. L. (2014). The dynamics and regulators of cell fate decisions are revealed by pseudotemporal ordering of single cells. *Nature Biotechnology*, 32, 381–386.
27. Zheng GX, Terry JM, Belgrader P, Ryvkin P, Bent ZW, Wilson R, Ziraldo SB, Wheeler TD, McDermott GP, Zhu J, et al. (2017). Massively parallel digital transcriptional profiling of single cells. *Nature Communications*, 8, 14049. doi: 10.1038/ncomms14049.
28. 10x Genomics. (2018). Chromium Single cell gene expression user guide (CG00052). Rev B.

47 Preparation of *Streptomyces* RNA for Transcriptomics

Balaji Muralikrishnan

47.1 INTRODUCTION

Transcriptomics is an evolved version of microarray analysis except for the fact that it has roped in the next-generation sequencing to provide a holistic pattern of gene expression. Interestingly, the earlier generation of microarray technique restricts the number of transcripts to the number of probes provided in the cassette while the next-generation sequencing provides information on novel transcripts that were not considered in the former technique. Transcriptomics in *Streptomyces* is vital to understand its secondary metabolite production machinery which should be helpful to resurrect the lost interest in Waksman's platform of drug discovery. Apart from this, *Streptomyces* originally served as a model organism for understanding the genetics and morphological development of *Actinobacteria*. Identification of how the biosynthetic gene clusters function and understanding its metabolic wiring is essential to unearth the secret regulatory genes that orchestrate the functioning of these clusters to render valuable natural products. Nevertheless, the use of transcriptomics and the scientific knowledge that can be gained through transcriptomics is left to novel methods and analysis. To achieve this, quality RNA sample preparation is crucial. In this chapter, we describe the protocols to isolate RNA from *Streptomyces*; deplete the ribosomal RNA and concentrate the mRNA and non-coding long RNA; and convert them to cDNA (cDNA synthesis), and sequence them using the Illumina platform of next-generation sequencing. Finally, check for the quality of reads before analysis.

47.2 MATERIALS AND METHODS

47.2.1 RNA Isolation

Lysis buffer prepared in DEPC-treated water (250 mM sodium acetate, 10 mM EDTA and 1% Triton X-100 at pH 5), liquid nitrogen, mortar and pestle, TRIzol, phenol:chloroform:isoamylalcohol (25:24:1), 70% ethanol, DEPC-treated water, DNase I (NEB, England).

47.3 PROTOCOL

1. Suspend the mycelium (100 mg, wet weight) in lysis buffer (1 mL) and add 5–10 mL of liquid nitrogen to freeze the cell mass. Using an autoclaved mortar and pestle, grind it to paste and allow it to turn back into liquid form.

   (The next step has to be performed immediately, and unless specified, all reactions have to be maintained on ice. Otherwise, it may lead to degradation of RNA.)
2. Extract the sample with 2 mL of TRIzol (Invitrogen) and vortex gently for 30 seconds and incubate at room temperature for 5 minutes.
3. Add 400 μL of chloroform, vortex the mixture for 30 seconds, allow it to incubate at room temperature for 5 minutes, and centrifuge at 12,000 rpm for 5 minutes at 4°C.

   (Now, the mixture separates into three layers, the bottom phenol layer, an interphase and a top aqueous layer. Insufficient TRIzol added to the sample could harbor DNA

DOI: 10.1201/9781003398400-47

contamination in the aqueous layer. Therefore, the addition of the same proportion mentioned is advisable).

4. Transfer the aqueous layer to a fresh DEPC-treated Eppendorf. Add 1 mL of ice-cold isopropanol and incubate at –20°C for 20 minutes and centrifuge at 12,000 rpm and 4°C.
5. Remove the supernatant and wash the RNA pellet with 75% ethanol (200 µL) at 12,000 rpm for 5 minutes. Repeat this step again.
6. Remove the washed ethanol and air dry the pellet to avoid ethanol contamination.
   (*Do not completely dry the pellet; otherwise, it would not dissolve in the water*).
7. Add 50–100 µL of sterile water and aspirate gently to dissolve RNA (Figure 47.1).
   (*Place it on ice if you are proceeding with the cDNA preparation immediately; otherwise, always store the RNA at 80°C*).

47.4 RIBOSOMAL RNA DEPLETION

Depletion of ribosomal RNA (16S, 23S, and 5S RNA) is very critical for transcriptomics as their presence hinders the efficiency of identifying the whole transcriptome of a bacteria. We have used a kit (RiboMinus™ Bacteria 2.0 Transcriptome Isolation Kit, Invitrogen) that depletes the ribosomal RNA by using specific prokaryotic ribosomal RNA probes which capture the former and a purified mRNA aliquot is achieved. The steps followed are given below.

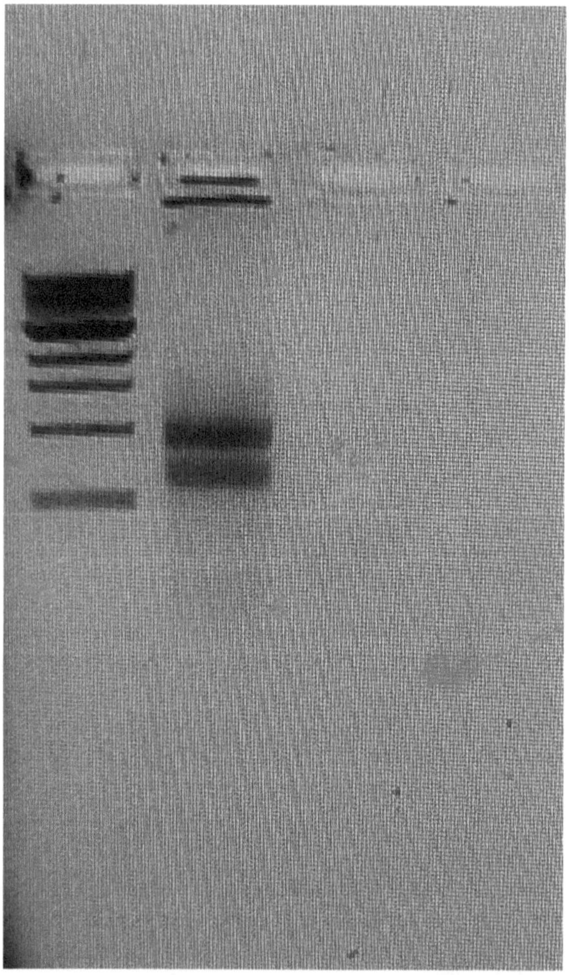

FIGURE 47.1 Isolation of RNA from *Streptomyces* sp.

47.5 PREPARATION OF BEADS

1. Take the magnetic bead solution and vortex it completely. Transfer 500 μL of the solution to a clean RNase-free Eppendorf tube and place it on the magnetic stand.
2. Allow the magnetic beads to adhere to the bottom and aspirate out the top liquid layer out. Now, add DEPC-treated water (200 μL) to the beads, vortex and repeat the above steps to keep the beads clean for RNA hybridization.
3. Next, after sufficient washing (at least 5–7 times), add 1X hybridization buffer (200 μL) to the beads and keep it at 37°C for 5 minutes for activation.

47.6 PREPARATION OF RNA-PROBE COMPLEX AND ITS REMOVAL

1. Mix the RNA sample with the probes given in the kit and incubate for 15 minutes at 37°C in a Polymerase Chain Reaction (PCR) tube. Centrifuge the mix for 4 minutes at 9,000 rpm at 4°C and place the tube in a magnetic stand and wait until the solution becomes clear.
2. Transfer the supernatant to the RNA-binding bead solution provided in the kit. Allow it to bind for 15 minutes at room temperature with gentle intermittent aspiration using a micropipette.
3. Keep the beads in the magnetic stand to remove the supernatant. Now, add 1 mL of 100% ethanol to wash the beads to remove any contaminants.
4. Next, allow it to dry to get rid of ethanol and add preheated 10–12 μL RNase-free distilled water to release the RNA from the beads.
5. Again, place the tube in the magnetic stand to remove only the supernatant which has the RNA. *The quality of the RNA sample can be assessed using the Qubit RNA fluorescence assay kit. The sample should not contain the ribosomal RNAs in it.*

47.7 CDNA SYNTHESIS

iScript™ cDNA Synthesis Kit (Bio-Rad) provides both random primers as well as oligo dT primers for the synthesis of cDNA. OligodT primers are used for cDNA synthesis.

Instructions for cDNA synthesis can be followed according to the protocol provided by the manufacturer. Changes are to be made only if there is a change in the concentration of input RNA. Once cDNA is synthesized, the same can be used for next-generation sequencing using either Illumina platform, Nanopore technology (Oxford) or Ion torrent (Thermo Fisher Scientific) platforms.

47.8 CONCLUSION

Streptomyces and other Actinobacteria are excellent sources of novel secondary metabolites. Understanding their gene regulation is an important aspect to develop better strains as well as device novel strategies for enhanced production. Transcriptomics is a powerful technique that uses next-generation sequencing to analyze holistic gene expression. For this, the preparation of quality RNA is indispensable. This protocol portrays an effective method to isolate RNA and remove the unwanted ribosomal RNA before next-generation sequencing.

48 A Protocol for CRISPR-Cas9-Mediated Knockout Construction in Actinobacteria

Khushal Khambhati, Gargi Bhattacharjee, and Vijai Singh

48.1 INTRODUCTION

Clustered regularly interspaced short palindromic repeats (CRISPR) have been recognized as part of the bacterial and archaeal adaptive immune system. The CRISPR-Cas (CRISPR-associated proteins) system has been broadly classified as Class I and Class II categories, which further include several types and subtypes. Depending on the CRISPR system, the nuclease encoded by their respective cluster could target either the DNA or RNA. For instance, the nuclease categorized within type VI is known to target RNA (Chaudhuri et al., 2022). The basic biology of the CRISPR-Cas system has been well adapted as RNA-guided endonuclease for nucleic acid detection and genome engineering purpose. The CRISPR-Cas system has brought a feasible and easy-to-use solution for detecting or engineering gene of interest with high specificity (Xu and Li, 2020). Due to such properties, the CRISPR-Cas systems have emerged as a revolutionizing technology for biomedical, industrial, and basic research (Ding et al., 2020).

Out of all the Cas nuclease, Cas9, a single effector endonuclease belonging to type II CRISPR-Cas system, was the first one to be used as a programmable RNA-guided nuclease for targeting the gene of interest. Apart from Cas9 endonuclease, a synthetic guide RNA (sgRNA) and protospacer adjacent motif (PAM) on the target sequence are required for successfully implementing Cas9 as a tool for genome engineering purpose (Khambhati et al., 2020). sgRNA is required to guide the Cas9 for specifically targeting the sequence of interest. sgRNA encompasses guide RNA (gRNA) scaffold region whose sequence changes depending on the Cas9 orthologue being used. The scaffold region of the gRNA helps Cas9 to attain an active state from an auto-inhibited state (Khambhati et al., 2020). The sgRNA is designed in a way that the first 12 nucleotides of the same are complementary to the seed region of the target sequence. This first 12 nucleotide sequence is one of the factors that determines the specificity of Cas9 toward the target sequence. The target sequence should be adjacent to the PAM. PAM is a specific 2–6 nucleotide region that is recognized by Cas9 endonuclease. PAM is required by Cas9 to demonstrate its lytic activity (Khambhati et al., 2020). Cas9 will swift form an inactive state to an active state once it forms a ribonucleoprotein complex with sgRNA and Cas9 endonuclease would form a complex with each other resulting Cas9 to be in an active state. This Cas9 would identify PAM that is in the targeted sequence. Once PAM is recognized, Cas9 would unwind the DNA toward the 3′ end of the PAM, resulting in opening up the seed and distil sequence of the targeted region. Given that the first 12 nucleotides of sgRNA are complement to the seed sequence, the Cas9 would perform its endonuclease activity. It is followed by a double-strand break and activation of non-homologous end joining (NHEJ) DNA repair mechanism, which may lead to knockout preparations. However, if replacement of the nucleotide is

 DOI: 10.1201/9781003398400-48

desired in the targeted region, then the donor DNA needs to be supplied which would facilitate homology-directed repair mechanisms (Khambhati et al., 2020).

The chapter aims to provide a protocol for Cas9-mediated knockout preparation in Actinobacteria. *S. coelicolor* A3 is an Actinobacteria and the best-studied actinomycete. *S. coelicolor* A3 is known to produce Actinorhodin (ACT) 1. ACT1 is a blue-pigmented antibiotic belonging to the class of aromatic polyketides (Okamoto et al., 2009). The current chapter takes the example of ACT1 biosynthetic gene cluster as the target for Cas9 nuclease to demonstrate knockout construction in *S. coelicolor* A3.

48.2 MATERIALS AND REAGENTS

48.2.1 Plasmids

- pCRISPR-Cas9-ScaligD (Addgene)

48.2.2 Enzymes

- T4 ligase (NEB)
- NcoI (NEB)
- SnaBI (NEB)

48.2.3 Molecular Biology Kits

- Plasmid MiniPrep Kit (Monarch)
- PCR Purification Kit (GeneJET)

48.2.4 Other Reagents and Antibiotics

- Apramycin (Sigma)
- Chloramphenicol (Sigma)
- Kanamycin (Sigma)
- Nalidixic acid (Sigma)

48.2.5 Cells

- Mach1™-T1 R *E. coli* (Life Technologies, UK)
- *E. coli* ET12567/pUZ8002
- *S. coelicolor* A3

48.2.6 Media

- Luria-Bertani (LB) medium
- Cullum agar medium
- ISP2 media

48.3 SPACER DESIGN FOR SGRNA EXPRESSION

CRISPy-web (http://crispy.secondarymetabolites.org/) is a web tool that is used to design sgRNA for Cas9 nuclease (Blin et al., 2016). The first step is to select "start" after uploading the *S. coelicolor* A3 genome sequence in genebank format using the "Browse" tab at its home page. This would provide information embedded in the genebank file such as organism name and sequence length. In the "Select target" box, enter the nucleotide range, locus tag, or cluster number of which knockout is desired. For the current example, enter 5,496,473–5,567,376, as it is the base pair rang in which ACT1 cluster is located in *S. coelicolor* A3 genome (Blin et al., 2016). The scan for potential spacer begins once the Find Target button is clicked. The spacer is the sequence that exactly matches with the target region. Spacer is also the sequence from which nucleotides complementary to the target region in gRNA arise through transcription. The output of the query will be a table consisting of columns such as the start and end of the potential target sequence, spacer sequence, strand, PAM site, and number of mismatch. The server also provides an option to zoom in on the gene of interest of which knockout is desired (Blin et al., 2016).

48.4 CONSTRUCTION OF SGRNA-EXPRESSING PLASMID

1. pCRISPR-Cas9-ScaligD is the vector with Cas9 gene under the control of tipA promoter. tipA is a thiostrepton inducible promoter which also requires thiostrepton responsive activator (TipA) to induce the expression. The gene for TipA is present in the genome of most of the *Streptomyces*, including *Streptomyces coelicolor* A3 (Tong et al., 2015).
2. In order to use pCRISPR-Cas9-related plasmid for knockout construction, the only requirement is to insert the spacer upstream to gRNA scaffold. This is achieved using forward and reverse primers. The sequence of the forward and reverse primer should be CATGCCATGGN_{20}GTTTTAGAGCTAGAAATAGC and ACGCCTACGTAAAAAAAGCACCGACTCGGTGCC, respectively. The N_{20} in the forward primer is the spacer sequence identified using CRISPy-Web tool (Tong et al., 2015). Using the primer mentioned before and pCRISPR–Cas9-ScaligD as template polymerase chain reaction (PCR) is performed. For 50 µL of PCR reaction, 20 ng of plasmid DNA is used. Ten microliters of high fidelity Phusion buffer (5X) is added, 1 mM of dNTP is used as the substrate for polymerization, and 0.5 µM of each primer is added. At the end, 1U enzyme and 1.5 µL DMSO are added (Tong et al., 2018). The reaction tube is tapped with fingers a couple of times. The reaction mixture is spun down and the tube is placed in a thermocycler. The PCR conditions are as follows: Initial denaturation at 98°C for 30 seconds, followed by 35 cycles of denaturation at 98°C for 10 seconds, annealing for 30 seconds (the temperature needs to be optimized), and extension at 72°C for 10 seconds. Once 35 cycles of denaturation, annealing, and extension terminate, one cycle of final extension at 72°C for 5 minutes is conducted. The PCR product is analyzed on 2% agarose gel (Tong et al., 2018). Once the size of the product is conformed, the PCR product is purified using the GeneJET PCR purification Kit. The concentration and purity are measured using nanodrop.

1. The purified product obtained is digested using NcoI and SnaBI. Then, ligation of NcoI- and SnaBI-digested PCR-purified product to NcoI and SnaBI are pre-digested pCRISPR-Cas9-ScaligD vector. The ligation is carried out using 100 ng of digested vector in such a way that for one molecule of vector DNA there should be three molecules of insert DNA (Tong et al., 2018). Ten microliters of ligation is carried out using 1 µL of 10× T4 ligase buffer and 1U of T4 DNA ligase at 16°C for 12 hours.

3. The ligated product is transformed in chemical competent Mach1™-T1 R *Escherichia coli* cells using the heat shock method. Ten microliters of ligation reaction is incubated with 50 µL of chemical competent cells for 30 minutes and then subjected to heat shock at 42°C for 45 seconds (Tong et al., 2018). The cells are incubated on ice for 2 minutes followed by the addition of 1 mL LB broth. The cells are incubated at 37°C for 1 hour. The transformed cells are grown in apramycin (50 µg/mL) containing LB agar plates as the vector consists of apramycin-resistant gene as the selection marker (Tong et al., 2018). The conformation of the insert into the vector could be done using colony PCR and Sanger sequencing.

48.5 CONJUGATION OF *E. COLI* ET12567/PUZ8002 WITH *S. COELICOLOR* A3 FOR PLASMID TRANSFER AND CAS9-MEDIATED KNOCKOUT PREPARATION

1. Transform the chemical competent *E. coli* ET12567/pUZ8002 cells with sgRNA and Cas9-expressing plasmid using the heat shock method. The cells are grown for 24 hours at 37°C in LB agar containing apramycin (50 µg/mL), kanamycin (50 µg/mL), and chloramphenicol (25 µg/mL) as a selective environment. Individual colonies are picked and inoculated in 10 mL LB broth containing the same selective environment (Tong et al., 2018).
2. The cells are suspended in 1 mL of LB broth after washing them twice with LB broth without antibiotics. Fifty microliters of *S. coelicolor* A3 spore are mixed with 100 µL of re-suspended *E. coli* ET12567/pUZ8002 culture. The culture is plated over Cullum agar plate and incubated at 30°C for 16 hours (Tong et al., 2018). Flood the plate with 1 mL of sterile distilled water containing 1 mg of nalidixic acid and apramycin. The exconjugants are allowed to grow for 3–4 days at 30°C. Ten to twenty exconjugants are picked and co-cultured by streaking and allowed to grow at 30°C for 4–5 days in ISP2 plates containing thiostrepton (1 µg/mL), apramycin (50 µg/mL), and nalidixic acid (50 µg/mL) (Tong et al., 2018).
3. Mycelia are scratched from the clone using a sterile toothpick and inoculated into 10 µL of dimethyl sulfoxide (DMSO) using PCR tubes. The tubes are heated at 100°C under shaking conditions followed by a vortex for 1 minutes at room temperature. The tubes are spun down at the top speed offered by the instrument and then 2 µL of supernatant is used as template for PCR to conform if the desired knockout is achieved (Tong et al., 2018).
4. In the current case, the chosen cluster for knockout construction produces pigmented antimicrobial. Thus, if the enzyme targeted in the cluster is essential for pigment production and is dysfunctional with NHEJ repair mechanism, then the cells should not produce pigmented antibiotics.

48.6 CONCLUSION

Over the past years, several programmable genome-engineering tools have been developed such as TALEN and ZFN. However, out of all of them, CRISPR-Cas9 has emerged as an easy-to-use and a highly specific programmable genome engineering nuclease. Pieczynski et al. (2019) also demonstrate the same, where they include CRISPR-Cas9-based knockout construction as a part of undergraduate laboratory experience and the students were able to generate the knockout of targeted gene with ease. The current chapter provides a brief description of the protocol that is used for Cas9-based knockout construction in actinobacteria. The protocol has been described taking the example of *S. coelicolor* A3 as a host. However, CRISPR-Cas9 has also been used to generate knockout in numerous other organisms. The Cas9-based technology is just not limited to knockout construction but have also been much explored for gene replacement, knockdown via CRISPR interference (CRISPRi), and upregulating gene expression via CRIRPRa technology (Yang et al., 2021). In the future, not only Cas9 but also other Cas proteins of different CRISPR systems will also play an important role in shaping basic and biomedical science.

ACKNOWLEDGMENTS

Khushal Khambhati and Gargi Bhattacharjee acknowledge the Indian Council of Medical Research of the Government of India for financial assistance in the form of a Senior Research Fellowship (File No. AMR/Fellowship/36/2022-ECD-II and File No. 5/3/8/41/ITR-F/2022-ITR, respectively). Khushal Khambhati and Gargi Bhattacharjee also acknowledge Indrashil University for providing infrastructural facilities.

CONFLICT OF INTEREST

The authors have no conflict of interest to declare that is relevant to the content of this protocol chapter.

DATA AVAILABILITY

No original data has been generated for the current chapter.

REFERENCES

Blin, K., L. E. Pedersen, T. Weber, and S. Y. Lee. 2016. CRISPy-web: An online resource to design sgRNAs for CRISPR applications. *Synth Syst Biotechnol* 1 (2):118–121. doi: 10.1016/j.synbio.2016.01.003.

Chaudhuri, A., K. Halder, and A. Datta. 2022. Classification of CRISPR/Cas system and its application in tomato breeding. *Theor Appl Genet* 135 (2):367–387. doi: 10.1007/s00122-021-03984-y.

Ding, W., Y. Zhang, and S. Shi. 2020. Development and application of CRISPR/Cas in microbial biotechnology. *Front Bioeng Biotechnol* 8: 711. doi: 10.3389/fbioe.2020.00711.

Khambhati, K., N. Gohil, G. Bhattacharjee, and V. Singh. 2020. Development and challenges of using CRISPR-Cas9 system in mammalians. In *Genome Engineering via CRISPR-Cas9 System*, ed. V. Singh and P. K. Dhar, 82–93. Academic Press.

Okamoto, S., T. Taguchi, K. Ochi, and K. Ichinose. 2009. Biosynthesis of actinorhodin and related antibiotics: Discovery of alternative routes for quinone formation encoded in the act gene cluster. *Chem Biol* 16 (2): 226–236. doi: 10.1016/j.chembiol.2009.01.015.

Pieczynski, J. N., A. Deets, A. McDuffee, H. L. Kee. 2019. An undergraduate laboratory experience using CRISPR-cas9 technology to deactivate green fluorescent protein expression in Escherichia coli. *Biochem Mol Biol Educ* 47 (2):145-155. doi: 10.1002/bmb.21206.

Tong, Y., P. Charusanti, L. Zhang, T. Weber, and S. Y. Lee. 2015. CRISPR-Cas9 based engineering of actinomycetal genomes. *ACS Synth Biol* 4 (9): 1020–1029. doi: 10.1021/acssynbio.5b00038.

Tong, Y., H. L. Robertsen, K. Blin, T. Weber, and S. Y. Lee. 2018. CRISPR-Cas9 toolkit for actinomycete genome editing. *Methods Mol Biol* 1671: 163–184. doi: 10.1007/978-1-4939-7295-1_11.

Xu, Y., and Z. Li. 2020. CRISPR-Cas systems: Overview, innovations and applications in human disease research and gene therapy. *Comput Struct Biotechnol J* 18: 2401–2415. doi: 10.1016/j.csbj.2020.08.031.

Yang, Y., Q. Feng, K. Hu, and F. Cheng. 2021. Using CRISPRa and CRISPRi technologies to study the biological functions of ITGB5, TIMP1, and TMEM176B in prostate cancer cells. *Front Mol Biosci* 8: 676021. doi: 10.3389/fmolb.2021.676021.

49 Design of gRNA for DNA-Directed RNA Polymerase Beta Subunit (*rpoB*) Gene in Actinomycetes

Naveenkumar Venkatesan

49.1 INTRODUCTION

Actinomycetes are fungi-like bacteria present in the soil. Though they are similar to fungi, they are not affected by antifungal agents [1,2]. They are noted for their filamentous growth pattern. Some actinomycetes resemble molds as they undergo asexual spore formation for reproduction which is carried externally. They are found in both fresh water and in marine water [3]. They are gram positive and have high guanine-cytosine (GC) content than any other bacteria [4]. They play a major role in agriculture by decomposing organic matter of dead organisms, helping the plants to take up the molecules. Some live by having a symbiotic (mutualistic) relationship with plants whose roots permeate through the soil. They fix nitrogen for the plants in exchange of some saccharides from them for energy [5]. They also inhibit the growth of several plant pathogens [6]. They are responsible for the earthy odor of the soil as they produce geosmin, a by-product [7]. Natural compounds produced by actinomycetes are widely used in the pharmaceutical business. Over the past 25 years, more than 50% of all anti-infective and anti-cancer chemicals have been derived from or are natural products of actinomycetes. For the manufacture of clinically necessary antibiotics, strain improvement for overproduction and the discovery of new antibiotics are significant [8]. The *rpoB* gene, encoding the beta-subunit of RNA polymerase which is a key enzyme for bacteria and it is a DNA-dependent RNA polymerase (RNAP), catalyzes the transcription of DNA into RNA. Tanaka et al. demonstrated that mutations in the *rpoB* gene effectively enhanced the production of antibiotics in actinomycetes by activating the secondary metabolite biosynthetic gene clusters, which are "silent" or poorly expressed under ordinary culture conditions [8].

Cas13 is an RNA-guided RNase belonging to the class 2, type VI CRISPR-Cas system and divided into four subtypes (A-D) [9].

49.2 AIM

In this experiment, we will be designing the guide RNA (gRNA) for cas13 enzymes against mRNA of *rpoB* gene in Actinomycetes.

49.3 OBJECTIVES

- Retrieval of mRNA sequence from NCBI.
- Design of spacer sequence.
- Direct Repeats-DR combining to spaces sequence.
- Addition of T7 promoter sequence at the 5′ end of gRNA for IVT (or)
- Direct gRNA synthesis as oligos.
- *In vitro* and *In vivo* cleavage

DOI: 10.1201/9781003398400-49

49.4 Things to Know 49.4.1 gRNA

- The DR in the gRNA should be specific for Cas13.
- A gRNA for Cas13 is a small RNA with a conserved sequence (Direct Repeats-DR) at 5′ end and a spacer sequence at 3′ end.
- Cas13 binds to DR and the spacer binds to the complementary target sequence.
- The spacer sequence of your gRNA should be complementary to your target RNA.
- A non-targeting gRNA control should be included in your experiments.

49.4.2 Difference of Cas13 gRNA from Cas9

- Cas9 requires gRNA with tracr-RNA and crRNA, whereas Cas13 requires shorter single gRNA.
- Cas13 does not need PAM sequence next to the target sequence but prefers protospacer flanking sequence (PFS) so that we can target any region of RNA [10].
- Cas13 only binds and cut ssRNA. RNA has a secondary structure and generates loops and dsRNA. It is essential that your gRNAs for Cas13 target the loops or single-stranded regions of the RNA and exhibit collateral RNase activity [11].

49.4.3 Designing of Cas13 gRNA

1. First choose the desired Cas13 protein. It could be Cas13 a, b, c, d, etc., because each cas13 type binds to its own sequence of the DR. Cas13a binds to a different DR than other cas13 types.
2. Check the bacterial strain for Cas13. Because Cas13a from one bacteria will recognize only its own specific DR and will not recognize the DR from other bacteria.

Cas13 Protein Type	Species	DR Sequence	Reference
LwCas13a	*Leptotrichia wadei*	GATTTAGACTACCCCAAAAACGAAGGGGACTAAAAC	[12]
PspCas13b	*Prevotella* sp.	GTTGTGGAAGGTCCAGTTTTGAGGGGCTATTACAAC	[13]
RxCas13d	*Ruminococcus flavefaciens XPD3002*	CCCCTACCAACTGGTCGGGGTTT	[14]

3. After deciding the Cas13 and DR sequence, then design the spacer sequence.

49.4.4 Designing of Spacer Sequence

1. To get the RNA sequence, go to the NCBI database (https://www.ncbi.nlm.nih.gov/gene) and select the gene of interest, find the reference transcript sequence, and download the FASTA sequence.
2. Example: Retrieve the Actinomycete NRRL B-1347 DNA-directed RNA polymerase beta subunit (rpoB) gene, (GenBank: KJ996304.1)
3. To design cas13 gRNA, one can use Sanjana Lab cas13design gRNA design tool (https://cas13design.nygenome.org) [15,16].
4. Select the design custom gRNAs option.
5. Upload your sequence as a FASTA file or paste it if it is shorter than 500 bases.
6. Click the "design guides" button, and a table with the predicted gRNAs will be generated or sent to you by email.

7. The gRNA sequence will be shown as below:
8. The table will show the spacer RNA sequences, the region of the crRNA targets, predicted efficacy score, quartile value (Q4-high efficacy, Q1-low efficacy), and potential off-target.
9. Choose the gRNA with the highest quartile value and guide score with less or no potential off-targets by number of mismatches (number of sequences in the transcriptome with 0, 1, or 2 mismatches).
10. Synthesis the gRNA by IVT or as synthetic oligo or clone in a plasmid.

49.4 IN VITRO CLEAVAGE

1. Incubate the gRNA along with Cas13 for 10 minutes at 25°C.
2. Incubate with Target sequence for 10 minutes at 37°C.
3. Cas13 will exhibit collateral cleavage after binding to the target sequence; this can be detected by adding external probe.

49.5 IN VIVO CLEAVAGE

Get the plasmid from Addgene by searching Cas13 gRNA.

Select the plasmid for bacteria and the plasmid should have a U6 promoter, the DR specific for each Cas13 protein, a restriction enzyme site to allow insertion of the crRNA, and a transcription termination site.

After transformation, the testing can be done by measuring the RNA levels of the transcript of rpoB 48–72 hours after transformation with Cas13 and the gRNA. RT-qPCR is the best and easiest way to do it. If your gRNAs work, you will see a reduction in the transcript levels.

REFERENCES

1. Pepper IL, Gentry TJ. Chapter 4 – Earth Environments. In: Pepper IL, Gerba CP, Gentry TJ, editors. *Environmental Microbiology* (Third Edition) [Internet]. San Diego: Academic Press; 2015. pp. 59–88. Available from: https://www.sciencedirect.com/science/article/pii/B9780123946263000041
2. Lechevalier HA, Lechevalier MP. Biology of Actinomycetes. *Annu Rev Microbiol.* 1967;21(1):71–100. Available from: https://doi.org/10.1146/annurev.mi.21.100167.000443
3. Williams ST, Wellington EMH. Actinomycetes. In: *Methods of Soil Analysis.* John Wiley & Sons, Ltd; 1983. pp. 969–987. Available from: https://onlinelibrary.wiley.com/doi/abs/10.2134/agronmonogr9.2.2ed.c45
4. De Simeis D, Serra S. Actinomycetes: A Never-Ending Source of Bioactive Compounds-An Overview on Antibiotics Production. *Antibiotics.* 2021;10(5):483. Available from: https://www.ncbi.nlm.nih.gov/pmc/articles/PMC8143475/
5. Kamal R, Gusain YS, Kumar V. Interaction and Symbiosis of AM fungi, Actinomycetes and Plant Growth Promoting Rhizobacteria with Plants: Strategies for the Improvement of Plants Health and Defense System *Int J Curr Microbiol App Sci.* 2014;3(7):564–585.
6. Goodfellow M, Williams ST. Ecology of Actinomycetes. *Annu Rev Microbiol.* 1983;37(1):189–216. Available from: https://doi.org/10.1146/annurev.mi.37.100183.001201
7. Gerber NN, Lechevalier HA. Geosmin, an Earthy-Smelling Substance Isolated from Actinomycetes. *Appl Microbiol.* 1965;13(6):935–938. Available from: https://www.ncbi.nlm.nih.gov/pmc/articles/PMC1058374/
8. Tanaka Y, Kasahara K, Hirose Y, Murakami K, Kugimiya R, Ochi K. Activation and Products of the Cryptic Secondary Metabolite Biosynthetic Gene Clusters by Rifampin Resistance (rpoB) Mutations in Actinomycetes. *J Bacteriol.* 2013;195(13):2959–2970. https://doi.org/10.1128/JB.00147-13
9. O'Connell MR. Molecular Mechanisms of RNA Targeting by Cas13-containing Type VI CRISPR-Cas Systems. *J Mol Biol.* 2019;431(1):66–87. https://doi.org/10.1016/j.jmb.2018.06.029
10. Abudayyeh O. et al. C2c2 Is a Single Component Programmable RNA-Guided RNA-Targeting CRISPR Effector. *Science.* 2016;353(6299):aaf5573.
11. Bandaru S. et al. Structure-Based Design of gRNA for Cas13. *Sci Rep.* 2020;10:11610.
12. Abudayyeh O. RNA Targeting with CRISPR-Cas13. *Nature.* 2017;550(7675):280–284.

13. Cox D. et al. RNA Editing with CRISPR-Cas13. *Science*. 2017;358:1019–1027.
14. Konermann S. et al. Transcriptome Engineering with RNA-Targeting Type VI-D CRISPR Effectors. *Cell*. 2018;173(3):665–676.e14. https://doi.org/10.1016/j.cell.2018.02.033
15. Wessels HH. et al. Massively Parallel Cas13 Screens Reveal Principles for Guide RNA Design. *Nat Biotechnol*. 2020;38:722–727.
16. Guo X. et al. Transcriptome-Wide Cas13 Guide RNA Design for Model Organisms and Viral RNA Pathogens. *Cell Genomics*. 2021;1:100001.

50 Acute and Chronic Toxicity of Bioactive Compounds from Actinobacteria

M. Joyce Nirmala, Uma Kizhuveetil, Ipsita Debnath, and Ramamurthy Nagarajan

50.1 INTRODUCTION

Bioactive compounds are chemical components, mostly of actinomycetes origin, which have capabilities above and beyond their nutritional values (Teodoro 2019). They are compounds which are used to confer immunity, work as antimicrobial or anticancer agents, and can counteract many metabolic syndromes. Though found in minute quantities, these bioactive compounds are being extracted and used in multiple formulations against various diseases. Spices such as cinnamon, turmeric, pepper, and thyme possess many excellent bioactive compounds which have a wide spectrum of activities (do Nascimento et al. 2020). Actinomycetes include a large number of organisms, amongst which are the pathogenic species such as *Streptomyces* sp. and *Propionibacterium* sp. that are known to cause infections of skin, bone, lungs, and blood (Kapadia et al. 2007; Perry and Lambert 2011). However, *Streptomyces* sp. is among the richest sources of a variety of antibiotics (Selim 2021), resulting in a high degree of resistance against many antimicrobial drugs. Spice extracts have been shown to possess wide-spectrum antimicrobial activity and are used in multiple drug formulations against these infections.

One major requirement of any drug preparation is that it must be shown to be safe for human use before obtaining approval for marketing. For this, the drug formulations have to be tested first *in vitro* and then *in vivo* in rodents such as mice and rats, and then non-rodents such as rabbits, before being approved for clinical testing on humans. The testing procedure depends on the type of formulation and may include topical preparations such as gels and ointments, as well as oral intake preparations such as tablets and intravenous preparations. The testing methods vary according to the type of preparation, as well as the duration of drug exposure. Toxicity studies involve the treatment of these rodent or non-rodent species with varying concentrations of the drug formulations in order to determine their adverse effects. Acute toxicity studies involve single or multiple drug administration within a period of 24 hours and are used to determine the short-term side effects, whereas chronic toxicity studies are conducted over a longer period of time. The protocol for both remains almost the same, with variations in the number of doses as well as the period of observation and are detailed in this chapter.

50.2 MATERIALS REQUIRED

Animal house, selected species of animals (rats, mice, rabbits), cannula, stomach tube, razors (according to the mode of drug administration), and the drug being tested.

50.3 ACUTE TOXICITY TEST

Acute toxicity test consists of two parts: limit test and main test (OECD 2022a). Limit test is used for preparations which are relatively non-toxic in nature, such as drugs prepared from food materials

DOI: 10.1201/9781003398400-50

and other sources approved as safe for consumption. The main test needs to be conducted only on the observance of toxicity in the limit test.

50.3.1 Method

a. Limit test for determination of acute toxicity
- Determine the weight of the rats. (Normally, female rats are preferred owing to their higher sensitivity to many drugs. Male rats are used for testing drugs for gender-specific diseases related to males.)
- Calculate the volume of the drug preparation required for 2,000 mg/kg of the rat.
- Keep the animals on fasting (no food but water provided) overnight. This fasting time changes for each animal and should be considered accordingly.
- Administer the calculated dose of the drug to one rat. The drug may be fed through a cannula for the intravenous route or a stomach tube for the oral route of drug administration. For dermal administration, marked skin areas where fur has been shaved off may be used. United States Food and Drug Administration (US FDA) guidelines require mandatory determination of intravenous toxicity for all routes of drug administration.
- Observe for 48 hours.
- In case of death of the rat, continue for the main test as LD50 values are less than 2,000 mg/kg.
- If the rat survives, dose four more rats sequentially at intervals of 48 hours.
- The LD_{50} value can be confirmed to be higher than 2,000 mg/kg if there are three or more surviving rats after 48 hours of drug administration.
- The rats must be kept under continued observation, and any delayed toxicity up to 12 days needs to be reported.
- For drugs which are considered safer, a limit test at 5,000 mg/kg may be performed. However, this is discouraged owing to the unnecessary harm to animals and is to be done only if there might be a specific regulatory need.

b. The main test for the determination of acute toxicity
- This is done using an up-and-down procedure according to the current Organization for Economic Co-operation and Development (OECD) guidelines.
- The requirements of fasting for the animals remain the same as the limit test. However, the number of rats required varies until the LD_{50} value is determined.
- If the limit test was performed and the rat died, reduce the dose by a factor of 3.2.
- After dosing by the selected mode of administration, observe the rat for 48 hours.
- In case of death, reduce the dose further by a factor of 3.2 and continue the above procedure.
- For drugs which are not known to be safe, the test is done from a lower dose after obtaining an approximate LD_{50} value through *in vitro* and *in-silico* toxicity tests. The dose may be either decreased (upon the death of the rat at the current dose) or increased (upon the survival of the rat) by a factor of 3.2 and continued sequentially as described above.

50.3.2 Notes

- Healthy adult young animals which are nulliparous (females which have not given birth to any offspring, dead or alive) and non-pregnant have to be used for all studies. Their weight should be within ±20% of the previously dosed animals.
- Animals should be kept under lab conditions for at least five days prior to drug administration to get acclimatized and should be less than nine weeks old at the start of the study.

- The dose should be fixed at a constant volume for each administration.
- For testing of multiple doses, the drug administration may be done as many times as required within a period of 24 hours from the first drug administration.
- The dose progression factor of 3.2 is based on half-log units.

The LD_{50} value may be calculated using the maximum likelihood method using appropriate software. A list of common software (OECD, 2018b) used for the determination of LD_{50} values is available on the OECD website.

50.3.3 Results

The data need to be logged in a tabular form as shown below with the number of animals being limited to a maximum of 15. All excluded animals may be denoted by an identity E as shown in Table 50.1.

The LD_{50} value may be calculated by entering the same data in the software or by reporting above a certain value (if only limit tests are used). The drug could be classified based on the LD_{50} values obtained as:

- <5 mg/kg – Extreme;
- 5–50 mg/kg – High;
- 50–500 mg/kg – Mild;
- 500–5,000 mg/kg – Slight;
- 5,000–15,000 mg/kg – Non-toxic; and
- >15,000 mg/kg – Comparatively harmless.

50.4 CHRONIC TOXICITY TESTING

Chronic toxicity testing is needed in order to determine the side effects resulting from repetitive exposure to a particular drug formulation over a long duration of time, which should be a significant fraction of the lifespan of the animal being tested. Results of chronic toxicity studies provide information regarding the side effects arising from long-term exposure, the organs which are affected, the dose-response relationship, and other necessary data which can be used to determine the benchmark dose and a no-observed-adverse-effect level. The protocol predominantly covers the oral route of administration.

TABLE 50.1
Data Interpretation for Acute Toxicity Testing

Test Sequence	Include/ Exclude	Dose (mg/kg) (Mode of Administration, e.g., Oral)	Short-Term Outcome (First 48 hours)	Long-TermOutcome (Successive 12 days)	Reference
1	I	175	O	O	
2	I	550	O	X	
3	I	1,750	X	X	
4	I	550	O	O	(OECD 2022a)
5	I	1,750	X	X	
6	I	550	O	O	
7	I	1,750	X	X	
8	E	-	-	-	

I, included; E, excluded; O, no response; X, response.

50.4.1 Method

- After the data from the acute toxicity studies are obtained, further toxicity assessment is done through the 28-day (OECD 2018a) and 90-day (OECD 2014) repeated dose toxicity studies.
- Animals of both sexes are used for chronic toxicity studies. A minimum of ten animals are to be used for the 28-day study. However, the minimum number required for the 90-day toxicity study is 20.
- No fasting is required as the doses are given through gavage (stomach/nose tube) or through food/water.
- Three test groups with different doses, as well as a control group, are to be used for the toxicity studies. The doses are calculated using the data from the acute toxicity studies with a preferably even variation of two- to ten-fold between the doses, depending on the test substance's characteristics (OECD 2022a).
- Prior to dosing the animals, detailed clinical, ophthalmological, and behavioural data need to be observed and recorded.
- The doses need to be administered daily throughout the study period (28/90 days) at the same time.
- Clinical assessments need to be done every day with detailed assessments every week.
- All changes in external appearance, movement, and behaviour of all the groups are to be observed and recorded.
- In the pre-final week of the study, reactivity to stimuli also needs to be recorded.
- At the closure of the test period, necessary haematology, pathology, clinical biochemistry, and histopathology analyses need to be performed on all the animals.
- The chronic toxicity studies are done sequentially, first 28 days, then 90 days, and if necessary, depending on the morbidity of the animals being tested, 12 months (OECD 2018b).
- The procedure, requirements, result, and analysis are similar for all studies.

50.4.2 Results

The test report should include all the details of the test substance, vehicle, animals, conditions, endpoints, assessments, and a detailed discussion of the results leading to a final documented conclusion.

ACKNOWLEDGEMENT

We gratefully acknowledge the support of Pratiksha Trust, Bangalore towards this research.

REFERENCES

do Nascimento, L.D., de Moraes, A.A.B., de Costa, K.S., Pereira Galúcio, J.M., Taube, P.S., Costa, C.M.L., Cruz, J.N., de Aguiar Andrade, E.H., de Faria, L.J.G. 2020. Bioactive Natural Compounds and Antioxidant Activity of Essential Oils from Spice Plants: New Findings and Potential Applications. *Biomolecules* 10 (7): 988. doi:10.3390/biom10070988

Kapadia, M., Rolston, Kenneth V.I., Xiang, Y.H. 2007. Invasive *Streptomyces* Infections: Six Cases and Literature Review. *American Journal of Clinical Pathology* 127 (4): 619–24. doi:10.1309/QJEBXP0BCGR54L15

OECD. 2014. *Guidance Document 116 on the Conduct and Design of Chronic Toxicity and Carcinogenicity Studies, Supporting Test Guidelines 451, 452 and 453: Second Edition*. OECD Series on Testing and Assessment. OECD. doi:10.1787/9789264221475-en

OECD. 2018a. *Test No. 408: Repeated Dose 90-Day Oral Toxicity Study in Rodents*. OECD Guidelines for the Testing of Chemicals, Section 4. OECD. doi:10.1787/9789264070707-en

OECD. 2018b. *Test No. 452: Chronic Toxicity Studies*. OECD Guidelines for the Testing of Chemicals, Section 4. OECD. doi:10.1787/20745788

OECD. 2022a. *Test No. 425: Acute Oral Toxicity: Up-and-Down Procedure*. OECD Guidelines for the Testing of Chemicals, Section 4. OECD. doi:10.1787/9789264071049-en

Perry, A., Lambert, P. 2011. *Propionibacterium acnes*: Infection beyond the Skin. *Expert Review of Anti-Infective Therapy* 9 (12): 1149–56. doi:10.1586/eri.11.137

Selim, M.S.M., Abdelhamid, S.A. and Mohamed, S.S. 2021. Secondary Metabolites and Biodiversity of Actinomycetes. *Journal of Genetic Engineering and Biotechnology* 19 (1): 72. doi:10.1186/s43141-021-00156-9

Teodoro, A.J. 2019. Bioactive Compounds of Food: Their Role in the Prevention and Treatment of Diseases. *Oxidative Medicine and Cellular Longevity* 2019: 3765986. doi:10.1155/2019/3765986

51 Production, Extraction, Purification, and Characterization of Bioactive Derivatives of Phthalate Ester from Marine *Streptomyces* sp. VITJS4

C. Subathra Devi, S. Jemimah Naine, M. Maneesha, Sanjeev K. Ganesh, Kothari Neeti, and V. Mohanasrinivasan

51.1 INTRODUCTION

Actinomycetes are saprophytic, gram-positive prokaryotes, which are abundantly present in the soil. These organisms create primary and secondary metabolites, also known as bioactive substances that are extremely potent, physiologically active, and continue to be valuable sources for the development of new medicines. A wide variety of antibiotics, such as aminoglycosides, glycosides, anthracyclines, peptides, and polyenes, are produced by the majority of actinomycetes. Over 70% of the earth's surface is encompassed by oceans, and creatures that inhabit earth developed from the sea. According to experts, some marine environments, coral reefs and the deep-sea floor, exhibit more biological richness than tropical rainforests. Actinomycetes habitat distribution in aquatic ecosystems is mostly unresolved [1]. Marine actinomycetes have piqued the interest of researchers due to their metabolic uniqueness and distinctive physiological capabilities as potential sources of antitumor compounds and pharmacological properties. There are various bioactive compounds produced by marine actinobacteria. The bioactive compound, 1,2-benzene dicarboxylic acid, mono(2-ethylhexyl) ester, belongs to the derivatives of phthalate ester and has been isolated from the marine *Streptomyces* sp. VITJS4 (NCIM No. 5574). It showed antitumor activity against liver cancer cells (HepG2) [2]. Phthalate esters, generally known as phthalates, are a group of chemicals that are frequently used as plasticizers. These phthalic acid esters are routinely added to polymers to increase their transparency and flexibility. They are employed in the production of PVC products, including vinyl flooring, shower curtains, toys, medical devices, and many consumer items. The benzene dicarboxylic acid and two side chains make up the fundamental chemical building block of phthalates [3]. Due to their industrial origins, these substances are regarded as environmental contaminants that pose health risks. The production of these chemicals by plants, bacteria, actinomycetes, fungi, fern, higher plants, and mammals claimed to be gradually altering this notion [4]. Different phthalate esters generated from organisms, including a variety of species such as algae, bacteria, actinomycetes, fungi, ferns, higher plants, and mammals tend to have antimicrobial, antitumor, larvicidal, antidiabetic, immunostimulant, and anti-inflammatory activity. Through current biotechnological techniques, the bioactivities of these diverse phthalate esters may be potentially valuable for human wellbeing [3]. This chapter highlights the extraction and purification methods for phthalate esters from marine actinobacteria.

DOI: 10.1201/9781003398400-51

51.2 METHODOLOGY

51.2.1 *Streptomyces* sp. VITJS4 Spore Preparation

- Spore preparation: Inoculate the cells in the sporulation media composed of starch casein agar.
- Growth conditions: Incubate the plates at 28°C for 7 days.
- Spore collection: Add 1 mL of distilled water over the plate to the spore suspension. Preserve the spores in 25% glycerine solution under –20°C [1].

51.2.2 Inoculum Preparation

- Seed culture (Inoculum) preparation: Inoculate the spore suspension in 100 mL of starch casein broth.
- Growth conditions: Incubate the inoculated media for 7 days at RT [5].

51.2.3 Production

- Preparation of production medium: Prepare 100 mL of starch casein broth (soluble starch 10 g/L, casein 1 g/L; in addition, add beef extract [nitrogen source] and glucose [carbon source] to the medium, 25% marine water and 25% marine soil extract at a pH of 7.2).
- Inoculation: Inoculate 2% (v/v) of the seed culture in the production medium.
- Incubation: Incubate the *Streptomyces* sp.VITJS4 in the production medium for 7 days in room temperature at 200 rpm [5].

51.2.4 Extraction

- Filter the culture broth through Whatman no.1 filter paper.
- Place the broth for 20 minutes at 5,000 rpm at 4°C.
- Add ethyl acetate to the broth and mix thoroughly.
- Repeat it thrice.
- Using rotary evaporator, remove the solvents.
- Keep the extract at 4°C till further use [1].

51.3 CHARACTERIZATION OF BIOACTIVE COMPOUNDS

51.3.1 GC-MS Analysis

- System: GC SHIMADZU QP2020
- Conditions: Column Elite-1 fused silica capillary column (30* 0.25 mm ID* 1 EM df) of 100% dimethyl polysiloxane
- Operation: Electron impact mode at 70 eV
- Carrier gas: Helium (99.999%)
- Flow: 1 mL/min
- Injection volume: 0.5 EI
- Injector temperature: 250°C
- Ion-source temperature: 280°C
- Oven temperature: Temperature has to be increased from 110°C to 280°C, by gradually increasing the temperature in every 10 minutes. Finally, after the temperature has reached 280°C, the last cycle has to be run for 9 minutes, and at 70 eV, mass spectra should be measured.
- Database for peak Identification: NIST08 and WILEY8 database. The retention time for phthalate ester is 22.014 [5,6] (Figure 51.1).

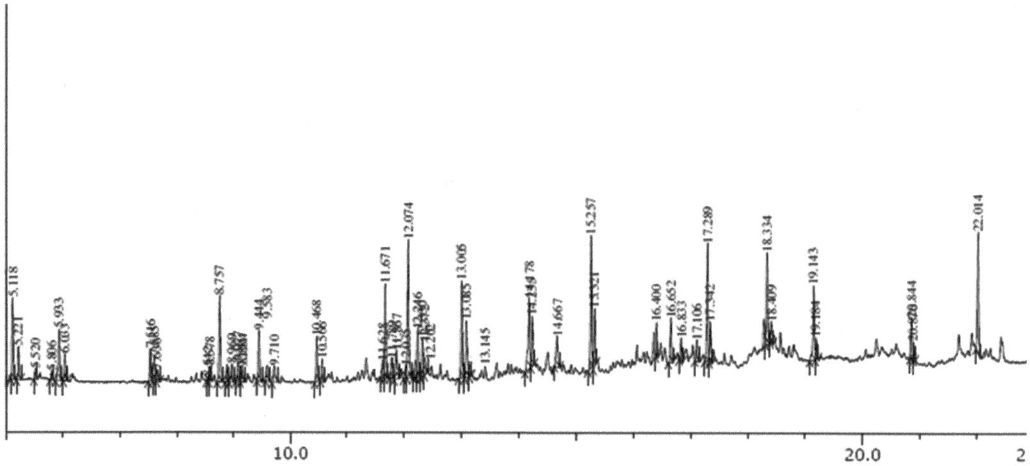

FIGURE 51.1 GC-MS chromatogram of *Streptomyces* sp. VITJS4 extract.

For VITJS4, 52 peaks are observed in GC-MS, in which the major compounds are 1,2 benzenedicarboxylic acid, mono(2-ethylhexyl) ester (22.04%), phenol, 2,4-*bis*(1,1-dimethylethyl)-(12,07%), 1-Nonadecene.

51.3.2 Purification

Column chromatography was used to purify the potent VITJS4 isolate by using silica gel, and a mixture of chloroform, ethyl acetate, and methanol was used for eluting out the fractions. From the elution, six fractions will be obtained, for which thin layer chromatography was performed to identify the purity of the fractions. For the mobile phase, ethyl acetate and chloroform are used in a 3:2 ratio. The spots obtained on TLC have to be placed on a silica plate and developed using the same solvents. After that, spots can be visualized using a TLC chamber. For the color change to take place, a mixture of 5% sulfuric acid in ethanol and 0.36% vanillin in 3.6% sulfuric acid in methanol was prepared and sprinkled on the plates and then kept at 100°C [7].

51.3.3 Structure Elucidation of the Purified Compound

a. **Spectroscopy:**
   - Determine the absorption spectrum of the compound using Shimadzu UV-160 A at a absorption range of 200–400 nm [8] (Figure 51.2)

b. **FT-IR Analysis**
   - Record the FT-IR spectra of the purified compound using Thermo Nicolet, Avatar 370 spectrometer over the 4,000–400 cm^{-1} range at the resolution of 4 cm^{-1}
   - Measure the maximum source aperture and the infrared spectra of the compound [9] (Figure 51.3)

The peaks in FT-IR profile are seen at 1,581, 1,462, 1,402, 1,382, 1,288, 1,126, 1,074, 1,009, 846, 887, 771, and 742 cm^{-1}, where the compound is a α, β unsaturated carboxylic acid.

The hydroxyl groups in FT-IR are seen at 3,417.86–2,547.97 region and for acid carbonyl at 1,641 cm^{-1}.

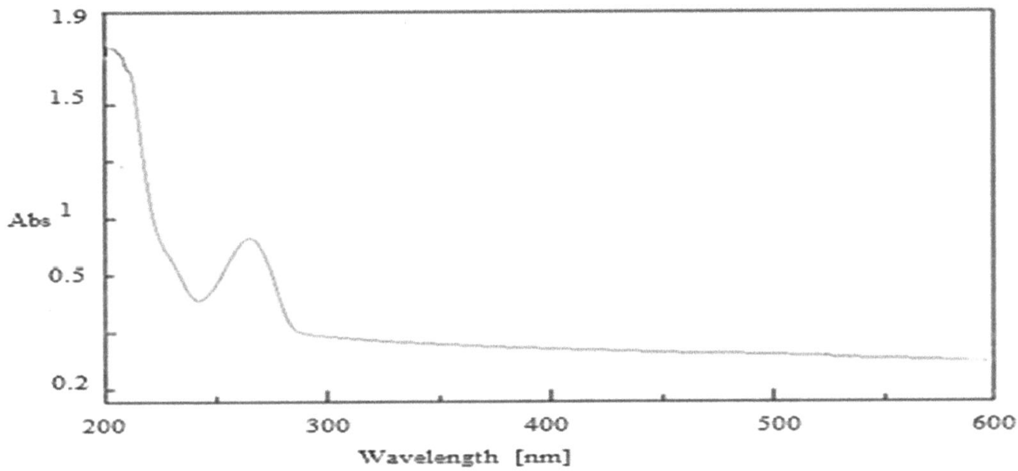

FIGURE 51.2 UV spectrum of phthalate ester.

51.3.4 NMR & EI-MS

- Determine the structure of the pure compound by NMR analysis.
- Run 1 H NMR at either 300 or 400 MHz and 13C NMR at 75 MHz using the solvent signal as reference.
- Perform High-resolution Electron Impact Mass Spectroscopy (EI-MS) [10] [Shimadzu, EI method] (Figures 51.4–51.6).

The proton peak for 1 H and 13 C NMR spectra are obtained at 7.80 ppm, and carbon signals at 132.01. At 10.22 ppm, a carboxylic group was seen and carbon signals are seen at 167.67 ppm. From this, the structure can be elucidated. The retention time for 1,2-benzenedicarboxylic acid, mono(2-Ethylhexyl) ester in EI/MS spectrum is 22.014 and peak at *m/z* 167 [M]+.

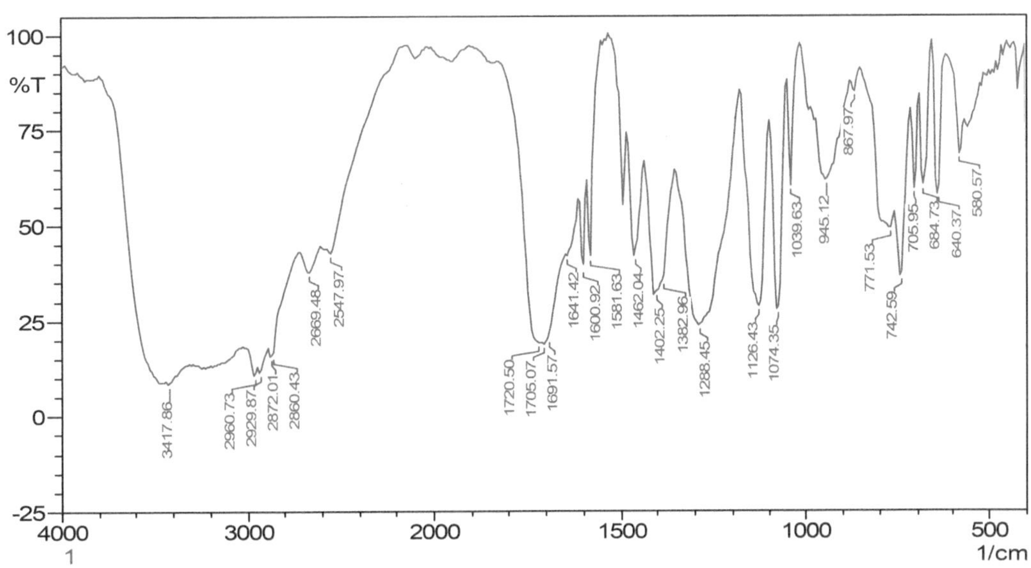

FIGURE 51.3 FT-IR profile of phthalate ester.

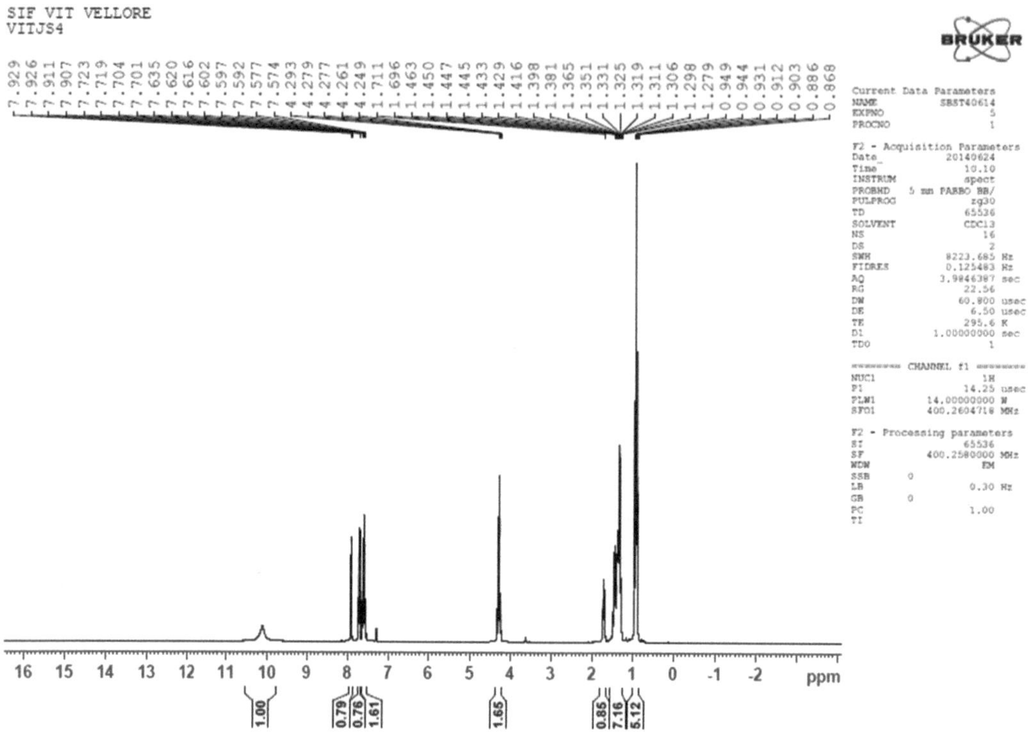

FIGURE 51.4 ¹H NMR spectrum of phthalate ester.

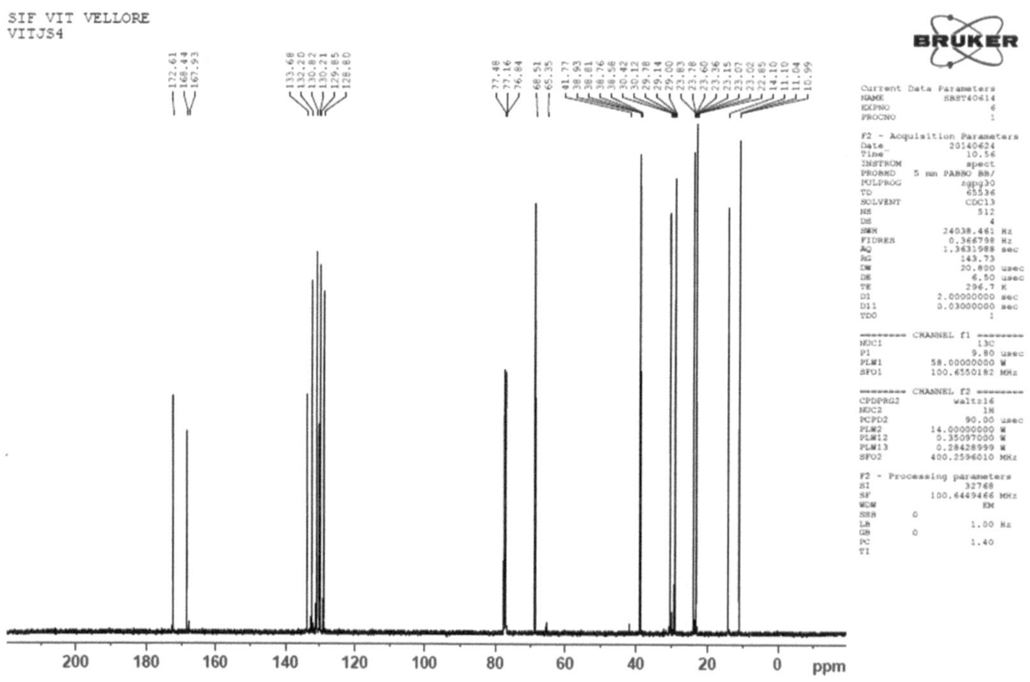

FIGURE 51.5 ¹³C NMR spectrum of phthalate ester.

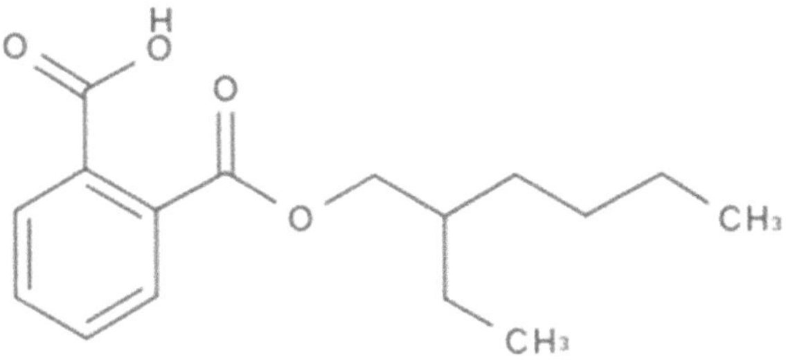

FIGURE 51.6 Structure of 1,2-benzene dicarboxylic acid, mono(2-ethylhexyl) ester.

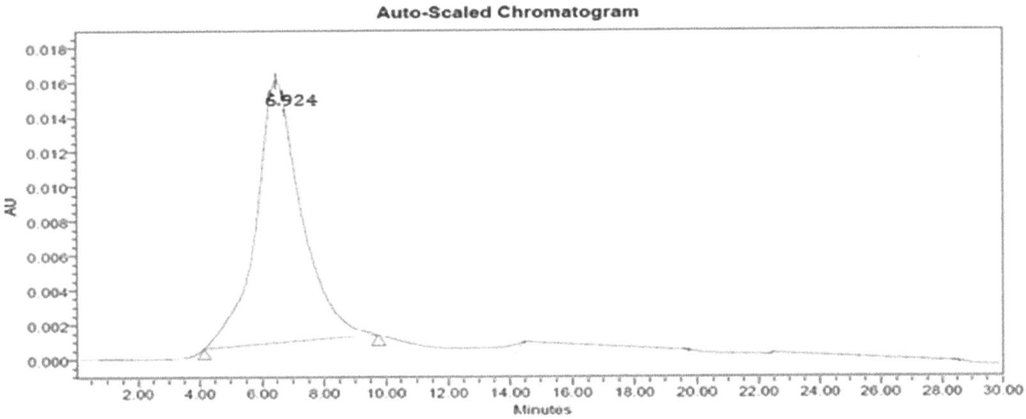

FIGURE 51.7 HPLC chromatogram of phthalate ester.

51.4 DETERMINE THE CONCENTRATION OF EXTRACTED BIOACTIVE COMPOUND

Protocol for the analysis of phthalate ester using HPLC [11].

- LC-10 AT vp model HPLC
- Column: 250 × 4.60 mm Rheodysne column (C-18)
- Mobile phase: Methanol: water (95:5)
- Flow rate : 1 mL/min
- Injection volume: 25 µL [Mix the sample with solvent in the ratio of 50:50 and filter using Millipore filter before injection]
- UV detection: 310 nm
- Approximate retention time: 6.9 (Figure 51.7)

REFERENCES

1. Selvakumar JN, Nasimunislam N, Babu V, Mohanasrinivasan V, Subathra Devi C. 2012. Isolation of soil actinomycetes inhabiting amirithi forest for the potential source of bioactive compounds. *Asian Journal of Pharmaceutical and Clinical Research.* 5 (4): 189–192.
2. Selvakumar JN, Chandrasekaran SD, Doss GPC, Kumar TD. 2019. Inhibition of the ATPase domain of human topoisomerase IIa on HepG2 cells by 1, 2-benzenedicarboxylic acid, mono (2-ethylhexyl) ester: molecular docking and dynamics simulations. *Current Cancer Drug Targets.* 19: 495–503. https://doi.org/10.2174/1568009619666181127122230.

3. Roy RN. 2020. Bioactive natural derivatives of phthalate ester. *Critical Reviews in Biotechnology*. 40 (7): 913–929. https://doi.org/10.1080/07388551.2020.1789838.
4. Ortiz A, Sansinenea E. 2018. Di-2-ethylhexylphthalate may be a natural product, rather than a pollutant. *Journal of Chemistry*. 2018. 1–7. https://doi.org/10.1155/2018/6040814
5. Jemimah Naine S, Subathra Devi C, Mohanasrinivasan V, George Priya Doss C. 2016. Bioactivity of marine *Streptomyces* sp. VITJS4: Interactions of cytotoxic phthalate derivatives with human topoisomerase II α – An *in silico* molecular docking analysis. *Interdisciplinary Sciences: Computational Life Sciences*. 10 (2): 261–270. https://doi.org/10.1007/s12539-016-0187-2.
6. Jemimah Naine S, Subathra Devi C, Mohanasrinivasan V. 2015. Bioprospecting of marine derived *Streptomyces spectabilis* VITJS10 and exploring its cytotoxicity against human liver cancer cell lines. *Pharmacognosy Magazine*. 11: s469–s473. https://doi.org/10.4103/0973-1296.168974.
7. Duraipandiyan V, Sasi AH, Islam VH, Valanarasu M, Ignacimuthu S. 2010. Antimicrobial properties of actinomycetes from the soil of Himalaya. *Journal of Medical Mycology*. 213: 6–10. https://doi.org/10.1016/j.mycmed.2009.11.002.
8. Jemimah Naine S, Subathra Devi C, Mohanasrinivasan V, George Priya Doss C, Thirumal Kumar D. 2016. Binding and molecular dynamic studies of sesquiterpenes (2R-acetoxymethyl-1,3,3-trimethyl-4t-(3-methyl-2-buten-1-yl)-1t-cyclohexanol) derived from marine Streptomyces sp VITJS8 as potential anti-cancer agent. *Applied Microbiology and Biotechnology*. 100(6): 2869–2882. https://doi.org/10.1007/s00253-015-7156-2.
9. Fiedler HP. 1993. Biosynthetic capacities of actinomycetes. Screening for secondary metabolites by HPLC and UV-visible absorbance spectral libraries. *Natural Product Letters*. 2: 119–128. https://doi.org/10.1080/10575639308043796
10. Boussaada O, Ammar A, Saidana D, Chriaa J, Chraif I, Daami M, Helal AN, Mighri Z. 2008. Chemical composition and antimicrobial activity of volatile components from capitula and aerial parts of Rhaponticum acaule DC growing wild in Tunisia. *Microbiological Research*. 163: 87–95. https://doi.org/10.1016/j.micres.2007.02.010.
11. Armarego W, Perrin DD. 1997. *Purification of Laboratory Chemicals*. Butterworth Heinemann, Oxford, UK, 4th edition. https://doi.org/10.3390/21000152.

Index

Note: **Bold** page numbers refer to tables and *italic* page numbers refer to figures.

FSC
www.fsc.org
MIX
Papier | Fördert
gute Waldnutzung
FSC® C083411

Kolibri 360 GmbH
Ferdinand-Jühlke-Straße 7
99095 Erfurt, Deutschland
produktsicherheit@kolibri360.de